# Barrier Engineering

This book aims to provide a systematic approach to the design, assessment, operation, and maintenance of safety barriers that are used for preventing accidents and protecting humans, equipment, and the environment.

*Barrier Engineering: Models and Methods for Technical Safety* is based on the philosophy of risk management, providing a thorough guide on identifying, analyzing, designing, operating, and maintaining safety barriers. It presents general theories, models, and both qualitative and quantitative analysis approaches, addressing both design and operational challenges for technical and non-technical barriers. The focus is on analyzing and evaluating the effectiveness and performance of technical safety barriers to ensure the functional safety of complex systems. This book also introduces the concepts of barrier security, applications of artificial intelligence, resilience, and sustainability considerations in safety barrier engineering and management. PowerPoint slides and a solutions manual are available for facilitating teaching and self-learning.

This book can be used as a textbook for master-level students in process and machinery safety, industrial and systems engineering, and management, and it is also an invaluable reference for risk analysts and engineers in complex system design, operation, and maintenance.

# Barrier Engineering
## Models and Methods for Technical Safety

Yiliu Liu

CRC Press
Taylor & Francis Group
Boca Raton London New York

CRC Press is an imprint of the
Taylor & Francis Group, an informa business

Designed cover image: Shutterstock - Alohaflaminggo



First edition published 2025
by CRC Press
2385 NW Executive Center Drive, Suite 320, Boca Raton FL 33431

and by CRC Press
4 Park Square, Milton Park, Abingdon, Oxon, OX14 4RN

*CRC Press is an imprint of Taylor & Francis Group, LLC*



ISBN: 978-1-032-15775-7 (hbk)
ISBN: 978-1-032-15776-4 (pbk)
ISBN: 978-1-003-24563-6 (ebk)

DOI: 10.1201/9781003245636

Typeset in Times
by KnowledgeWorks Global Ltd.

Access the Support Material: www.Routledge.com/9781032157757

*To Han Han, Wuge, and Zege*

# Contents

# Preface

Safety of complex sociotechnical systems is critical to humans and environments, and it has been a key component in achieving the sustainability development goals (SDGs) of the United Nations, for example, SDG 7 – providing affordable and clean energy, SDG 8 – promoting decent work, and SDG 9 – building resilient infrastructure. Safety barriers refer to those physical and non-physical means in different applications for ensuring safety by preventing the occurrences of hazardous events and mitigating their consequences.

The idea to write a book about safety barriers was first inspired by a review paper the author wrote and published in *Journal of Loss Prevention in the Process Industries* in 2020. Despite widespread recognition of their importance and the abundance of literature available, a dedicated textbook specifically focusing on safety barriers is still absent. Several on-shelf books have concentrated on safety design or some particular types of barriers, such as safety-instrumented systems, but at least in the opinion of the author, it is still meaningful to lift eyes at a higher level to explore the common features of the different technical and managerial approaches aimed at similar protective functions. Students can benefit from such a broader viewpoint. Furthermore, distinguishing safety barriers from the other safety assurance methods is also interesting to the designers, industrial facility owners, and risk analysts who need to manage safety issues well but with budgetary considerations.

Barrier engineering is a relatively new concept, and it can be roughly defined as efforts of identifying, analyzing, designing, operating and maintaining safety barriers. General theories, models, and quantitative analysis approaches are included in this textbook as the prerequisite of introducing detailed state-of-the-art methods. Different from the previous books focusing on design of single barriers, the author analyzes and evaluates the effectiveness and performance of safety barriers under a broader and more complex context. This book connects the separated domains of barrier engineering and incorporates new, research-based knowledge. The discussion extends beyond design to the operational phases and includes both technical and non-technical barriers. In addition, security of barriers and barrier functions for resilience and sustainability are also explored in this textbook.

This work can be used as a textbook for graduate students and a reference for reliability, safety, and maintenance engineers in different industries. Students in other engineering disciplines will also be the readers of this book since it provides a fresh perspective on accident prevention and protection of equipment, people, and the environment.

This book can be regarded as a summary of the research works that I have conducted with my colleagues and students in the last 15 years. I am deeply grateful to my mentor, Professor Marvin Rausand, who guided me during my PhD, postdoc research, and my early professional career. He is always my model example for being a researcher and a teacher. His several books on system reliability, risk assessment, and safety-critical systems also act as the foundations and the most important references of this book. My heartfelt thanks also go to my colleagues at the Norwegian

University of Science and Technology, Professor Mary Ann Lundteigen, Professor Stein Haugen, Professor Jørn Vatn, and Dr. Hui Jin, and my former PhD students – Aibo Zhang, Lin Xie, Yixin Zhao, Abraham Ahmed Jigar, Shengnan Wu, and Tiantian Zhu, whose exceptional works contribute to the core contents of this book.

**Yiliu Liu**
*10.08.2024, Trondheim, Norway*

# About the Author

Yiliu Liu is a professor in the Department of Mechanical and Industrial Engineering, at the Norwegian University of Science and Technology (NTNU). His main research interests include safety barriers and safety-critical systems, risk management, system reliability, and resilience engineering. By the end of 2024, he has published more than 200 papers on international journals and conferences. He is the co-chair of Sweden-Norway joint section of IEEE Reliability Society, the Editor-in-Chief for *International Journal of Reliability and Safety*, the subject editor of *Process Safety and Environmental Protection*, and the director of International RAMS (reliability, availability, maintainability, and safety) master program at NTNU. He also manages several national and international research projects with the focused applications in traditional and renewable energies and transportations.

# 1 Introduction of Safety Barriers

## 1.1 INTRODUCTION

The term "safety barrier" was originally used in road transportation sectors to refer to physical defenses made of wood, concrete, or metal. Positioned in the middle of or along roads, these barriers are used to prevent vehicles from leaving the road or colliding with oncoming traffic in the opposite lane. As safety science and safety engineering have evolved, the concept of safety barriers has been broadened and used in different applications. For instance, safety barriers appear in the energy-and-barrier model or hazard-and-barrier model introduced by Gibson (1961), in the research article of Haddon (1973), and in the management oversight and risk tree (MORT) technique, developed by Johnson (1975). MORT can be regarded as one of the earliest systematic approaches to using safety barriers for identifying and analyzing potential safety hazards in complex systems.

The introduction of new instrumentation, control technologies, and materials leads to the development of equipment featuring advanced safety barriers such as safety sensors, emergency shutdown systems, and safety interlocks. These barriers are designed and installed to prevent accidents and mitigate their consequences across various industrial sectors. On the other hand, since industrial safety has been relying on the protection function of safety barriers, it has become apparent that numerous severe accidents occur due to the absence or failure of safety barriers. For example, in the 2010 Deepwater Horizon oil spill, the primary cause was the failure of technical safety barriers on the rig, including the blowout preventer (BOP), which is installed to stop the flow of oil in the event of a well blowout. Failure to activate BOP leads to the escape of oil and gas, thus resulting in a devastating explosion and oil spill.

As such, safety barriers are critical safety engineering approaches in industrial sectors. They protect human lives and the environment and play a vital role in achieving sustainable development goals. In this chapter, we will provide an introduction to safety barriers, explore their general concepts, and discuss their role within the overall risk management framework. Additionally, we will examine various types of safety barriers and investigate the differences between safety design and emergency response programs.

## 1.2 RATIONALE FOR SAFETY BARRIERS

In common sense, safety is understood as a state where humans and the environment are free from any harm, such as injuries, fatalities, or pollution. These harms may result from hazardous events or specific conditions, and safety is achieved by avoiding or mitigating these hazardous events or conditions that pose threats to

DOI: 10.1201/9781003245636-1

human well-being and the environment through appropriate measures. Researchers have developed several models for understanding the relationships between hazards, safety, measures to achieve safety, and other relevant elements.

### 1.2.1 Hazard-and-Barrier Model

The *hazard-and-barrier model* (also known as the *hazard-barrier-target model* or *energy-and-barrier model*) incorporates the concept of safety barriers into accident analysis and prevention. This model was first proposed by Gibson (1961) and then was used by the U.S. Department of Energy (DOE) in the guidance document of hazard-and-barrier analysis (US DOE, 1996). As depicted in Figure 1.1, the classic hazard-and-barrier model consists of three fundamental elements: *hazard*, *asset*, and *barrier* (or *energy*, *target*, and *barrier* in the literature of energy-and-barrier models).

For the element on the left side of the hazard-and-barrier model, the *hazard*, we can adopt the definition developed by Rausand and Haugen (2020) to define it as follows:

- *Hazard*: A source or condition that has the potential to cause harm, either on its own or when combined with other factors.

In this model, a hazard is often associated with a certain type of energies. For example, the kinetic energy of a moving vehicle and the thermal energy from a fire in Figure 1.1 are typical hazards. Some hazards are not directly related to energy, such as risk-taking driving behavior and lack of training, but they can still be analyzed using the hazard-and-barrier model. In the guidance released by U.S. DOE, 72 different hazards are identified and organized in 15 categories (US DOE, 1996).

On the right side of Figure 1.1, we can see the element of *asset*. In this book, the asset, or more specifically, the asset of interest (AOI) for an organization, is defined as follows:

- *Asset (AOI)*: A physical or non-physical entity that has potential or actual value to an organization and is being protected or is desired by the organization to be protected.

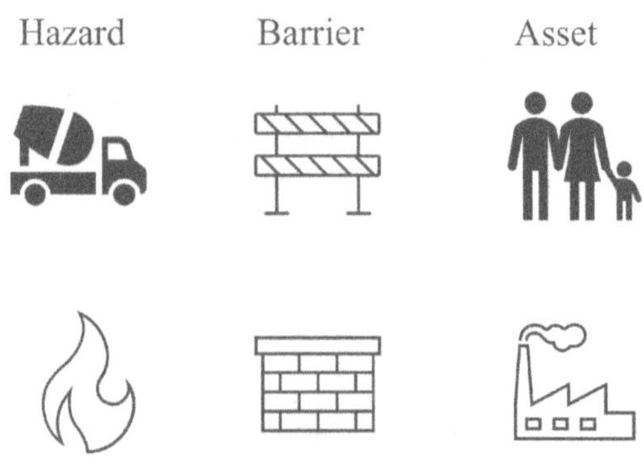


**FIGURE 1.1** Illustration of the hazard-and-barrier model.

For example, in Figure 1.1, pedestrians are the asset for the human society as an organization, and equipment in a factory is the asset of its owner. An asset typically has economic value or serves a vital function for the organization that aims to safeguard it. If the asset is damaged, the organization will experience a loss or regret. However, in some instances, compliance with regulations or fulfilling social responsibilities may also mandate the protection of an asset. When protective measures are employed, the asset can be referred to as the *asset under protection* (*AUP*) in this book.

According to IEC62443-1-1 (2009), assets are classified into three types: physical, logical, and human. Physical assets may include equipment and facilities owned by a specific organization or natural environments valuable to humanity. Logical assets include confidential data, proprietary knowledge, and other intellectual properties. Human assets refer to groups targeted by a specific organization or any potential victims who could be harmed by hazards. Another common classification method categorizes assets into three groups: people, environment, and property. People may include employees, customers, visitors, or others who might be victims of harm. The environment covers both natural and built environments, including elements such as air quality, water quality, soil quality, and other environmental factors. Property, in this context, is related to the physical equipment and facility of an organization, such as buildings and machines.

Then, a barrier, or safety barrier, is placed in the middle of the hazard-and-barrier model, or between the hazard and asset, as depicted in Figure 1.1. The figure demonstrates the function of a barrier that is to avoid the contact between the hazard and the asset or to stop the flow of energy to the target. In other words, a barrier is the measure that is adopted by the organization to protect its AOI.

### 1.2.2 Bow-Tie Model

Another model explains safety by illustrating the process that an asset is harmed by a hazard and considering the ways to avoid such a harm. Hazards are omnipresent, but they do not necessarily pose a threat to an asset until some specific conditions or events, particularly unexpected ones, act as triggers, potentially releasing energy toward the asset. On the other hand, the impact of harm on an asset can vary significantly, from very minor to extremely severe. A limitation of the hazard-and-barrier model lies in its focus on a single event, a single hazard, and also implies a specific harm. In practical risk management of industries, *the bow-tie model*, as the diagram shown in Figure 1.2, that resembles a bow tie, traditionally worn as formal neckwear by men, is often employed to involve more than one hazard and different severities of harm and illustrate the progress from hazards to harms.

In such a model, several events might occur during the progression from a hazard to the actual harm on an asset, including the following:

- A *pivotal event*, also called a *hazardous event* or a *top event* (especially in the fault tree analysis, see Chapter 5), is the event that can harm an asset. Commonly, it can be referred to as an incident or negation, such as the failure of a critical component, leakage of poisonous gas, and a car collision. The pivotal event is a crucial linkage between hazards and assets, marking

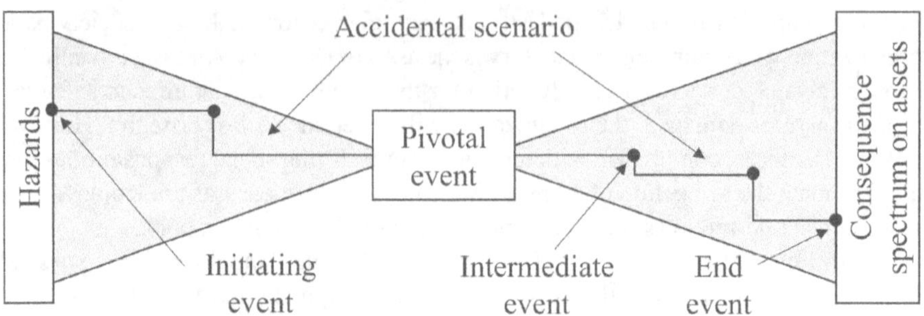


**FIGURE 1.2** A bow-tie diagram.

the moment when a potential hazard becomes an actual threat. Identifying and determining the nature of the pivotal event is a significant task of most risk analysis efforts. A pivotal event is always abnormal and negative, and thus, prevention is important. In case a pivotal event occurs, it can lead to various types and severities of harms/consequences to the asset.

- An *initiating event*, as the starting point in the analysis based on a bow-tie model, can be the first event that activates a potential hazard, such as high-speed driving in the following Example 1.1. The initiating event can make a situation or condition hazardous or has a potential to release energy. However, it must be noted that an initiating event can be an abnormal event or a normal event in daily life or in the operation of a technical system. For instance, high-speed driving is normal if within the speed limit, but it has the potential to develop into a pivotal event, i.e., a collision.
- An *end event* on an asset presents the final consequence or outcome resulting from a pivotal event. This event can either harm the asset or have no impact. The severity of the end event depends on the nature of the pivotal event, the condition and properties of the asset, and potential measures to respond to the pivotal event. In some instances, an asset remains indefinitely in a state following harm, for instance, a person may be dead after a collision, or a machine might be broken after a severe failure. This state is referred to as an *end state*, and in such a case, the end event is the transition of the asset into this state.
- An *intermediate event* describes any occurrence between the initiating and pivotal event or between the pivotal and end events. Such an event is also called an *enabling event*. The occurrence of an intermediate event can alter the outcome of its preceding event. It is important to note that specific conditions can also change outcome of an event.

The sequence of these events from a hazard to the harm on an asset can be regarded as an *accidental scenario*, with all events directing the scenario toward a worse outcome regarded as *hazardous events*. In practice, it is challenging to identify the initiating event of an accidental scenario, since most normal behaviors and operations can evolve into a pivotal event under certain conditions. Another challenge is identifying the end event when developing an accident scenario from a conceptual stage, given the difficulty in estimating the potential long-term impacts of hazardous events, such as environmental effects and impacts on future generations.

The bow-tie diagram can illustrate all the above events associated with hazards and assets, as well as accidental scenarios, facilitating risk analysis (see CCPS and EI (2018)). The pivotal event is in the center of the bow-tie model, and the paths extending from left to right represent accidental scenarios, while the points turning the paths are events. The far left points depict the initiating events coming from hazards, while the far right points are for the end events. It is important to note that in this model, all end events are considered within the spectrum of consequences, each associated with a specific level of harm to the asset.

### Example 1.1:

### *Accidental Scenario in the Collision of Two Cars*

Consider an accidental scenario involving two cars on a highway: The trailing car was traveling close to the speed limit, while the leading car was gradually decelerating. The driver behind did not notice the deceleration and maintained his speed, resulting in a collision that slightly injured both drivers.

Based on the terminology introduced in this section, the elements of this accident, corresponding to those in Figure 1.2, include the following:

- *Hazard*: The kinetic energy of the moving cars (particularly the trailing car) and the potential chemical energy from the fuel, which could lead to fire or an explosion.
- *Asset*: The drivers and the cars themselves. Additionally, highway infrastructure such as the road surface and guardrails, and even the flow of traffic, which is crucial for transportation management, can also be considered assets.
- *Initiating event*: This can be different depending on the focus of the analysis. For example, if the aim is to identify potential human errors and enhance driver training, the initiating event could be the deceleration of the leading car. On the other hand, if the goal is to explore all potential accident types to improve highway safety, driving on the highway might be considered the initiating event.
- *Pivotal event*: The collision is the pivotal event in this example. The sequence leading up to the collision includes driving on the highway, the deceleration of the leading car, and the failure of the trailing driver to recognize this deceleration.
- *End event*: In an actual accident report, the end event is typically recorded as injuries to the drivers and significant damage to the cars. However, when assessing potential outcomes of a collision, accidental scenarios may range from more optimistic to more pessimistic outcomes. For instance, delayed rescue due to heavy traffic or a remote location can escalate minor injuries to severe ones, potentially leading to amputation or fatalities.
- *Intermediate event*: In the accidental scenario where amputation is the end event, and the events that the drivers are injured, and the rescue is delayed are intermediate events. ■

Such events and the associated accident scenarios can be involved in a bow-tie model as shown in Figure 1.2.

### 1.2.3 Definition of Safety Barrier

Based on the hazard-and-barrier model and the bow-tie model, we can analyze how hazards threaten an asset and understand how safety barriers protect this asset. To make a definition of safety barrier, we can roughly define safety at first, as follows:

- *Safety (1)*: A state of an asset free from a harm.

Then, a safety barrier, simply referred to as a barrier, serves to shield an asset from hazards or ensure safety of the asset. For instance, examples in Figure 1.1 illustrate how barriers maintain asset safety in the presence of hazards: A guard fence acts as an upper barrier, safeguarding pedestrians from an out-of-control vehicle, while a firewall in the lower part halts the spread of fire to a factory. Such safety barriers exist in our daily lives and across various industries, with their forms tailored to specific applications. A barrier might be a tangible object or an intangible strategy. For example, in Example 1.1, the braking light of the car serves as an informational barrier by signaling the trailing driver to decelerate. Although the braking light does not physically absorb kinetic energy, it enhances safety by conveying vital information, thus qualifying as a part of a safety barrier.

In MORT, Johnson (1975) described a safety barrier as a physical or procedural measure designed to channel energy appropriately and control unwanted releases. Given the diverse ways in which assets can be protected, we define a safety barrier for use in this book as follows:

- *Safety barrier*: An entity and/or method introduced to protect an asset from harm caused by existing and potential hazards.

It should be noted that a single safety barrier can have multiple functions related to asset protection in the presence of hazards. Within this text, the term "function" means an activity performed by an item, typically engineered with a specific purpose for the human-made objects. A function is generally expressed as a verb followed by a noun, such as "to release pressure", "to contain fire", or "to reduce temperature". When assessing the activity of a safety barrier in shielding an asset from harm, we refer to this as a *barrier function*. Based on the hazard-and-barrier model in Figure 1.1, barrier function is defined as follows:

- *Barrier function (1)*: The function of shielding an asset from the damage by a specific form of energy.

According to the bow-tie model, when a barrier is effectively deployed, it performs its barrier function by interrupting the progression of an accidental scenario, thus preventing the originally anticipated end event. For example, in the model of Figure 1.2, six events are in the given accidental scenario, including an initiating event, a pivotal event, an end event, and three intermediate events. Five barriers can thus be interposed between the six events, as illustrated in Figure 1.3. When the first barrier is in use, it

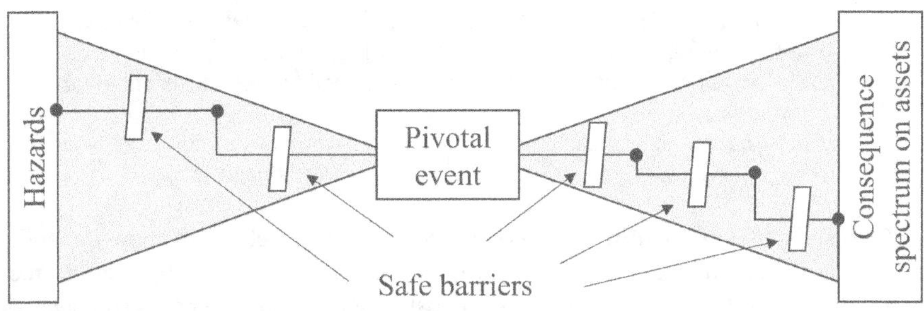


**FIGURE 1.3** A bow-tie diagram with safety barriers.

stops the accidental scenario immediately just after the initiating event. If this barrier does not work well or is not effective, the second barrier can be utilized to stop the development of the accidental scenario from the first intermediate event to the pivotal event. The other three barriers on the right side of the pivotal event can also stop the accidental scenario before it finally progresses to the end event.

Thus, the bow-tie model is actually an informative method for visualizing the roles or functions of safety barriers in protecting the asset by avoiding severe harms. We can also rely on the bow-tie model to identify and roughly analyze potential safety barriers. Barrier management based on the bow-tie model will be introduced in Section 2.4.

### Example 1.2:

#### *Accidental Scenario in the Collision of Two Cars (Continue…)*

Revisiting Example 1.1, let us re-analyze the two-car collision using the model depicted in Figure 1.3. The sequence of events in this hypothetical accident scenario progresses as follows (from left to right):

- *Initiating event*: The front car begins to decelerate.
- *Intermediate event 1*: The following driver fails to brake in time.
- *Pivotal event:* A collision between the two cars occurs.
- *Intermediate event 2*: The driver in the following car is hit hard during the collision.
- *Intermediate event 3*: The trailing driver's left leg is severely crushed, and circulation is compromised, leading to potential tissue death.
- *End event*: Due to delayed rescue and prolonged lack of medical attention, the driver ultimately loses his left leg.

Given this scenario, we can identify five potential safety barriers that could be strategically placed between these events, as illustrated in Figure 1.3:

- *Safety barrier 1*: An automatically activated brake light on the front car that alerts the following driver.
- *Safety barrier 2*: An automatic braking system that detects the proximity of the car ahead and decelerates accordingly.

- *Safety barrier 3*: An airbag at the driver seat that is activated upon collision.
- *Safety barrier 4*: An automatic door-unlocking feature that activates in a collision, allowing the driver to escape and facilitating quicker medical intervention.
- *Safety barrier 5*: A clear emergency lane that ensures rapid access for rescue services, crucial in preventing irreversible damage. ■

In this discussion, we utilize the hazard-and-barrier model and the bow-tie model to introduce fundamental concepts regarding safety barriers. It is important to recognize, however, that these two models are just selected from numerous frameworks used to comprehend the dynamics of accidents and the function of barriers. For instance, when a barrier has a flaw or fails to function as intended, it creates a "hole" that allows a pathway for the hazard to reach the asset. This aspect of the model aligns with Reason's Swiss cheese model (Reason, 1990) used in epidemiological accident analysis. In the Swiss cheese model, slices of cheese represent the barriers, and the holes within these slices symbolize the weaknesses of barriers.

We cannot cover all models that incorporate the concepts of hazard, safety, and safety barriers. For readers interested in more accident models and the integration of safety barriers within these frameworks, Chapter 8 of the book by Rausand and Haugen (2020) is an excellent resource.

## 1.3 BARRIER IN RISK MANAGEMENT

The safety barriers depicted in Figure 1.1 play a critical role in separating the assets from hazards, thereby mitigating or controlling the risk of harm. We have referenced the term "risk" several times in Section 1.2, and thus it is essential to define relevant terminologies and understand the role of safety barriers within the broader risk management framework.

It is important to note that this book does not deeply investigate risk science or risk philosophy. Instead, we just offer a concise and straightforward introduction to risk management concepts, avoiding complex discussions on terminology. For those interested in exploring risk further, the books by Aven (2015) and Rausand and Haugen (2020) are recommended.

To introduce the concept of risk, we rely on some existing definitions that, despite the lack of universal consensus, have been widely used.

According to Kaplan and Garrick (1981), risk is defined as follows:

- *Risk (1)*: The combined answer of three questions related to the asset of our interest: (1) What can go wrong? (2) What is the likelihood of that happening? And (3) What are the consequences?

In a more recent study, Aven and Renn (2009) describe risk in terms of uncertainty and the relevant knowledge:

- *Risk (2)*: Uncertainty about and severity of a certain of events and their consequences, influenced perception of people based on their observations and/or causal knowledge about the event.

In this book, we consider risk as the likelihood of any hazardous event occurring and the potential harm it poses to the safety of an asset. We also think that risk is intricately linked to the level of uncertainty surrounding the event, as well as our knowledge of it.

According to the international standard for risk management, ISO31000 (2018), *uncertainty* refers to the state of having insufficient information to understand an event, including its likelihood or consequences. Uncertainties are generally categorized into two types: *epistemic uncertainty* and *aleatory uncertainty*. The previous type comes from the lack of knowledge about the objects of study, while the latter type arises from natural variability and randomness. The epistemic uncertainty can be reduced with the increase of knowledge, but the aleatory uncertainty is generally irreducible. As an illustration, the spread of fire is heavily influenced by wind speed and direction at the time of occurrence. However, even though we can gather knowledge about the most likely wind conditions at a specific location over an extended period of observation and study, the exact wind speed and direction at a specific future moment are unpredictable. Consequently, it is also uncertain whether an asset at a certain place will be damaged by fire.

Given the existence of aleatory uncertainty, absolute safety, as defined earlier, may be unattainable. Therefore, a more reasonable approach is to define safety in terms of managing and controlling risk:

- *Safety (2)*: A state of an asset with an acceptable level of risk concerning specific hazards.

To determine the safety of an asset, it is necessary to acquire relevant risk information and assess it accordingly. *Risk assessment* is the endeavor of identifying all hazards to a specific asset, analyzing the risk information, and evaluating the risk level of the asset. Based on this information, we can engage in a systematic process of understanding and controlling the risk of the asset of interest, as *risk management*. IEC31010 (2019) is an international standard of risk management, and it has outlined the main steps of risk management as shown in the framework of Figure 1.4.

Here are brief descriptions of the main steps in risk management:

1. *Establishment of context*: This first step in risk management is to define the objectives for managing risk and identify the assets that require protection. At this stage, it is also necessary to determine the scope by identifying the system, which spatially includes hazards, assets, and barriers, as well as the process, which temporally encompasses potential accidental scenarios and protecting actions. In this book, the term "*process*" is defined as follows:
   - *Process*: A series of steps or actions by humans, technical systems, or both. From the perspective of risk management, any deviation of the process that is brought by a hazard may be a hazardous event resulting in a harm to the asset.

   For example, operating a chemical plant or driving a car from place A to place B are considered processes. In the study of a process, hazards and

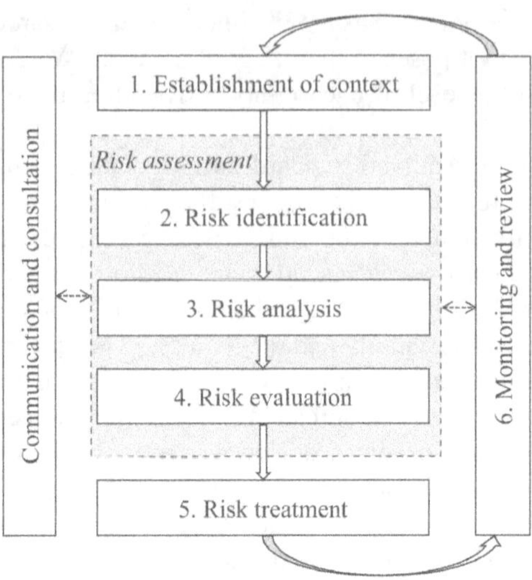


**FIGURE 1.4** General risk management framework. (Adapted from IEC31010 (2019))

assets can be either included in the process or external to it. For instance, while driving, the car and the driver are assets within the process, whereas pedestrians are external assets, because many of their decisions do not directly change their risk level, though their presence may influence the decision of the driver.

After the process is identified, the next task is to locate the asset of interest that needs protection. In most risk management practices, the focus is usually on a single asset or a group of similar assets. This approach is taken because if multiple assets are considered, the methods and resources to be used to satisfy their safety requirements can vary significantly.

Within a process, a range of accidental scenarios can be identified considering varying *deviations* from the expected performance and magnitudes of these deviations. From such a perspective, the process can be regarded as a *collection of scenarios*. If there is no deviation within the process, the asset remains unthreatened. The actions within the process may be executed by an individual, a group of people, or one or more technical systems. When the performance of a technical system is influenced by human, social, or organizational factors, it is termed a sociotechnical system.

We assume to undertake *general* or *strategic risk management* in the early phase before the process starts operating. In such kind of risk management, timeliness is not a primary concern. Once the process has started, for example, a technical system has been in operation, *operational risk assessment or dynamic risk assessment* methods can be employed to gather risk information and support decision more quickly. However, this chapter does not focus on the operational phase or these particular methodologies. Readers interested in dynamic risk assessment are encouraged to consult the book by Paltrinieri and Khan (2016).

2. *Risk identification*: We need to identify the potentially pivotal events that can harm the identified asset, along with the associated accidental scenarios, to address the first question in the definition of risk by Kaplan and Garrick (1981). As we know, the elements of an accidental scenario typically include hazards, initiating events, intermediate events, and end events. Ideally, each hazardous event that might endanger the asset should be thoroughly studied. However, it is impractical to identify all such events affecting the asset in a single risk identification session. Therefore, it is necessary to prioritize and focus on those events that are most representative and likely to occur. In addition, it is important that the pivotal event is described with sufficient specificity. For instance, a vague term like "gas leakage" is less informative, whereas "leakage of gas in Tank X" provides more clarity and is more useful for the following risk analysis.

   We can use $E$ to denote a hazardous event that may harm the asset and use $R(S|E)$ to denote the risk of system $S$ with respect to the event $E$.
3. *Risk analysis*: In this step, we need to answer the second and the third questions in the definition of risk by Kaplan and Garrick (1981), specifically the likelihood and the consequences of a hazardous event. To obtain such information, a lot of works are needed. Both qualitative and quantitative methods are helpful, some of which will be explored in later chapters of this book.

   For quantitative risk analysis, *probability* is used to denote the likelihood of an event, where probability values range from 0 to 1. In practical risk management, probability is often reflected by frequency (especially for those we have observations) and can be categorized into several levels, such as $\leq 0.01$/year, $0.01 - 0.1$/year, $0.1 - 1$/year, and $\geq 0.1$/year. For a deeper understanding of the terms *likelihood*, *probability*, and *frequency*, readers can consult Chapter 2 of the book by Rausand and Haugen (2020). We will introduce the mathematical foundation of quantitative risk analysis in Chapter 5.

   Regarding consequences, a hazardous event can lead to a range of potential outcomes, which can be described as a consequence spectrum and represented mathematically as a vector:

$$c = [c_1, c_2, \ldots, c_n] \tag{1.1}$$

   These consequences can represent various severities of harm to the asset. For instance, in Example 1.2, the consequences of a two-car collision range from slight injuries to severe injuries and even fatalities. In practice, consequences are classified into categories such as catastrophic, critical, major, and minor damage to differentiate levels of impact.

   Consequently, both probability ($p$) and consequence ($c$) are incorporated into the risk formula for the asset, expressed as $R[S|E(p,c)]$.
4. *Risk evaluation*: Risk evaluation is conducted based on the results of step 3. Given that safety is often regarded as an acceptable level of risk, it is

| Likelihood/Consequence | 1. Minor | 2. Major | 3. Critical | 4. Catastrophic |
|---|---|---|---|---|
| 1. Rare ($< 0.0001/y$) | Low | Low | Low | Medium |
| 2. Unlikely ($0.0001 - 0.001/y$) | Low | Low | Medium | High |
| 3. Possible ($0.001 - 0.01/y$) | Low | Medium | Medium | High |
| 4. Likely ($0.01 - 0.1/y$) | Medium | Medium | High | High |
| 5. Often ($\geq 0.1/y$) | Medium | High | High | High |

**FIGURE 1.5** An example of risk matrix.

necessary to compare the estimated risk against a predetermined threshold to determine if it has been sufficiently mitigated.

Numerous tools are now available for presenting and evaluating the results of risk analysis. One such tool is the *risk matrix*, illustrated in Figure 1.5, where the likelihood of a hazardous event occurring is plotted on the horizontal axis and the consequences are plotted on the vertical axis. This matrix categorizes risk using five levels of likelihood and four levels of consequence. The combination of these factors divides risks into three broad categories: low, medium, and high. This visual representation aids in quickly evaluating and ranking the risks associated with different scenarios and thus facilitates decision-making regarding risk management strategies.

Another commonly used risk index is the *risk priority number* (RPN). In its simplest form, the RPN of a hazardous event can be calculated using a *multiplication rule*, where it is determined by multiplying the probability of the event by its consequence. These factors are often quantified through objective or subjective numerical scales, such as the 1–5 scale used in Figure 1.5 for classifying likelihood:

$$\text{RPN} = p \cdot c \tag{1.2}$$

In some contexts, the *addition rule* is also applied, whereby the RPN is calculated as the sum of the probability and consequence:

$$\text{RPN} = p + c \tag{1.3}$$

Regardless of the calculation method, a higher RPN indicates a higher-risk level, raising a need for more attention and possibly immediate action. For a more in-depth discussion on the applications and implications of RPN, readers can refer to Chapter 6 of the book by Rausand and Haugen (2020).

The overall process of risk analysis and risk evaluation is called *risk assessment.*

5. *Risk treatment*: After the risk evaluation, decision-making is required on how to manage the evaluated risk. This process is called *risk-informed*

*decision-making*. Generally, there are three strategic options available based on the assessed levels of risk:

- *Avoidance*: When the risk is evaluated as extremely high, it is reasonable to avoid the activity or reject the system design that generates such a risk. This approach is particularly applicable when the potential outcomes are deemed unacceptable or when the risk poses severe threats to critical assets.
- *Acceptance*: When the risk is evaluated as very low, meaning that the hazardous event is highly unlikely to occur, or its consequences are negligible, the risk can be accepted or tolerated. In such cases, the asset is considered almost safe, and no further action is necessary.
- *Treatment*: When an activity is not avoidable but its risk on the asset is not ignorable, it is essential to implement risk treatment measures. One common measure is *risk transfer*, which involves shifting the financial burden of risk to another party. For example, a company may transfer the financial risk of a product liability claim to an insurance company. However, transferring risk is not always applicable or ethical, especially when human life or the environment is at stake. Thus, another effective measure is *risk reduction*, which involves decreasing either the likelihood of the pivotal event or mitigating its potential impact on the asset. This can be achieved through various controls, safety enhancements, or process improvements.

6. *Monitoring and review*: This step is to verify that the implemented risk management actions are being executed effectively and that the associated risks remain controlled within acceptable levels. It is essential to continually reassess the risks and the effectiveness of the mitigation measures, especially in response to any changes in potential operations, newly identified hazards, or experiences from others. Regular reviews help to ensure that the risk management strategies are relevant and are able to continue to provide the intended level of protection against risks.

In risk management, particularly within high-risk industries such as aviation, nuclear power, and oil and gas, the principle of *ALARP* (*as low as reasonably possible*) is frequently employed to determine whether a risk, possibly with specific treatments, is acceptable. This principle anchors the definition of safety under the context of risk management, suggesting that risks should be reduced as far as reasonably practicable, taking into account both the costs and the feasibilities of implementing risk reduction measures. ALARP acknowledges the reality that not all risks can be completely eliminated, and therefore, a compromise must be found between reducing risks and maintaining practicality.

Safety barriers are a typical method of risk reduction, especially when risks cannot be adequately controlled through standard technical system designs or typical procedural safeguards. The adoption of safety barriers is rooted in a *fail-safe* philosophy, which means that if a negative event occurs (fails), whether it is an abnormal initiating event or a pivotal event, these barriers can maintain the safety of the asset, or keep it within an acceptable risk level. This is aligning with what Figure 1.3

depicts. It also means that risk assessment needs to consider not only the probability and consequence of a hazardous event, but the efficacy of existing and potential barriers on preventing and mitigating the event.

Mathematically, we denote a barrier (or barriers) as *B*, anticipating that the barrier will be effective enough to ensure the following inequality is satisfied:

$$R\left[S|E(p,c),B\right] < R\left[S|E(p,c)\right] \tag{1.4}$$

This indicates that the risk level of an asset when a safety barrier is in place should be lower than the risk level in its absence.

We can then see the expanded risk management framework in Figure 1.6, which builds upon the framework presented in Figure 1.4, with emphasizing the use of barriers as the central strategy for risk reduction.

In the risk analysis phase, in addition to conducting probability and consequence analysis of the hazardous event, it is important to identify and analyze existing safety barriers. This approach provides a comprehensive understanding and a more

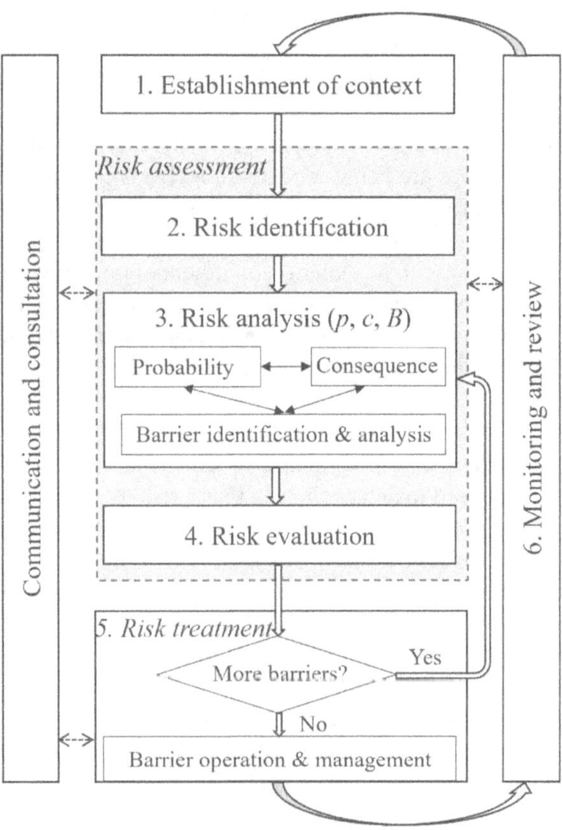


**FIGURE 1.6** Risk management framework in consideration of safety barriers.

accurate assessment of the actual risk to the asset. By examining the current safety barriers in place, it is also possible to identify any gaps or weaknesses that may increase the likelihood or severity of the hazardous event.

Following this, the risk treatment phase requires answering a key question based on the collected, analyzed, and evaluated risk information: Are additional barriers needed to ensure the asset being at an acceptable risk level?

If the answer is Yes, it is necessary to return the risk analysis phase to involve more barriers and then repeat the risk analysis and evaluation. Methods for introducing new barriers may include designing new physical or procedural barriers, implementing additional training for personnel, or enhancing the safety culture. This process should be iterative, continuing until the estimated risk to the asset is reduced to an acceptable level.

Conversely, when the answer is No, the focus shifts to managing and operating the existing barriers effectively. This includes ensuring proper operations, conducting regular inspections and maintenance of barriers, and providing regular training for relevant personnel. The purpose is to ensure the barriers to perform their required functions effectively, thereby maintaining the risk to the asset at an acceptable level.

The discussions on barrier engineering in this book primarily support risk analysis and risk treatment. The framework outlined in this subsection will be consistently followed throughout the book.

## 1.4 CLASSIFICATIONS OF SAFETY BARRIERS

Introducing classifications of safety barriers has significant value in understanding the functions and operational mechanisms of barriers. Moreover, these classifications can serve as guiding frameworks for identifying potential barriers, and thus some overlaps between certain groups within specific classifications can be accepted. The following subsections outline a variety of classifications for safety barriers.

### 1.4.1 Classification Based on Roles of Barriers in Risk Management

This classification of safety barriers is according to their roles in risk management, for understanding how barriers function along with an accidental scenario. A commonly used risk assessment framework is the bow-tie model, which effectively illustrates two main types of barriers (Liu, 2020; Sklet, 2006), as depicted in Figure 1.3. It can be found that barriers positioned on the left side of the pivotal event in the bow-tie model are designed to prevent the accidental scenario from developing from hazards to the pivotal event. These are referred to as *proactive barriers*, *frequency-reducing barriers*, or *preventive barriers*. Considering any hazardous event can be regarded as the pivotal event of study in a certain bow-tie model, such kind of safety barriers can be defined as follows:

- *Preventive safety barriers*: Safety barriers that are used to prevent the occurrence of a hazardous event or at least reduce the occurring probability/frequency of the event.

For example, in the context of car collisions, preventive barriers include the rear braking light on the front car and the automatic braking system and distance tracking system in the following car.

Conversely, barriers located on the right side of the pivotal event in the bow-tie model aim to stop the accidental scenario progressing from the pivotal event to severe consequences. These are known as *reactive barriers*, *consequence-reducing barriers*, or *mitigative barriers*. For a generally hazardous event, such kind of barriers is defined as follows:

- *Mitigative safety barriers*: Safety barriers that are used to mitigate the consequence/harm of the hazardous event to the asset.

In the case of car collisions, mitigative barriers include safety seat belts, airbags, and automatic door unlocking after a collision.

Expanding on this classification, Reason (1997) classified safety barriers into several categories, including (1) barriers for understanding and recognizing hazards, (2) barriers guiding safe operation, (3) barriers providing alarms, (4) barriers restoring the asset to a safe state, (5) barriers interposing between hazards and assets, (6) barriers containing and eliminating hazards, and (7) barriers providing escape and rescue.

Similarly, Rathnayakaa et al. (2011) identified seven types of barriers specific to process industries in the event of a release and categorized them as follows: (1) human factor barrier, (2) management and organizational barrier, (3) release prevention barrier, (4) dispersion prevention barrier, (5) ignition prevention barrier, (6) escalation prevention barrier, and (7) damage control emergency management barrier. The first three categories function as preventive barriers, while the latter four serve as mitigative barriers. Sevcik and Gudmestad (2014) proposed a similar method by categorizing safety barriers as prevention barriers, detection and control barriers, mitigation barriers, and emergency responses.

This classification highlights the two crucial roles that safety barriers play within a comprehensive risk management framework. When examining safety barriers in a system, it is important to assess whether both types of barriers, preventive and mitigative, are feasible to implement, or if one type has been overlooked.

### 1.4.2 Classification Based on Barrier Modality

The classification of safety barriers based on their modality or format is essential for understanding how the barrier function is realized. In this book, we categorize safety barriers into entitative and non-entitative types:

- *Entitative safety barriers*: Tangible physical entities that are designed and constructed to enhance safety of assets.
- *Non-entitative safety barriers*: Measures that are not physical entities but play an important role in enhancing safety of assets.

Entitative barriers can be further classified as physical barriers, technical barriers, and informative barriers:

- *Physical safety barriers*: These include any physical structure capable of performing barrier functions. Examples of physical safety barriers include the following:
  - Physical facilities located in a plant or other energy-related places such as fences, guardrails, barricades, walls, and barriers made of concrete, metal, or other durable materials.
  - Most personal protective equipment, such as helmets, safety glasses, gloves, thick-soled boots, and earplugs for hearing protection.
- *Technical safety barriers*: These are equipment that the use of various technologies to automatically perform at least some of barrier functions. Examples of technical safety barriers include the following:
  - Instrumentation systems, such as shutdown valves controlled by programmable logic controllers (PLCs).
  - Sensing systems, such as gas and fire alarms.
- *Informative safety barriers*: These entities are designed to provide critical safety-related information to individuals to prevent the occurrence of hazardous or to mitigate their consequences. Examples of informative safety barriers include the following:
  - Safety signs, such as warning signs, hazard symbols, and instructional signage that convey important information about potential dangers and necessary precautions.
  - Alarms, including audio or visual alarms that notify individuals of immediate or imminent dangers.
  - *Emergency information*, which includes evacuation maps and emergency exit signs that guide individuals to safety in crisis situations.

In this book, we classify informative safety barriers as the entitative ones because we assume that the safety information used to alert operators is provided by a tangible entity, which, like other physical barriers, requires operational and maintenance plans. However, some literature regards informative barriers as non-entitative, arguing that it is the information itself, not a physical object, that performs the barrier function.

It is important to recognize that informative barriers often need to work together with other types of barriers or require active engagement from individuals to effectively maintain safety. For instance, the effectiveness of a speed limit sign depends on drivers complying with the displayed speed limits. This interdependence highlights the need for a holistic approach to barrier engineering, where informative barriers are integrated with technical and organizational measures to ensure comprehensive protection.

It also should be noted that when we refer to "technical safety barriers" in this book, we typically consider the equipment that can take all required actions to ensure safety. However, if the equipment merely provides alarms or information and requires manual intervention, it is classified as an informative safety barrier. For instance, gas detectors with alarms and braking lights are considered informative safety barriers in this book, despite their reliance on digital sensing technologies.

The term "*safety-critical systems*" is more commonly used to describe what are essentially technical safety barriers in the context of high-risk industries such as nuclear power (NEA, 2003). In this book, we need to distinguish safety-critical systems from another term "safety-related systems", with the following definitions:

- *Safety-critical systems*: These are technical systems whose primary functions are to act as barriers. The failure of these systems when hazards exist may directly result in harm to the asset they are designed to protect. However, if there is no hazard, the failures of safety-critical systems do not impact the functionality of the asset.
- *Safety-related systems*: These are technical systems that perform primary functions other than acting as barriers but have the capability to protect the asset. The failure of these systems may lead to a loss of functionality of the asset, regardless of the presence of hazards.

We can view a safety-related system as an integral part of the *inherent safety design* of the larger system in which it is located. For example, in the event of a car collision, the airbag system is classified as a safety-critical system due to its direct role in mitigating harm, while the steering system is considered a safety-related system, as it primarily serves another function but contributes to overall vehicle safety. The relationship between barriers and safety design will be explored further in Section 1.5.

Safety-critical systems frequently employ instrumentation and digital technologies to detect abnormalities and analyze situations. The international standard for process industries, IEC61511 (2016), uses the term *safety-instrumented system (SIS)* to describe safety-critical systems equipped with instrumentation technologies. While in IEC61508 (2010), the term of *electrical/electronic/programmable electronic (E/E/PE) safety-related systems* is aligning with the concept of safety-critical systems discussed in this book. The primary focus of this book is on safety-critical systems and their roles in ensuring safety in various industries.

On the other hand, non-entitative barriers can be further classified as follows:

- *Procedural safety barriers*: These refer to specific job procedures and tasks that guide operations to prevent or reduce the risk of harm to the asset of interest. Examples of procedural safety barriers include the following:
  - Action limitations, such as exposure limits and permit-to-work systems, which require obtaining authorization from a designated authority before work can begin.
  - Lockout/tagout (LOTO), as a procedure designed to ensure that equipment is de-energized and isolated before maintenance or repair work is conducted.
- *Organizational safety barriers*: These encompass safety policies and strategic measures implemented at the organizational level to reduce risks. They

are also known as *administrative safety barriers*. Examples of organizational safety barriers include the following:

- Safety policy as formal documents that specify the guidelines and requirements for maintaining a safe working environment.
- Safety committees in a company for identifying and addressing safety issues within the organization and management commitment.

- *Perceptional safety barrier*: These barriers involve the culture established within an organization and other perceptual factors that influence workplace safety. Examples of perceptional safety barriers include the following:
  - The recognition of safety priority at all organizational levels.
  - Safety training and education, and other ongoing programs that reinforce the importance of safety, ensuring it is a constant focus.
  - Smooth communication channels ensuring that safety information is communicated effectively to all organizational members, facilitating a well-informed and responsive workforce.

In some software programs, specific codes or modules are introduced and implemented to avoid dangerous operations or tolerate faults, and these are considered conducing barrier functions. In this subsection, such software barrier functions also belong to non-entitative barriers. They can either be regarded a separate subcategory, or grouped under the procedural safety barriers, because it is reasonable to treat the codes or modules as specific tasks of the program.

Distinguishing these barriers from the intrinsic characteristics of an organization can be challenging, as many factors influence both safety and operational efficiency. To practically assess a barrier, one should consider whether it introduces additional workload or cost compared to a scenario where safety is not emphasized. More examples of these non-entitative barriers can be explored in Examples 1.3 and 1.4.

**Example 1.3:**

*Examples of Non-Entitative Barriers in a Process Plant*

In the chemical industry, plants often rely on the following non-entitative safety barriers.

Organizational safety barriers:

- Strong commitment to safety from top management;
- Regular safety audits and inspections to identify and mitigate potential hazards;
- Comprehensive safety policies and guidelines that are frequently reviewed and updated;
- Established communication channels specifically for reporting and addressing safety concerns;
- Continuous monitoring and evaluation of safety performance metrics;

Procedural safety barriers:

- Detailed standard operating procedures for the safe handling, storage, and disposal of hazardous materials;
- Mandatory process hazard analysis (PHA) before initiating hazardous operations;
- A permit-to-work system that mandates formal authorization before undertaking hazardous tasks;
- LOTO procedures to ensure equipment remains de-energized during maintenance;
- Well-documented emergency response plans and procedures;
- Procedures for incident reporting and investigation to determine root causes and prevent recurrence.

Perceptional safety barriers:

- Regular safety training and education for all staff;
- A culture that promotes accountability and responsibility for safety at all levels;
- Positive reinforcement and recognition for safe practices;
- An organizational culture that encourages reporting and learning from mistakes and incidents. ■

## Example 1.4:

### *Examples of Non-Entitative Barriers Related to Driving*

The following non-entitative safety barriers can help prevent injuries and fatalities in car collisions.

Organizational safety barriers:

- Regulations and laws mandating the use of seat belts, with penalties for non-compliance;
- Mandatory vehicle maintenance programs to ensure all cars meet safety standards;
- Routine performance checks on drivers conducted by traffic police;

Procedural safety barriers:

- A standardized driver license issuing process that includes rigorous testing;
- Pre-driving safety checks to ensure vehicle roadworthiness;

Perceptional safety barriers:

- Driver education programs that emphasize safe driving practices, defensive driving, and hazard awareness;
- A driving culture that discourages risky behaviors such as distracted driving, speeding, and driving under the influence of alcohol or drugs. ■

It should be noted that various researchers have proposed classifications similar to those discussed. For example, Sklet (2006) categorized safety barriers into three main types: physical barriers, technical barriers, and human and/or operational barriers. He further divided technical barriers into Safety instrumented system (SIS), other technology safety-related systems, and external risk reduction facilities. In addition, Hollnagel (2004) divides safety barriers into four categories: material barriers (corresponding to physical barriers), functional barriers (which overlap with both procedural and technical barriers), symbolic barriers (equivalent to informative barriers), and immaterial barriers, which encompass procedural, organizational, and cultural barriers. These classifications help to illustrate the diversity of approaches and mechanisms that can be used to enhance safety across various domains.

### 1.4.3 Classification Based on Operational Features

Safety barriers can also be classified according to their operational features, which describe how they function in response to specific conditions (Lugauer et al., 2016; Rausand, 2014):

- *Passive safety barriers*: These barriers operate inherently once installed, without requiring any external command, action, or energy input. They are also referred to as *constant barriers*.
- *Active safety barriers*: These barriers need to be activated by a command, action, or some form of energy to perform their functions in response to certain events. They are also known as *on-demand barriers*.

Examples of passive safety barriers include helmets and other personal protective equipment, safety defenses and bollards that prevent vehicles from colliding with buildings, fire-rated walls and doors, and safety nets designed to catch falling objects. It can be found that all these examples are entitative physical barriers.

Non-entitative barriers can also be passive in nature. For instance, when guidelines and rules are established, they can continuously govern behavior and operations without needing further input or activation. Once training and educational programs are conducted, they can leave a lasting awareness or mindset that influences behavior automatically.

On the other hand, examples of active safety barriers include fire and gas detectors in a house, anti-lock braking system (ABS), electronic stability program (ESP), and airbag system in a car, and pressure relief valves in the process industries.

In some studies, such as those by Kang et al. (2016), active barriers are also termed *positive barriers*. Most technical safety barriers or safety-critical systems belong to this category and typically include three subsystems: a sensor, a logic solver or analyzer, and an actuator, which together perform the full barrier function.

**Example 1.5:**

*A Matrix of Barrier Classification*

Using the above classifications, we can construct a matrix with three dimensions: role in risk analysis, barrier modality, and operational feature. This matrix can help locate various safety barriers in the context of a car collision scenario:

- *Driver training and test*: Classified as a preventive, perceptual, and passive barrier;
- *Speed limit on the road*: Classified as a preventive, informative, and passive barrier;
- *Braking light in the front car*: Classified as a preventive, informative and active barrier;
- *Airbag system*: Classified as a mitigative, technical, and active barrier. ■

This matrix is also useful for identifying potential barriers in accidental scenarios, helping to visualize how different barriers function and interact within an overall safety system.

### 1.4.4 Classification Based on Functional Sequence

Safety barriers are often implemented in multiple layers to protect valuable assets, with each layer designed to act in sequence to mitigate risks. These barriers can be classified based on their order in response to a hazard:

- *Primary safety barrier*: This is the first line of defense, positioned closest to the hazard. It is the most frequently used preventive barrier, designed to stop the development of a hazard to a hazardous event.
- *Secondary safety barrier*: This barrier is used when the primary barrier is bypassed or fails to function as expected. It serves as the next layer of protection, positioned to act after the primary barrier.
- *Ultimate safety barrier*: This is the last layer of protection, positioned closest to the asset. It is used only when all previous barriers have failed or been bypassed, providing the final attempt to protect the asset.

Between the primary and the ultimate barriers, there may be additional layers (including the secondary safety barrier) referred to as *intermediate safety barriers*. Typically, if a barrier is closer to the asset and farther to the hazard, it is less demanded.

It is needed to clarify that the terms "closer to" the hazard or asset refer to the logical sequence in which the barriers operate, rather than their physical proximity. These barriers serve various roles within the sequence. The secondary barrier generally experiences fewer demands and less force than the primary barrier due to the initial mitigation effects of the primary barrier. However, if the secondary barrier fails after the primary barrier has been breached, the resulting damage can often be significantly more severe. If the ultimate barrier is breached or fails, the asset remains unprotected and is likely to be damaged.

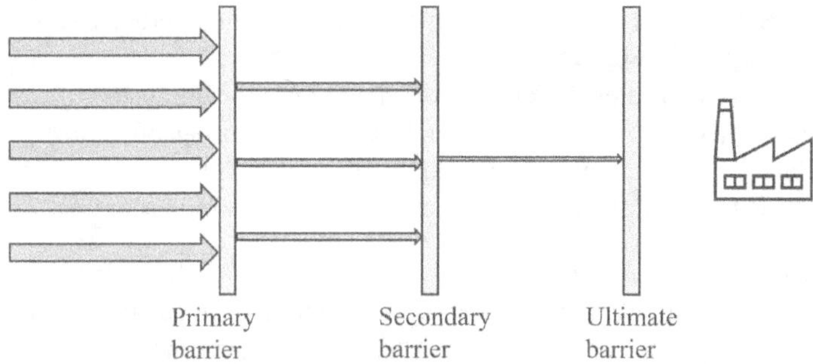


**FIGURE 1.7** Example of safety barriers in sequence.

For example, consider the car collision scenario: If we view a strong hit on the driver as the end event, then the speed limit might serve as the primary barrier, the braking light of the car in front as the secondary barrier, and the airbag system as the ultimate barrier.

Figure 1.7 illustrates the relationships and sequences of different types of safety barriers. The arrows indicate the direction of hazard demands, and the width of the arrows represents the strength of these demands. While the figure shows three layers of barriers, in practice, there can be additional layers, such as a seat belt, which might act as a tertiary barrier in a collision scenario, positioned before the airbag system.

### 1.4.5 Haddon's Classification

In his classical article of countermeasures (safety barriers), William Haddon proposed ten strategies based on the premise that accidents are caused by harmful energy when effective barriers are absent between the energy source and the asset (Haddon, 1973). Haddon suggests identifying countermeasures or safety barriers in three distinct phases and from ten different perspectives:

*Preinjury phase (Before the pivotal event)*:

1. Barriers that eliminate energy concentration, for example, using weather forecasts to avoid driving during severe weather or opting not to use flammable gases.
2. Barriers that limit the amount of energy, such as speed limit signs to discourage speeding (limiting kinetic energy) or reducing pressure within containers.
3. Barriers that prevent the uncontrolled release of energy, including ABSs or anti-corrosion coatings on tank walls (preventing chemical energy).

*Injury phase (from the pivotal event to consequence)*:

4. Barriers that reduce the rate or spatial distribution of released energy, such as studded tires (reduce kinetic energy by improving grip) or shutdown valves (cut off the flow of a hazardous substance with energy).
5. Barriers that spatially or temporally separate the asset from the energy source, such as proximity alarms or maintaining safe distances between fuel stations and neighboring buildings.

6. Physical barriers that isolate the asset from the energy, like mid-road barriers or firewalls.
7. Barriers that alter the quality of the energy, including modifications to the contact surface or basic structure, such as airbags in vehicles.
8. Barriers that enhance the resistance of asset to damage, such as personal protective equipment like helmets or safety training programs.

*Postinjury phase (consequence):*

9. Barriers that limit the progression of damage, for example, automatic unlocking of vehicle doors or fire suppression systems.
10. Barriers that ameliorate the damage, such as onboard first aid kits or repair and reconstruction efforts.

This strategic framework serves as a practical classification of safety barriers, and it actually can be used as a checklist for safety barrier identification. It is advisable to formulate specific questions for each type of safety barrier, like "Can the amount of energy be limited?" to facilitate brainstorming and discussions with domain experts aimed at enhancing safety.

### 1.4.6 Other Classifications

Recently, new classification approaches have been proposed that reflect evolving trends in barrier technology and research. For example, Pitblado et al. (2016) have introduced categories based on the operational constancy of safety barriers:

- *Static safety barriers*: These barriers maintain a consistent performance level, upheld through regular predetermined inspections and maintenance routines.
- *Dynamic safety barriers*: These barriers are characterized by performance that may degrade over time or under specific conditions.

The same authors (Pitblado et al., 2016) also proposed another classification based on the types of failures that safety barriers aim to prevent, which directly or indirectly influence the progression of accidental scenarios:

- *Immediate safety barriers*: These barriers are designed to respond to immediate failures that have a direct impact on causing accidents.
- *Temporal safety barriers*: These barriers address latent failures or hidden defects within the system that may not immediately cause harm but can allow accident scenarios to develop over time.

Based on the strategies employed by safety barriers to manage energy or hazards, barriers can be categorized into three types:

- *Hazard-containing barriers*: These safety barriers are designed to confine and control the flow of energy or hazardous elements, so as to protect the asset. For example, a firewall that stops the spread of flames and heat to critical properties.

- *Hazard-diluting barriers*: These safety barriers reduce the intensity of energy or hazard as it propagates. An example is a ventilation system that expels smoke or toxic gases from a confined space.
- *Hazard-transferring barriers*: These safety barriers either change the nature of hazardous energy to make it safe or redirect the energy flow away from the asset of interest. A lightning rod is a prime example, attracting and conducting lightning strikes away from structures and safely grounding the powerful electrical energy.

In the context of managing dependent failures within complex systems, Xie et al. (2018) classified safety barriers into three further categories: *safety barriers against individual failures*, *safety barriers against common cause failures*, and *safety barriers against cascading failures*. This classification will be explained in detail and discussed further in Chapter 10.

More comprehensive classifications can be explored in the review paper by Yuan et al. (2022), which provides an extensive overview of safety barrier categorizations.

## 1.5 SAFETY BARRIER AND INHERENT SAFETY DESIGN

In Section 1.3, we discussed that safety barriers are specific types of risk reduction measures, but it is not feasible to regard all risk reduction measures as barriers. For most sociotechnical systems, main risk reduction measures include inherent or integrated safety design features and (add-on) safety barriers (Kjellen, 2007; Liu, 2020; Yuan et al., 2022).

A sociotechnical system can be designed to be *inherently safe*, and this method is also known as *human-centered design* or *ergonomic design* when its primary focus is on human safety. In such a system, the safety or barrier functions are integrated into the essential functions of the system itself. For example, in the environments where movement and human activity are frequent, such as in schools, hospitals, and office spaces, the risk of injury from sharp corners is significant. To address this, furniture and equipment can be designed with rounded corners, which inherently reduces the likelihood of injuries from collisions. This design feature seamlessly integrates safety into the daily operations without requiring additional safety measures or behavior changes from the users. On the other hand, braking system is a good example of integrating safety function into a system. The primary role of a braking system is to control speed, but it acts as a crucial safety measure during critical moments.

In the framework of layer of protection analysis (LOPA, CCPS (2001)), which is widely utilized in the process industries, the concept of a "layer of protection" is closely aligned with that of a safety barrier. LOPA also includes inherent safety design, treating it as a distinct layer of protection. For example, a braking system can be categorized under LOPA as a basic control process system (BCPS), as an integral part of the system. Meanwhile, safety barriers are treated as additional protective layers beyond the BCPS. Further details on LOPA will be explored in Chapter 2.

There are several design principles that contribute to making a sociotechnical system inherently safe, as detailed in the work of Khan and Amyotte (2003):

- *Simplification*: Simplifying the design and structure of a system involves removing unnecessary complexity, reducing vulnerable components, enhancing control, and streamlining processes. Complexity is often the root cause of many safety issues within a system. As complexity increases, the system becomes harder to understand, troubleshoot, and maintain. Simplification helps to reduce the number of potential failure points, thereby decreasing the likelihood of malfunctions. Moreover, a simpler system is easier to diagnose and rectify, leading to quicker resolution of safety issues. However, it is important to note that oversimplification may lead to unintended safety risks, such as diminished system capabilities or increased vulnerabilities in certain scenarios.
- *Minimization*: This principle involves reducing the quantity of hazardous materials, energy, and other inherent hazards during the design phase. For example, toxic materials should be avoided, and heavy substances should be minimized in structures located in earthquake-prone areas. Minimizing hazardous materials can be realized by substituting dangerous materials with safer alternatives, such as replacing lead-based solder with lead-free solder in electronics. Another effective strategy is to limit the exposure of the asset to these hazardous materials and energy sources.
- *Safety margin (Overdesign)*: A safety margin is the calculated disparity between the capability of an item and the demands expected of it. Designing with a substantial safety margin ensures that components are more reliable, can withstand unexpected high loads, and are less likely to fail. Approaches to achieving a sufficient safety margin include using stronger and more durable materials, as well as incorporating redundant structures. The implications of safety margin will be further discussed in Chapter 3.
- *Fool-proof*: This principle indicates that a system should be designed to minimize the potential for user error, accommodating users who may lack specific training or knowledge. For example, user interfaces should be designed intuitive and easy to understand, reducing the complexity of operations, and enhancing the resilience of system to user mistakes. Ensuring that systems are fool-proof helps prevent misuse and enhances overall safety.

Designers need to be conscientious about safety regulations, especially when developing complex technical systems. Human-centered design and ergonomics play the key roles in enhancing safety and safeguarding human operators and users. These approaches need safety experts to combine engineering principles with the insights from psychological sciences (Kjellen, 2007). For instance, incorporating easily accessible emergency shut-off switches and designing equipment that minimizes the risk of musculoskeletal injuries can enhance operational safety through ergonomic design.

However, the focus of this book is on add-on safety barriers, as opposed to inherent safety design elements. To distinguish between these two approaches, we describe safety barriers as "add-on" because they are not directly related to the essential functions of the system or asset they protect. Essentially, these installed safety barriers should not impact the performance or functionality of an asset, provided that the asset is not subjected to hazards. Typically, safety barriers are implemented only when hazards cannot be completely eliminated or sufficiently mitigated by the inherent design characteristics of the system. This strategy is in line with the principle of simplification in design.

As we have discussed, in certain scenarios, safety barriers are introduced based on a fail-safe principle:

- *Fail-safe*: A system design principle where the system, as the asset of interest, can sojourn or transit to the safe state in the event of a specific failure or incident.

In practice, it may not always be essential to distinguish between safety barriers and integrated safety design when the primary objective is to minimize risks and protect assets. However, the clarification provided above is important to understand that safety barriers possess unique characteristics compared to other elements within a technical system. These barriers have specific operational modes and design requirements that will be investigated in this book. By exploring these unique aspects of safety barriers, we can obtain a deeper insight into how they can be designed and used to enhance system safety. The article by Kjellen (2007) offers additional insights into the relationship between inherent safety design and safety barriers, discussing how to balance these two critical approaches to safety engineering.

## 1.6 SAFETY BARRIER AND EMERGENCY RESPONSE

While it may not always be essential to distinguish between safety barriers and *emergency response programs*, clearly defining these concepts can enhance overall safety management and help identify appropriate strategies in various scenarios.

Safety barriers are measures designed to either reduce the occurring likelihood of a hazardous event or mitigate its impact (avoid an accident). In contrast, emergency response programs are the measures prepared to respond to an accident or emergency (caused by a hazardous event) once they have occurred. These programs involve detailed procedures and resources aimed at managing the immediate consequences of an accident. Examples of emergency response programs include firefighting procedures, evacuation drills, and emergency medical responses.

A key difference between safety barriers and emergency response programs is their timing relative to an accident or harm, which is represented as the end event in the consequence spectrum of a bow-tie model. The measures implemented before harm occurs are classified as safety barriers, whereas the measures applied after harm are considered emergency responses. In the accident model proposed by Sevcik and Gudmestad (2014), the authors identify three critical points in the accident chain: the hazard (initiating event), critical deviation (pivotal event), and accident (end event). Methods employed after an accident occurs are categorized as emergency responses.

However, distinguishing between safety barriers and emergency responses can be complex in practice, as it is often challenging to define precisely what constitutes an end event, as discussed in Section 1.2. In many instances, what is considered the pivotal event might be treated as an accident, and all mitigative measures placed after this point in the bow-tie diagram are viewed as components of the emergency response program.

Another factor that distinguishes safety barriers from emergency response programs is their integration within the system they protect. Safety barriers are typically installed as part of the system that they aim to protect, even if they do not directly relate to the main system function. In contrast, emergency response programs often rely on external resources or need to be initiated externally. For example, in car collisions, medical rescue teams and firefighters are not considered safety barriers installed in cars and even not in the transportation system. From an operational standpoint, emergency response programs are always active or on-demand, only respond when needed.

## 1.7 MORE TERMINOLOGIES

To fully comprehend the concept of safety barriers, it is essential to have a foundational understanding of the various terminologies commonly used across different sectors and disciplines to describe them. These terms are often used interchangeably, which sometimes can lead to confusion and misunderstandings. Therefore, before the detailed discussion of safety barriers in this book, a brief overview of the fundamental concepts and definitions is necessary.

Some of the common terminologies that are similar to or relevant to safety barriers include the following:

- *Risk reduction measure*: As discussed earlier in Section 1.2, this term can be any measure that either reduces the likelihood of a pivotal event or mitigates the consequences of such an event. Risk reduction measure has a broader scope than that of a safety barrier, including methods used in inherent safety design and organizational/operational management.
- *Defense*: According to Harms-Ringdahl (2009), this term includes both physical safety barriers as hard defense and non-physical barriers such as regulations, procedures, and training, as soft defense. From this perspective, the concept of defense in this book covers the same scope as that of a risk reduction measure.
- *Safeguard*: A safeguard is a measure or system that is implemented to prevent or mitigate a hazard. Within this book, a safeguard is synonymous with a safety barrier, and in some interpretations, a safety barrier is considered a type of safeguard that prevents hazardous events from progressing beyond a certain point.
- *Protection layer or protective layer*: This term originates from LOPA (CCPS, 2001), where a protection layer is defined as a device, system, or human action that reduces the likelihood and/or severity of a specific loss event. Protection layer has similarities with safety barriers, but unlike safety barriers, layers in LOPA are classified according to their physical proximity to the asset, which is not typically the analytical approach used with safety barriers.
- *Safety indicator*: This involves a technical parameter, or a measure related to human or organizational behavior that serves to assess and monitor the safety performance of a system, process, or organization. Examples of

safety indicators include the number of accidents, incidents or near-misses, the frequency of safety inspections, and the availability and usage of safety barriers.

Next, we can introduce additional terms for use in the analysis of safety barriers within the framework of risk management. Up to now, the concept of safety barriers has evolved beyond the simple hazard-barrier model, leading to a redefinition of the barrier function:

- *Barrier function (2)*: The function of reducing the occurrence probability of a hazardous event and/or mitigating its consequences.

As discussed in Section 1.4, certain barrier functions of technical elements, especially those involving informative functions, require interaction with human behaviors to effectively ensure asset safe. For example, the rear braking light of the front car provides a barrier function in preventing collisions. However, its role is limited to alerting the trailing driver of a potential hazard. It is then the responsibility of the trailing driver to promptly notice the warning and take appropriate action to prevent a collision. In this scenario, the barrier function of the braking light is activated, but it is insufficient for ensuring safety alone. The overall barrier function in this case is divided into several subfunctions, with the braking light fulfilling just one aspect. Consequently, we also need to introduce the following:

- *Sub-barrier function*: A part of a barrier function that normally involves detecting hazards, analyzing the situation, preventing a hazardous event, or mitigating its consequences.

A single barrier subfunction can be further broken down into sub-subfunctions if detailed analysis is required. Thus, a barrier function can be decomposed into subfunctions, forming a hierarchical or tree structure as illustrated in Figure 1.8. This process of breaking down the overall barrier function into sub-barrier functions is known as *functional analysis*. A *functional tree* can be constructed by questioning *how* a function is realized, with the answers forming its subfunctions. This methodical breakdown helps clarify the roles and effectiveness of different safety mechanisms within a comprehensive risk management strategy.

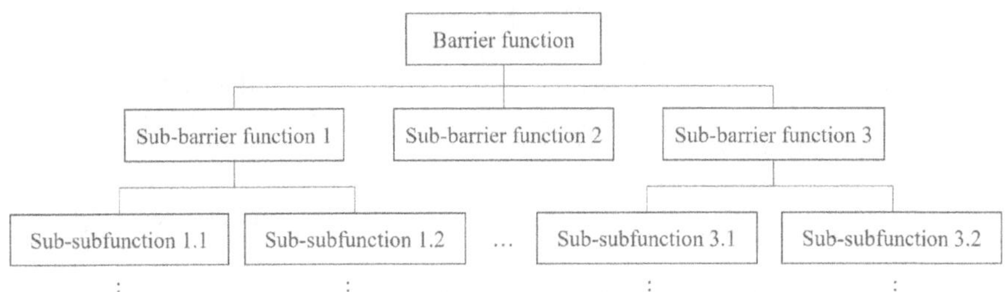


**FIGURE 1.8** Example of a functional tree for safety barrier.

**Example 1.6:**

### *Barrier Function Decomposition of Emergency Shutdown*

Consider the critical barrier function of an emergency shutdown system in a process plant, which is activated to stop or shut down the process under abnormal conditions. This barrier function comprises several sub-barrier functions:

- *Detecting the abnormal condition*: This involves the continuous monitoring and collection of process data to identify any deviations from normal operating parameters.
- *Analyzing collected data*: Once data is gathered, it is analyzed to determine the severity of the deviation and the necessary control actions to be taken.
- *Implementing the shutdown action*: Based on the analysis, the system executes the shutdown procedure to safely halt the process and prevent any further escalation of the abnormal condition.

Furthermore, the sub-barrier function of detecting the abnormal condition can be broken down into multiple sub-subfunctions, which include the following:

- Detecting high temperature.
- Detecting high pressure.
- Detecting loss of power.

When a barrier function is achieved by a technical system, a social system, or a combination of both, this system is referred to as a *barrier system*:

- *Barrier system*: A group of elements that is configured to perform specific barrier functions and to interact with other systems, including the asset of interest.

In this book, the concept of a barrier system aligns with that of technical safety barriers. Thus, in subsequent chapters, references to barrier systems specifically exclude organizational or procedural barriers. A barrier system may incorporate mechanical, electro-mechanical, and/or software elements to execute its barrier functions. According to Yuan et al. (2022), a barrier system should be capable of independently performing a complete safety function. Such a barrier system can be broken down into several subsystems:

- *Barrier subsystem*: A group of elements within a barrier system that deliver at least one barrier subfunction, or a part of the barrier function needed by the system.

A barrier subsystem can then be further decomposed into barrier components:

- *Barrier component*: An element that delivers some aspect of the barrier function. One or more barrier components can make up a barrier subsystem. ■

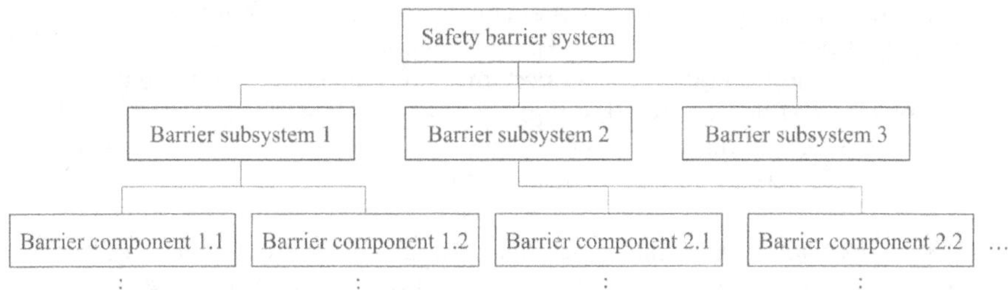


**FIGURE 1.9** Example of a physical decomposition of a safety barrier system.

In some studies, a barrier component is also referred to as a *unit.* Throughout most of this book, a barrier subsystem consists of several barrier components that share the same or similar barrier function. If a certain number of these components are operational, the function of the subsystem can be realized. Moreover, a barrier component is considered the smallest unit meaningful for performance assessment and is subject to maintenance, namely it is the maintainable unit.

Figure 1.9 illustrates the hierarchical structure of a barrier system and its components. The process of decomposing a system into components is known as *system structural analysis.*

### Example 1.7:

#### *Physical Decomposition of an Emergency Shutdown System*

Consider the emergency shutdown function described in Example 1.6, which is performed by an emergency shutdown system. This system acts as a technical safety barrier and comprises three subsystems: a sensing subsystem for data collection, a logic solver subsystem for evaluating whether conditions are abnormal, and an actuating subsystem for executing the shutdown action. Within the sensing subsystem, three sensors operate as barrier components. The logic solver subsystem and the actuating subsystem are equipped with one or two programmable logic controllers (PLCs) and two shutdown valves, respectively. These two shutdown valves are often referred to as two channels within the actuating subsystem.

Readers can go to Figure 3.1 to see the system configuration, and further details about how this system operates will be discussed later in the book.

To streamline terminology in this book, we frequently use the term "barrier system" to describe a system or a subsystem. For example, two shutdown valves together may form a system, which we refer to as a barrier system for simplicity, though it technically constitutes a barrier subsystem that requires sensors and PLCs to perform the full barrier function.

Additionally, we will use the following two terms for clarity:

- *Item*: This term is used as the general description for a system or any part of a system.
- *Element*: This term is used to describe a part of an item when necessary.

Based on these definitions, a barrier component can be referred to as an item or as an element within the item of a barrier system.

It also should be noted that the performance of a barrier component, while performing its barrier function, is not solely determined by its own design characteristics. It is also influenced by various other factors, which are known in barrier engineering as performance-influencing factors:

- *Performance-influencing factor (PIF)*: An internal or external factor that can influence the performance of a barrier or a barrier element. ■

Examples of PIFs include maintenance methods, effectiveness of maintenance, frequency of regular tests and inspections, operational strategies, and training programs. These factors can significantly influence the reliability and effectiveness of barrier systems in preventing or mitigating hazards.

## 1.8 SCOPE OF THIS BOOK

Based on the definitions provided, here we provide a summary of the scope and approach of barrier engineering:

- *Barrier engineering*: The process of designing, analyzing, and implementing both entitative and non-entitative safety barriers to prevent hazardous events or mitigate their consequences, thereby maintaining the risk to the asset of interest at an acceptable level.

Barrier engineering principles are applicable across various industries, including manufacturing, transportation, energy, and healthcare, where safety risks are significant, and the consequences of accidents can be severe.

The remainder of this book details barrier engineering and management, organized as follows:

- Chapter 2 introduces barrier management methods, focusing particularly on non-technical safety barriers and providing a comprehensive overview of strategies and practices.
- Chapter 3 discusses the design and operational characteristics of technical safety barriers, including an exploration of their potential failure modes.
- Chapter 4 covers hazard identification methods and qualitative risk analysis techniques, providing readers with the tools needed to identify potential failures in safety barriers and assess associated risks.
- Chapter 5 presents quantitative analysis methods and models used in safety barrier management, along with their foundational probability theories. This chapter builds the mathematical groundwork necessary for the analyses conducted in subsequent chapters.
- Chapter 6 outlines the main performance measures for safety barriers, establishing the criteria for their evaluation.
- Chapter 7 discusses the data analysis methods in barrier engineering, some of which support the quantitative analysis models introduced in Chapter 5.

- Chapter 8 applies the methodologies introduced in Chapter 5 to analyze the performance of barriers across different configurations and operational strategies.
- Chapter 9 focuses on the effects of various testing and maintenance strategies on barriers and assesses the time-dependent performance of safety barriers, considering phenomena of degradation.
- Chapter 10 offers models and algorithms for evaluating dependent failures within a safety barrier system and interdependent failures between safety barriers and other system elements they protect.
- Chapter 11 explores digitalization issues in barrier engineering and the potential applications of artificial intelligence technologies in this field.
- Chapter 12 describes security issues related to safety barriers and explores how the principles of barrier engineering can be applied to enhance security assurance.
- Chapter 13 investigates the relationship between barrier engineering and sustainability, focusing on designing and operating barriers sustainably and using safety barriers to improve sustainability.
- Chapter 14 extends the concept of safety barrier to resilience enhancer and discusses foundational topics related to this novel concept.

With these chapters, this book aims to provide a comprehensive exploration of safety barrier design, operation, management, making them essential for professionals and academics interested in advanced safety and risk management.

## REFERENCES


Aven, T. (2015). *Risk Analysis.* John Wiley & Sons.

Aven, T., & Renn, O. (2009). On risk defined as an event where the outcome is uncertain. *Journal of Risk Research, 12*(1), 1–11.

CCPS. (2001). *Layer of Protection Analysis: Simplified Process Risk Assessment.* American Institute of Chemical Engineers.

CCPS, & EI. (2018). Bow Ties in Risk Management: A Concept Book for Process Safety. American Institute of Chemical Engineers.

Gibson, J. J. (1961). The Contribution of Experimental Psychology to the Formulation of the Problem of Safety—A Brief for Basic Research. *Behavioral Approaches to Accident Research, 1,* 77–89.

Haddon, W. J. (1973). Energy damage and the ten countermeasure strategies. *Human Factors: The Journal of the Human Factors and Ergonomics Society, 15*(4), 355–366.

Harms-Ringdahl, L. (2009). Analysis of safety functions and barriers in accidents. *Safety Science, 47*(3), 353–363.

Hollnagel, E. (2004). *Barriers and Accident Prevention.* Ashgate.

IEC31010. (2019). Risk management – Risk assessment techniques. Geneva, Switzerland: International Electrotechnical Commission.

IEC 61508. (2010). Functional Safety of Electrical/Electronic/Programmable Electronic Safety-Related Systems, International Electrotechnical Commission, Geneva, Switzerland.

IEC 61511. (2016). Functional Safety - Safety Instrumented Systems for the Process Industry Sector, International Electrotechnical Commission, Geneva, Switzerland.

IEC62443-1-1. (2009). Industrial communication networks – Network and system security. In *Part 1-1: Terminology, Concepts and Models.* Geneva, Switzerland: International Electrotechnical Commission.

ISO31000. (2018). Risk management – Guidelines. Geneva, Switzerland: International Organization for Standardization.

Johnson, W. G. (1975). MORT: The management oversight and risk tree. *Journal of Safety Research*, *7*(1), 4–15.

Kang, J., Zhang, J., & Gao, J. (2016). Analysis of the safety barrier function: Accidents caused by the failure of safety barriers and quantitative evaluation of their performance. *Journal of Loss Prevention in the Process Industries*, *43*, 361–371.

Kaplan, S., & Garrick, B. J. (1981). On the quantitative definition of risk. *Risk Analysis*, *1*(1), 11–27.

Khan, F. I., & Amyotte, P. R. (2003). How to make inherent safety practice a reality. *The Canadian Journal of Chemical Engineering*, *81*, 2–16.

Kjellen, U. (2007). Safety in the design of offshore platforms: Integrated safety versus safety as an add-on characteristic. *Safety Science*, *45*, 107–127.

Liu, Y. (2020). Safety barriers: Research advances and new thoughts on theory, engineering and management. *Journal of Loss Prevention in the Process Industries*, *67*, Article 104260. https://doi.org/10.1016/j.jlp.2020.104260

Lugauer, F., Stiehl, T., & Zaeh, M. (2016). Functional safety of hybrid laser safety systems – How can a combination between passive and active components prevent accidents? *Physics Procedia*, *83*, 1196–1205.

NEA. (2003). *Engineered Barrier Systems (EBS) in the Context of the Entire Safety Case.* Nuclear Energy Agency.

Paltrinieri, N., & Khan, F. I. (2016). *Dynamic Risk Analysis in the Chemical and Petroleum Industry: Evolution and Interaction with Parallel Disciplines in the Perspective of Industrial Application.* Butterworth-Heinemann.

Pitblado, R., Fisher, M., Nelson, B., Fløtaker, H., Molazemi, K., & Stokke, A. (2016). Concepts for dynamic barrier management. *Journal of Loss Prevention in the Process Industries*, *43*, 741–746.

Rathnayakaa, S., Khan, F., & Amyotte, P. (2011). SHIPP methodology: predictive accident modeling approach. Part II. Validation with case study. *Process Safety and Environment Protection*, *89*(2), 75–88.

Rausand, M. (2014). *Reliability of Safety-Critical Systems: Theory and Applications.* John Wiley & Sons.

Rausand, M., & Haugen, S. (2020). *Risk Assessment: Theory, Methods and Applications.* Wiley.

Reason, J. (1990). *Human Error.* Cambridge University Press.

Reason, J. (1997). *Managing the Risks of Organizational Accidents.* Ashgate.

Sevcik, A., & Gudmestad, O. T. (2014). *A systematic approach to risk reduction measures in the Norwegian offshore oil and gas industry.* The 9th International Conference on Risk Analysis and Hazard Mitigation, New Forest, UK.

Sklet, S. (2006). Safety barriers: Definition, classification, and performance. *Journal of Loss Prevention in the Process Industries*, *19*(5), 494–506.

US DOE. (1996). Hazard and Barrier Analysis Guidance Document. U.S. Department of Environment.

Xie, L., Lundteigen, M. A., & Liu, Y. (2018). Common cause failures and cascading failures in technical systems: Similarities, differences and barriers. In S. Haugen, A. Barros, C. Gulijk, T. Kongsvik & J. E. Vinnem (Eds.), *Safety and Reliability – Safe Societies in a Changing World.* CRC Press.

Yuan, S., Yang, M., Reniers, G., Chen, C., & Wu, J. (2022). Safety barriers in the chemical process industries: A state-of-the-art review on their classification, assessment, and management. *Safety Science*, *148*, 105647. https://doi.org/10.1016/j.ssci.2021.105647

# 2 Barrier Management and Non-Technical Barriers

## 2.1 INTRODUCTION OF BARRIER MANAGEMENT

### 2.1.1 Barrier Management and Barrier Engineering

Barrier management is recognized within the offshore industries as a key component of risk management. According to the Norwegian Petroleum Safety Authority (PSA, 2017), barrier management is a systematic and continuous process that involves the identification, establishment, and maintenance of barriers. This process ensures that the barriers in use remain relevant, effective, and robust. In the earlier version of barrier management report by the same institution (PSA, 2011), barrier management includes the management activities related to processes, systems, solutions, and measures. Additionally, a report of DNV (2014)[1] links barrier management to various aspects of improving safety, including safety culture, operational risk management, and organizational learning.

Generally speaking, the term "management", as defined by Henri Fayol (1841–1925), includes several tasks: Planning, organizing, commanding, coordinating, and controlling. Therefore, barrier management in this book is defined as:

- *Barrier management*: The integration of planning, organizing, and coordinating the possible resources and relevant elements to achieve and control the required barrier functions for protecting the asset of interests.

This definition highlights the importance of coordinating interactions between elements of safety barriers and other factors related to the asset. In this context, the approach of executing barrier management is referred to as a *barrier management strategy* or simply a *barrier strategy*.

On the other hand, as defined in Chapter 1, barrier engineering primarily focuses on the design and implementation of physical and technical safety barriers (entitative barriers). This can be considered a part of the broader barrier management framework. Additionally, the aim of barrier engineering is to establish the necessary measures to mitigate risks, while safety barrier management ensures these measures are effective and feasible, given organizational goals and resource constraints. Thus, barrier engineering and barrier management are interdependent, each bolstering the effectiveness of the other.

In subsequent sections of this book, we will explore detailed aspects of barrier engineering for technical safety barriers, including modeling, quantitative methods for barrier design and analysis, and the inspection and maintenance of barriers. However, this chapter will concentrate on the comprehensive process of barrier management, addressing both technical and non-technical aspects.

DOI: 10.1201/9781003245636-2

### 2.1.2 Principles of Barrier Management

There are some general principles of barrier management that should be noted.

- *Integrating with risk management*: Barrier management should be systematically integrated with risk management to enhance safety measures effectively (Johansen & Rausand, 2015). As discussed in Chapter 1, safety barriers have a crucial role within the overall risk management framework. The bow-tie model introduced earlier highlights the significance of prioritizing preventive or proactive safety barriers on the left side of the pivotal event. These barriers aim to eliminate or reduce the occurrence likelihood of a hazardous event and are generally more effective. This chapter introduces a barrier management framework aligned with risk management principles, illustrating how insights from risk analysis can guide the decision-making in barrier management.
- *Keeping independence and diversity of safety barriers*: Safety barriers are expected to operate and behave independently from other parts of the asset and from other barriers to ensure that a failure in one does not compromise the performance of others or the asset as a whole. To this end, barriers should be designed and operated using diverse materials, techniques, design methods, operational modes, and maintenance strategies within a unified management framework. By relying on a variety of approaches, each barrier can function differently and effectively even in the face of similar hazards and is less likely to suffer from the same failure causes. This approach, known as *defense in depth* (Holmberg, 2017), provides multiple layers of protection and reduces the likelihood of simultaneous failures. Defense in depth will be explored further in Section 2.3.
- *Encouraging interdisciplinary collaboration*: Effective barrier management requires collaboration across various disciplines, as illustrated in Figure 2.1. Corporate management sets the strategic direction and policies, aligning risk management with the organization's culture and values while ensuring the allocation of necessary resources for barrier management. It is important for corporate management to establish a structured approach and

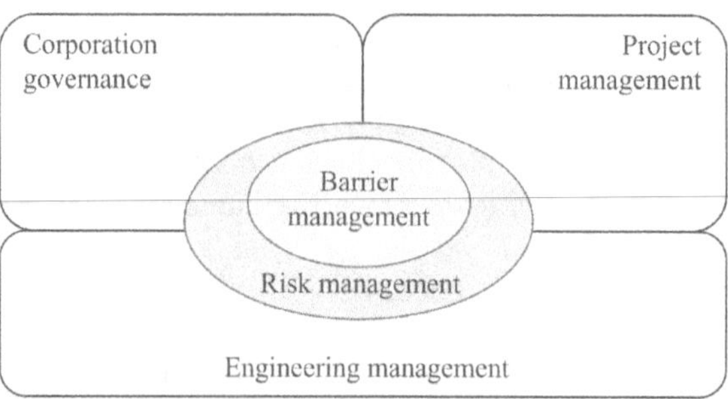


**FIGURE 2.1** Barrier management as the intersection of multiple disciplines.

provide adequate resources to support the development and implementation of barrier management. Project management plays a vital role in integrating barrier management throughout the project lifecycle, from planning to decommissioning. This integration involves identifying potential hazards and risks, developing risk mitigation strategies, and establishing suitable barriers to manage these risks effectively. Engineering management is essential for the design, implementation, and maintenance of safety barriers, particularly technical barriers. Responsibilities of engineering management include developing and testing barriers to verify their suitability, regularly monitoring their performance, and carrying out any necessary upgrades or modifications to maintain their effectiveness.

- *Streamlining communication*: Seamless communication and collaboration across all management areas are critical for the successful implementation of barrier management. By fostering a collaborative environment to identify, assess, and manage risks, organizations can proactively address potential communication issues. This ensures the highest level of safety and protection for their assets and stakeholders, maintaining an efficient and effective risk management process.

## 2.2 BARRIER MANAGEMENT MATRIX

Based on Figure 2.1, it is meaningful to explore how barrier management can be integrated with corporate and project management over time. Such analysis can clarify the roles of various personnel and departments within an organization in managing safety barriers. For example, the organization responsible for overseeing safety barriers can be in various forms, such as a small team, a company, an industrial sector, or even a national entity. In this book, we typically focus on an industrial corporation that operates a sociotechnical system, ensuring the safety of assets connected to the system, including employees, customers, nearby residents, machinery, facilities, and the environment.

Effective safety and barrier management of sociotechnical systems within a corporation can generally be conducted at two levels: governance and execution level. The tasks at the governance level involve developing long-term barrier strategies and allocating and managing resources related to barriers. The execution level, on the other hand, is directly responsible for the practical realization and implementation of these barriers as safety measures. In this context, barrier engineering, as discussed in this book, primarily pertains to activities at the execution level.

In addition, the project management process for ensuring safety in sociotechnical systems can be divided into three distinct phases: the planning phase, during which risks and corresponding barriers are identified; the operating phase, which begins when the asset and barriers are operational and during which the barriers are implemented, monitored, and maintained; and the responding phase, which is triggered by the occurrence of a hazardous event and necessitates actions to mitigate its impacts.

Considering these two levels of corporate management with the three phases of project management, we can construct a matrix with six sectors/stages. This

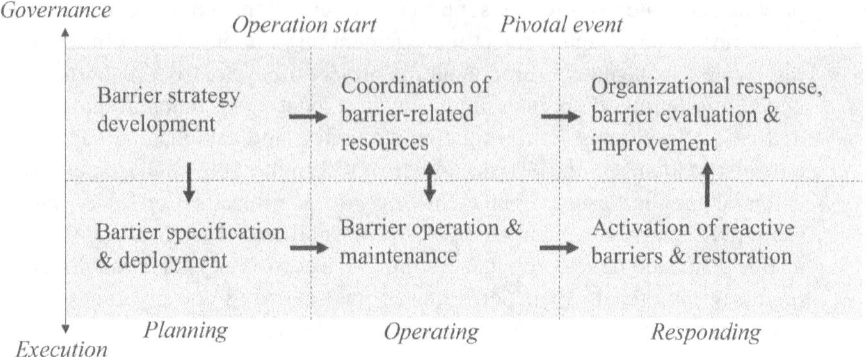


**FIGURE 2.2** Barrier management at different levels and at different phases.

organizational structure is depicted in Figure 2.2, and the details of each stage are described as follows.

### 2.2.1 Stage I: Barrier Strategy Development

This stage, located at the intersection of the governance level and the planning phase, focuses on the strategic tasks such as setting long-term goals for barrier management and establishing performance requirements for barriers. It also involves analyzing available and potential resources for barrier management and developing a safety culture and barrier management regulations within the organization.

The performance requirements for barriers should be manageable, meaning that they need to be practical, feasible, and achievable. This implies that barriers should be straightforward to install and maintain, and their performance requirements should be achievable with the available resources and expertise in the organization. Too complex performance requirements might hinder the effectiveness of safety barriers in preventing hazards and risks. Meanwhile, too high or strict performance requirements may pose significant challenges for the organization, potentially leading to safety hazards or compliance issues.

Overall, the objective in this stage is to build a high-reliability organization (HRO), as detailed in the article by Rochlin (1996), characterized by a strong safety culture, risk-awareness, well-structured organizational frameworks, and efficient procedures that support and enhance barrier management.

### 2.2.2 Stage II: Barrier Specification and Deployment

This stage, located at the execution level and within the planning phase, is for identifying and analyzing potential hazards and risks. This essential step establishes the foundation of the overall barrier management system for the organization.

In addition to risk identification, this stage assesses the inherent safety design of sociotechnical systems, which includes the interactions of people, machines, and processes that collectively produce goods and services. Evaluating the inherent safety design is vital for determining the required extent of risk reduction. The subsequent task is to identify the most suitable safety barriers for implementation, which includes evaluating the effectiveness of different barrier options. This evaluation often draws on lessons learned from existing applications, accident reports, and best practices within the industry. Various qualitative and quantitative analysis methods discussed later in Chapters 5–8 are utilized in this stage.

Once appropriate safety barriers are identified, the next step includes their deployment and installation at proper locations. The installation process requires careful planning and execution to ensure that the barriers are deployed in a manner of effectively mitigating the identified risks. Following installation, it is necessary to assess the effectiveness of these safety barriers within the context of the overall barrier management system of the organization.

### 2.2.3 Stage III: Barrier-Related Resource Allocation

This stage, positioned at the governance level and operating phase, involves coordinating efforts across different departments within the corporation to allocate necessary resources, such as personal protective equipment, safety equipment, and the labor of safety experts and technicians, as well as other relevant consumable materials. These resources are essential for the smooth operation of various safety barriers.

Management also needs to monitor the effectiveness of organizational, perceptual, and procedural barriers to ensure they are integrated with the corporation's culture and effectively interact with other parts of the organization.

In this stage, conducting a safety-economy analysis is very helpful to establish an optimal balance between safety and production objectives, considering that all barriers are with costs. For example, in the offshore industry, the deployment of safety barriers, including passive safety barriers like fire and blast walls and active technical barriers such as emergency shutdown systems and pressure relief systems, can lead to increased capital expenditures and affect platform size and weight. It is therefore essential to carefully evaluate which safety barriers are necessary and how they should be deployed to minimize both capital and operational costs while maintaining risks at acceptable levels. This approach requires a thorough analysis of the costs associated with safety barriers in relation to the potential risks and hazards they mitigate.

It should be noted that Stages II and III often run concurrently but at different management levels. In practice, Stage III has interactions with both Stages II and IV.

### 2.2.4 Stage IV: Barrier Operation and Maintenance

This stage, located at the execution level and operating phase, primarily focuses on the daily operation of safety barriers. One critical task is to ensure that all informative barriers, such as warning signs, labels, and manuals, are current, clearly visible, and comprehensible to both employees and visitors. Implementing a well-structured and consistent method for displaying these barriers is essential to prevent confusion and ensure the correct interpretation of crucial safety information.

Physical barriers, such as fences, walls, or access control systems, require regular inspections to detect any damage or defects that might compromise their functionality. On the other hand, technical barriers need more frequent or continuous monitoring to guarantee their status and availability. Regular tests and inspections are important to keep these barriers functioning well and prepared to activate in an emergency. Additionally, regular maintenance tasks, including calibration and part replacement, are also vital to maintain these barriers functioning.

*Preventive maintenance* is a key component of effective barrier management and engineering, with the aim of preventing failures before they occur. This may include proactive measures such as replacing worn-out parts before they fail or changing filters before they become clogged. On the other hand, *corrective maintenance* becomes necessary when a physical or technical barrier fails to perform as expected. Such maintenance is expected to restore the barrier to its original operational state, potentially involving part replacement, equipment repair, or system reconfiguration. We will discuss more details of barrier maintenance in Chapter 9.

### 2.2.5 Stage V: Barrier Activation and Restoration

This stage is at the execution level and the responding phase, focusing on the activation of safety barriers to stop the development of an accidental scenario. The emphasis here is on technical- and active barriers, specifically designed to respond to specific hazards or emergency situations. Failure to activate these barriers will lead the accidental scenario to the worst consequences and perhaps severe damage.

Once a hazardous event is contained, it is crucial to assess the condition of the barrier and, if possible, restore its function to an as-good-as-new state.

Additionally, this stage involves conducting a post-incident review to evaluate the effectiveness of the activated barriers and identify the areas to improve. This review may lead to modifications in the barriers to enhance their responsiveness to the specific hazards or emergency scenarios they are designed to address.

### 2.2.6 Stage VI: Organizational Response, Barrier Evaluation, and Improvement

This stage is at the governance level and the responding phase, with tasks centered around the necessary responses of the organization when a hazardous event occurs. Responsibilities of this stage include disseminating alarm information, organizing the evacuation of employees and nearby residents, and isolating high-risk areas to prevent further damage.

After the emergency response, the organization needs to evaluate the performance of the overall barrier management system, which includes both entitative and non-entitative barriers. This evaluation should be conducted promptly to capture lessons learned and implement necessary changes to the barrier management system. The aim is to identify weaknesses within the existing barrier system and provide recommendations for future enhancements. These enhancements could include modifications to the design or placement of barriers, the addition of technical barriers, and further training for employees on barrier management procedures.

It is evident that management at the execution level primarily focuses on entitative barriers and is closely linked to barrier engineering. The objective of this level is to ensure these barriers are functioning effectively and efficiently, providing the necessary protection. On the other hand, the governance level has a broader scope, dealing with all types of barriers and considering how barrier management integrates with the business objectives and regular operations of the organization. Barrier management at the governance level does not involve the detailed activities related to entitative barriers.

Figure 2.2 illustrates the relationships between different stages and the sequence of tasks. Stage I provides guidelines for the work in Stages II and III. Stage II is followed by Stage IV in chronological order. Stages III and IV interact extensively, as the allocated resources can determine the approaches to operation and maintenance of barriers, and feedback from these activities guides adjustments in resource allocation. Stage IV influences Stage V, while Stage III impacts Stage VI. Stage VI includes a comprehensive evaluation of the performance of the barriers and barrier management, requiring inputs from Stage V.

## 2.3 ACCIDENT MODELS AND BARRIER MANAGEMENT METHODS

Barrier management should be a continuous and consistent process throughout the entire lifecycle of a process or a system. In this section, we will provide an overview of the commonly used models, methods, and approaches in generic barrier management.

Kjellen (2007) has outlined the essential tasks in this process, which include the establishment and implementation of barrier management, management during operation, monitoring, and risk management. All effective barrier management methods and frameworks incorporate these tasks. The choice of approach for managing safety barriers heavily depends on how to understand the relationships among safety, risks, and accidents and the models that are used to describe such relationships. The following accident models, as detailed by Rausand and Haugen (2020), are widely used and particularly relevant to barrier management:

- *Energy flow models or energy-barrier models*: As introduced in Chapter 1, these models propose that most accidents involve some form of energy. Safety barriers in these models are designed to either halt the flow of energy toward an asset or isolate the energy source from the asset. The bow-tie model, which illustrates the fundamental concept of a barrier based on energy flow principles, is an example of this model type.
- *Causal sequencing models*: These models regard an accident as a series of discrete events that occur in a specific temporal sequence, where each event triggers the next. A well-known example is Heinrich's classical domino model (Heinrich, 1931), which likens the accident causation to a chain reaction similar to falling dominoes. Each domino represents a contributing/hazardous event to an accident. In this analogy, a safety barrier is visualized as either increasing the gap between the dominos or securing a domino in place. While this model effectively highlights the role of safety barriers, it may not suit detailed analytical needs. Readers can easily find images and examples of this model on the internet.

- *Epidemiological accident models*: These models explain an accident as the result of multiple interacting factors and events, drawing an analogy to disease spread. In such models, the hazardous events do not need to follow a strict order; there are also no necessarily direct causal links between the events. An example of this type is the Swiss Cheese model, where holes in the layers represent latent failures aligning to cause an accident, as depicted in Figure 2.3 in Section 2.5.
- *Systematic control models*: These models attribute failures to a loss of control within a sociotechnical system or an organization. Examples include Perrow's Normal Accident Theory (Perrow, 2011) and Leveson's Systems-Theoretic Accident Model and Processes (STAMP) (Leveson, 2004). These models are contributive not only for designing organizational and procedural barriers but also for understanding the operational principles and potential failures of technical safety barriers. However, systematic control models are typically not employed in conjunction with quantitative analysis methods, limiting their intensive usage in some analytical contexts.

These accident models serve as the cornerstone for risk and barrier analysis. It is important to note that while there are numerous accident models and methods within each category described above, this book does not cover all of them due to their similarities in barrier analysis or limited usage in barrier analysis. For readers interested in a deeper exploration of accident models, the textbook by Rausand and Haugen (2020) is recommended.

Drawing from the models discussed, several methods and approaches have been developed to guide the analysis and management of safety barriers. These include:

- *Bow-tie analysis*: Previously introduced in Chapter 1, this method utilizes the energy-barrier model to visually map out the pathways from hazards to consequences and the barriers that prevent them.
- *Layer of protection analysis (LOPA)*: This method combines elements from the energy-barrier and epidemiological accident models to evaluate and enhance the effectiveness of safety barriers.
- *Barrier and operational risk analysis (BORA)*: Rooted in the causal sequencing model, this approach focuses on identifying and managing sequential accident events.
- *Management oversight and risk tree (MORT)*: Based on a tree diagram, MORT is an analytical tool that will be discussed alongside fault tree analysis in Chapter 4.

Following this introduction, we will explore these established methods for barrier analysis and management, such as bow-tie, LOPA, and BORA. We will also introduce a new framework designed to streamline barrier analysis, engineering, operation, and management, which will serve as the basis for the subsequent chapters of this book.

## 2.4 BOW-TIE-BASED BARRIER MANAGEMENT

We have previously introduced the general structure of the bow-tie model and its key definitions while discussing the concept of safety barriers in Chapter 1. This section will detail how barrier management is practically implemented using the bow-tie model.

The main steps of barrier management based on the bow-tie model are as follows:

**Step 1: Determining asset and the pivotal event**

The first step is to identify the asset of interest and establish the pivotal event within a bow-tie diagram. The pivotal event is selected for its potential to significantly impact or harm the identified asset. Positioned at the center of the diagram, the pivotal event is the focus of the analysis, chosen based on its severity and direct threat to the asset of interest. The pivotal event is inherently hazardous but is not considered the final or ultimate event. It is also an event that is not typically expected under normal operating conditions and can be mitigated through effective preventive measures.

**Step 2: Identifying hazards and events that lead to the pivotal event**

Once the pivotal event has been determined, the next step is to identify the causes or hazards that can lead to its occurrence. These hazards can range from highly likely to remotely possible. It is recommended to compile a comprehensive list of potential causes and prioritize those that are most likely to occur. The bow-tie analysis team should organize causal analysis discussions, utilizing methods such as cause-and-effect diagrams or root cause analysis (Emslie, 2007), to thoroughly reveal all hazards that may lead to the pivotal event. Additionally, it is essential to pinpoint factors or conditions that increase the occurrence likelihood of the pivotal event. Hazard identification methods to be introduced in Chapter 4 are used in this step.

**Step 3: Identifying the consequential events arising from the pivotal event**

The next step moves to the right side of the pivotal event, where the task involves compiling a list of all potential consequences that may arise from the pivotal event. The bow-tie analysis team then needs to determine whether these consequences are static end states or events if they potentially escalate into further outcomes. It is essential to assess if the immediate consequences of the pivotal event could evolve into more severe outcomes under certain conditions or over time. Identifying the ultimate consequences is often complex; however, the analysis should conclude with the best available information and within the boundaries of manageable control.

**Step 4: Constructing the bow-tie diagram with accidental scenarios**

Using the identified hazards, hazardous events, and potential consequences, we can construct accidental scenarios that connect these elements within a comprehensive bow-tie diagram, as illustrated in Figure 1.2. On the left side of the pivotal event, the fault tree method to be introduced in Section 4.7 is commonly applied to visualize and analyze the interconnections among hazards, hazardous events, contributing factors, and the pivotal event. This method provides a systematic way to assess and quantify risk and, thus, the effectiveness of barriers. A detailed discussion of the fault tree method will be provided later in this book.

On the right side of the pivotal event, the event tree model can be used to present how different consequences evolve based on specific conditions or changes over time. This model helps in understanding the sequence and impact of potential outcomes. An in-depth exploration of the event tree model will be outlined in Section 5.10.

**Step 5: Identifying safety barriers**

This step is to identify existing safety barriers along with accidental scenarios. These identified safety barriers serve two main purposes: preventing the progression from a hazard to the pivotal event and responding effectively to the pivotal event to mitigate the occurrence of severe consequences. In addition, identification should consider entitative barrier elements, such as engineering controls, safety equipment, and infrastructure, as well as those non-entitative methods like procedural controls, job safety guidance, and organizational factors. Integrating these elements into the bow-tie diagram helps visualize their functions.

**Step 6: Risk and safety barrier assessment**

A thorough evaluation on the effectiveness and integrity of each safety barrier is important to confirm its capability to interrupt the progression of accidental scenarios. This assessment can utilize quantitative analysis methods, as discussed in Chapters 5–8, to analyze the performance of safety barriers and the likelihood and severity of potential consequences resulting from the pivotal event. This analysis helps determine the associated level of risk to the asset of interest.

**Step 7: Monitoring, testing, and continuous improvement**

If the risk to the asset associated with the pivotal event is regarded as acceptable, indicating the adequacy of current safety barriers, the operational phase may begin. This step requires ongoing monitoring of the barrier performance to ensure they can function as intended. Establishing a feedback loop for continuous review and enhancement of barrier management is crucial. The methods to be presented in Chapter 9 are specifically designed to aid in this ongoing task of continuous improvement.

Conversely, if the assessed risk level is unacceptable, it is necessary to enhance existing barriers or add new ones. Then, we need to revisit Step 5 and iterate the process until the risk is reduced to an acceptable level.

## 2.5 LAYER OF PROTECTION ANALYSIS

### 2.5.1 Description of the Method

The concept of safety barrier is partly derived from LOPA, a simplified risk assessment framework that is widely used in the process industries (CCPS, 2001). LOPA continues to be an important tool in barrier management, mainly used to analyze the effectiveness of barriers in risk reduction. It is also recognized in IEC61511 (2016), particularly focusing on instrumented barrier systems.

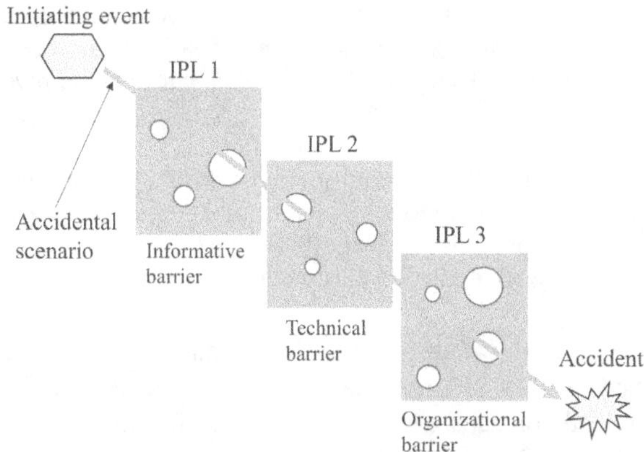


**FIGURE 2.3** An illustrative example of the model for layer of protection analysis.

LOPA works on a specific initiating event within a single scenario, which may develop through a sequence of events under given certain enabling conditions, leading to a significant consequence. As discussed in Chapter 1, within LOPA, each safety barrier is referred to as an *independent protection layer* (IPL). To illustrate the underlying philosophy of LOPA, Figure 2.4 demonstrates how three different types of safety barriers, functioning as IPLs, are strategically positioned between the initiating event and the potential accident or other severe consequence impacting the asset.

LOPA aligns with the approach of defense-in-depth (Reason, 1997), which illustrates that a severe consequence on the asset only occurs if all barriers fail. Reason (1997) developed the famous *Swiss Cheese Model* as an analogy for this approach, as shown in Figure 2.3. In this model, each slice of cheese represents a barrier, and a hole in a slice represents a failure in that barrier. Consequently, only when a hazard successfully passes through the holes in all the slices, the locations of which are different, it can result in a harm to the asset. This model illustrates the importance of having multiple layers of independent protection to effectively mitigate risks.

### Example 2.1:

### *LOPA for an Accident of Car Collision*

We can use the car collision example from Chapter 1 to demonstrate the concepts in LOPA. For instance, over-speeding on a highway can serve as the initiating event leading to a collision, which is considered an accident in this context.

To identify potential IPLs that can prevent car collisions, we can refer to Figure 2.3. Three possible IPLs include:

*IPL 1*: Speed sign and alarm. Roadside speed signs and in-car alarms or GPS alerts can notify the driver if they are exceeding speed limits. This IPL serves as an informative barrier because it provides critical information to the driver.

*IPL 2*: Automatic deceleration. If sensors detect over-speeding and a short distance to the car in front, the automatic braking function of the trailing vehicle should be activated to decelerate the car to help prevent a collision. This IPL is a technical safety barrier that utilizes sensing and control technologies.

*IPL 3*: Safety action of the driver. The driver should be conscious of their speed and the proximity to the car in front and respond by decelerating to avoid a collision. This IPL depends on safe driving practices reinforced through training, categorizing it as an organizational barrier. ■

It is important to acknowledge that each IPL may have its own vulnerabilities (holes in the cheese), such as driver carelessness or inexperience, alarms that are unclear or unnoticed, or sensor malfunctions due to environmental factors like strong sunlight. LOPA aims to address these vulnerabilities by proposing solutions to strengthen the IPLs and prevent hazards from bypassing the safety barriers.

### 2.5.2 Procedure of LOPA

Several guidelines have been published to support conducting LOPA, such as those by the Center for Chemical Process Safety (CCPS, 2001). The procedures outlined below are derived from these industry standards and exhibit some overlap with the barrier management strategies discussed in Section 2.2.

The main steps of LOPA include:

**Step 1: Establishment of an accidental scenario**

The first step is to identify the initiating event of the accidental scenario, which should be an abnormal occurrence, according to LOPA. In this book, while an initiating event can be either normal or abnormal, for LOPA it is explicitly abnormal. The scenario completes with an end event that results in significant harm to the asset, referred to as a "consequence of concern" in LOPA. It is essential to define the scope of this consequence, which may include injuries, fatalities, or environmental damage.

**Step 2: Risk acceptance analysis**

This step assesses the level of risk that is considered acceptable for the identified initiating event. Essentially, this involves determining "how safe is safe enough". Given that the accidental scenario has already been identified with its consequences in mind, the focus here is on establishing the acceptable frequency of these events and the occurrence rate of the consequence of concern. A risk matrix is often employed to visually represent the links between the frequency of events and their consequences and evaluate them.

**Step 3: Cause and condition analysis**

The objective of this step is to identify the causes and conditions that lead to the initiating event and any subsequent events within the accident scenario. HAZOP (Hazard and Operability Study) and HAZID (Hazard Identification Study) are commonly employed to identify these causes and conditions, as detailed in Chapter 4. LOPA specifically necessitates the analysis of causes and enabling conditions, which are essential for the development of the initiating event, and *condition modifiers*, which affect the frequency of subsequent events and the probability of significant consequences.

**Step 4: Identification and implementation of IPLs**

This step focuses on identifying and implementing potential IPLs that are independent of each other. The critical questions to answer in this step include:

- What types of protection layers are necessary for the accidental scenario?
- Which protection layers have been involved in the accidental scenario?
- Among these layers, which meet the criteria to be considered an IPL?

Step 4 should follow the principle of independence in barrier management, ensuring that each IPL is capable of interrupting the accidental scenario on its own. This approach emphasizes the need for each layer to function effectively on its own to mitigate risks.

**Step 5: Assessment of IPLs**

Each IPL must effectively respond to the accident scenario. This step is to evaluate the efficacy of each IPL in preventing events or mitigating condition modifiers, as well as assessing the likelihood of an IPL's failure to respond. Key questions to address in this step include:

- How significantly should/can each IPL reduce risk?
- Is the implementation of this IPL necessary?
- Is this IPL performing as expected, and what are its integrity targets?
- How significantly should/can all the IPLs collectively reduce risk?

Details on the quantitative measures and methods for such assessments will be discussed in Chapter 7.

**Step 6: Re-evaluation of risk**

After assessing all IPLs, the frequency with which the consequence of concern occurs as the end event of the accident scenario should be re-evaluated to ensure it remains below the acceptable frequency. This step revolves around the question:

- Is there a need to introduce an additional IPL?

If the answer is yes, it is important to determine the extent of further risk reduction required, then return to Step 3 to analyze additional causes of the events and to identify and implement new IPLs.

For documentation purposes, a simple worksheet as shown in Figure 2.4 can be used to record information used in a LOPA analysis. This aids in maintaining clear and organized records of the process.

| Accident | Accident scenario | TEF | UEF | IPL 1 | MEF 1 | RRF 1 | IPL 2 | MEF 2 | RRF 2 |
|---|---|---|---|---|---|---|---|---|---|
| Car collision | The driver behind is driving fast and he/she does not notice the deceleration of the car in front | $10^{-7}$ | $10^{-3}$ | Overspeed alarm | $10^{-4}$ | 10 | Automatic braking system | $10^{-8}$ | $10^{4}$ |

**FIGURE 2.4** A simplified generic example of worksheet in LOPA.

In such a LOPA table for car collision, the columns serve the following functions:

- *Column 1 – Accident description*: Specify the type of accident being analyzed, such as a car collision.
- *Column 2 – Accident scenario description*: Provide details on the accidental scenarios, initiating events, and proximate causes, such as the events and conditions leading up to a car collision.
- *Column 3 – Tolerable event frequency (TEF)*: Indicate the maximum acceptable frequency of the hazardous event occurring. For example, in the given table, the value of $10^{-7}$/hour suggests that a typical driver has about a 0.5% chance of experiencing a car collision during 50 years of driving.
- *Column 4 – Unmitigated event frequency (UEF)*: Represent the frequency of the hazardous event occurring without any protective measures. While the example provides a rough estimate, in practice, this value is often derived from historical data. It is not easy to provide an accurate UEF value because most existing sociotechnical systems have been equipped with protection measures.
- *Column 5 – First IPL (IPL 1) description*: Provide detailed description on the first IPL in the accidental scenario, such as an overspeed alarm triggered by sensors detecting high speed and proximity to another vehicle.
- *Column 6 – Mitigated event frequency by IPL 1 (MEF 1)*: List the estimated frequency of the hazardous event after the application of IPL 1, providing a specific risk reduction metric.
- *Column 7 – Risk reduction factor of IPL 1 (RRF 1)*: Evaluate the effectiveness of IPL 1, calculated as the ratio of UEF to MEF 1. It can also be expressed as a relative risk reduction. In this case, RRF is calculated by UEF divided by MEF 1, but it also can be calculated in other ways, for example, as a relative risk reduction (UEF – MEF 1)/UEF.
- *Column 8 – Second IPL (IPL 2) description*: Describe the second layer of protection, such as an automatic braking system that activates to slow the vehicle when necessary.
- *Column 9 – Mitigated event frequency by IPL 2 (MEF 2)*: Present the frequency of the hazardous event after both IPLs are applied, further reducing the risk.
- *Column 10 – Risk reduction factor of IPL 2 (RRF 2)*: Evaluate the effectiveness of IPL 2, determined by dividing MEF 1 by MEF 2 in the above table, indicating a further reduction in risk due to the second IPL.

The complexity and length of a LOPA worksheet vary based on the number of involved IPLs. Often, additional columns are required to describe the integrity of technical IPLs, such as an automatic braking system in the case of car collision, because their activation without failure cannot be guaranteed.

In the analysis of complex technical systems, the LOPA worksheet may feature more extensive columns to accommodate detailed assessments. This is explored in greater depth in Chapter 15 of the book by Rausand and Haugen (2020). Additionally,

LOPA is frequently integrated with HAZOP tables to enhance the identification and evaluation of hazards, as illustrated in Figure 4.6. We will discuss HAZOP in Section 4.5.

**Example 2.2:**

*Frequency of Car Collision*

Following Example 2.1, let us consider two IPLs to prevent car collisions. Assume the probability of collisions is 0.001 when only qualified drivers are in control and there are no alarms or mitigation measures. Then, estimate the failure probability of a well-trained driver noticing the speed sign and sounding an alarm (IPL 1) to be no more than 10%. Additionally, the failure frequency of the automatic braking system (IPL 2) is estimated to be $10^{-4}$ per hour, or approximately one failure every 300,000 km of driving.

It is important to distinguish between probability and frequency, though both can be used for rough estimates in this context. The frequency of a car collision, considering the two IPLs, can be calculated as $0.001 \times 0.1 \times 10^{-4} = 1.0 \times 10^{-8}$/hour.

This calculation suggests that if an individual drives about 1,000 hours per year over 50 years, there is approximately 1 in 2,000 chances of experiencing a car collision in their lifetime. This estimation, while rough and perhaps too optimistic, can be used to illustrate the potential effectiveness of IPLs. The assumption here is that the automatic braking system, when functioning, can completely prevent collisions.

This case study is designed for illustrative purposes to demonstrate how risk assessments with barriers compare against tolerable risks for car collisions. If the calculated frequency of collisions is higher than the acceptable level, it may be necessary to introduce additional IPLs or enhance the effectiveness of the existing ones. ■

### 2.5.3 Advantages and Limitations of LOPA

LOPA is widely applied since it has several advantages:

- *Clear hazard understanding and IPL identification*: LOPA aids in identifying and clarifying the specific IPLs needed to manage particular events. This ensures that all necessary measures are in place, utilizing available resources effectively.
- *Optimal allocation of resources*: By evaluating the relevance and effectiveness of each IPL, LOPA helps prioritize resource allocation to the most effective layers, optimizing resource use and potentially saving costs.
- *Structured and comprehensive approach*: LOPA provides a structured framework for risk and barrier management, documenting each decision and its rationale. This process ensures comprehensive consideration of all relevant factors and IPLs, demonstrating how risks are reduced to acceptable levels through preventive and mitigative measures (Torres-Echeverria, 2016).
- *Standardization*: As a method compliant with IEC 61511, LOPA is standardized, making it accessible and implementable across different stakeholders

and organizations. It also provides semi-quantitative, risk-based responses to critical questions in barrier management, offering decision-makers a clear, concise summary of the risks involved.

On the other hand, the limitations of LOPA also should be noted:

- *Subjectivity and data quality dependence*: LOPA relies on expert judgment, particularly regarding the effectiveness of non-technical IPLs, resulting in a level of subjectivity that can lead to inconsistent outcomes. Moreover, the quality of the data utilized in LOPA analyses can greatly influence the accuracy and reliability of the results.
- *Time-consuming and limited scope*: LOPA can be a time-consuming process, especially for complex systems, as it typically focuses on analyzing one hazardous event at a time. This narrow scope may limit its application in broader decision-making related to system design or operational strategies, where a comprehensive assessment of multiple critical hazards and protective measures is necessary.
- *Overlook of dependent failures among protection layers*: LOPA generally assumes the independence of protection layers and does not consider common cause failures or cascading failures among these layers. For example, in Example 2.1, the assumption that IPL 1 and IPL 2 are completely independent may be inaccurate, as lack of training of the driver could lead to disregarding overspeed alarms, impacting both IPLs.

## 2.6 BARRIER AND OPERATIONAL RISK ANALYSIS

### 2.6.1 Introduction of the BORA Method

The BORA framework was developed as part of a Norwegian research project. This method was initially applied to analyze risks associated with hydrocarbon leaks, with the aims to offer a detailed, quantitative model for assessing barrier performance (Vinnem et al., 2004). Over the past decade, the applications of BORA have broadened to include accident prevention across various industries, such as oil and gas, offshore operations, and others.

BORA focuses on the operational phase, integrating the management of production operations and the handling of preventive and mitigative safety barriers. The objective of developing this method is to address the shortcomings of traditional risk assessments, which often lacked comprehensive analysis on the operational issues related to barriers and assets. It serves as an overall framework for qualitative, semi-quantitative, and quantitative barrier analysis, considering risk influence factors (RIFs) related to technical, human, operational, and organizational aspects of assets.

BORA is often used together with other analytical methods like barrier block diagrams, event trees, fault trees, and risk influence diagrams (Aven et al., 2006). These methods will be discussed briefly in the following subsection and explored in detail with respect to quantitative analysis in Chapter 5.

### 2.6.2 Procedure of BORA

According to literature (Aven et al., 2006; Sklet et al., 2006; Vinnem et al., 2004), the procedure of BORA includes the following steps:

**Step 1: Establishment of a risk analysis model with safety barriers for an accident**

BORA typically utilizes a *barrier block diagram* to present accidental scenarios and assess the effectiveness of safety barriers in influencing these event sequences. A simple example of this can be seen in Figure 2.5, which outlines various potential accidental events and the roles of safety barriers.

The barrier block diagram comprises an initiating event, several barrier functions, and the outcomes of the event sequence, with the accident being one of the potential outcomes. Each block within the diagram symbolizes a safety barrier function, with two potential outputs: a horizontal arrow pointing right signifies the successful operation of the barrier, while a vertical arrow pointing downward indicates a failure.

It is important to acknowledge that the diagram presented in Figure 2.5 illustrates only a segment of a comprehensive barrier block diagram. This particular example concludes with a car collision, identified as the critical/pivotal event. However, the ultimate worst-case scenario could involve fatalities. Further details on constructing a complete barrier block diagram are available in the BORA Handbook (Seljelid et al., 2007).

**Step 2: Performance analysis of safety barriers**

For a quantitative assessment, it is essential to model and measure the performance of safety barriers quantitatively. Chapter 6 elaborates on various performance measures for safety barriers. Within the BORA

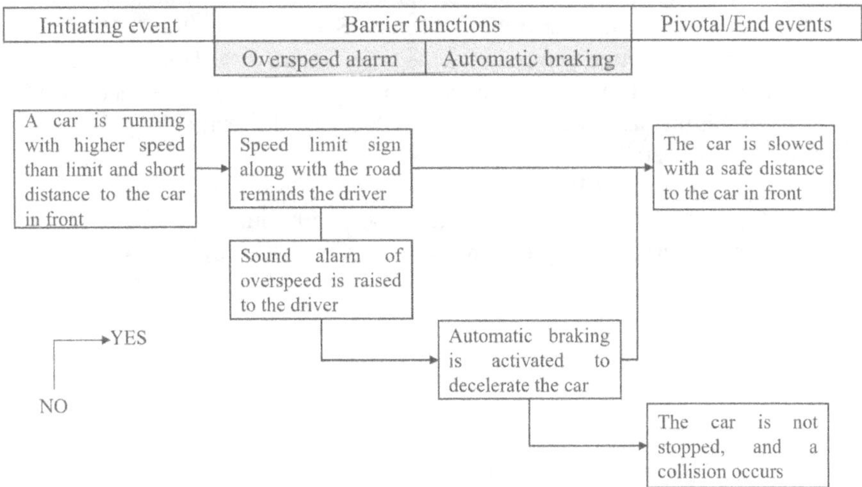


**FIGURE 2.5** A simple example of barrier block diagram.

framework, the fault tree method is recommended for performance analysis. In this method, the failure of a safety barrier is treated as the "Top event" in a fault tree. A comprehensive overview of fault tree analysis is provided in Section 5.4.

**Step 3: Obtaining the reference information**

The fault tree method requires the information of frequencies or probabilities of the initiating event and other intermediate events within the accidental scenarios. Plant-specific data is typically preferred for determining the frequencies of specific events, but in many cases, industry average data, or generic data, may be more readily available from databases such as OREDA (2009) for offshore industries and HIAD for hydrogen facilities (see Galassi et al. (2012)). If neither plant-specific nor generic data is available, it becomes necessary to rely on expert judgment to provide reference frequencies or probabilities to the different events.

**Step 4: Identification of RIFs and development of risk influence diagrams**

The objective of this step is to identify the RIFs that affect the occurrence of each event in an accidental scenario, including the initiating event and the instances where a safety barrier fails. RIFs differ from the PIFs of a safety barrier, which are defined in Section 1.7 as the factors that can affect the performance of a barrier or a barrier element. In this context, PIFs may be considered a subset of RIFs.

A top-down approach is recommended for identifying the causes and influencing factors of an event. By questioning why an event occurs, the analyst can pinpoint relevant factors. These factors and their relationships to the event can then be visually mapped using a risk influence diagram. An illustrative example of this process is shown in Figure 2.6, detailing the event where "the driver does not notice the sound alarm of overspeed".

In the process industries, RIFs can be systematically classified into five categories: human factors, task-related factors, technical factors, administrative factors, and organizational factors (Seljelid et al., 2007). Here, we do not discuss more about the methods for identifying RIFs. For a more comprehensive framework and additional details, readers are encouraged to consult the articles by Aven et al. (2006) and the textbook by Rausand and Haugen (2020).

**Step 5: Assessment of RIFs**

This step is to analyze the RIFs associated with each specific event and benchmark them against industry averages. According to Aven et al.

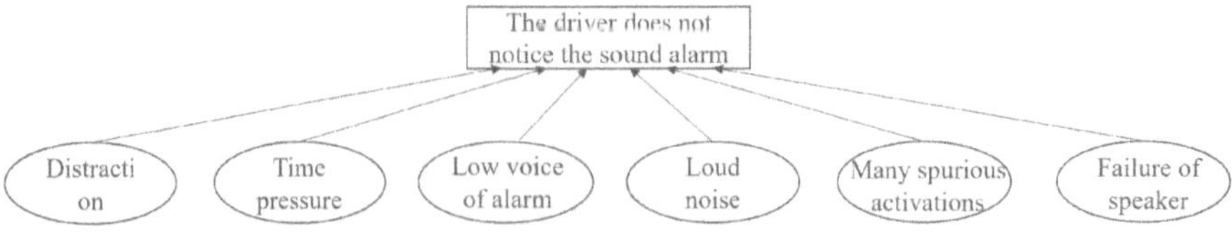


**FIGURE 2.6** An example of risk influence diagram.

**TABLE 2.1**
**A Scheme of Scoring RIFs**

| Score | Descriptions |
|---|---|
| A | Status corresponds to the best standard in the industry |
| B | Status corresponds to a level better than the industry average |
| C | Status corresponds to the industry average |
| D | Status corresponds to a level slightly worse than the industry average |
| E | Status corresponds to a level worse than the industry average |
| F | Status corresponds to the worst practice in the industry |

*Source:* Aven et al. (2006) and Vinnem et al. (2004).

(2006), a six-level scoring scale ranging from A to F is utilized, as detailed in Table 2.1. This scoring system quantifies the frequency or probability of an event in a specific case relative to the industry average for similar events. Detailed methodologies and calculations for this analysis are available in the BORA Handbook (Seljelid et al., 2007).

Alternatively, a simpler five-grade scale can be employed to evaluate each RIF. A score of 1 indicates that the RIF's status corresponds to a significantly lower risk level than the industry average. A score of 3 suggests that the risk level is comparable to the industry norm, and a score of 5 implies that the RIF's status represents a significantly higher risk level than the industry average.

Additionally, in Step 5, the RIFs are weighted according to their contributions, as determined by the analysis. Each RIF can be assigned a relative weight on a scale from 1 to 10, which are then normalized to ensure that the sum of the weights for all RIFs related to a specific event equals 1. The criticality of a RIF is assessed by multiplying its status score by its normalized weight. This process is dynamic, and the weights assigned to RIFs may change over time as the system continues to operate.

Information sources for scoring and weighting RIFs include accident investigation reports, system audit or review reports, and work group meetings that incorporate expert judgment.

**Step 6: Calculation of risks of the specific events**

This step involves calculating the frequencies of specific events in accidental scenarios, taking into account the statuses of RIFs and the performance of safety barriers, as well as the probabilities of various outcomes. Event tree analysis (ETA) is employed for this purpose (see Section 5.10). The calculation results can be updated as new data pertaining to events, barriers, or RIFs becomes available. Initially, BORA offers only a preliminary estimation of risks at the beginning of system operation, making this approach semi-quantitative. However, when additional data is collected, these risk estimations can be further refined and precise.

**Example 2.3:**

*A Simple Example of Calculating the Frequency of a Specific Event*

As an example, let us use Figure 2.6 to illustrate Step 5 of BORA by roughly scoring the RIFs for a scenario involving a young driver listening to loud music while driving. Employing the earlier mentioned 5-grade scale, we can assign scores as follows: "Distraction" receives a 5, "Time Pressure" a 4, "Loud Noise" another 5, while the remaining three factors each score a 3.

Next, we assign relative weights to the six factors depicted in Figure 2.6, from left to right, as: 8, 4, 8, 10, 6, and 4. The sum of these weights is 40. Consequently, the normalized weights for these factors are 0.20, 0.10, 0.20, 0.25, 0.15, and 0.10, respectively.

Assuming that the probability of a well-trained and careful driver failing to notice an alarm is 0.1 (based on industry averages), we can now calculate the probability of the young driver, who is distracted by loud music, missing an alarm as follows:

$$\frac{5\times0.20+4\times0.10+3\times0.20+5\times0.25+3\times0.15+3\times0.10}{3}\times0.1=0.133$$

The denominator used in this calculation is 3, because a score of 3 represents equivalence with the industry average within the framework of evaluating RIF status. The calculated result suggests that the probability of such a driver ignoring the alarm is 33% higher than the industry average. It is important to note that this example is provided only for illustrative purposes only and does not align exactly with the methodologies recommended by the BORA Handbook (Seljelid et al., 2007), which is available for open access online for those interested in further exploration. ■

### 2.6.3 Advantages and Limitations of BORA

The advantages of BORA include:

- *Systematic identification of RIFs*: BORA offers valuable insights into RIFs and provides a systematic method to identify these factors. This approach is useful in finding suitable solutions to enhance barrier performance and reduce risks.
- *Integrated framework*: BORA integrates a barrier analysis and management framework with established risk management principles, incorporating well-known risk analysis methods such as fault tree analysis and ETA. This integration can facilitate comprehension among stakeholders from diverse backgrounds.
- *Dynamic risk assessment*: BORA supports the continuous update of risk assessments in response to the implementation of new barrier functions or the availability of new information.

The limitations of BORA include:

- *Assumption of independence*: BORA uses risk influence diagrams, which are based on the assumption that RIFs are independent, but it is often impractical. For instance, in Figure 2.6, factors such as distraction and loud noise may be interconnected, as drivers are likely to be more distracted when music volume is high. This limitation could be addressed by adopting Bayesian networks to account for the interdependencies among factors.

- *Subjective judgments*: BORA heavily relies on subjective judgments to determine the statuses, weights of RIFs, and their criticalities, often without a robust justification process.
- *Ambiguity in definitions*: BORA does not provide a clear definition of RIFs, which are typically considered factors that may influence the occurrence likelihood of an event but are not necessarily performance influencing factors of a safety barrier (Teng, 2010). Some RIFs may not be directly relevant to barrier management.

## 2.7 RISK-INFORMED BARRIER MANAGEMENT

According to PSA (2017), the essential tasks in the barrier management process include hazard identification, barrier function identification, barrier element identification, performance requirement establishment, and barrier performance maintenance. While in the guidance provided by Hauge and Øien (2016), barrier analysis is structured into three steps: barrier function analysis, barrier element analysis, and requirement analysis.

In light of this literature, we propose a framework titled risk-informed barrier management (RIBM). This framework adapts the barrier analysis, engineering, and management processes to suit two distinct contexts: initiating barrier management for new projects or systems and integrating barrier management into existing projects or systems.

### 2.7.1 Barrier Management in the New Context

For newly established companies or organizations, as well as those introducing new technologies, assets, or projects, it is crucial to integrate a barrier strategy with their operations from the beginning. The main steps in such a barrier management process are depicted in Figure 2.7. Here, the solid lines with single arrows indicate actual transformations between steps in barrier management, while the dotted lines with double arrows represent communications and potential transformations among steps. The main steps include:

**Step 1: Identification of assets and hazards**

The first step is to identify the critical assets of the company or organization and to assess the potential hazards and threats that could impact these assets. The tasks include clarification of the process potentially threatening the assets (study scope) and detailed examinations of the inherent design, structure, and operational mode of these critical assets with a focus on safety. It is also essential to recognize the vulnerabilities of these assets to various threats. Methods for hazard identification will be introduced in Chapter 4.

**Step 2: Integration of the barrier philosophy into risk management**

The second step is to incorporate the barrier philosophy into the existing risk management framework of the process under study. This philosophy

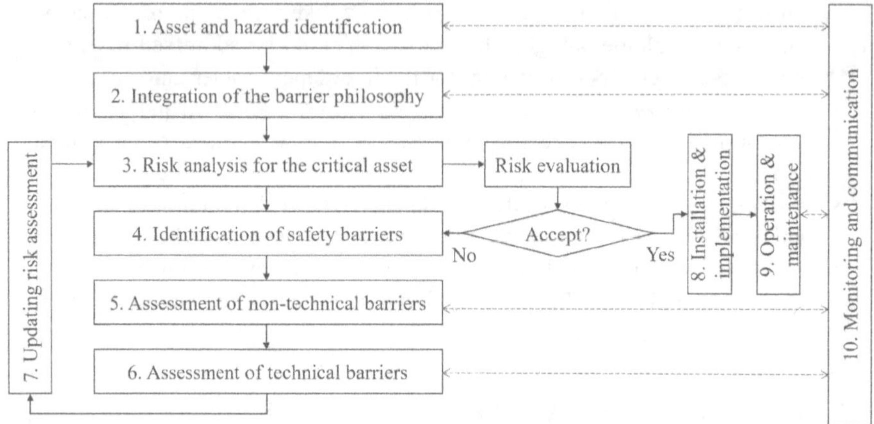


**FIGURE 2.7** Flowchart of the RIBM process in a new context.

includes recognizing the "fail-safe" principle and ensuring that severe consequences from abnormal situations or initiating events can be mitigated through the implementation of additional equipment and measures. This approach can be further explained in Figure 2.8, which presents a perspective on how barriers are employed to reduce the risk to an asset to an acceptable level or lower. This step aims to clearly define a specific risk assessment procedure for the asset of interest, outline associated risk treatment approaches, and formulate an appropriate strategy for integrating and managing barriers. It is advantageous for the risk management model to visualize safety barriers and their status clearly. Later in this chapter, we will introduce some methods for integrating barriers into risk management.

**Step 3: Risk assessment of the critical assets**

The third step is to assess the likelihoods and consequences of the hazardous events on each asset of interest, in qualitative, semi-quantitative, and quantitative ways. After the most significant events for each critical asset

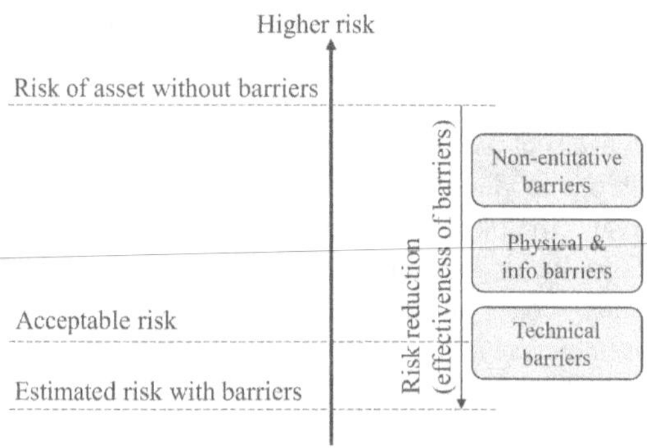


**FIGURE 2.8** Illustration of the barrier philosophy.

are identified based on the assessment, the next task in this step is to identify the significant accidental scenarios related to the events, from the threats to consequences. The risk picture generated based on the works above is regarded by PSA (2017) as the foundation of further barrier analysis. In other words, the work of this step is to determine the position of the upper line "Risk of asset without barriers" in Figure 2.8. Methods for conducting likelihood and consequence analyses will be introduced in Chapter 5.

**Step 4: Identification of various safety barriers**

The fourth step is to identify the potentially procedural, organizational, and cultural barriers, as well as physical, technical, and informational barriers that could be implemented around each asset. In this step, it is important to define the expected barrier functions and then identify the available barrier functions and elements, including both tangible and intangible aspects. If certain essential barrier functions are missing, it becomes necessary to propose new methods and solutions that align with the available resources and safety requirements. Methods for identifying barriers will be covered in Chapter 4, with a discussion on the properties of non-technical safety barriers to follow later in this chapter.

**Step 5: Assessment of non-technical safety barriers**

In the fifth step, the effectiveness of non-technical safety barriers (including non-entitative and entitative ones) is analyzed and evaluated. The effectiveness of these barriers is measured by the degree of risk reduction they provide, mostly in the qualitative way or with mathematical formulas for this evaluation provided in Chapter 6. This evaluation may also include reviews of accident reports. Unlike technical barriers, non-technical barriers are presumed to function without failures, thus making effectiveness a sufficient measure of their functionality.

**Step 6: Assessment of technical safety barriers**

The next step focuses on the analysis and evaluation of technical safety barriers. Here, effectiveness alone is not a sufficient metric since these barriers may fail to perform their intended functions. Integrity, defined as the probability that a barrier will perform as required, is a crucial performance measure for technical safety barriers. This step involves determining the design and operational parameters necessary for the effectiveness and integrity of barriers.

Methods for analyzing the integrity of technical barriers will be introduced in Chapter 6 and further explored in subsequent chapters. The impact of operational and maintenance strategies on the functionality and performance of technical barriers also needs to be examined. It should be noted that an important task in both Steps 5 and 6 is the safety-economy analysis, which assesses whether the barriers are cost-effective in terms of expenditure. This analysis will be discussed in Section 2.8.

**Step 7: Updating risk assessment**

After assessing different types of safety barriers, the results are utilized to update the risk assessment for each asset of interest. In Step 7, the initial evaluations from Step 3 are modified using the new information obtained in Steps 4–6, leading to a recalculated and re-evaluated risk for each critical asset. If the newly estimated risk still exceeds the acceptable level, it becomes necessary to revisit Steps 4–6. This iterative process continues until the risk, with the addition of more or newly designed barriers or new barrier operational strategies, is reduced to a tolerable level. Steps 3–7 collectively comprise the barrier assessment process.

Steps 3–7 collectively comprise the *barrier assessment* process.

**Step 8: Installation and implementation of safety barriers**

Once the risk level is found acceptable, the subsequent step is the design, procurement, installation, and initiation of the necessary safety barriers. It is also important to consider the layout of these barriers within their intended context. Factors, such as the physical environment, potential hazards, and operational requirements, are critical to ensure that the barriers are positioned optimally to effectively mitigate risk. Safety barriers should be implemented at the beginning of the operation of the asset to ensure that the measures are fully integrated into the operational process and that all personnel are acquainted with the barriers and their functions.

**Step 9: Operation and maintenance of safety barriers**

Regular inspections and tests are necessary to identify potential faults and degradation in the barrier systems and components. Appropriate diagnostic and preventive maintenance strategies must be carefully planned and executed to prevent failures in technical barriers. These activities and the corresponding analyses are referred to as *dynamic barrier management* (Pitblado et al., 2016). The impact of inspections and tests on the performance of technical safety barriers will be further explored in Chapter 9, with an in-depth examination of degradation issues.

**Step 10: Monitoring and communications for improvement**

This step focuses on developing guidelines for monitoring barrier performance and establishing effective communication channels to facilitate continuous improvement. This involves setting up and maintaining a robust system for reporting issues related to physical and technical barriers, as well as human errors, organizational weaknesses, and cultural factors that could potentially impact the effectiveness of the risk management framework and barrier strategy. Effective communication is crucial to ensure all stakeholders involved in barrier management are aware of potential problems and are actively engaged in the improvement process. This may include implementing regular reporting and feedback mechanisms, along with training and awareness programs to prompt a culture of continuous improvement and a shared commitment to safety.

It is evident that information plays a critical role in decision-making across all steps. The role of information in barrier management will be further discussed in Section 2.9.

Steps 1–8 should be implemented at least during the system design and project preparation phase, while Steps 9 and 10 occur during the system and project operation phase. These steps may provide feedback that necessitates revisiting Steps 1–7. Risk information for the asset can be periodically updated through Steps 8 and 9, a process referred to as *dynamic risk analysis*. Readers may refer to the book edited by Paltrinieri and Khan (2016) for relevant models and analysis methods.

For a clearer understanding of the roles of different components of barrier management within the overall risk management framework, readers are encouraged to compare Figures 2.7 and 1.6.

### 2.7.2 Integration of Barrier Management with the Existing Context

For projects or systems already in operation, the barrier management process may differ slightly from those being newly implemented. To develop or enhance the existing barrier management strategy and integrate it seamlessly, the following steps should be included:

**Step 1: Identification of assets and hazards**

For an operating project or system, critical assets are already identified, and thus a focused summary of hazards associated with these assets can be obtained.

**Step 2: Evaluation of the current risk management framework and barrier philosophy**

This step is to evaluate the existing barrier philosophy and its clarity within the current context and determine if the general procedures of barrier management and types of barriers are compatible with the ongoing risk management framework.

**Step 3: Risk assessment of the critical assets**

Utilizing existing risk assessment reports can save time and resources. It is important to evaluate whether the current barriers adequately reduce the risk of asset to the acceptable level. If the risk has been sufficiently low, the process may advance directly to Step 8; otherwise, proceed to Step 4.

**Step 4: Identification and implementation of new safety barriers**

Given that new safety barriers are needed, this step assesses the potential for new protective approaches, technologies, and training programs that could enhance safety barrier functions. It is also essential to consider the need for new regulations to mitigate human-related errors and hazardous events.

**Step 5: Improvement of non-technical safety barriers**

The aim of this step is to improve the effectiveness of non-technical barriers in risk reduction. This may involve optimizing barrier layouts, refining barrier management practices, renewing physical barriers, and revising existing safety training programs.

**Step 6: Improvement of technical safety barriers**

Focusing on increasing both the effectiveness and integrity of technical barriers, this step can implement various improvement approaches, including upgrading technical systems, addressing causes of failure to boost reliability, redesigning barrier structures, and enhancing inspections, tests, and preventive maintenance.

**Step 7: Updating risk assessment**

This step is similar to Step 7 for new contexts, re-assessing risks based on newly implemented or improved barriers.

**Step 8: Installation and implementation of safety barriers**

With the involvement of new barriers, it is important to evaluate their impacts on the effectiveness of existing barriers.

**Step 9: Operation and maintenance of safety barriers**

It is essential to ensure that current operation and maintenance strategies are cost-effective while maintaining acceptable risk levels. Adjusting maintenance practices, test intervals, and other operational parameters can enhance the integrity of technical barriers and improve overall barrier effectiveness.

**Step 10: Monitoring and communications for improvement**

This step is similar to Step 10 for new contexts, focusing on developing guidelines for monitoring barrier performance and establishing effective communication channels to facilitate continuous improvement.

### 2.7.3 Advantages and Limitations of RIBM

The RIBM method proposed in this book is a structured approach to enhancing safety and risk management in both new and existing systems. Here are some advantages and limitations of the BIBM method:

- *Aligning with risk management framework*: RIBM offers a structured methodology for identifying, assessing, and managing risks through various types of safety barriers, in line with the procedure of risk management in ISO31000 (2018). This systematic approach helps organizations to ensure that more potential risks are considered and addressed appropriately.
- *Incorporation of risk Information*: By using risk-informed decisions, RIBM ensures that the management of safety barriers is directly tied to the actual risk levels associated with different assets and operations. This alignment helps in prioritizing actions based on the most significant risks.
- *Flexibility for different contexts*: RIBM is adaptable to both new projects and existing systems. It provides guidelines for integrating barrier management strategies effectively regardless of the project stage.

On the other hand, the limitations of RIBM include:

- *Dependence on accurate risk assessment*: The effectiveness of the RIBM method heavily relies on the accuracy of the initial risk assessments. Inaccurate or incomplete risk evaluation can lead to inappropriate safety barrier choices and risk management strategies.
- *Slow response to emergent risks*: Given its structured nature, RIBM may be less agile in responding to emergent risks that require immediate action, especially in highly dynamic environments where risks can evolve rapidly.

## 2.8 DECISION-MAKING IN BARRIER MANAGEMENT[2]

### 2.8.1 Types of Decision-Makings in Barrier Management

We consider decisions in barrier management as those that can impact the function and performance of safety barriers, thereby influencing the risk to an asset by either reducing it or not. Decisions in this context always include choosing whether or not to undertake specific actions. Examples of such decisions in barrier management include determining the overall barrier strategy, specifying barrier equipment, allocating budgets for barrier maintenance, and manually activating reactive barriers.

According to how a decision is reached, decision-making can be generally classified as (Zhu, 2023):

- *Bounded rational decision-making*: A decision-maker strategizes to generate multiple alternatives and finds the optimal choice by comparing all the alternatives in a systematic way (March, 1994). This process demands considerable effort in collecting and interpreting information, reasoning, and deliberation, assuming that the decision-maker is always rational. Bounded rational decision-making can be *data-driven* or *expert-based*. The information used in decision-making may be straightforward, complete, and easily understood, or it may be ambiguous, incomplete, and difficult to interpret. In situations with high uncertainty, such as developing barrier strategies, a process known as *sensemaking* is necessary. Sensemaking involves identifying patterns, integrating information, and interpreting the significance of events to support decision-making.

  In risk management and barrier management, decisions made using comprehensive information about risks and dedicated analysis of potential consequences and uncertainties related to safety barriers are examples of bounded rational decision-making. In literature (Cepin, 2010; Ersdal & Aven, 2008), such a decision is called as *risk-informed decision-making* (RIDM), whose outcome may influence the risk to the asset.
- *Rule-based decision-making*: A decision-maker relies on a predefined set of rules or criteria to act. This approach only requires information about rules, typically established from past experiences, best practices, and lessons from accidents. Rule-based decision-making is often applicable in situations where there is a high degree of certainty and decisions must be made swiftly. In barrier management, this method is commonly used in determining barrier layout and operation approaches.
- *Recognition-primed decision-making*: Here, the decision-maker utilizes prior experiences, skills, and knowledge to recognize and diagnose problems, identify patterns, and select actions that have been effective in similar past situations. This process is typical in high-stress and time-critical environments, and decision-making behaviors are by professionals, such as firefighters using hoses, water, and foam to suppress fires (emergent activation of safety barriers).
- *Intuition-based decision-making*: This approach relies on the decision-maker's intuition or "gut feeling" to make a decision automatically without

conscious thought (Kahneman, 2011). It is rapid and draws on past experiences, patterns, and environmental cues. Intuition-based decision-making is often used in situations with a high degree of uncertainty. Although its quality can vary, especially when data or necessary information is incomplete or unavailable, educated intuition-based decision-making can share some similarities with recognition-primed decision-making as both are results of accumulated knowledge and experience (Salas et al., 2010). However, general (less mature) intuition-based decision-making may be prone to biases, errors, and heuristics, potentially leading to suboptimal decisions.

The amount and quality of information required by the above four decision-making approaches are very different. The rule-based approach requires complete and clear information about the relevant rules, while bounded rational decision-making or RIDM needs more detailed information about risks compared to the other methods. At a higher level of barrier management, such as in the development of barrier strategies and the specification of barriers, where time and resources are less constrained, RIDM is the recommended approach as it allows for more comprehensive data collection and analysis. However, under increased time pressure and with limited information, it may be natural and necessary to shift to rule-based, recognition-primed, and intuition-based approaches successively. An empirical study in the oil and gas industry shows that middle managers use rule-based decision-making more often, even though 80% of their decision activities are associated with risk management (Rezvani & Hudson, 2021).

The selection of decision-making approaches is influenced by the characteristics of the *risk-related decision problems*, in which decision-makers need to consider uncertainties and potential adverse outcomes associated with each choice. In literature (NASA, 2010; Zhu, 2023), a risk-related decision problem can be observed and evaluated from the following perspectives to make a decision on which decision-making approaches can be used:

- *Criticality*: Criticality indicates the potential change in risk that results from implementing a decision. It considers the objective of reducing risk, the gap between the current state and the objective, and the proximity to hazards. The level of criticality influences the number of resources allocated for information gathering and analysis for decision-making. For decisions with high criticality, relying on intuition-based approaches is generally discouraged.
- *Uniqueness*: Uniqueness describes how distinct a decision problem is in comparison to the others. High uniqueness typically means less available information and limited experience. When dealing with a high unique decision problem, it is challenging to rely on the recognition-primed approach. In some cases, unique decision problems may also complicate the identification of applicable rules.
- *Complexity*: Complexity is related to the amount of information within the problem statement, the structure of the statement, the number of uncontrollable variables involved in decision-making, the interactions among controllable variables, and their dynamics. It is difficult to appropriately use

recognition-primed and intuition-based approaches to deal with complex decision problems due to the intricate nature of the variables involved.

- *Uncertainty*: Uncertainty arises from a deficiency of information, which can also complicate the decision-making process. However, uncertainty can also be related to the unpredictable outcomes of a decision. When outcomes are highly uncertain, decision-makers might prefer the rule-based approach, as it provides a structured framework for decision-making in the face of ambiguity.
- *Emergency*: Emergency reflects the time available from the decision to a potential hazardous event, or from one decision point in an accidental scenario to the next. In other words, it measures how much time in maximum can be tolerated before a necessary action must be taken. In emergent situations, there is limited time for a thorough information collection and analysis, and thus the bounded rational approach is less feasible. In the bow-tie model, decisions related to preventive barriers are generally less urgent, whereas decisions concerning mitigative barriers are more time-sensitive.

In subsequent sections of this book, the focus will be primarily on the RIDM methods and models, which are crucial for making key decisions in the development, design, deployment, and maintenance of barrier strategies.

### 2.8.2 Risk-Informed Decision-Making

In this subsection, we will briefly discuss the related concepts and main steps of RIDM, and its integration into barrier management. However, we will not further investigate the specific models and algorithms used for decision-making. Readers interested in decision-making theory and related methods are encouraged to refer to sources such as the book by Kochenderfer (2015).

First, we can define RIDM based on literature (NUREG-1855, 2009) as:

- *Risk-informed decision-making (RIDM)*: An approach where decision-makers consider and interpret risk information along with other factors in their decision-making processes.

This definition emphasizes the role of human judgment in decision-making and suggests that such decisions can encompass multiple objectives. It is important to differentiate RIDM from *risk-based decision-making* (RBDM), where decisions are made solely based on the results of risk assessment.

In the context of barrier management, RIDM is primarily applied to the tasks of barrier identification and analysis within risk analysis. The outcomes of RIDM also influence the operational and management aspects of barriers. According to NASA (2010), RIDM typically involves three steps, as illustrated in Figure 2.9:

**Step 1: Identification of decision objectives and alternatives**

The decision-making process may target a single objective or multiple objectives. Sometimes, a primary decision objective can be broken down into several subobjectives. For example, the objectives of installing gas and

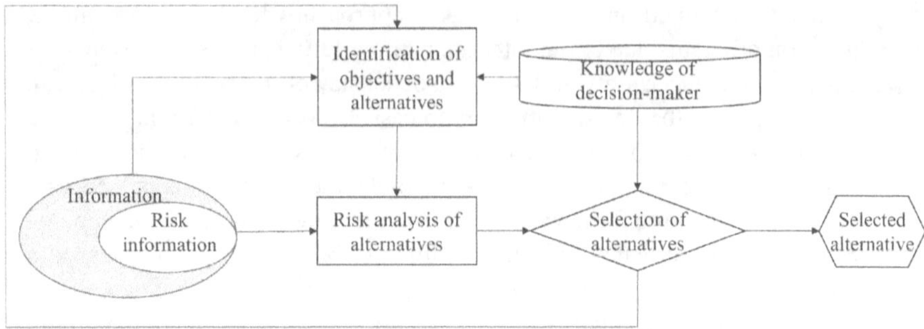


**FIGURE 2.9** Elements of a risk-informed decision-making process.

fire detection systems in a chemical plant might include protecting the facilities, preventing injuries and fatalities among operators, reducing production downtime, and minimizing environmental pollution. These objectives might align or conflict with each other, making it likely that an option may favor some objectives but be less effective for others. Additionally, various constraints, such as costs associated with deploying the barrier system, may impede the achievement of these objectives.

The identification of decision objectives and constraints should be grounded in a thorough understanding of the expectations and requirements of *stakeholders*. According to NASA (2010), a stakeholder is an individual or organization affected by the decision's outcome, which can be potential accident victims, operators of technical barriers, vendors, etc.

An option or *decision alternative* refers to the combination of variables that can be controlled in barrier design, operation, and management. For example, this can involve the choices such as the types and numbers of sensors in technical barriers, the layout of barriers, the materials and structures of physical barriers, test intervals, and maintenance policies.

Performance measures should then be defined as the extent to which a decision alternative meets the performance objectives. For instance, risk reduction to a certain asset is a typical performance measure for a safety barrier. Detailed discussion on performance measures for safety barriers will be provided in Chapter 6.

This book will not extensively cover issues and approaches related to multiple objective decision-making but will have short introductions on the most commonly used methods, such as the analytic hierarchy process (AHP), in Chapter 13.

**Step 2: Risk analysis of the decision alternatives**

In RIDM, it is important that the decision-maker can quantitatively assess the performance measures of each decision alternative, taking into account significant uncertainties. This is because in the presence of uncertainties, the actual outcome of a decision alternative will be just one within a range of possible outcomes (NASA, 2010). The performance objective can only be addressed based on the occurrence or nonoccurrence of various events and conditions. For instance, technical systems are inherently imperfect, and failures in operation are inevitable. Therefore, it is essential to analyze the unreliability and unavailability of technical safety barriers, and these factors must be incorporated into performance measures. The concept of integrity is utilized for this type of risk analysis, which will be discussed in detail in Chapter 7.

**Step 3: Risk-informed alternative selection**

In this step, the decision-maker needs to compare the decision alternatives based on their outcomes and risk analysis results. Selection criteria must be established to choose the most reasonable alternative that satisfies the performance objectives. It is important to carefully weigh the advantages and disadvantages of each alternative, ensuring alignment with constraints such as budget and acceptable risk levels. It is also worth noting that there may be situations where no alternative can provide an optimal solution. In such cases, the decision-maker should seek new alternatives and re-evaluate the available options.

Within the context of barrier engineering and management, a performance commitment is expected once a decision alternative is chosen. This commitment should involve a quantified performance measure value that meets the criteria for acceptable risk, which can later be evaluated and compared with the actual outcomes of the decision. The effectiveness of a decision will be judged based on such comparisons.

Further details on the RIDM process are available in the openly accessible NASA report (2010).

Historically, the responsibility of making decisions in RIDM belongs to humans. However, in recent decades, with the advent of advanced automatic control technologies and the development of artificial intelligence (AI) in industries, programmable logic controllers and computers are increasingly taking on more significant roles. It is therefore vital to understand the mechanics of RIDM and the metrics that quantify its performance to make the process comprehensible for computers and to automate these processes where possible.

### 2.8.3 Risk Information in Decision-Making and Barrier Management

In RIDM, the general function of information is to create a state of knowing (knowledge) for decision-makers and to reduce the uncertainty of actions (Zhu, 2023). In such a context, we can adapt the definition of information by Zins (2007):

- *Information*: Facts and ideas that are accessible and comprehensible to someone at a given moment.

Information may originate as data, but it becomes meaningful after it is processed, organized, and structured to provide insights. It can then be communicated to relevant parties, including individuals or artificial decision-makers, facilitating informed decision-making.

In risk management and barrier management, risk information serves as a critical foundation for identifying, assessing, and managing risks as a subset of information. This term is defined within this book as:

- *Risk information*: Facts related to accidental scenarios and perceptions of risks that are accessible to someone at a given moment.

Specifically, the risk information needed in a RIDM process includes:

- Details about safety issues or the current state of identified hazards, such as the severity of the problems and the time required for decision-making.
- Likelihoods of hazardous events in various scenarios.
- Identified RIFs and other contextual factors, such as the conditions under which safety barriers operate.
- Constraints on reasonable alternatives, including the availability of necessary resources like financial support, time, space, skills, and compliance with national and sectoral regulations.
- Stakeholder-related information, including their risk preferences, prioritized barrier types, their objectives, and their significance in decision-making.
- Predicted outcomes and effects of these alternatives, such as changes in risk levels, benefits or costs, and the potential for new hazards.

The role of risk information in a RIDM process is illustrated in Figure 2.9. It forms the backbone of effective decision-making by providing the necessary details about potential hazards, their impacts, and the effectiveness of safety barriers.

The value of risk information can be evaluated by the difference it makes in decision outcomes with and without it, particularly in terms of risk reduction. Generally, the value of risk information is linked to its relevance, quality, and usefulness:

- *Relevance*: Does the information address the decision-making problem directly?
- *Quality*: Is the information reliable and valid, and can it be interpreted accurately?
- *Usefulness*: Does the information provide actionable insights that can inform decisions?

The impact of risk information on a decision-maker depends significantly on their level of knowledge. While information is external, knowledge is within the brain of a person, encompassing the understanding, skills, and expertise an individual has gained through learning, experience, and practice. In many cases, presenting the same information to different decision-makers may not yield the same knowledge in their brains. For instance, in risk management, risk information typically includes uncertainties about specific events, and people's perception of these uncertainties is significantly influenced by the information they have obtained. A well-informed decision-maker can utilize this information to make precise evaluations and choices, while a less informed individual may not fully grasp the implications or importance of the same information.

Moreover, the effectiveness of risk information is contingent upon its timely delivery to the appropriate decision-maker. If crucial information is delayed or misdirected, its utility diminishes, potentially leading to suboptimal decisions based on outdated or incomplete data.

## 2.9 HUMAN FACTORS IN BARRIER MANAGEMENT

In the last section of this chapter, we have investigated the vital roles that humans play in barrier management, highlighting both the opportunities and challenges these roles present.

Generally, humans interact with safety barriers in the following ways:

- *Designers and operators*: Humans are responsible for the design, evaluation, installation, operation, and maintenance of barriers and barrier strategies.
- *Assets under protection*: Humans are the primary assets that barriers aim to protect, underscoring the human-centric focus of safety measures.
- *Integral components*: Humans often act as components of barrier systems, interacting dynamically with other elements.
- *Sources of hazard*: Humans can also introduce hazards that need to be mitigated through effective barrier strategies.

While the first two roles are straightforward and require little elaboration, this subsection will focus on the complex interactions involving humans as active components of barriers, as well as the potential for human actions to disrupt barrier functions.

It is important to recognize that organizational, procedural, and perceptual barriers are deeply interacted with humans, as the philosophy of these barriers is to avoid any human behaviors that can jeopardize the safety of assets. In the evaluation of these non-entitative barriers, it is necessary to consider the questions from the following perspectives:

- *Role and responsibility*: What are the different roles of individuals in performing and managing organizational, procedural, and perceptual barriers?
- *Competence requirement*: What are the requirements for knowledge, expertise, and skills of the personnel with the specific roles?
- *Training requirement*: What training programs are necessary to acquire the essential knowledge and expertise, particularly in cases where new recruitment is not feasible?
- *Performance assessment*: How can the performance of individuals be effectively evaluated in barrier management?
- *Influence factor*: What factors can positively or negatively influence the performance of personnel?

While some physical and technical barriers in the process industries, such as firewalls and emergency shutdown valves, are passive or fully automated, others require human interaction for effective operation. For example, technical and informative barriers often need humans to act as actuators, controllers, or backups. In scenarios where the automatic braking function fails, manual intervention to apply the brakes becomes essential. Similarly, when a gas detector triggers an alarm upon detecting hazards, it is necessary for human operators to manually close the valve to halt the gas flow and activate the ventilation system, effectively engaging with the barrier equipment.

To enhance barrier management, it is necessary to understand how human characteristics can help in developing and operating more robust and responsive barrier systems. For example, when humans are a part of safety barriers, they have some strengths:

- Humans can adapt to changing circumstances and make adjustments, e.g., on process operation and barrier activation in response to unexpected events.
- Humans are capable of making rapid decisions based on intuition and judgment, drawing from their experiences and knowledge to assess situations instantaneously. This capability is invaluable in emergency situations where standard procedures may not exist.
- Humans have the ability to coordinate actions and collaborate effectively, which is important when external assistance is required to enhance barrier functions.

However, it is not uncommon for a barrier system to fail to function as expected due to humans' unintentionally or even intentionally ignoring alarms or being unable to perform barrier-related actions effectively. In addition to the predictable physical limitations of humans in terms of strength, endurance, and sensory capabilities, the following human limitations should also be recognized in barrier management:

- Humans are prone to errors arising from fatigue, distraction, excessive workload, and insufficient knowledge, among other factors. Errors can also come from misunderstandings, misinterpretations, or incomplete transmission of information during communications. Such errors can compromise the effectiveness of barriers and increase the risk of accidents.
- Personal biases and unfounded assumptions can affect decision-making processes, introducing mistakes and inconsistencies in subjective judgments and actions. This can lead to inconsistencies in barrier management practices. Some human errors are from non-compliance with established rules due to lack of awareness, negligence, or intentional shortcuts.
- Humans exhibit performance variations due to differences in training, knowledge gaps, and personality traits. This leads to significant disparities in hazard detection, perception of consequences, and emergency responses. Consequently, the same barriers may function effectively with some individuals and less so with others.

To address these challenges, organizations need to develop strategies that leverage human strengths while mitigating limitations through targeted training, well-designed and standardized procedures, ergonomic design, and technological supports. These efforts also necessitate continuous evaluation and improvement of barrier management practices, especially as personnel responsible for operating and managing barriers change.

## NOTES

1. Det Norske Veritas (DNV) is an international accredited registrar and classification society headquartered in Høvik, Norway.
2. This section is motivated by the PhD thesis of Tiantian Zhu (2023) at Norwegian University of Science and Technology.

## REFERENCES

Aven, T., Sklet, S., & Vinnem, J. E. (2006). Barrier and operational risk analysis of hydrocarbon releases (BORA-release): Part I. Method description. *Journal of Hazardous Materials*, *137*(2), 681–691.

CCPS. (2001). *Layer of Protection Analysis: Simplified Process Risk Assessment*. American Institute of Chemical Engineers.

Cepin, M. (2010). Probability safety assessment and risk-informed decision-making. In P. Tsvetkov (Ed.), *Nuclear Power*. IntechOpen.

Øie, S., Wahlstrøm, A. M., Fløtaker, H., & Rørkær, S. (2014). *Good Practices—Barrier Management in Operation for the Rig Industry*. DNV GL.

Emslie, S. (2007). *Root Cause Analysis – Application Guidelines*. Ministry of Health (MoH), UK.

Ersdal, G., & Aven, T. (2008). Risk informed decision-making and its ethical basis. *Reliability Engineering & System Safety*, *93*(2), 197–205.

Galassi, M. C., Papanikolaou, E., Baraldi, D., Funnemark, E., Håland, E., Engebø, A., Haugom, G. P., Jordan, T., & Tchouvelev, A. (2012). HIAD – Hydrogen incident and accident database. *International Journal of Hydrogen Energy*, *37*(22), 17351–17357.

Hauge, S., & Øien, K. (2016). *Guidance for Barrier Management in the Petroleum Industry*. SINTEF Technology and Society, Trondheim, Norway.

Heinrich, H. W. (1931). *Industrial Accident Prevention: A Scientific Approach*. McGraw-Hill.

Holmberg, J.-E. (2017). Defense-in-depth. In N. Moller, S. O. Hansson, J.-E. Holmberg, & C. Rollenhagen (Eds.), *Handbook of Safety Principles*. John Wiley & Sons.

IEC61511. (2016). Functional safety – Safety instrumented systems for the process industry sector. Geneva, Switzerland: International Electrotechnical Commission.

ISO31000. (2018). Risk management – Guidelines. Geneva, Switzerland: International Organization for Standardization.

Johansen, I. L., & Rausand, M. (2015). Barrier management in the offshore oil and gas industry. *Journal of Loss Prevention in the Process Industries*, *34*, 49–55.

Kahneman, D. (2011). *Thinking, Fast and Slow*. Farrar, Straus and Giroux.

Kjellen, U. (2007). Safety in the design of offshore platforms: Integrated safety versus safety as an add-on characteristic. *Safety Science*, *45*, 107–127.

Kochenderfer, M. J. (2015). *Decision Making Under Uncertainty: Theory and Application*. The MIT Press.

Leveson, N. (2004). A new accident model for engineering safer systems. *Safety Science*, *42*(4), 237–270.

March, J. G. (1994). *Primer on Decision Making: How Decisions Happen*. The Free Press.

NASA. (2010). *Risk-Informed Decision Making Handbook*. NASA/SP-2010-576 Version 1.0, NASA Headquarters, Washington, DC.

NUREG-1855. (2009). *Guidance on the Treatment of Uncertainties Associated with PRAs in Risk-Informed Decision Making.* U.S. Nuclear Regulatory Commission, Office of Nuclear Regulatory Research.

OREDA. (2009). *Offshore Reliability Data* (5th ed.). OREDA Participants.

Paltrinieri, N., & Khan, F. I. (2016). *Dynamic Risk Analysis in the Chemical and Petroleum Industry: Evolution and Interaction with Parallel Disciplines in the Perspective of Industrial Application.* Butterworth-Heinemann.

Perrow, C. (2011). *Normal Accidents: Living with High-Risk Technologies – Updated Edition.* Princeton University Press.

Pitblado, R., Fisher, M., Nelson, B., Fløtaker, H., Molazemi, K., & Stokke, A. (2016). Concepts for dynamic barrier management. *Journal of Loss Prevention in the Process Industries, 43*, 741–746.

PSA. (2011). *Regulations Relating to Management and the Duty to Provide Information in the Petroleum Activities and at Certain Onshore Facilities.* NASA/SP-2010-576 Version 1.0, NASA Headquarters, Washington, DC.

PSA. (2017). *Principles for Barrier Management in the Petroleum Industry.* Petroleum Safety Authority, Stavanger, Norway

Rausand, M., & Haugen, S. (2020). *Risk Assessment: Theory, Methods and Applications.* Wiley.

Reason, J. (1997). *Managing the Risks of Organizational Accidents.* Ashgate.

Rezvani, Z., & Hudson, P. (2021). How do middle managers really make decisions within the oil and gas industry? *Safety Science, 139*, 105199.

Rochlin, G. I. (1996). Reliable organizations: Present research and future directions. *Journal of Contingencies and Crisis Management, 4*(2), 55–59.

Salas, E., Rosen, M. A., & DiazGranados, D. (2010). Expertise-based intuition and decision making in organizations. *Journal of Management, 36*(4), 941–973.

Seljelid, J., Haugen, S., Sklet, S., & Vinnem, J. E. (2007). *Operational Risk Analysis – Total Analysis of Physical and Non-Physical Barriers – BORA Handbook.* Preventor.

Sklet, S., Vinnem, J. E., & Aven, T. (2006). Barrier and operational risk analysis of hydrocarbon releases (BORA-release): Part II: Results from a case study. *Journal of Hazardous Materials, 137*(2), 692–708.

Teng, B. (2010). *Assessing Risk and Prevent Accidents in Complex System.* Norwegian University of Science and Technology.

Torres-Echeverria, A. C. (2016). On the use of LOPA and risk graphs for SIL determination. *Journal of Loss Prevention in the Process Industries, 41*, 333–343.

Vinnem, J. E., Aven, T., Hauge, S., Seljelid, J., & Veire, G. (2004). Integrated barrier analysis in operational risk assessment in offshore petroleum operations. In C. Spitzer, U. Schmocker, & V. N. Dang (Eds.), *Probabilistic Safety Assessment and Management* (pp. 620–625). Springer.

Zhu, T. (2023). *Information and Decision-Making for Major Accident Prevention: A Concept of Information-Based Strategies for Accident Prevention* Norwegian University of Science and Technology.

Zins, C. (2007). Conceptual approaches for defining data, information, and knowledge. *Journal of the American Society for Information Science and Technology, 58*(4), 479–493.

# 3 Technical Safety Barriers

## 3.1 INTRODUCTION

Following the introduction of barrier management and non-entitative safety barriers, this chapter transitions to technical safety barriers or safety-critical systems. When these systems incorporate instrumentation technologies, they are termed "safety instrumented systems", as defined by the international standard IEC61511 (2016).

As discussed in Chapter 1, in the process industries and many other industries, safety is commonly defined as a state where the risk has been reduced to and is maintained at a level that is as low as reasonably practicable. Considering a complex socio-technical system as the asset of interest, its overall safety is dependent on several factors: design characteristics, effectiveness of management, skills and experiences of operators, stability and predictability of operational conditions, and the reliability of technical safety barriers. This brings us to a definition of functional safety:

- *Functional safety*: The aspect of overall system safety that relies on the proper operation of technical safety barriers.

In many industrial sectors, including nuclear, oil and gas, automotive, aviation, and medical, functional safety is not just a legal requirement but also a moral obligation to ensure the safety of the end-users and operators. Functional safety raises requirements to the design, development, testing, and certification of safety-critical systems and their components to ensure compliance with essential safety standards and regulations. Before introducing quantitative methods in functional safety, in this chapter, we start from a discussion on the general design properties of these technical safety barriers.

## 3.2 ELEMENTS IN TECHNICAL BARRIERS

In this section, we present the general structure and elements of technical safety barriers, along with their voting mechanisms.

### 3.2.1 Subsystems of a Technical Barrier

A technical safety barrier or safety-critical system capable of performing a complete barrier function typically consists of three subsystems. These are presented in the case of an emergency shutdown system designed for a pressurized vessel, as shown in Figure 3.1.

- *Sensing subsystem*: This subsystem involves one or more sensors that observe or monitor the condition of the asset, and they aim to promptly detect any abnormal situations or undesired events. The subsystem serves

DOI: 10.1201/9781003245636-3

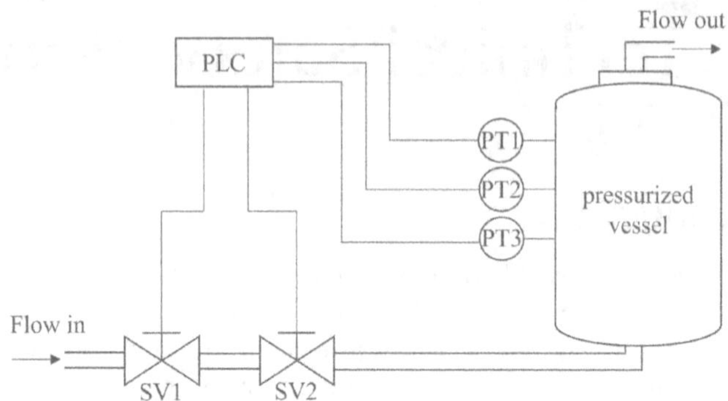


**FIGURE 3.1** An illustrative example of three subsystems of an emergency shutdown system.

as the input element of a technical barrier, transmitting signals to the controlling subsystem. In the example shown in Figure 3.1, the sensing subsystem includes three pressure transmitters (PT1, PT2, and PT3).

- *Controlling subsystem*: This subsystem contains at least one controller, such as a programmable logic controller (PLC), logic solver, or computer, responsible for processing the information relayed by the sensing subsystem. It acts as the decision-maker within the technical barrier, analyzing the data to determine necessary safety actions and communicating commands to the actuating subsystem. In the example shown in Figure 3.1, the controlling subsystem is represented by a safety PLC.
- *Actuating subsystem*: This subsystem includes one or more physical components, often referred to as *actuators* or *final control elements* in safety literature (Rausand, 2014), to execute the safety actions requested by the controlling subsystem. These components are crucial in converting the decisions of the controlling subsystem into actionable steps that prevent or mitigate harm to the asset. In the example shown in Figure 3.1, the actuating subsystem comprises two solenoid valves (SV1 and SV2).

The operational principle of the safety-critical system in Figure 3.1 is as follows: Three transmitters continuously monitor the gas pressure within the vessel. If the pressure exceeds acceptable limits, the transmitters send this information to the PLC, which analyzes the information and then issues commands if necessary to the two shutdown valves. The solenoid valves respond by closing actions, thereby stopping the flow of incoming gas through the pipeline and preventing any further rise in vessel pressure.

There are generally two approaches for activating the actuating subsystem in such technical safety barriers:

- *Energize-to-trip*: In this approach, the actuating elements remain inactive during normal operations and are activated to perform the barrier function when supplied with electricity, hydraulic pressure, or pneumatic pressure.

For example, a hydraulic gate valve that closes to stop fluid flow when hydraulic pressure is applied.
- *De-energize-to-trip*: Here, the actuating elements are energized during normal operations and are triggered to perform the barrier function when the energy supply is removed. For example, a solenoid valve that remains open, allowing fluid flow while electrically powered, and automatically closes when the electricity supply is discontinued.

A technical safety barrier is always regarded as an active barrier (see Section 1.4.3), even when it is operated with the de-energy-to-trip approach, since its barrier function is implemented based on a command.

### 3.2.2 Alarm as a Safety Barrier

It is important to note that systems possessing only one or two of these subsystems can also function as technical safety barriers or safety-critical systems. A sensing subsystem alone may trigger alarms to warn nearby operators to evacuate if the pressure becomes excessively high. Such a sensing and alarm system, despite the lack of actuators, is undeniably critical to safety and is the study object in this chapter.

Alarm systems incorporate the same three subsystems as other technical safety barriers. In these systems, the actuating subsystem is specifically designed to issue alert or warning signals to personnel. Due to its working mechanism, this subsystem generally demonstrates higher reliability and experiences fewer failures than other types of actuators. Nevertheless, it is important to understand that an alarm system alone is not sufficient to completely eliminate risks to assets. It depends on human intervention, either to evacuate or to implement additional safety measures. Consequently, the analysis and management of alarm systems are slightly different from those applied to other technical safety barriers.

According to ANSI/ISA-18.2 (2016), the tasks of an alarm system include the detection of an abnormal condition, the notification and communication of this condition to human operators, and the documentation of state changes within the alarmed system. There are several potential challenges associated with these alarm functions that could impact their effectiveness in performing necessary barrier functions, including the following:

- *No or late alarms*: Alarms may fail to raise or may be delayed, which can be attributed to a variety of factors such as technical malfunctions in alarm components or improper configuration of alarm thresholds. For example, if ultrasonic sensors use the same alarm threshold for detecting both hydrogen and natural gas leaks, a minor hydrogen leak might not trigger the alarm despite its significant ignition risk due to the wider ignition range of hydrogen.
- *Alarm flooding*: This issue occurs when an excessive number of alarms are triggered in a short period, overwhelming human operators. This flood makes it difficult to discern and prioritize critical alarms from less urgent ones. Alarm flooding often results from overly conservative alarm settings,

where even minor deviations in parameters can trigger a barrage of alarms, creating a chaotic environment. This can lead to alarm fatigue, causing the operator to possibly turn off the alarm.

- *Standing alarms*: Sometimes alarms remain active for an extended period without being acknowledged or addressed, typically due to alarm failures, equipment malfunctions, or operator inattention. This issue can make it harder to respond properly, sometimes resulting in the alarm being turned off.
- *Low alarm prioritizing*: This situation results from assigning lower priorities to alarms considered less critical or urgent. Although this strategy can be effective in reducing alarm flooding and preventing operator overload, it can inadvertently lead to neglect of alarms that, while seemingly less urgent, may still require prompt attention and action.

Since alarms require human intervention, a common issue is that operators or users may fail to take appropriate actions or respond effectively. This can be due to a lack of training, insufficient understanding of the alarm systems, or complacency. As a result, alarms might be ignored, overlooked, or mishandled, leading to potential safety risks, damage to equipment, or inefficiencies in operations.

Therefore, alarm management, as a specific area of barrier management, needs careful preparation, planning, and implementation to enhance the effectiveness of alarm systems and minimize risks to assets. For guidelines on alarm system design and management, international standards such as ANSI/ISA-18.2 (2016) and IEC62682 (2014) are very useful. Readers interested in techniques and data analysis methods in alarm management can refer to the chapter by Yang et al. (2022).

## 3.3 DESIGN AND DEVELOPMENT OF TECHNICAL BARRIERS

### 3.3.1 Design Requirements

Requirements for a technical system can be categorized as functional- and non-functional ones. A *functional requirement* defines what the system is supposed to do. As discussed in Chapter 1, the functional requirement of a technical safety barrier is to reduce the risk to the asset. The following design requirement is crucial:

- *Specificity*: This refers to the need for a barrier to be tailor-made to deal with a particular hazard or hazardous event. In most cases, a barrier designed specifically for a particular hazard tends to be more effective in mitigating that hazard than a general-purpose barrier. For instance, a medical device like an implantable defibrillator must be specifically designed to meet the unique medical needs and conditions of the patient. It needs to detect and respond to life-threatening cardiac events accurately while avoiding false alarms or unnecessary interventions.

On the other hand, a *non-functional requirement*, also called a *performance requirement*, relates to how well a system performs its functions. Here are some common performance requirements for technical systems, including barrier systems:

- *Reliability*: The capability of a barrier to perform its required function consistently without failure for a specified period under defined environmental and operational conditions. In this book, reliability is mostly distinct from availability by focusing on the capability to operate successfully up to the point of any potential failure.
- *Availability*: The capability of an item to be in a state ready for use at a given time or over a specified period. Compared with reliability, availability is applied to repairable items and includes considerations of scenarios post-failure and the impacts of downtime. The availability of technical safety barriers is often described in terms of safety integrity, which will be explored in Chapter 6.
- *Maintainability*: The easiness that an item can be maintained or restored to working condition after a failure. Maintainability is determined by various aspects, such as the simplicity of detecting and locating failures, the rapidity of performing the needed maintenance tasks, and the efficiency of the verification process and restart.

Moreover, researchers have proposed more dedicated requirements for the design of technical safety barriers (e.g., see Hollnagel (2004) and Johansen and Rausand (2015)):

- *Adequacy*: This requirement ensures that a safety barrier reduces the risk to the asset to an acceptable or at least to the expected level. If the barrier is found to be inadequate, additional barriers should be implemented. In Chapter 6, we will use "effectiveness" as a measure of adequacy, which also qualifies as a performance requirement.
- *Independence*: This principle dictates that each safety barrier should be designed and operated without reliance on other barriers or the asset it protects. This ensures that the failure of one barrier is expected not to compromise the performance or reliability of others. More importantly, in a hazardous event that may harm the asset, the functions of safety barriers should not be simultaneously weakened.
- *Auditability*: This includes the ability to track and verify that a barrier has been designed, installed, and maintained in accordance with established safety standards and regulations. In safety-critical sectors, auditors may review the designs of safety barriers to ensure compliance with regulatory demands and industry benchmarks. Such reviews typically include examining design documents, inspecting installations, and evaluating barrier performance over time. Auditability also provides valuable data that can be used to enhance the design and maintenance of safety barriers in the future.

Chapter 6 will provide a comprehensive overview of performance measures used in the quantitative analysis of safety barriers, along with the algorithms tailored for such analysis. It is important to recognize that the performance discussed so far is based on customer or management requirements, which may not always correspond to the performance of a prototype developed according to the design or the actual

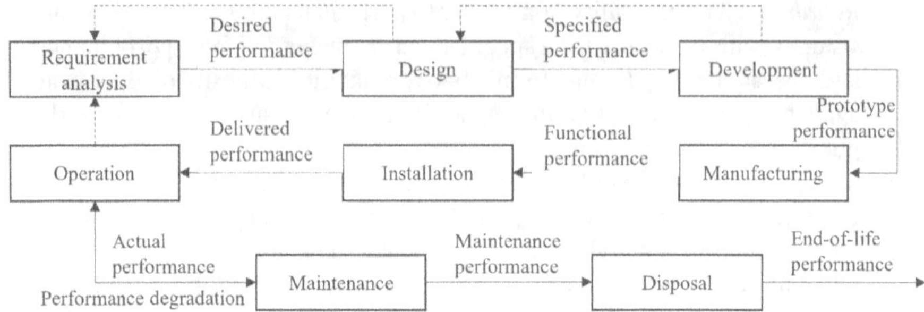


**FIGURE 3.2** A simplified lifecycle model with performance outputted from each phase.

performance of the product during its operation. Figure 3.2 illustrates the different phases in the lifecycle of a technical product/system, highlighting the specific performance terminologies associated with each phase. The dotted lines in the figure represent the flow of information.

### 3.3.2 Generic Design Rules and Methods for Safety Barriers

When considering the requirements for technical safety barriers, it is beneficial to establish a set of general rules and methods early in the early design process for satisfying the requirements with relatively low cost. In Section 1.5, we have presented a collection of design rules for inherent safety, which are also applicable to safety barriers as both are technical systems. The following guidelines provide a selection of generic rules and their corresponding methods for designing technical safety barriers:

- *Design for adequacy and reliability of barriers*: This can be achieved through *overdesign*, namely introducing a larger safety margin by using materials with higher strength, durability, or resistance. This creates a buffer against wear, fatigue, or unexpected loads. Redundancy can enhance reliability and will be discussed further in the upcoming subsection.

  Another approach in design for reliability is *robust design*, which involves developing a barrier system capable of withstanding variations in operating and environmental conditions, such as extreme temperatures, humidity, and vibrations.

  Additionally, conducting a comprehensive identification of potential failures and hazards during the system design phase is also important for improving reliability. We will introduce some relevant methods in Chapter 4.
- *Design for maintainability and availability of barriers*: Maintainability can be improved by all the methods that shorten any step in the whole maintenance process and thus improve availability. For instance, integrating an automatic diagnosis module within the technical barrier can facilitate the detection and localization of failures. The system design should allow easy access for maintenance crews, with components designed for more efficient removal and replacement.

Other approaches used in design for maintainability include *modular design* that minimizes maintenance impact on functioning parts and standardized verification for technologies in barriers. *Testability* is also a critical aspect of maintainability, which specifies whether the status of a barrier or its components can be determined promptly and accurately. Approaches of improving barrier testability include making test results easily observable and understandable, ensuring compatibility with various testing tools, and achieving consistent test outcomes.

- *Design for independence*: To achieve a high level of independence, safety barriers should be designed utilizing diverse barrier modalities. This involves using different types of technologies, approaches, or components within each barrier function to ensure redundancy and minimize the risk of common-cause failures that occur on several barriers or barrier components at the same time (these failures will be further discussed in Chapter 10). Additionally, implementing physical separation between barriers can provide an extra layer of protection and reduce the risk of simultaneous failures across multiple barriers.

### 3.3.3 Redundant Design of Safety Barriers

One key method to enhance the reliability and availability of technical barriers is to incorporate a safety margin through with *redundancy* or a *redundant structure*. In the example depicted in Figure 3.1, both the sensing and actuating subsystems include multiple components, which constitute redundant structures. In this book, we use the term "channel" to describe either a single component or a combination of several components that work together to perform a specific function within a redundant structure.

The benefits of redundancy can be found even without quantitative analysis, since having multiple channels allows the same function to be performed by different components. In the event of a failure in one channel, the others can continue to operate, thus maintaining system functionality.

Take the example from Figure 3.1, where two valves are installed to manage the pipeline flow. If one valve fails to close, the other remains functioning, ensuring that the flow can still be stopped. Consequently, as long as either SV1 or SV2 is functioning, the barrier function is maintained.

Generally, we can implement redundancy in system design using the following approaches (for example, considering two channels in a system):

- *Hot redundancy*: Also known as *active redundancy*, this type of redundancy is commonly used in many critical applications, where all channels are always active or in-use. There is no distinction between the primary and backup channels. Both channels run in parallel, continuously receiving data and performing identical tasks. In the event of a failure, the functioning channel continues to operate, ensuring uninterrupted service.
- *Cold redundancy*: Also called the *standby redundancy*, there is a backup channel that remains idle and turned off (de-energized) until the moment

when the primary channel fails. In this case, the backup channel does not monitor or process data but rather just sitting ready to be powered up. When the primary channel fails, the cold backup channel is powered on and started up from scratch.

- *Warm redundancy*: This type of redundancy combines aspects of hot- and cold redundancy. The backup channel is running and receiving data but is not processing it to the same extent as the active channel, so as to save energy or computation resource. In the event of a failure, the warm backup channel is brought up to full capacity, taking over from the failed one that was previously operating. This approach provides a balance between cost and speed/success of recovery.

We will discuss modeling and reliability analysis of various redundant structures in more detail in Chapters 5 and 8.

As depicted in Figure 3.1, the three pressure transmitters employ hot redundancy, as all of them simultaneously receive data from the vessel. The two valves, however, can be set up with either hot- or cold redundancy, depending on specific requirements. For instance, with cold redundancy, SV1 serves as the primary channel and SV2 is activated only if SV1 fails to close due to unforeseen circumstances.

It is important to note that our discussion has so far focused only on hardware redundancy. However, redundancy can also be implemented through software redundancy, such as using multiple error-checking codes, or temporal sequencing, referring to scheduling multiple, staggered backups to ensure data integrity.

Redundancy is actually conflicting with another design rule for safety – simplification, because a redundant structure has more components and perhaps more interactions, making a system more complex. In practice, it is necessary to find an optimal design in consideration of different rules. Further discussions on the types, potentials, and pitfalls of redundancy are available in the paper by Yellman (2006).

Coming back to a barrier system, consider a sensing subsystem with $n$ identical channels, one extreme situation is that the subsystem is functioning only when all the $n$ channels are functioning, and we can call this case as an *n-out-of-n voting structure*. The opposite extreme situation is that the subsystem is functioning as long as at least one channel is functioning, and we call this case as a *1-out-of-n voting structure*. A more general case is that the subsystem is functioning when at least $k$ of the $n$ channels are functioning, and this is called the *k-out-of-n voting structure*. We will further discuss the voting mechanisms in a system with several identical channels and voted structures of technical barriers in Chapter 8.

### 3.3.4 Development and Reliability Test

Following the design phase, a *test-analyze-and-fix* (TAAF) process is necessary to evaluate whether prototype barrier products meet the designated functional and non-functional requirements. Performance of a prototype barrier product, especially those aspects related to reliability, is often focused in such tests, which are thus called *reliability tests*. The primary goal of reliability tests is to develop exceptionally reliable products that minimize the incidence of unexpected failures. Specifically, the tests conducted during the development phase prior to mass production can be classified into three distinct groups based on their respective objectives:

- *Failure identification test*: These tests aim to uncover the failure modes of an item, identify their causes, and understand their relationships with the design and applied stresses. The objective is to remove the defects associated with the identified failure modes and avoid these failure modes in operation.
- *Environmental stress test*: These tests evaluate the effects of environmental conditions on an item. The objective is to determine the operational envelope, beyond which failures are not regarded as unexpected events on such as an item.
- *Reliability estimation test*: These tests are conducted to estimate reliability parameters, with the objective of verifying if the reliability requirements are met. If the reliability parameters based on the performance observed during a test do not meet the expected or required levels, it is crucial to find the causes and make necessary design modifications. Another objective is to compare different design options in terms of reliability.

These laboratory tests can be conducted under stress conditions similar to those of the real applications. However, they are more frequently carried out under conditions where stresses exceed the nominal values. This approach is adopted because most modern products, especially those such as safety barrier components that are designed for high reliability, are difficult to evaluate for long-term performance and failure modes during normal operations within a limited timeframe. Applying higher stresses accelerates the occurrence of failures and reduces the time required to collect necessary data, and such an approach is *accelerated test*. Stresses in these tests can include electrical, vibrational, thermal, and moisture levels, among others (MIL-HDBK-781A, 1996).

One of the commonly employed reliability testing approaches used by manufacturers to identify failures and reveal the operational envelope is the *highly accelerated life test* (HALT). This method involves exposing product prototypes to stresses that surpass their design limits, thereby revealing potential failure modes. HALT is conducted in a dedicated testing chamber equipped to simulate various types of stresses, such as extreme temperatures, thermal cycles, and vibration. The items under test are placed in the chamber and progressively subjected to different stressors until they reach their destruction limit. Following HALT, root cause analysis is undertaken for revealing failure causes, and improvement analysis is implemented for eliminating similar failures in the future. Readers interested in further information and case studies on HALT can refer to McLean (2009).

On the other hand, acceleration in reliability estimation tests can be achieved through the following methods:

- *Over-stressing*: The tested items are operated under stresses that are higher than usual but still within the design specification limits. This approach assumes that such conditions will not introduce new failure modes but can make the items more vulnerable and fail in a shorter time.

- *Intensive usage*: The usage frequency of the tested items is increased, potentially leading to failures in a reduced amount of calendar or clock time.
- *Threshold tightening*: In some cases, failures are defined as events where performance exceeds a certain threshold. In accelerated tests, this threshold can be adjusted downward to reduce the testing duration.

Further details about accelerated testing methodologies are available in the international standard IEC62506 (2013). In Chapter 7, we will discuss the analysis methods used to interpret the data collected from these reliability estimation tests.

## 3.4 MANUFACTURING AND INSTALLATION OF TECHNICAL BARRIERS

This book does not focus on the manufacturing and installation phases. Instead, it offers a concise introduction to these essential steps in the lifecycle of a technical safety barrier (refer to lifecycle model in Figure 3.2), with providing relevant literature for further exploration.

### 3.4.1 Reliability Test and Quality Assurance

The manufacturing process for equipment with barrier functions does not significantly differ from that of other technical products. A critical aspect of manufacturing process is ensuring that mass-produced items meet the same performance standards as the prototypes. Manufacturing inherently varies due to the reasons such as improperly adjusted machines, human errors, or defective materials (Montgomery, 2012). To maintain the required performance, it is vital to operate the manufacturing process consistently, with minimal variations. *Statistical process control* (SPC) serves as an effective toolkit for analyzing and mitigating these variations. For a more detailed exploration of SPC, readers are encouraged to refer to Montgomery (2012).

The *highly accelerated stress screen* (HASS) is a technique similar to HALT but is specifically designed for use during the manufacturing and production stages. HASS is employed to detect shifts and weaknesses inherent to the manufacturing process, often applying stress levels beyond those typically encountered in the field to reveal any manufacturing defects. The stress thresholds identified in HALT guide the stress levels used in HASS. Another relevant concept is the *highly accelerated stress audit* (HASA), which utilizes a statistical sample rather than a 100% screening like HASS. HASA offers a more cost-effective alternative for large-scale product assessments, considering the constraints of testing duration and resources. Further details on HASS and HASA can be found in Hobbs (2000) and McLean (2009).

Many products, such as electronic components and sensors integral to technical safety barriers, are susceptible to early failures, often known as *infant mortality*, which can result from defects during the manufacturing process. *Burn-in testing* is often employed to mitigate these early failures before the components are operational. The purpose of burn-in testing is to detect and remove defective items, thereby significantly enhancing the average reliability of the delivered products. Theoretically, with a sufficiently extended burn-in period, the components are largely assured to be free from early failures. More insights into burn-in testing techniques and analysis algorithms can be found in the PhD thesis by Ye (2011).

Before the installation of technical barriers, it is essential to develop a comprehensive installation plan. This plan should include the sequence of installation steps, resource allocation, logistical arrangements, and coordination with other construction or installation activities on-site. It is important to prioritize relevant safety protocols, secure the necessary permits, and obtain any required approvals specific to the installation of technical safety barriers. A comprehensive assessment of the installation site is also vital to identify any potential obstacles, hazards, or unique requirements that might affect the installation process. Throughout the installation process, strict adherence to manufacturer guidelines and safety regulations is paramount.

### 3.4.2 Technical Verification

Another critical task before delivery is conducting *acceptance tests* (primarily field tests) and technical *verification* to confirm the functionality and performance of produced items. In various industrial sectors, there are several guidelines and standards for technical verification, such as NORSOK S-001 (NORSOK, 2008), DNV-OSS-304 (2011), and API-RP-17Q (2018). According to these documents, verification in such a context can be defined as follows:

- *Verification*: The process of testing and examining equipment to confirm its capability to meet specified requirements.

This verification process is typically carried out independently, often by a third party distinct from both the manufacturer and the user. Engaging in verification allows for the barriers to be tested against simulated demands or real-world scenarios, evaluating their responses to ensure compliance with the operational parameters and performance criteria set forth in the specifications.

Given that the verification object is a technical safety barrier, the general verification process can include the following steps:

**Step 1: Specification of asset and hazards**

This first step focuses on identifying and clarifying the asset that the barrier will protect, along with any potential hazards it may face. This involves collecting data on the operational environment of the asset and the barrier, expected performance of the asset, and potential risk influence factors that could compromise safety. While risk assessment results provided by the barrier supplier or the user can be used, to keep independence of verification, a third party may conduct a new risk assessment.

**Step 2: Scoping the verification project**

This phase defines the scope of the verification project by establishing the objectives of the verification activities and identifying which activities in the product lifecycle should be considered. It also involves determining the acceptance criteria for verification and identifying available resources. DNV-OSS-304 (2011) has highlighted that verification can be conducted at different levels: Low-level verification or brief verification is for the well-known equipment and the situation where consequences of failures are low or moderate; and high-level verification or thorough verification is for innovative designs or the situation where consequences of failure are severe.

**Step 3: Developing a verification plan**

Developing a verification plan involves outlining the methodology and tools required for conducting the verification. This plan sets the timeline, assigns responsibilities to team members, and specifies the testing and evaluation methods to be used, ensuring all aspects of the barrier are comprehensively covered. The following verification works are needed for different activities in a product lifecycle of a technical safety barrier and other equipment:

- Reviews on system and detailed design, including drawings and design principles, design procedures and cost estimates.
- Audits on project and quality management, such as project quality management documentation, quality control procedures, and simulation and laboratory testing models, with attendance at tests for the most critical components.
- Reviews of manufacturing, fabrication, and inspection procedures, and manufacturing specifications, along with audits of quality records.
- Reviews of operations management systems and inspection plans, including attendance during start-up, critical inspections, and activities related to modification and repair.
- Audits of human resources and employee training.

**Step 4: Implementing verification**

This step is the actual execution of the verification plan, which includes conducting the aforementioned reviews and audits on both documentation and through real-world observations to assess the functionality and integrity of the safety barrier. This phase is critical for identifying any discrepancies, weaknesses, or failures in the barrier system.

**Step 5: Evaluating and reporting**

The final step focuses on analyzing the data gathered during the testing phase, evaluating the performance of the safety barrier against the specified requirements. The results are then compiled into a detailed report that provides insights into the verified barrier, highlights any areas of concern, and recommends improvements if necessary. This report is crucial for stakeholders to make informed decisions regarding the safety barrier's deployment and operation.

The guidance and standards mentioned above can be references for more detailed information of technical verification.

## 3.5 OPERATION OF TECHNICAL BARRIERS

In most production systems, items are considered to be functioning when they are actively operating. However, barrier engineering often employs a standby or dormant mechanism to minimize disruptions to the normal operations of the asset being protected, to limit degradation of the barriers, and to reduce energy consumption. It is essential to understand the operational modes of safety barriers before conducting further quantitative analysis.

### 3.5.1 Demand and Operational Mode

Most technical safety barriers are active barriers, meaning that they require a specific energy input or a command to be activated. This input or command is referred to as a *demand*:

- *Demand*: An event or condition that necessitates the activation of a safety barrier to perform its barrier function.

A demand may also be referred to as a *deviation* or a *process upset*. Events such as the pivotal or hazardous ones in an accidental scenario can generate demands on associated safety barriers. If a technical safety barrier is functioning properly, it is expected to respond to this demand by performing its barrier function, thereby eliminating or at least mitigating the potential harm related to the demand. Conversely, if a technical safety barrier is faulty, is unavailable, or is ineffective, the demand will bypass the barrier to proceed to the next barrier or potentially harm the asset if the barrier is the ultimate barrier.

In consideration of demands, safety barriers generally operate in one of two modes:

- *Continuous mode*: In this mode, a barrier is designed to operate continuously, without any dormant state. It actively manages any deviations related to the asset and interrupts an accidental scenario when it develops to its position.
- *Demand mode*: In this mode, a barrier remains in a standby or dormant state during normal operations and is activated only by specific demands to perform its barrier function. In this operational mode, the barrier appears transparent in case of no demand, taking no action unless a deviation occurs, and its intervention is required.

It is important to note that when classifying safety barriers according to their operational modes, passive barriers are regarded as functioning in continuous mode, whereas active barriers operate in demand mode.

In certain safety-critical systems, subsystems may operate in different modes simultaneously. For example, in the emergency shutdown system depicted in Figure 3.1, the sensors and PLC operate continuously, constantly collecting data and processing calculations, while the valves operate in a demand mode and are activated only when the pressure in vessel exceeds a specified threshold. In practical analysis, it is also reasonable to consider that the sensing and controlling subsystems might operate in the demand mode, as their barrier functions become critical under high-pressure conditions.

The demand mode of operation for safety barriers can be further categorized into two distinct types based on the frequency of activation demands:

- *Low-demand mode (1)*: The frequency of activation demands is relatively low compared to the frequency of tests and inspections conducted on the safety barrier.
- *High-demand mode (1)*: The frequency of activation demands is relatively high compared to the frequency of tests and inspections.

This categorization provides a basic differentiation between the two modes of operation. According to IEC61508 (2010), a practical guideline for this classification is that if the frequency of demands to activate a safety barrier does not exceed once per year, it is considered a low-demand mode; otherwise, it is categorized as a high-demand mode. This method offers a straightforward way to categorize operational demands, and the rationale behind this classification will be explored in detail in Chapter 8.

In most situations, a demand can be viewed as a shock of negligible duration, implying that the demand is considered finished once the barrier function has been performed. However, the duration of a demand does need to be considered in certain contexts. For example, when assessing the effectiveness of a ventilation system in mitigating a gas leak, the event that gas concentration is higher than a certain value is regarded as the demand. The barrier system needs some time to ventilate the flammable gas before its concentration in a room drops below the threshold. Here, we define *demand duration* as follows:

- *Demand duration*: The time span from the arrival of a demand, when a barrier function becomes necessary, to the moment when the demand ceases and the barrier function is no longer required.

Figure 3.3 illustrates some examples of different operational modes of safety barriers. As we will discuss in Chapter 6, different metrics can be employed to evaluate the performance of safety barriers both upon and during the demands.

When considering scenarios with noticeable demand duration, an active safety barrier typically operates in two distinct states: *dormant state* and *demanding state.* In the dormant state, the barrier remains inactive or on standby, not performing its designated barrier function, and consequently experiences the minimal stress. On the other hand, during the demanding state, the barrier is actively engaged in performing its intended barrier function and is subjected to higher levels of stress.

### 3.5.2 Tests and Diagnosis

For the safety-critical systems operating in the demanded mode, especially for those in the low-demand mode, tests and inspections are necessary to examine their operational conditions and to verify that these systems are functioning properly. It is

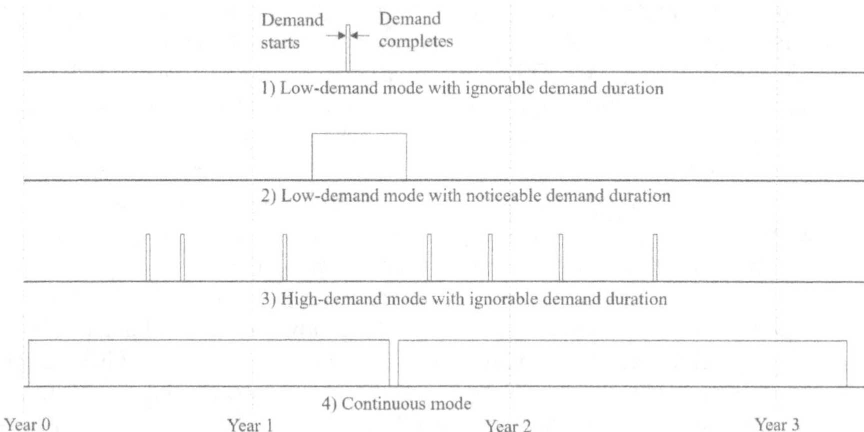


**FIGURE 3.3** Four examples of different operational modes of safety barriers.

important to note that the tests and diagnosis we discuss here are conducted on the safety-critical systems themselves, rather than on the assets they protect. There are two main types of tests used for safety barriers:

- *Diagnostic test*: Also known as a *diagnosis*, this test utilizes a built-in diagnostic module or feature within a safety-critical system to identify specific types of failures or abnormal conditions. A diagnostic test can be performed in real time, allowing for immediate detection of failures as they occur. These tests are non-intrusive and do not disrupt normal production activities, which is why they are often referred to as *online tests*.
- *Proof test*: This test is conducted to confirm that a safety-critical system can effectively perform its barrier function as intended. Proof tests are scheduled regularly, occurring at fixed intervals, denoted as $\tau$ in this book, existing between two consecutive tests. Because these tests can interrupt normal production or reduce productivity, they are typically classified as *offline tests*.

By continuously monitoring performance and identifying any deviations from expected behavior, diagnostic tests play a critical role in preventing catastrophic failures and minimizing system downtime. For example, in automobiles, diagnostic tests can display results through indicators on the dashboard. If there is an issue, such as a malfunction in the anti-lock braking system (ABS), the corresponding diagnostic module will automatically detect the fault and alert the driver by activating the relevant indicator icon.

There are two widely used techniques for conducting diagnostic tests: *reference diagnosis* and *comparison diagnosis*:

- *Reference diagnosis*: This technique involves measuring specific characteristics of the barrier, such as current, voltage, pressure, and temperature, and comparing these measurements to predetermined reference values to identify any abnormalities. It is typically used for simpler systems where reference values are well-established and reliable indicators of system health.

- *Comparison diagnosis*: This more complex technique involves comparing data from two or more components of a barrier to detect discrepancies. It is particularly useful in complex systems where no well-established reference values exist, and analyzing data from multiple sources is necessary to reveal faults. For example, within the electronic control unit of a vehicle, comparison diagnosis can be used to assess data from various sensors to identify inconsistencies that could indicate a malfunction.

It is important to note that diagnostic tests do not always reveal all potential failures in safety-critical systems. For example, while a sensor malfunction in an emergency shutdown system might be immediately detected if it reports incorrect altitude readings, a failure in a valve that prevents it from closing a pipeline tightly may not be detectable through these tests. The extent to which faults in a safety-critical system can be identified through diagnostic tests is referred to as the *diagnostic coverage* (a formal definition and additional details are available in Chapter 9). Therefore, even with diagnostic modules installed, proof tests are still useful to ensure the overall reliability and functionality of safety-critical systems.

Proof tests are important for the industries where safety is critical. The primary goal of a proof test is to confirm that a barrier system or component adheres to required safety standards and can function as expected under real-world conditions. Reflecting on the definition of demand in the last subsection, a proof test can be seen as a specific, planned type of demand imposed on a safety-critical system. In other words, a proof test is a deliberate attempt to demand the barrier function of the system to examine its availability.

For a barrier system with a redundant structure, when one barrier channel is undergoing testing, another channel can maintain the required barrier function. In the case of a single barrier component, such as a shutdown valve, a common approach is to temporarily bypass the barrier function during the test to prevent an accidental activation. For example, a signal might be sent to the actuator to confirm it can receive the signal and then execute a full operational test of the barrier function. During the proof test, any visible damage is also inspected. Once testing is complete, the bypass is removed, and the normal operation of the barrier is restored.

Lundteigen and Rausand (2008a) have analyzed and compared diagnostic tests and proof tests, summarizing their differences in Table 3.1.

The appropriate interval for proof tests can be determined by considering various factors, such as the criticality of the barrier, the cost and downtime associated with testing, and the performance requirements of the barrier. For example, in a process plant, the actuating subsystem of a shutdown valve system might require a proof test every 12 months, while the sensing subsystem might only need testing every 24 months.

Regarding the extent of coverage provided by proof tests, they can be categorized into two types:

- *Full proof test (FPT)*: This comprehensive testing approach is designed to uncover all potential faults within a safety-critical system. An FPT subjects the barrier system to a series of demanding conditions or even extreme

**TABLE 3.1**
**Comparison of Diagnostic Test and Proof Test**

| | Diagnostic Test | Proof Test |
|---|---|---|
| Execution | Fully automatic | Often require human involvement |
| Effect on production | Do not need process shutdown | Require process shutdown |
| Frequency | Very frequent, in the range of every millisecond and to every few hours | Less frequent, in the range of every few months to every few years |
| Coverage | Some failures, dependent on types of barrier components | All failures in the ideal situation |
| Barrier state after test | Not necessarily restore to an as-good-as-new condition | As-good-as-new |

*Source:* Adopted from Lundteigen and Rausand (2008a).

conditions, simulating different process deviations and activation scenarios to ensure thorough evaluation.

- *Partial proof test (PPT)*: This testing approach focuses on identifying specific faults within particular components or subsystems of a safety-critical system. A PPT targets a specific part of the system, exposing it to the abnormal operating conditions to test its response and robustness.

Although FPTs are typically comprehensive, they are also time-consuming and resource-intensive. Consequently, PPTs are employed to minimize asset downtime and reduce costs while still maintaining adequate safety levels. For example, consider the activation of a shutdown valve. If the valve moves and stops the movement shortly after, this can be verified through a PPT. This PPT confirms that parts of the safety-critical system can activate as required, such as ensuring that there is no high friction between the stem and the seals, and the valve stem is able to move. However, this PPT does not assess whether the valve can fully close the pipeline without any leakage. The advantage of using a PPT in this scenario is that it allows the barrier to be tested with minimal disruption to normal production, meaning the flow in the pipeline does not need to be entirely halted. Nonetheless, it is crucial to understand that a PPT alone is insufficient for comprehensive safety assurance and should be supplemented by an FPT to detect any potential leaks and ensure the overall safety of the process plant in the real event of a shutdown demand.

Regarding test results, diagnostic tests are considered partial tests since they only evaluate specific functions of a safety-critical system. However, it is important to note that diagnostic tests do not simulate real-world demands. Moreover, the objectives and techniques of diagnostic tests differ from those of PPTs, which verify only parts of the ability of a safety-critical system to perform its intended functions. Thus, in the remainder of this book, we clearly differentiate between diagnostics and PPTs as distinct testing approaches.

In upcoming discussions, particularly in Chapter 8, we will explore availability analysis of safety-critical systems based on regular FPTs. Then, in Chapter 9, we will examine the impacts of different types of tests and their testing intervals on the performance of safety-critical systems.

### 3.5.3 Maintenance

After a proof test reveals a fault, a following-up *repair* can be conducted to eliminate the revealed fault, and such a method is also called *corrective maintenance*. In the basic analysis of safety-critical systems, we assume that only the corrective maintenance approach is used. Another assumption is that corrective maintenance is perfectly executed, meaning that after the repairs, the state of a barrier component or system is restored to an "as-good-as-new" state. In Chapter 9, we will explore and discuss more maintenance strategies.

## 3.6 FAILURES OF TECHNICAL BARRIERS

Given the unique operational modes and testing requirements of technical safety barriers, it is essential to assess the specific failure modes of these systems. This section will provide an overview of key concepts related to failure, followed by an exploration of the main failure types commonly observed in safety-critical systems.

### 3.6.1 Concepts Related to Failure

First, we can define a failure based on literature (IEC61511, 2016; Rausand et al., 2020) as follows:

- *Failure*: Termination of the ability of an item to perform as required.

Failure represents an event wherein the condition of an item shifts from being able to perform as required to not being able to meet the performance requirement. Failures can occur in various ways. For example, they may occur suddenly and unexpectedly, like a light bulb failing due to a short circuit. On the other hand, from the perspective of mechanics, failure can be viewed as an occurrence that the working load on an item is higher than the capacity or strength of the item.

Additionally, a failure may result from a progressive degradation process that reaches a critical point or threshold. Degradation is a natural phenomenon for most hardware items, due to the stress on the component or from a normal aging. Mechanically, the strength of an item tends to diminish over time while the workload might stay consistent or even escalate. When the strength falls below the operational load, the item fails, as illustrated in Figure 3.4. It is important to recognize that the decrease in strength is not always predictable, and the workload might also vary, making the occurrence time of such a failure random.

The aforementioned threshold can be quantitatively defined, such as the minimum tread depth required for winter tires in certain countries. While tire wear progresses gradually, tires are considered failed when the tread depth dips below this specified threshold, despite possibly remaining functional on non-icy roads. Similarly,

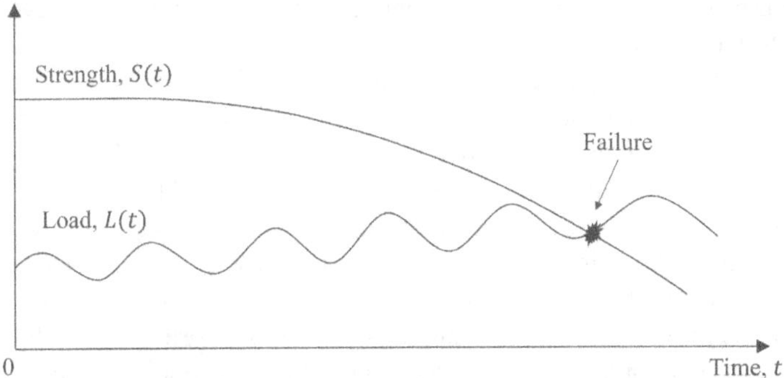


**FIGURE 3.4** Occurrence of a failure in a continuous degradation process.

safety mechanisms like shutdown valves must close within a designated time frame. Initially, a new valve on a pipeline might close promptly as needed. Over time, however, particulate matters may accumulate between the valve and the interior surface of pipeline, impeding movement of the valve. To prevent accidents, a closure time threshold of 10 seconds may be established. If the valve shuts within this period, it functions correctly; if closure takes longer, excessive gas may flow downstream, increasing the risk of an explosion. In this scenario, failure is defined by the valve's inability to close within the required timeframe.

It is important to acknowledge that failures triggered by exceeding a threshold in a gradual process may not be immediately apparent. For instance, the accumulation of particles affecting closure times might not be noticeable until the movement of the valve significantly slows. Regular inspections and proof testing of the equipment are essential to detect potential failures before they escalate to the critical levels.

After a failure occurs, the item enters a state known as a *fault* or *faulty state*. It is important to clarify the distinction between failure and fault by defining fault as follows:

- *Fault*: A state in which an item cannot perform as required.

This definition implies that the item resides in a fault state following the event of a failure, and if the item is repairable, it may return to its functioning state after undergoing repair or restoration.

In this book, we also use the concepts of failure cause, failure mode, and failure effect, with their definitions as follows:

- *Failure cause*: The reason why a failure occurs.
- *Failure mode*: The manner in which an item fails.
- *Failure effect*: The consequence after a failure occurs.

These three terms are important in identifying hazards that result in a hazardous event, understanding what to be observed in tests and operations, and analyzing the consequences of a hazardous/pivotal event within a risk analysis framework using a

Bow-Tie approach. Here, the pivotal event in a bow-tie diagram is always identified as the failure of an item.

Failure cause encompasses a variety of factors, conditions, and events. These causes can be the inherent malfunction, flawed designs, or manufacturing defects of the failed item, which are considered *failure mechanisms*. Additionally, external factors such as deliberate sabotage, extreme weather conditions, natural disasters like earthquakes, or operational changes, such as increased usage of machinery over a short period due to a large order, can also be failure causes.

Failure causes can be categorized into *proximate causes* and *root causes*. Proximate causes are the immediate factors, conditions, and events presenting just before a failure, directly contributing to its occurrence. Conversely, root causes are the deeper underlying factors, conditions, and events that lead to the proximate causes. Within a Bow-Tie model, root causes can be considered hazards and initiating events, while proximate causes are the events immediately preceding the pivotal event (on the left side of the model).

Failure mode describes the manner in which a failure occurs and is observed. However, the description of a failure mode does not include the causes of the failure. For example, the "failure to close as demanded" is a failure mode of a shutdown valve, whereas "a broken spring in the actuator" might be identified as the proximate cause and "lack of inspection and preventive maintenance" as the root cause.

The failure effect of a general item typically involves downtime, economic losses, and impacts on human safety and the environment. The failure effects of a technical safety barrier, however, can vary depending on whether there is a threat to the asset or a requirement to activate the barrier. These various failure effects of technical safety barriers will be discussed in the following subsection.

### 3.6.2 Classifications of Failures

IEC61511 (2016) classifies failures of technical systems into two categories based on different causes:

- *Random failure*: This type of failure occurs randomly as a result of one or more random events or potential degradation mechanisms where the timing when the item's strength falls below the operational load is uncertain.
- *Systematic failure*: This failure type occurs consistently under specific conditions, often due to existing faults in the system design, manufacturing, operation, documentation, or other performance-related activities.

Chastain-Knight (2023) has provided several examples of systematic failures within technical safety barriers, such as deficiencies in management systems, biases in tools and assessment methods, and reliance on overly optimistic data. Systematic failures can be addressed by identifying and removing faults through modifications in design, manufacturing, operating procedures, and documentation. It is almost impossible to completely prevent random failures in a technical system, but their occurrence likelihood can be minimized, and their impacts can be mitigated.

For safety-critical systems, failures can be categorized based on different testing approaches discussed in Section 3.4, particularly in terms of their detectability:

- *Detected failure*: A failure that can be immediately identified after it occurs, either by the diagnostic module installed in the safety-critical system or through manual observation.
- *Undetected failure*: A failure in a safety-critical system that remains unnoticed by the diagnostic module and observation. The faulty state of the safety-critical system due to such a failure is expected to be revealed in a proof test.

Another method to classify failures of safety-critical systems considers the severity of the failure effects:

- *Dangerous failure*: A failure that can result in a serious harm on the asset that is protected by the safety-critical system. This type of failure compromises or disables the barrier function of the safety-critical system when it is needed.
- *Safe failure*: A failure that does not lead to serious harm to the asset that is protected by the safety-critical system. In many cases, such a failure does not prevent the barrier system from performing its function.

Considering these two classifications for the failures of safety-critical systems, we can adopt a matrix approach as illustrated in Figure 3.5. This results in four distinct types of failures:

- *Dangerous detected (DD) failure*: This type of failure hinders or disables the barrier function and is immediately detectable upon occurrence.
- *Dangerous undetected (DU) failure*: This type of failure hinders or disables the barrier function but is not immediately detectable.
- *Safe detected (SD) failure*: This failure does not affect the barrier function and is immediately detectable upon occurrence.
- *Safe undetected (SU)*: This failure does not affect the barrier function and remains undetected initially.

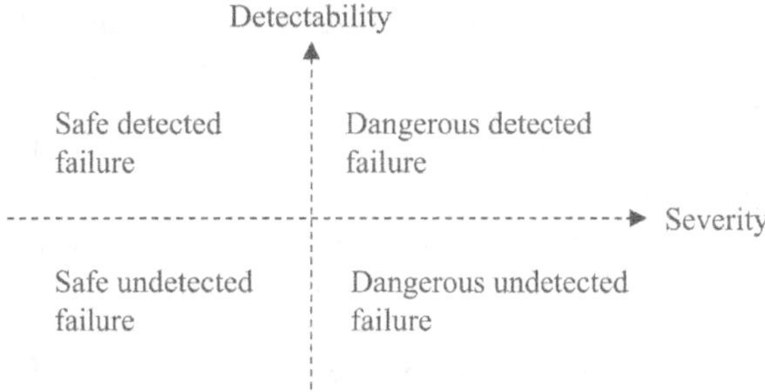


**FIGURE 3.5** Classification matrix of failures of safety-critical systems.

**Example 3.1:**

*Typical Failures of a Shutdown Valve*

In this example, we analyze the potential failures of the shutdown valve depicted in Figure 3.1, serving as the actuator of a safety-critical system. We categorize these failures into distinct types based on their impacts and detectability:

- *Failure to close*: This failure occurs when the valve is unable to move into a position that stops the flow in the pipeline. Potential causes include a broken spring, swollen stem seals, or debris and sand lodged in the valve cavity. This failure compromises the valve's barrier function and is undetected until there is a demand to shut down, classifying it as a DU failure.
- *Spurious closure*: This failure means that the valve closes the pipeline unnecessarily. It may occur due to an incorrect power shutoff, operating under a de-energize-to-trip principle. We will explore this failure type in greater detail in the next subsection. While such a failure can disrupt normal operations and lead to economic losses, it generally does not cause severe damage to adjacent equipment or pose risks to operators. Since the closure is immediately observable, this failure is categorized as an SD failure.
- *Leakage during normal operation*: This type of failure means that gas leaks from the valve into the environment. Possible causes include cracks or seal defects. If gas detectors are installed, this failure can immediately be detected. Given that gases in the pipeline are often flammable, explosive, or toxic, leakage poses significant risks to operators and property, making this a DD failure.
- *Failure to open*: In this scenario, the shutdown valve can close as required but fails to reopen after the downstream pressure has decreased. A potential cause can be that the gate was skewed during closure. Although this failure leads to unexpected production downtime, it typically does not pose additional hazards to the process plant, similar to the spurious closure. However, this failure is only noticeable after an attempt to reopen the valve, and it cannot be detected by the diagnostic module, thus classifying it as an SU failure. ■

### 3.6.3 Spurious Activation

The spurious trip of a shutdown valve, as described in Example 3.1, is a significant and specific issue for safety-critical systems that can lead to partial or complete shutdowns of production processes and result in economic losses (Lundteigen & Rausand, 2008b). It is therefore beneficial to further investigate the nature of such failures in this subsection.

Safety-critical systems are designed based on the fail-safe principle, which ensures that these systems activate in response to hazards, abnormal conditions, or failures within the asset they protect. However, when the condition of the asset is inaccurately recognized or analyzed, it can lead to an erroneous action by the actuator. This, in turn, can result in *spurious activation* of a safety-critical system. Namely, a barrier function is triggered by a false demand. In addition, spurious activations may also come from issues within the actuating subsystems or failures in their components, which could be due to factors like electromagnetic interference, mechanical wear and tear, or human errors.

To provide a clearer understanding, we can define spurious activation of a safety-critical system formally, drawing from standards such as CEN ISO/TR12489 (2016) and Jigar et al. (2016) as follows:

- *Spurious activation*: The activation of a safety-critical system or its parts in an untimely manner.

The term "untimely" is a crucial keyword in this context as it describes the distinction between a barrier function that is appropriately activated at the right moment and one that is mistakenly triggered without an actual demand. In essence, a spurious activation occurs when the barrier function is executed without a valid need.

Spurious activations of safety-critical systems are common, and such events can be observed in daily life and in various industries. For instance, the automatic braking system in vehicles may activate undesirably if it mistakenly identifies a benign object as a vehicle ahead, leading to an unnecessary slowdown. Another example involves the protective switch on ovens designed to trigger when excessively high temperatures are detected, perhaps from leaving a pan unattended on the stove. However, if the activation threshold is set too low, the switch may trigger during normal cooking activities like frying or wok cooking, inadvertently shutting off and locking the oven.

According to Lundteigen and Rausand (2008b), spurious activations of a safety-critical system typically are mainly due to two causes:

- An internal failure within the actuating subsystem or its ancillary components.
- An inaccurate response by the sensing subsystem to the condition of the asset.

In many contexts, spurious activations are considered safe failures because most of them do not impair the barrier function of a safety-critical system. However, such activations can sometimes adversely affect asset safety. For instance, in the case of an automatic braking system, if the sudden reduction in speed catches the drivers of both the vehicle and the one following by surprise, it may lead to a collision. Similarly, a spurious closure of a shutdown valve may lead to over-pressurization of upstream equipment and loss of containment.

A tragic illustration of the potential consequences of spurious activation is the Ethiopian Airlines Flight 302 disaster in 2019. The Maneuvering Characteristics Augmentation System (MCAS) of the aircraft, which was designed to stabilize flight, was mistakenly activated, causing the plane to pitch downward. Despite the pilot has tried to counteract this, the MCAS persistently forced the nose down, leading to a fatal crash. The Ethiopian Civil Aviation Authority's investigation concluded that "Repetitive and un-commanded airplane-nose-down inputs from the MCAS due to erroneous AOA input, and its unrecoverable activation system which made the airplane dive...was the most probable cause of the accident". While the interaction between the pilots and the MCAS system during the incident remains a subject of ongoing discussion, the spurious activation of the MCAS was undeniably a significant factor.

In the study by Jigar et al. (2016), terminating a barrier function earlier than expected is also considered a spurious activation. However, in this book, we consider such an event as a dangerous failure rather than a spurious activation. We understand the earlier stop of a barrier function is understood as an issue related to the durability of a safety-critical system, which will be discussed in detail in Chapter 6.

Designing and operating safety-critical systems require a subtle balance between preventing dangerous failures and reducing spurious activations, but these goals can sometimes conflict. For example, as depicted in Figure 3.1, incorporating two shutdown valves in a redundant configuration enhances the availability of the subsystem and improves safety by guarding against dangerous failures, such as the scenario where no valve closes the flow upon demand. However, this redundancy may increase the likelihood of spurious activations since a false trigger in either valve could unnecessarily halt the flow. Similarly, setting a lower threshold for a gas detector increases its sensitivity, making it more capable to detect even minimal gas concentrations and thus enhancing asset safety. Nonetheless, this high sensitivity may lead to more frequent alerts, including some that are unnecessary, potentially leading to disruptions and annoyance.

It is also important to recognize that although spurious activation is categorized as a safe failure within a safety-critical system, not all safe failures qualify as spurious activations. Quantitative analysis of spurious activations will be introduced in Chapter 8.

## 3.7 INTERNATIONAL STANDARDS OF TECHNICAL BARRIERS

There are numerous sector-specific, national, and international standards and regulations that specify the design, functionality, and performance requirements for safety-critical systems and technical safety barriers. Compliance with these standards is crucial for the designers and operators of safety-critical systems to ensure safe operation. In this section, we will briefly introduce some widely recognized international standards that have been in use in recent decades. These standards provide a framework for developing, testing, and maintaining safety-critical systems to ensure that they meet the requirements related to safety.

It should be noted that some of the standards mentioned below may not be the latest versions or may even be outdated when this book is in use, since all the international standards are updated time by time.

### 3.7.1 IEC 61508

IEC61508 (2010), developed by the International Electrotechnical Commission (IEC), is a generic international standard for the functional safety of technical safety barriers, referred to as electrical, electronic, and programmable electronic safety-related systems (Safety E/E/PES) within the standard. This standard offers a comprehensive framework to ensure that Safety E/E/PESs or safety-critical systems are designed, implemented, operated, and maintained to achieve the safety integrity levels (SILs) required for their intended applications.

The standard is divided into seven parts. Parts 1–3 outline the normative hardware and software requirements for Safety E/E/PESs. Part 4 provides definitions, while Parts 5–7 offer guidelines on techniques and measures for system development, as well as examples.

A significant innovation in IEC 61508 is its definition of safety lifecycle requirements and guidance for the systematic development of Safety E/E/PESs or safety-critical systems. The overall safety lifecycle, encompassing 16 steps, is depicted in Figure 3.6. These steps are generally organized into three phases: Phase I – definition and requirement, Phase II – specification and realization, and Phase III – operation and maintenance.

In Phase I, the focus is on the asset, with key activities including identifying threats and risks to the asset, as well as defining the safety requirements necessary for determining appropriate risk reduction measures. The focus of Phase II shifts to the Safety E/E/PES, alongside other risk reduction measures, with the main efforts in designing, constructing, and validating the E/E/PES to ensure the asset is protected and that its safety requirements are met. Phase III concentrates on the interactions between the Safety E/E/PES and the asset, with the main task being to maintain the performance of the Safety E/E/PES at the required level.

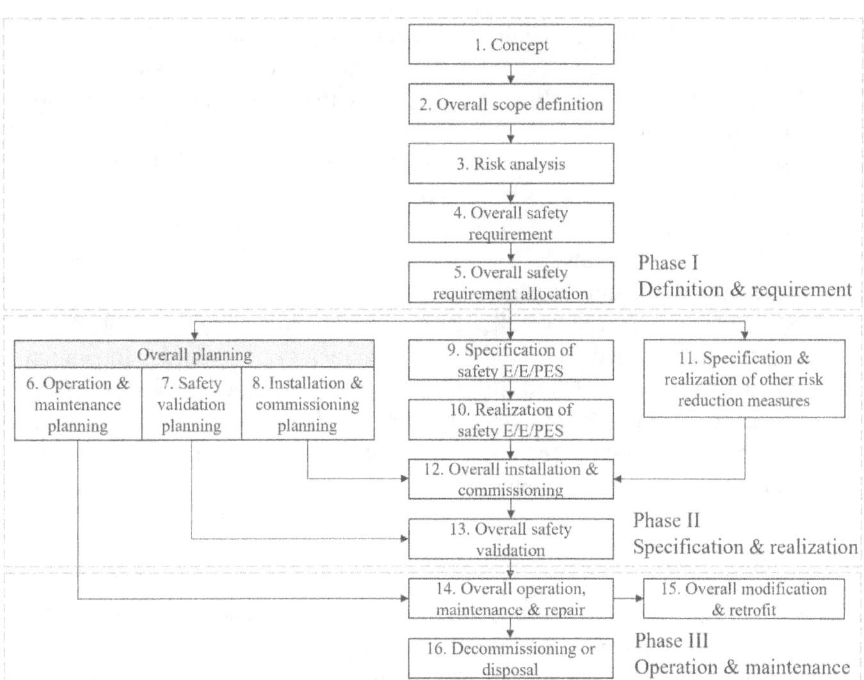


**FIGURE 3.6** The overall safety lifecycle. (Adapted from Figure 2 in IEC 61508-1.)

IEC 61508 is an expansive standard applicable to a wide range of fields, including power generation, process industries, medical devices, automotive, and railway systems, among others. The sector-specific standards to be discussed in the following subsections align with the generic requirements of IEC 61508.

### 3.7.2 IEC 61511

IEC61511 (2016) is titled as "Functional safety – Safety instrumented systems for the process industry sector", as an international standard that provides a framework for the management of functional safety in the process industries. In this standard, the term "safety instrumented system" (SIS) is the synonym of technical safety barrier, as it focuses on safety equipment that utilizes instrumentation technologies.

IEC 61511 proposes the requirements for the specification, design, installation, operation, and maintenance of SISs intended to mitigate process hazards in industries such as chemical, oil and gas, and pharmaceuticals. The standard is based on the risk reduction philosophy and offers guidance on selecting SISs and determining the necessary SILs for these systems. It also provides guidelines for developing safety plans, verifying SIS performance, and managing SISs throughout their lifecycles to ensure their reliability and effectiveness.

Additionally, several other standards and guidelines in the oil and gas and offshore industries relate to functional safety and technical safety barriers, including the following:

- ISO13702 (2015) "Petroleum and natural gas industries – Control and mitigation of fires and explosions on offshore production installations – Requirements and guidelines" developed by International Organization for Standardization (ISO).
- ISO10418 (2019) "Petroleum and natural gas industries – Offshore production installations – Process safety systems".
- API RP 14C (API, 2007) "Analysis, design, installation, and testing of basic surface safety systems for offshore production platforms", developed by American Petroleum Institute.
- NORSOK S-001 (NORSOK, 2008) "Technical safety", a Norwegian standard developed by Norwegian Technology Standards Institution (NORSOK). This standard, together with ISO 13702, describes the safety aspects related to specific safety systems and barriers, the necessary utilities for these systems, their functional requirements, and their interfaces with other safety systems and barriers.

### 3.7.3 IEC 62061

IEC62061 (2021), titled as "Safety of machinery – Functional safety of safety-related electrical, electronic, and programmable electronic control systems", provides guidance on the development and implementation of technical safety barriers for machinery. This includes machinery used in industrial, commercial, and public environments.

Similar to the IEC 61508, IEC 62061 covers the entire lifecycle of technical barriers designed to protect machinery. This includes the phases such as design, development, validation, installation, operation, maintenance, and decommissioning. The standard is widely recognized and utilized by machinery manufacturers, system integrators, and end-users globally.

### 3.7.4 IEC 62279 AND EN 50129

Both standards are pivotal for the functional safety of railway applications. IEC62279 (2015), titled as "Railway applications – Communication, signaling, and processing systems – Safety-related electronic systems for signaling", provides a framework for the development and assessment of safety-related electronic systems that serve as technical barriers, including communication, signaling, and processing systems in the railway sector. The standard provides detailed requirements for the development and assessment of these systems, including requirements for system architecture, software development, verification and validation, and documentation.

EN50129 (2018) is a European standard that adapts the principles of IEC 62279 specifically for use within Europe. It outlines requirements for the acceptance and approval of safety-related electronic systems in railway signaling. The standard specifies the integrated design, implementation, commissioning, operation, and maintenance of safety-related systems. EN 50129 proposes requirements for both safety-related hardware and software, and it places greater emphasis on risk assessment and management than IEC 62279.

### 3.7.5 ISO 26262

ISO26262 (2011) "Road vehicles – Functional safety" is adopted from IEC 61508, tailored specifically with slight modifications to fit the automotive industry and the applications related to vehicles running on roads. This standard applies to technical safety barriers that incorporate one or more electrical/electronic (E/E) techniques and are installed in passenger cars with a maximum gross vehicle mass of up to 3,500 kg.

ISO 26262 provides a framework for developing these technical safety barriers and offers guidelines and processes that encompass hardware, software, and their integration. It also defines the roles and responsibilities of various stakeholders involved in the development process, including suppliers, original equipment manufacturers (OEMs), and regulatory bodies. The standard covers all electronic systems installed in road vehicles, such as advanced driver assistance systems, electronic control units, and sensors.

The residual risk after deploying technical safety barriers is addressed in ISO 26262 as the *Automotive Safety Integrity Level* (ASIL). ASIL A means the least stringent level, while ASIL D represents the most stringent level. The standard also emphasizes the importance of planning throughout the functional safety lifecycle and defines supporting processes such as requirements management, configuration management, and change management.

### 3.7.6 IEC 60730

The IEC 60730 series of standards are developed to ensure the safe operation of embedded control hardware and software in household appliances. Initially published in 1977, the original standard has evolved into a continuously updated series. For example, one of the most recent updates, IEC60730-1 (2022) "Automatic electrical controls – Part 1: General requirements" was released in 2022.

IEC 60730 encompasses a broad array of products, including household appliances, heating and cooling systems, and other electrical devices equipped with automatic controls. The standard specifies functional safety requirements for these automatic controls and safety-related systems. It also establishes criteria for protection against electric shocks and other hazards, in addition to outlining the requirements for testing and verification of the barrier functions of the control system. Compliance with IEC 60730 is often mandated by regulatory bodies and is recognized internationally as a benchmark of quality and safety for consumer products.

### 3.7.7 Compliance with Standards

Compiling an exhaustive list of all standards related to safety-critical systems and functional safety is challenging, considering that new standards are frequently developed and released. Generally, these standards can be categorized into three types:

- **Type A standards** provide foundational concepts, general principles, and guidelines for safety management applicable across all sectors. Compliance with Type A standards involves conducting risk assessments and developing comprehensive safety policies and management procedures. An example of a Type A standard is IEC31010 (2019).
- **Type B standards** provide generic requirements for safety design and ergonomics (Type B1), or for safety barriers (Type B2). IEC 61508 is an example of a Type B2 standard, and IEC 61511, along with other standards mentioned earlier, can also be considered Type B2 standards tailored to different industries. Compliance with these standards ensures that the equipment or processes meet specified safety requirements, which include designing, testing, verification, and documenting compliance.
- **Type C standards** specify product-specific requirements for safety barriers. For example, ISO22201-1 (2017), which addresses programmable electronic systems used for the functional safety of lifts and elevators, is a Type C standard.

To ensure compliance with the required standards in the design and operation of safety-critical systems, it is important to first evaluate if a Type C standard can be found suitable to the specific product. If such a standard exists, it is mandatory to comply with it. In the absence of a relevant Type C standard, we should refer to the applicable Type B standards and, as needed, Type A standards to guide our compliance efforts.

## REFERENCES

ANSI/ISA-18.2. (2016). Management of alarm systems for the process industries. Durham, NC, USA: International Society of Automation.

API-RP-17Q. (2018). *Subsea Equipment Qualification-Standardized Process for Documentation* (2nd ed.). American Petroleum Institute, Washington, DC, US.

API. (2007). API RP 14C – Analysis, design, installation, and testing of basic surface safety systems for offshore production platforms. Washington, DC: American Petroleum Institute.

Chastain-Knight, D. (2023). Systematic errors that compromise integrity of safety instrumented systems. *Process Safety Progress*, *42*(4), 1–10.

DNV-OSS-304. (2011). *Risk Based Verification of Offshore Structures*. D. N. Veritas.

EN50129. (2018). Railway applications – Communication, signalling and processing systems – Safety related electronic systems for signalling. Brussels, Belgium: European Committee for Electrotechnical Standardization (CENELEC).

Hobbs, G. K. (2000). *Accelerated Reliability Engineering: HALT and HASS*. John Wiley & Sons.

Hollnagel, E. (2004). *Barriers and Accident Prevention*. Ashgate.

IEC31010. (2019). Risk management – Risk assessment techniques. Geneva, Switzerland: International Electrotechnical Commission.

IEC60730-1. (2022). Automatic electrical controls – Part 1: General requirements. Geneva, Switzerland: International Electrotechnical Commission.

IEC61508. (2010). Functional safety of electrical/electronic/programmable electronic safety-related systems. Geneva, Switzerland: International Electrotechnical Commission.

IEC61511. (2016). Functional safety – Safety instrumented systems for the process industry sector. Geneva, Switzerland: International Electrotechnical Commission.

IEC62061. (2021). Safety of machinery – Functional safety of safety-related control systems. Geneva, Switzerland: International Electrotechnical Commission.

IEC62279. (2015). Railway applications – Communication, signalling and processing systems – Software for railway control and protection systems. Geneva, Switzerland: International Electrotechnical Commission.

IEC62506. (2013). Methods for product accelerated testing. Geneva, Switzerland: International Electrotechnical Commission.

IEC62682. (2014). Management of alarm systems for the process industries. Geneva, Switzerland: International Electrotechnical Commission.

ISO10418. (2019). Petroleum and natural gas industries – Offshore production installations – Process safety systems. Geneva, Switzerland: International Organization for Standardization.

ISO13702. (2015). Petroleum and natural gas industries – Control and mitigation of fires and explosions on offshore production installations – Requirements and guidelines. Geneva, Switzerland: International Organization for Standardization.

ISO22201-1. (2017). Lifts (elevators), escalators and moving walks – Programmable electronic systems in safety-related applications – Part 1: Lifts (elevators) (PESSRAL). Geneva, Switzerland: International Organization for Standardization.

ISO26262. (2011). Road vehicles – Functional safety. Geneva, Switzerland: International Organization for Standardization.

ISO/TR12489. (2016). Petroleum, petrochemical and natural gas industries – Reliability modelling and calculation of safety systems. Geneva, Switzerland: International Organization for Standardization.

Jigar, A. A., Liu, Y., & Lundteigen, M. A. (2016). Spurious activation analysis of safety-instrumented systems. *Reliability Engineering & System Safety*, *156*, 15–23. https://doi.org/10.1016/j.ress.2016.06.015

Johansen, I. L., & Rausand, M. (2015). Barrier management in the offshore oil and gas industry. *Journal of Loss Prevention in the Process Industries*, *34*, 49–55.

Lundteigen, M. A., & Rausand, M. (2008a). Partial stroke testing of process shutdown valves: How to determine the test coverage. *Journal of Loss Prevention in the Process Industries*, *21*(6), 579–588. https://doi.org/10.1016/j.jlp.2008.04.007

Lundteigen, M. A., & Rausand, M. (2008b). Spurious activation of safety instrumented systems in the oil and gas industry: Basic concepts and formulas. *Reliability Engineering & System Safety*, *93*(8), 1208–1217.

McLean, H. (2009). *HALT, HASS, and HASA Explained: Accelerated Reliability Techniques* (rev. ed.). ASQ Quality Press.

MIL-HDBK-781A. (1996). *Reliability Test Methods, Plans and Environments for Engineering, Development Qualification and Production*. US Department of Defense, Washington DC, USA.

Montgomery, D. (2012). *Statistical Quality Control: A Modern Introduction* (7th ed.). John Wiley & Sons.

NORSOK. (2008). NORSOK S-001 – Technical safety. Oslo, Norway: Norwegian Technology Standards Institution.

Rausand, M. (2014). *Reliability of Safety-Critical Systems: Theory and Applications*. John Wiley & Sons.

Rausand, M., Barros, A., & Hoyland, A. (2020). *System Reliability Theory: Models, Statistical Methods, and Applications* (3rd ed.). Wiley-Blackwell.

Yang, F., Wang, J., Asaadi, M., Hu, W., Wang, Z., & Zhang, Y. (2022). Alarm management techniques to improve process safety. In F. Khan, H. Pasman, & M. Yang (Eds.), *Methods in Chemical Process Safety: Methods to Assess and Manage Process Safety in Digitalized Process System* (Vol. 6, pp. 227–280). Elsevier.

Ye, Z. (2011). *Optimal Burn-in under Complex Failure Processes: Some New Perspectives*. National University of Singapore.

Yellman, T. W. (2006). Redundancy in designs. *Risk Analysis*, *26*(1), 277–286.

# 4 Hazard-barrier Identification and Qualitative Assessment

## 4.1 GENERAL APPROACHES OF HAZARD AND BARRIER IDENTIFICATION

Identifying hazards and their associated safety barriers within a process is an important step in risk management, as illustrated in Figure 1.6. The identification should be carried out in a structured way to provide a comprehensive list of hazards and barriers, and thus establish the foundation for more detailed subsequent analysis and evaluation. This chapter introduces general strategies of hazard and barrier identification, and then present several techniques and tools that are used in various industries for the identification and preliminary analysis of hazards and barriers.

Typically, the process of hazard and barrier identification does not start from scratch. Relevant data and information are firstly gathered and analyzed. This information is then systematically organized to facilitate the identification of potential hazards to the asset of interest, associated accidental scenarios, and the corresponding safety barriers. Consequently, the first step of a strategic approach for systematic hazard and barrier identification is to screen a variety of information sources and obtain useful information.

Information used in barrier engineering and management may be qualitative or quantitative and can be in various formats, such as text, numbers, diagrams, and videos. Sources of hazard and barrier information are also diverse, including suppliers, users, historical incidents, subject matter experts, engineering departments, online resources, scholarly literature, and national or sector-specific reports. Based on the different information resources, the general methodologies for identifying hazards and safety barriers are categorized accordingly.

- *Historical accident investigation*: This approach includes examining past accidents and incidents to reveal potential hazards, determine proximate and root causes and their contributing factors, and evaluate the effectiveness and vulnerabilities of existing risk reduction measures. By analyzing historical events, organizations can gain insights from prior mistakes, fostering proactive improvement strategies. This approach utilizes real-world cases to evaluate safety barriers and identify opportunities for improvement. Existing databases, such as MHIDAS, have been proved valuable in facilitating this investigative process.
- *Interview and survey*: This approach engages with professional stakeholders, such as system operators, designers, and maintenance engineers,

DOI: 10.1201/9781003245636-4

interviews can provide deep, insightful perspectives on safety risks and potential barriers. Interviews may vary in format, including in-person talks, telephone calls, or video conferences, and may be either structured or unstructured. On the other hand, questionnaires and surveys also serve as effective tools for gathering broader input on safety attitudes, behaviors, and perceptions from a larger cohort.

- *Field observation and information collection*: This approach includes the acquisition of data regarding hazards and safety barriers through direct observation or the analysis of video recordings and other documentation. Information gathering may be executed within an organization or alongside partners under specific agreements. Such an approach offers a more accurate view of existing safety conditions, highlighting those hazards that may have been overlooked. The insights derived from these observations are important in developing new safety barriers or improving existing ones.
- *Laboratory reliability tests*: This approach exposes the asset and relevant safety barriers to the controlled conditions to assess their operational thresholds, potential failure modes, and reliability parameters. Laboratory testing allows for the evaluation of safety barriers and equipment under extreme scenarios that may be impractical or rare in actual operational environments. Such tests are very useful in discovering the vulnerabilities of barriers systems and suggesting improvements. Moreover, laboratory scenarios are helpful for revealing the impacts of potential equipment failures and parameter deviations, aiding in the identification of necessary safety barriers.
- *Model-based analysis and simulation*: This approach employs accident models to simulate potential accident scenarios and then conduct quantitative analyses using a variety of data sources, including field and experimental data, supplemented by computer-generated data. The outcomes of these simulations provide critical insights into the effects of any failures or deviations on the normal operation of the process, and the functionality of safety barriers. The main benefits of such an approach are its cost-effectiveness and the controlled environment it offers. Simulations enable organizations to avoid the expenses and hazards linked to physical experiments or field tests, allowing for the rapid assessment of numerous scenarios and safety measures. However, it is crucial to corroborate simulation-generated data with empirical evidence to ensure its validity and reliability.

All the above approaches need to handle extensive data and require the transformation of raw data into meaningful information and actionable commands for barrier operations. In Chapter 7, we will further investigate the databases and data analysis techniques essential for processing and interpreting this data effectively.

In the remainder of this chapter, we will introduce several methods and tools for systematically identifying and documenting hazards and safety barriers. Most of these methods are also utilized for rough and qualitative risk and barrier assessments, as well as for the communications between different stakeholders.

Among these, the first three methods, hazard-barrier matrix, energy flow/barrier analysis, and safety-barrier diagram, are dedicated for barrier identification and analysis. These can be integrated into frameworks for barrier management, such as LOPA and BORA (see Chapter 2). The latter four methods: Hazard and Operability (HAZOP) analysis, Failure Mode and Effects Analysis (FMEA), Fault tree analysis (FTA), and Systems-Theoretic Process Analysis (STPA), are more general and comprehensive, involving overall risk identification and assessment where barriers are considered as part of the broader risk picture.

## 4.2 HAZARD-BARRIER MATRIX

The hazard-barrier matrix is an instrumental tool for identifying safety barriers and qualitatively analyzing their relationship with various types of hazards. Such an approach was initially proposed in a report of the U.S. Department of Energy (DOE, 1996) and has been referenced in the recent literature (Rausand & Haugen, 2020; Xie, 2022). Figure 4.1 demonstrates the use of the hazard-barrier matrix with an example involving a leakage at a gas fueling station. The matrix is divided into three main sections:

- *Hazard*: The first column of the matrix categorizes general hazard types, while the second column provides descriptions of specific hazards. Hazards are typically organized based on their sources, such as electrical, mechanical, or environmental. This categorization is the foundational step in constructing a hazard-barrier matrix.
- *Barrier*: Beginning from the third column, barriers are listed and can be grouped into various categories such as organizational, operational, informational, and technical barriers. Alternatively, barriers may also be grouped as either preventive or mitigative, depending on their function and the target audience of the matrix. The process of identifying relevant safety barriers across these categories constitutes the second phase of the hazard-barrier matrix methodology.

| Hazard category | Hazard description | Organizational barrier | | Procedural barrier | | Technical barrier | |
|---|---|---|---|---|---|---|---|
| | | Safety audit | Safety training | Regular maintenance | Standard fueling operation | Leakage sensor & alarm | Shutdown valve |
| Mechanical hazards | Corrosion of the tank wall | 1 | 2 | 3 | | 3 | 3 |
| | Incompatible material | 4 | 2 | | | 2 | 2 |
| Operational hazards | Low quality of installation | | 2 | 1 | | 2 | 2 |
| | Errors in fueling operation | | | | 4 | 4 | 4 |
| Environmental hazards | Ground movement due to drought | | | 3 | | 3 | 3 |
| | Movement of tanks by groundwater | | | 3 | | 3 | 3 |

**FIGURE 4.1** An illustrative example of hazard-barrier matrix.

- *Relevance of barrier to hazard*: The points where barriers and hazards intersect within the matrix denote their relevance. A simple symbol, like an "X", is commonly used to indicate this association. To facilitate further analysis, the potential effectiveness of each barrier against a specific hazard can be expressed using a numerical value or a color code. For example, with color codes, blue can be used to indicate full effectiveness, green is for high effectiveness, yellow is for moderate effectiveness, and orange is for ineffectiveness. In Figure 4.1, a scale from 1 to 4 is employed to rate the effectiveness of barriers, with 1 representing not effective and 4 signifying fully effective.

The matrix is a clear and organized way to visually represent the complex interplays between hazards and barriers, facilitating further analysis and decision-making regarding safety barriers. A hazard-barrier matrix serves several essential functions, including:

- *Identification and systematic listing of hazards and safety barriers*: The matrix offers a structured framework to categorize and describe hazards in detail. It facilitates the systematic classification of safety barriers into various groups, providing a comprehensive overview of the different measures in place. This structure helps ensure that different types of hazards are considered and that the corresponding barriers are documented.
- *Rough analysis of the existing safety barriers*: The matrix can support a preliminary assessment of the functionality and effectiveness of safety barriers in relation to the identified hazards. For example, the matrix can be used to perform simple calculations, such as summing the effectiveness values assigned to each barrier. This helps in assessing the significance and overall impact of each barrier in mitigating the associated hazards.
- *Identification of gaps in barrier management*: Analysis of the matrix can reveal gaps where specific hazards are not adequately addressed by existing safety barriers. If a hazard intersects with few or no effective barriers, it indicates a potential vulnerability in the safety strategy. Recognizing these gaps is important, because it can help to determine to introduce new barriers or strengthen the existing ones to better manage the risk. Furthermore, the simple risk assessment offers a semi-quantified evaluation of the risk level posed by each hazard, guiding further safety improvements and decision-making processes.

The hazard-barrier matrix is a straightforward tool that offers a relatively comprehensive view of the interrelations between hazards and safety barriers. It serves as an accessible starting point for barrier analysis and aids in effective communication among stakeholders. However, due to its inherent simplicity, there are several limitations that need to be considered:

- *Limited classification of safety barriers*: The matrix only allows for a single classification approach of safety barriers. For example, in Figure 4.1, barriers are categorized by their formats, but the matrix does not provide further differentiation regarding their roles in risk reduction, such as distinguishing

between preventive and mitigative functions. While the technical barriers in the last two columns of the matrix seem critical as they are relevant to all hazards, this relevance is obvious because they act as mitigative barriers in the event of a leakage. However, in barrier management, these barriers might not necessarily be more important than preventive barriers.
- *Insufficient utilization of risk assessment terminology*: The approach does not consistently use the formal terminology of risk assessment. It blurs the lines between hazards and hazardous events and fails to account for the diversity among assets. Thus, the hazard-barrier matrix may struggle to adequately assess whether the risk to a specific asset has been reduced to an acceptable level using the existing barriers.
- *Ignoring variations in hazard severity*: The matrix approach does not consider the varying severities of hazards. As a result, the simple summation of effectiveness values for a safety barrier may not always yield a meaningful result. A barrier can be effective against minor hazards but may fall short in dealing with more severe or critical hazards.

Given these limitations, it is important to supplement the hazard-barrier matrix with other risk assessment methodologies, such as HAZOP analysis, and quantitative analysis techniques.

## 4.3 ENERGY FLOW/BARRIER ANALYSIS

In consideration of the limitations of hazard-barrier matrix, the energy flow/barrier analysis (EFBA) emerges as a more versatile tool suitable for formal risk assessment terminology and offers a flexible method for evaluating the relationships between hazards and barriers.

EFBA examines the processes defined within barrier management frameworks, focusing particularly on harmful energy sources within these processes. This methodology identifies and analyzes barriers based on the paths of energy flow emanating from these sources, providing a dynamic perspective on how energy interacts with safety mechanisms in real-world scenarios.

EFBA has been used in system design, operation plan, procedure development, and accident prevention (Rausand & Haugen, 2020). An illustrative worksheet of EFBA is shown in Figure 4.2.

In the EFBA worksheet, it is essential to document each aspect related to the energy flows and barriers to provide a thorough analysis. The below is a breakdown of the information that should be included in each column:

- *Number (Column 1)*: This represents the unique reference number assigned to each energy source to facilitate clear identification and differentiation among them.
- *Study node (Column 2)*: This represents an energy source within the EFBA process. All potential harmful energy sources should be identified and listed here. Additionally, information about the type and quantity of energy, such as electrical, kinetic, potential, chemical, thermal, radiation, pressure, noise, vibration, etc., should be included in this column, as these

Study object: Hydrogen refueling station | Date:

Reference: XXX | Compiled by: Yiliu Liu

| No. | Energy source (type amount) | Hazard (energy pathway) | Affected asset | Barrier | Barrier effectiveness | Residual risk | Recommended action | Comments |
|---|---|---|---|---|---|---|---|---|
| (1) | (2) | (3) | (4) | (5) | (6) | (7) | (8) | (9) |
| 1 | 700-bar compressed gaseous hydrogen in tank (pressure, flammable and explosive gas, 149 L) | Leakage (flow by wind) | Facilities and people in the station | 1.1 Gas detector<br>1.2 Alarm<br>1.3 Shut-down valve | Fully effective | Low | Regular tests on the barrier equipment | |

**FIGURE 4.2** An illustrative example of an EFBA worksheet.

characteristics are critical in assessing risks and selecting appropriate barriers. If the study node is complex, sub-reference numbers should be assigned to the hazards, assets, and barriers associated with it.

- *Hazard (Column 3)*: This column documents all the possible energy flow paths that originate from each energy source and lead to assets of interest. These paths illustrate how harmful energy could potentially threaten assets. It is noted that multiple energy flows may target the same asset.
- *Asset (Column 4)*: This column provides a comprehensive list of assets that could be impacted by each listed energy source in Column 2. If applicable, the vulnerability of each asset should also be identified to tailor the barrier solutions effectively.
- *Barrier (Column 5)*: For new system/process designs, this column is used to identify potential barriers along each energy path from the source to the asset. If the EFBA is for accident investigation, we need to list the existing barriers and explain why they have failed to prevent the accident. This column should detail all barriers related to each study node and its associated hazards.
- *Effectiveness (Column 6)*: This column describes the effectiveness of the barriers listed in Column 5, either qualitatively or quantitatively. The effectiveness can be described qualitatively or quantitatively. For instance, in the illustrative example of Figure 4.2, a simple scale from 1 to 4 is used to indicate effectiveness. Additionally, the integrity of these barriers should be assessed. Methods for calculating these values can be found in Chapters 5 and 8.
- *Risk (Column 7)*: After implementing the barriers from Column 5, the residual risk is evaluated and documented in this column. This can be presented using a color code, descriptive statement, or numerical value, and may compare risks associated with the asset both with and without the barriers. This assessment helps determine the effectiveness and sufficiency of barriers in ensuring asset safety.
- *Improvement (Column 8)*: This column includes the recommended measures or actions to further enhance safety of the assets. These may include design modifications, the implementation of additional safety barriers, or the use of other techniques to better understand and mitigate hazards.

- *Comment (column 9)*: Any additional necessary information or comments can be listed in this column to provide further context or clarification regarding the analysis.

EFBA also has a variant known as energy trace and barrier analysis (ETBA), which is part of the management overview and risk tree (MORT) methodology. ETBA specifically focuses on detecting hazards related to different types of energies in a process and the barriers controlling these hazards. We will introduce ETBA along with MORT in Section 4.7.

EFBA is a practical and straightforward method, offering a common tabular working document that is used for communications between system designers, risk analysts, and management. EFBA is particularly valuable in accident investigations, as the detailed worksheet helps elucidate how energy flows can bypass existing barriers, leading to accidents.

## 4.4 SAFETY-BARRIER DIAGRAM

The safety-barrier diagram is a graphical tool designed for the systematic identification and analysis of barriers, supporting methodologies such as LOPA. This method was initially utilized in the process industry of Denmark and later formalized through the scholarly contributions of Duijm (2009) and his co-authors (Duijm & Markert, 2009).

The graphical presentation of a safety-barrier diagram is shown in Figure 4.3.

This basic model, with its fundamental elements, is similar to an accidental scenario in the bow-tie model. However, the safety-barrier diagram can extend from these basic representations by incorporating more complex logical relationships through the use of AND/OR gates, multiple inputs, and multiple outputs. This enhancement allows the safety-barrier diagram to reflect more intricate causal sequences between hazards, events, and consequences.

A more advanced example of a safety-barrier diagram is shown in Figure 4.4. This diagram includes an *AND-gate* and an *OR-gate* that are used to manage the logic of hazard and barrier interactions:

- *AND-Gate*: As located near Barrier 6 in the diagram, an AND-gate means that all input events linked to the gate must occur for leading to the output event. This gate is useful for the situations where multiple conditions must be satisfied together to trigger a hazardous event or to activate a barrier.
- *OR-Gate*: As located near Barrier 4, an OR-gate allows for any one of the input events associated with the gate to trigger the output event. This gate is typically used in the scenarios where any one of several different conditions can lead to a hazardous event or necessitate the activation of a safety barrier.

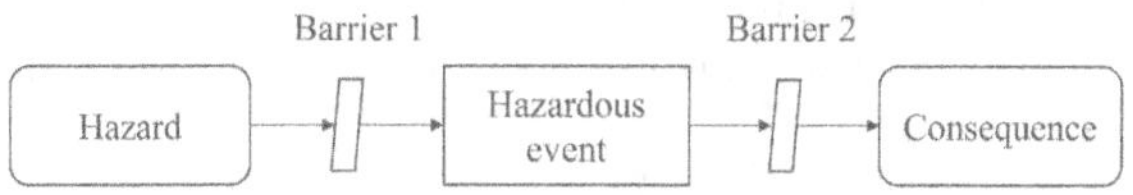


**FIGURE 4.3** A simple example of safety-barrier diagram.

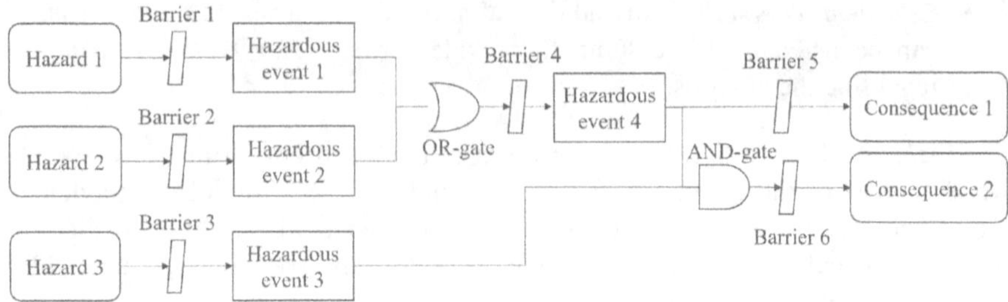


**FIGURE 4.4** An illustrative example of safety-barrier diagram for leakage in a hydrogen station.

These logical components enhance the utility of such a diagram by enabling it to model complex scenarios where multiple factors interact to impact safety outcomes. We will see more applications of AND-Gate and OR-Gate in the Section 4.7 when introducing FTA and associated quantitative analysis in Section 5.4.

### Example 4.1:

#### *Safety-Barrier Diagram Analysis*

This case study, using the information provided in Figure 4.1, considers three types of hazards at a gas fueling station, including the following:

- *Hazard 1*: Corrosion of the gas tank wall.
- *Hazard 2*: Improper operation by individual users during fueling.
- *Hazard 3*: Presence of ignition sources near the fueling area.

These hazards have the potential to develop into the following hazardous events:

- *Hazardous event 1*: Gas leakage from the tank due to corrosion.
- *Hazardous event 2*: Gas leakage from a nozzle or pipeline resulting from improper operation.
- *Hazardous event 3*: Occurrence of fire or sparking.

To prevent the occurrence of these hazardous events, the following safety barriers can be implemented:

- *Barrier 1*: Regular inspection and maintenance on the tank.
- *Barrier 2*: Standardized fueling operation guidelines displayed prominently near the fueling machines.
- *Barrier 3*: Safety signs prohibiting smoking and the use of open flames near the fueling area.

It is important to note that both Hazardous event 1 and Hazardous event 2 can result in the spread of flammable gas in the gas fueling station. Therefore, an

OR-gate is used to in the diagram to represent the logic that any input event can lead to the output event. This gives rise to another hazardous event:

- *Hazardous event 4*: Propagation of the flammable gas in the station with people.

Gas sensors can be employed to detect any gas leaks and activate alarms within the fueling station:

- *Barrier 4*: Gas leak sensors that trigger alarms to alert individuals in the station of a leak.
- *Barrier 5*: Alarms which facilitate the evacuation of the station, preventing exposure to potential fires.

When people hear the alarms, they should be able to evacuate the station. Therefore, a possible consequence in such a scenario could be:

- *Consequence 1*: Temporary downtime and economic loss for the fueling station, without any human injuries or fatalities.

However, if someone is smoking in the vicinity when a flammable gas leak occurs, there is a high risk of a pool fire. In such a scenario, both Hazardous event 3 (presence of an ignition source) and Hazardous event 4 (propagation of flammable gas) happen simultaneously. The AND-gate in the diagram shows that only when both these events occur together does the output event (pool fire) take place. When a fire occurs, people can immediately run behind a firewall for their safety. This introduces an additional safety barrier and another potential consequence:

- *Barrier 6*: A fire-resistant wall designed to withstand a pool fire for a specific duration, enhancing safety for individuals seeking refuge.
- *Consequence 2*: Damage to facilities due to the pool fire, but thanks to the fire-resistant wall, fatalities are avoided. ■

The safety-barrier diagram can enhance the understanding and management of risks in hazardous environments, with the following main functions:

- Identification of safety barriers in a specific accidental scenario.
- Presentation of the sequence of different barriers to be activated.
- Display of the logical dependencies between barriers and between hazards and barriers.
- Analysis of the adequacy of barriers for mitigating the identified hazards.
- Evaluation of the criticalities of barriers in consideration of various accidental scenarios.

A safety-barrier diagram typically incorporates data from risk assessment methods such as HAZOP or failure modes, effects, and diagnostics analysis (FEMDA), including the occurrence probabilities of events and the failure rates of barriers. By incorporating these inputs, the diagram enhances these risk assessment methods by focusing on the functions and criticalities of safety barriers within the overall risk management framework.

When analyzing a hazardous event, if multiple safety barriers are identified along the pathway leading to a specific consequence, intermediate events or conditions must be defined. This structure helps establish connections between the barriers, implying that the failure of a preceding barrier triggers the activation of the next.

Safety-barrier diagrams are designed to facilitate communication, even for those who are not experts. In addition, Duijm (2009) has highlighted several features of the safety-barrier diagram in barrier identification and analysis:

- *Uniqueness and conditions in one diagram*: Each barrier is unique and corresponds to specific events or conditions, with unique outputs. This singularity ensures that each barrier is represented only once, which is beneficial for precise specification and performance monitoring.
- *Decomposability*: No matter the complexity, a safety-barrier diagram can be decomposed into smaller, interconnected diagrams. This feature aids in detailed barrier analysis by allowing focus on specific parts of the system without losing the context of the whole.
- *Compatibility with FTA*: The qualitative methods and algorithms used in FTA can also be applied to models represented in safety-barrier diagrams. In practice, safety-barrier diagrams are often converted into fault trees for detailed quantitative analysis.

Relationships between safety barriers and their dependencies also can be analyzed using a safety-barrier diagram.

The Technical University of Denmark (DTU) has developed a dedicated software tool "SafetyBarrierManager", for modeling and analysis using safety-barrier diagrams. This tool provides robust support for practitioners looking to employ this method in their safety assessments. More information about this approach and the software tool is available at: https://safetybarriermanager.duijm.dk/wordpress/.

## 4.5 HAZID AND HAZOP

In the risk assessment of process industries, the Hazard and Operability Study (HAZOP) serves as an essential and comprehensive tool for identifying abnormalities in a process and their root causes. Similarly, the hazard identification (HAZID) study is widely employed to identify and analyze threats and hazards within a specific unit or workplace.

Both HAZID and HAZOP are introduced here in the same section due to their interconnected applications. These studies need to be performed by a qualified team comprising technical experts on the process, as well as personnel from operations, maintenance, and other relevant departments.

### 4.5.1 HAZID

HAZID is a general risk analysis tool, applicable to facilities, production processes, or machines. In a HAZID study, hazardous events are identified and preliminarily evaluated based on their causes, likelihood of occurrence, severity of consequences, and potential risk reduction measures.

The objectives of a HAZID study include the following:

- Identifying all the relevant hazards with a pivotal event(s) to the study node;
- Listing potential consequences to the asset;
- Recognizing existing safety barriers;
- Recommending potential safety barriers, including both preventive and mitigative ones;
- Providing guidance on the unit and process design and safety management.

HAZID aims to thoroughly examine all potential hazards throughout the design, installation, operation, and decommissioning phases. Upon completion, the study reports and worksheets of HAZID provide a long-term reference that can be consulted as needed.

An HAZID worksheet example for car collision (the study object is a driving process) is as shown in Figure 4.5.

The HAZID study, illustrated by the worksheet shown in Figure 4.5, is integral to the risk management process by effectively highlighting the hazardous events. Each hazardous event is designated as a *study node* in HAZID, with its reference number in Column 1 and detailed description in Column 2. The subsequent steps are to identify the causes (hazards in Column 3) and potential consequences (Column 4) of each event. The initial risk assessment, noted in Column 5, is calculated by evaluating the likelihood of the event and the severity of its consequences. Here, we can use the risk matrix and *risk priority number* (RPN) outlined in Section 1.3 as the straightforward assessment tools of risk.

Following this, the existing safeguards or safety barriers are listed in Column 6. These measures are then considered in the re-evaluation of the initial risk, and the revised and residual risk reflecting the implemented risk reduction measures, are detailed in Column 7. If the residual risk remains unacceptably high or insufficiently mitigated, Column 8 offers recommendations for further actions. These may include the introduction of various types of barriers or modifications to the process design. The final column specifies the responsible individual or department tasked with implementing these recommended actions.

Ideally, the HAZID study team should comprise three to five members, supplemented by a secretary to ensure thorough documentation. In a typical process plant

Study object: Car collision on a highway — Date:

Reference: XXX — Compiled by: Yiliu Liu

| No. | Hazardous event | Hazard | Consequence | Initial risk | Safeguard/ Barrier | Residual risk | Recommended action | Responsible |
|---|---|---|---|---|---|---|---|---|
| (1) | (2) | (3) | (4) | (5) | (6) | (7) | (8) | (9) |
| 1 | Car collision | High speed driving | Severe damage on cars and injuries of people | Intermediate | Speed limit sign, safe driving training | Moderate low | Electric alarm of short distance | |

**FIGURE 4.5** A simple example of HAZID worksheet.

scenario, the HAZID team would usually include a design engineer responsible for the study node, an asset or process manager, a safety expert, and a maintenance engineer or technician, all working collaboratively to ensure a comprehensive risk evaluation and mitigation strategy.

Generally, the HAZID study includes three structured steps:

1. *Identification of the hazard event*: This step is for determining the hazardous event, detailing its causes, and delineating the potential consequences.
2. *Risk estimation*: Here, the risk associated with the identified hazardous event is estimated. This estimation takes into account the effectiveness of existing barriers.
3. *Recommendations for additional actions and barriers*: Based on the risk estimation, this step is for proposing further measures to mitigate the risk, including additional barriers or procedural changes.

The HAZID study is foundational in risk assessment and barrier management activities and should be conducted as early as possible in the project lifecycle. However, it is important to revisit and revise HAZID results regularly or when changes in the process occur, such as during facility upgrades, redesigns, or in the aftermath of incidents like accidents or near misses.

In practice, HAZID worksheets can be integrated with other risk management tools, such as the Bow-Tie model, to enhance barrier management and risk analysis. For example, certain bow-tie software, like "BowTieXP", can automatically import data from HAZID worksheets, streamlining the risk analysis process.

The insights from HAZID studies are important for management to prioritize actions regarding barrier and safeguard strategies and to allocate resources effectively to critical areas. Additionally, HAZID helps identify situations where further in-depth studies, such as HAZOP, may be necessary.

While some sources treat HAZID as a component of Preliminary Hazard Analysis (PHA) or as a simplified PHA, this book does not explore PHA in depth. Those interested in further details about PHA may refer to the relevant technical standards and guidelines, such as MIL-STD-882E (2012).

### 4.5.2 HAZOP

#### 4.5.2.1 Introduction of HAZOP

HAZOP is a structured and rigorously documented method with the aim to identify the potential hazards in the operation of a facility or deviations from the design intention of a process. This method is widely applied across various sectors of the process industries, including chemical, pharmaceutical, oil and gas, and nuclear operations.

In conducting a HAZOP, the plant or process is segmented into several study nodes, each representing a distinct unit or component of the overall process. The selection of study nodes for a HAZOP requires judicious consideration of the balance between the thoroughness of the study and the constraints of time and cost, given the labor-intensive and time-consuming nature of the process.

The HAZOP examination of each node is facilitated through the use of several *guidewords* and *process parameters* during a series of structured meetings. Guidewords serve as benchmarks against the normative operational state, while process parameters act as quantitative or qualitative indicators of the current state. This combination of guidewords and parameters is designed to motivate individual thoughts and stimulate collective brainstorming within HAZOP meetings.

As standardized by IEC61882 (2016) and depicted in Table 4.1, guidewords express the typical operational conditions, while the process parameters in relevant industries commonly include the measurements such as flow, pH, temperature, pressure, and level. Additional parameters might include speed, frequency, composition, time, voltage, vibration, among others. These parameters often correlate with specific operational actions including maintenance, control, separation, reaction, activation or startup, inspection, etc. The synthesis of a guideword with a process parameter articulates a potential *deviation* in the process, formalized as:

$$\text{Guideword} + \text{Parameters} = \text{Deviation} \tag{4.1}$$

In the HAZOP methodology, the term "deviation" has the same meaning with the "pivotal event" used in Bow-Tie analysis. The incorporation of guidewords and process parameters aims to enhance the comprehensiveness of the analysis by enabling a systematic exploration of potential deviations from the intended design. However, it is important to recognize that not all hypothetical combinations of guidewords and

**TABLE 4.1**
**Typical HAZOP Guidewords and Examples of Deviation**

| Guideword | Meaning | Example of Deviation |
|---|---|---|
| No/Not | Complete negation of the design intent | No flow, no pressure, no reaction |
| More | Increase above the design intent | Higher temperature, lower pH value |
| Less | Decrease below the design intent | Lower temperature, higher level |
| As well as | Something else is also present even the design intent is achieved | Liquid flow as well as debris |
| Part of | Only some, rather than all, of the design intent is achieved | Partial control, partial flow |
| Reverse | Opposite of the design intent | Reverse flow |
| Other than | Something different substitutes the design intent | Unwanted/spurious activation, wrong composition |
| Early | Something happens earlier in the clock time than planed | Early activation, early started flow |
| Late | Something happens later in the clock time than planed | Late maintenance, late separation |
| Before | Something happens before it is expected in the sequence of order | Maintenance before inspection |
| After | Something happens after it is expected in the sequence of order | Inspection/examination after start-up |

parameters have meaningful insights. For example, notions such as "reverse pH value" or "other than level" are generally considered non-applicable and thus discarded.

Once deviations are identified, the next step of HAZOP is to analyze the associated risks and safety barriers. The objectives of a HAZOP study typically include the following:

- Identifying all the deviations of a process from its design intent and their causes.
- Detecting operability issues that arise from these deviations.
- Cataloging existing measures implemented to address these operability issues.
- Evaluating the adequacy of these measures and, if found insufficient, investigating the reasons for their inadequacy.
- Developing potential measures, safety barriers, and other corrective actions that can effectively address the identified issues.
- Monitoring the implementation of these barriers and recommended actions for efficacy and compliance.

In addition to the standard HAZOP, there are several specialized variants tailored to specific contexts, such as human HAZOP focusing on human errors, software HAZOP for identifying the issues in software development, and cyber HAZOP dealing with cyber threats. While these methods are outside the scope of this book, interested readers may refer to literature, such as the guidance book by Crawley and Brian (2015).

#### 4.5.2.2 Main Steps of HAZOP

The main steps of a standard HAZOP study include the following:

**Step 1: Preparing the project.**

The first step of a HAZOP project is to establish a HAZOP study team, define the scope (study object) and study objectives, and determine a systematic approach to guide the progression of this study. This stage also includes the allocation of adequate resources and the acquisition of essential information, to facilitate the smooth execution of subsequent steps.

When establishing a HAZOP team, a team leader should be appointed. The leader must have the relevant analysis experience and be familiar with the HAZOP procedure and the needed techniques. The main responsibilities of the team leader include organizing the team members to work toward a shared objective, coordinating the necessary resources for implementation of the study, and maintaining independence from the subject of the study to prevent any conflicts of interest.

A typical HAZOP team comprises 5–8 experts from different disciplinary backgrounds. Ideally, for example, a HAZOP study team focusing on a new chemical production process should include: a process designer, a process engineer responsible for developing piping and instrumentation diagrams (P&IDs), an electrical engineer who designs the associated electrical systems, control engineers tasked with designing or specifying the control systems, the operation manager overseeing the process, and a HAZOP secretary charged with documenting the meetings and managing relevant documentation.

To define the scope of HAZOP study, it is necessary to specify a particular component or section of the entire process as the study node, such as a pump or the segment extending from a pump to a separator. The design intent and normal operating conditions of the study node are also established during this phase.

Before the inaugural HAZOP meeting, all necessary information is expected to be in place. This includes P&IDs, equipment drawings, equipment layout, instrumentation specifications, control system logic diagrams, operational and maintenance guidelines, emergency response procedures, and applicable codes of practice. Given that certain documents, such as P&IDs, are only finalized after completing the process design, early lifecycle phase HAZOP meetings often rely on drawings from analogous systems or historical databases.

**Step 2: Identifying deviations.**

With the study object and its normal condition and design intent established in Step 1, the HAZOP team, led by the team leader, can start identifying the process deviations. Guidewords and process parameters (like in HAZID study) are used in this step to prompt rigorous expert discussions.

The team leader needs to stimulate the discussions among the experts with different backgrounds. For example, consider the scenario depicted in Figure 4.6, where the team leader should raise critical questions regarding the first process parameter "pressure" of the study node "pipeline":

- Is the combination of guideword "high" and the parameter "pressure" applicable?
- Could "high pressure" realistically occur as a deviation in the pipeline?

The findings from this step are systematically documented in the HAZOP worksheet, with filling in the Column 1 (reference number of the study node), Column 2 (description of the study node), Column 3 (guideword), and Column 4 (deviation).

**Step 3: Identifying causes of the deviations.**

To systematically uncover the origins of identified deviations, the HAZOP team leader needs to facilitate targeted inquiries, such as:

- Considering "high pressure" is identified as a deviation, what are the potential causes for its occurrence?

Study object: Gas transmission in a pipeline | Date:

Reference: XXX | Compiled by: Yiliu Liu

| No. | Study node | Guideword | Deviation | Causes | Consequence | Safety barriers | Risk | | | Improvements | Responsible | Comments |
|---|---|---|---|---|---|---|---|---|---|---|---|---|
| | | | | | | | F | S | RPN | | | |
| (1) | (2) | (3) | (4) | (5) | (6) | (7) | (8) | (9) | (10) | (11) | (12) | (13) |
| 1 | Pipeline (pressure) | More | High pressure | Intermediate | Potential explosion | Pressure transmitters, emergency shutdown system | 2 | 4 | 8 | Regular inspection and condition-based maintenances | Maintenance department | |
| | | Less | Low pressure | Corrosive rupture | Loss of gas flow | Pressure transmitters | 3 | 1 | 3 | Regular equipment maintenance | Maintenance department | |
| | | No | No pressure | Vent is wrongly open Loss of supply | Interruption of service | Pressure transmitters | 2 | 2 | 4 | Backup power supply | Utility department | |
| | Pipeline (composition) | As well as | Gas flow with impurities | Contamination during transportation, incorrect blending of gases | Reduced gas quality, increased emissions | Gas quality monitoring, proper handling and storage procedures | 3 | 1 | 3 | Separation system, regular equipment maintenance | Maintenance department | |

**FIGURE 4.6** An illustrative example of HAZOP worksheet.

Responses to this inquiry may vary widely, dependent on the operational complexities, the availability of detailed information, and the expertise of team members. It is important to document these potential causes comprehensively in Column 5 of the HAZOP worksheet to ensure all possible scenarios are explored and understood.

**Step 4: Identifying consequences of the deviations.**

In this step, the team leader guides the exploration of potential outcomes resulting from the deviations by proposing questions like:

- What are the possible consequences of "high pressure" within the system?
- Could these consequences disrupt normal production operations?
- Are these consequences hazardous to the pipeline, adjacent facilities, or personnel?

Such questions aim to clarify the direct and indirect impacts of the deviation, ensuring a thorough understanding of its potential to affect system integrity and safety. The responses are carefully recorded in Column 6 as the possible consequences of each identified deviation.

**Step 5: Identifying existing safety barriers.**

This step involves a detailed examination of preventative and mitigative measures existing within the process. Discussions are stimulated through questions such as:

- Is there an inherent mechanism or operational procedure that can prevent "high pressure"?
- Is there specific equipment installed within the system that can prevent "high pressure"?
- Are there sensors available that could detect "high pressure" immediately when the deviation occurs?
- Can the impact of "high pressure" be mitigated through certain mechanisms or equipment already present in the process?

These questions correspond to different categories of safety barriers, allowing the team to classify and evaluate each barrier according to its function and effectiveness. The findings from this step are entered into Column 7 of the HAZOP worksheet, with providing the detailed description of the existing barriers and their roles in maintaining operational safety.

**Step 6: Assessing risks of the deviations.**

With the causes, consequences, and existing safety barriers identified, the next step is to assess the risk associated with each deviation. Risk assessment in HAZOP studies often utilizes tools such as a risk matrix and a RPN to provide a semi-quantitative evaluation of the risks and the effectiveness of existing safety barriers.

Typically, the occurrence frequency of deviation is assigned a frequency class ranging from 1 to 5, according to a pre-established risk matrix as introduced in Section 1.3. Similarly, the severity of the consequences is classified on a scale from 1 to 4 within the same matrix. The RPN can then be calculated using Equation (1.2), offering a rough quantification of the risk level associated with each deviation.

Given that these estimations are rough and involve a degree of subjective judgment, it is important for the HAZOP team leader to prompt in-depth discussions among technical experts to validate the assessments.

The determined frequency of deviation, severity of consequences, and the corresponding RPN are systematically documented in Columns 8, 9, and 10 of the HAZOP worksheet, as illustrated in Figure 4.2.

**Step 7: Proposing improvements.**

Based on the results from the risk assessment, the HAZOP team leader then initiates the discussions for enhancing safety measures by proposing the questions such as:

- Can "high pressure" be prevented by incorporating additional operational steps or by installing new equipment?
- Given the potential and consequences of "high pressure", does it justify the additional investment in new equipment or design modifications?

This step is dedicated to the exploration of potential new safety barriers and other risk mitigation strategies. It is important to assess whether the cost of these new measures is justified by the expected enhancement in safety and operational efficiency. This evaluation helps in making informed decisions on implementing improvements.

The new safety measures are recorded in Column 11 of the HAZOP worksheet as proposed improvements. This documentation serves as a record for future reference and continuous improvement efforts.

**Step 8: Determining the follow-up approach and responsible persons.**

This step is critical for ensuring the actionable items identified during the HAZOP are effectively implemented and monitored. The responsibility for overseeing the development and implementation of each proposed improvement is assigned to specific individuals, teams, or departments. The name or designation of the responsible party is meticulously recorded in Column 12 of the HAZOP worksheet. This clear assignment of duties makes sure that each task has a specific person in charge, which is important for follow-up works.

Additionally, it is important to establish a robust method for tracking the effectiveness of the implemented improvements over time. This includes setting out detailed procedures for ongoing monitoring and assessment. The specifics of the follow-up approach, along with any relevant guidelines or additional pertinent information, should be documented in Column 13 as comments.

#### 4.5.2.3 Advantages and Limitations of HAZOP

The HAZOP method is widely recognized for its systematicity and comprehensiveness in hazard identification of analysis. By utilizing guidewords and process parameters, HAZOP has the potential to reveal all potential hazardous deviations of a study node. This method is particularly effective at identifying human errors and the errors that may arise from interactions between humans and machines, or among various machines themselves. Given the extensive adoption of HAZOP across various industries, technical experts are generally well-acquainted with the characteristics and features of this method.

However, the effectiveness of a HAZOP study can be heavily dependent on the experience and competence of the team leader and the expertise of the team members. There is also a risk that interactions between different study nodes may be overlooked. The HAZOP process requires a significant amount of detailed information, which may not always be available or provided at the necessary level of detail. Furthermore, the method can result in voluminous documentation if comprehensive records are maintained.

For readers who want to deepen their understanding of HAZOP, it is recommended to refer to the dedicated international standard IEC61882 (2016), which provides detailed guidance on conducting HAZOP studies.

To support the execution and management of HAZOP studies, several computerized tools have been developed, such as HAZOP Manager©,[1] and modules of some safety management programs. These tools are designed to streamline the process, enhance data management, and support the detailed analysis required in HAZOP studies. Utilizing these tools can significantly reduce the labor-intensive aspects of the process and improve the accuracy and efficiency of the analyses conducted.

## 4.6 FAILURE MODE AND EFFECT ANALYSIS

*Failure Mode and Effect Analysis (FMEA)* is a classical tool for systematically identifying failure modes of technical systems and conducting qualitative and semi-quantitative assessments. Originally developed in the late 1940s, FMEA was employed to enhance the reliability and safety of military hardware systems and was subsequently formalized and adopted widely in the 1970s. This technique is grounded in identifying where and how a product or process might fail, estimating the impact of different kinds of failures, and providing solutions to mitigate risk effectively. By systematically analyzing the possibilities of failures and their impacts, FMEA helps organizations proactively fix the parts of a system that could lead to future failures.

Since its inception, FMEA has evolved to include two significant variations that enhance its analytical capabilities:

- *Failure Mode, Effect, and Criticality Analysis (FMECA)*: This variation extends FMEA by incorporating a criticality metric, which aids in prioritizing failure modes based on their risk impact. The criticality assessment considers both the severity of the consequences and the likelihood of occurrence, providing a more robust framework for risk management.
- *Failure Modes Effects and Diagnostic Analysis (FMEDA)*: FMEDA expands the scope of FMEA by adding diagnostic capabilities to the analysis. This allows for the assessment of the ability of a system in detecting and responding to failures, which thus can be used for enhancing preventive maintenance strategies and improving system reliability.

Primarily applied within the discipline of reliability engineering, FMECA and FMEDA offer a more nuanced approach compared to HAZOP, which is generally for process hazard analysis. FMECA and FMEDA are tailored for the detailed analysis of hardware, making them important tools for sectors that rely heavily on complex machinery and systems, such as aerospace, automotive manufacturing, and electronics. These methodologies are particularly effective in identifying systemic vulnerabilities and improving the reliability and safety of engineering systems.

### 4.6.1 Failure Mode, Effect, and Criticality Analysis

FMECA is recognized as one of the most used tools in reliability engineering and risk assessment. The formulation of FMECA can be traced back to the U.S. military handbook (MIL-STD-1629A, 1980). While various variants of FMECA exist, we will focus on introducing the most relevant type: *product FMECA*.

#### 4.6.1.1 Objectives of FMECA

FMECA is typically integrated early in the design phase of system development across various industries, aiming to proactively address potential failures. Its main objectives are the following:

- Identifying potential failure modes for each component within a technical system as a product.
- Determining the likelihood of occurrence for each failure mode.
- Identifying the root causes of each failure mode and the underlying issues that may lead to failure.
- Assessing the impact of each failure mode on other components and the overall system.
- Evaluating the risk of each failure mode to gauge its severity and potential consequences.
- Identifying safety barriers and risk reduction measures associated with each failure mode.

Although we have mentioned in previous discussions that FMEDA expands upon FMEA by considering detectability, it is noteworthy that the easiness of detecting a failure mode is also taken into account in many FMECA projects.

It can be found that the hazardous event in FMECA is the failure of a component. We have defined the term "failure mode" in Chapter 3. Moreover, as discussed in Chapter 1, this term may pertain to an entire system, a subsystem, or a single component. To some extent, a failure mode illustrates the current state when a failure is observed. In contrast, a failure cause explains why the failure occurred in the past, and a failure effect pertains to the anticipated consequences of the failure in the future.

#### 4.6.1.2 Main Steps of FMECA

FMECA proceeds through the following steps (MIL-STD-1629A, 1980; Rausand et al., 2020), which correspond with the worksheet depicted in Figure 4.7:

**Step 1: Preparing the project.**

The first step of FMECA is similar with the preparation stage described in HAZOP, including establishing the study team, defining the scope and objectives of the FMECA study, and collecting necessary information.

Study object: Oil transmission in a pipeline | Date:
Study node: Centrifugal pump | Compiled by: Yiliu Liu
Reference drawing: XXX | Approved by:

| Description of study node | | | Description of failure | | | Effect of failure | | Risk | | | | Risk reduction measures/safety barriers | Responsible | Comments |
|---|---|---|---|---|---|---|---|---|---|---|---|---|---|---|
| Ref. | Function | Operational mode | Failure mode | Failure causes | Detection approach | Local effect | Global effect | F | S | D | RPN | | | |
| (1) | (2) | (3) | (4) | (5) | (6) | (7) | (8) | (9) | (10) | (11) | (12) | (11) | (12) | (13) |
| 1 | Deliver oil (liquid) as demand | Normal operation | No oil is delivered | Loss of power supply; Damage of motor; Control system failure | Diagnosis and observation immediately | Downtime of the pump and maintenance | Transmission stoppage | 1 | 4 | 1 | 4 | Backup power supply, monitoring motor performance | Facility department | |
| | | | Leakage of oil | Worn seals & gaskets; Loose connections | Diagnosis and condition monitoring | Reduced pump efficiency, | Loss of fluid, Pollution to environment | 2 | 3 | 2 | 12 | Regular check and replacement of seal and gaskets | Maintenance department | |
| | | | Overheating | Excessive load of bearing, Insufficient cooling | Condition monitoring | Increased wear; Damage to motor | Potential transmission stoppage | 3 | 2 | 2 | 12 | Regularly clean cooling passages | Maintenance department | |
| | | | Increased vibration | Shaft misalignment; Insufficient lubrication of bearing | Diagnosis and observation immediately | Damage to components in the pump | Reduced transmission | 3 | 1 | 2 | 6 | Standard procedure for alignment and verification; Regular check of lubrication oil | Engineering team, Maintenance department | |

**FIGURE 4.7** An illustrative example of FMECA worksheet.

However, it should be noted that the composition of the FMECA team differs slightly. Typically, a FMECA team consists of eight to ten members, led by a team leader who coordinates the analysis, facilitates communication, and monitors the study progress. The team is expected to include the following:

- *Domain experts*: Specialists with deep knowledge of the potential failure modes within the technical system.
- Design engineer of the system: She/he understands the design intent and specifications, offering insights into the system design and functional aspects.
- *Reliability engineers*: They are skilled in reliability engineering principles, statistical analysis, and risk assessment, and they can help quantify the likelihood and criticality of failure modes.
- *Quality control personnel*: They are responsible for identifying quality control issues during system manufacturing and ensuring that quality standards are met.
- *Operational personnel*: They are experienced with system behavior, and they can observe potential failure modes during operation and understand the impact of failures on system performance.
- *Safety specialists*: They are knowledgeable about safety regulations, standards, and risk management practices, helping evaluate associated risks.
- *FMECA secretary*: She/he documents the FMECA process and its results, assisting the team leader in administrative tasks.

Furthermore, the information required for FMECA differs from that used in HAZOP. The FMECA approach requires a detailed description of the system, including a Bill of Materials (BOM), which lists all components within the system comprehensively. The FMECA team must also prepare design and functional specifications, gather historical data on failures, understand operational conditions, and collect information about maintenance and inspection procedures, as well as safety and regulatory requirements.

**Step 2: Performing functional analysis.**

This step begins by identifying the primary function of the system, followed by detailing its operational modes and specifying the performance requirements for that function.

The system is then broken down into subsystems and components, similar to the hierarchical structure shown in Figure 1.9 for a barrier system. During this breakdown, the BOM for the system is utilized as a reference point. Further, the system function is decomposed into subfunctions and sub-subfunctions using the functional tree model illustrated in Figure 1.8. The interdependencies between components, subfunctions, and sub-subfunctions need to be thoroughly evaluated and documented. It is important to note that a single component may fulfill one or more sub-subfunctions, and conversely, a specific sub-subfunction may require multiple components for its realization.

A single component within the system may have multiple operational modes, such as running, switching, and standby. Each operational mode may be susceptible to specific failure modes. By documenting these operational modes, potential failure modes can be identified more effectively. If the operational modes significantly influence the frequency of a particular failure mode, they should be listed in a dedicated column on the FMECA worksheet. Otherwise, if the impact is less significant, the operational mode can be incorporated into the failure mode column itself.

In most product FMECA projects, a component acts as a study node or a unit and is assigned a unique identification number. All functions associated with that component are identified and listed. As discussed in Chapter 1, a barrier component is the lowest level item meaningful for performance assessment and is subject to maintenance activities.

**Step 3: Identifying failure modes and their causes for each component.**

This step focuses on determining the potential failure modes for each component based on its functions across different operational modes. A failure mode is typically described as the loss or degradation of the intended functions of a component. Performance degradation can also be categorized as a type of failure.

Similar to a HAZOP study, the FMECA team leader facilitates discussion by posing targeted questions to identify potential failures and their causes, such as, "What mechanisms can produce this failure mode?" This approach helps to stimulate thorough analysis and discussion among team members.

As depicted in Figure 4.7, some FMECA studies also incorporate the identification and documentation of detection methods for each failure mode. These detection methods are often considered part of the safety barriers. Common detection approaches include the following:

- *Operator observation and perception*: Operators can often notice clear signs of failure such as cracks, leaks, discoloration, or surface damage.

Their direct interaction with the system allows for immediate recognition of such issues.

- *Regular testing and inspection*: Conducted by trained technicians or engineers using specialized equipment, these routines are important for detecting not only apparent but also hidden failures, such as those in safety valves that might not be immediately evident.
- *Condition monitoring*: Monitoring critical parameters like temperature, pressure, vibration, or electrical signals can provide early indications of component deterioration or failure, facilitating proactive maintenance and repairs.
- *Automatic fault diagnosis*: Modern machinery may be equipped with diagnostic systems that automatically detect specific failures or malfunctions, generating error messages or indicators to alert operators or maintenance personnel.

**Step 4: Determining the effects of each failure mode.**

In this step, the FMECA worksheet, as shown in Figure 4.7, categorizes the impacts of each failure mode into *local effects* and *system effects*. Local effects refer to the consequences on adjacent components and the subsystem where the component is located, reflecting the immediate consequence within a smaller scope. On the other hand, system effects consider the broader influence or harm on the overall functionality of the entire system. In this step, it is necessary to provide descriptive text explain these effects, ensuring that both immediate and far-reaching consequences are clearly understood.

**Step 5: Assessing the risk related to each failure mode.**

This step evaluates the criticality of each identified failure mode. Using data from steps 2–4, each failure mode is scored based on its frequency, consequence, and detectability. Consistent with the methodology outlined in Section 1.3, a frequency class ranging from 1 (least frequent) to 5 (most frequent), and a consequence severity class from 1 (least severe) to 4 (most severe) are used. In the example, detectability is also classified into five groups, with group 1 indicating the highest detectability (easiest to detect) and group 5 the lowest (most difficult to detect). These classifications can ensure that higher scores correspond to greater risks. RPN can be calculated by multiplying the values of frequency, severity, and detectability, providing a quantitative measure of criticality of each failure mode.

**Step 6: Proposing risk reduction measures and safety barriers.**

Following the risk assessment, appropriate risk reduction measures and safety barriers are proposed to mitigate identified risks. Although the FMECA in Figure 4.7 does not detail changes in risk levels after the mitigation due to space constraints, the criticality matrix developed in Step 5 serves as a guideline for devising effective measures. Strategies might include reducing the frequency of failure occurrences, mitigating their consequences, or improving detectability. For example, installing new safety barriers can be an effective measure to reduce the impact of potential failures.

**Step 7: Determining the follow-up approach and responsible persons.**

As discussed in similar contexts previously, this step involves specifying follow-up actions and assigning responsibilities to ensure that all proposed measures are implemented and continuously monitored, similar to those in HAZOP. We do not repeat here.

#### 4.6.1.3 Advantages of Limitations of FMECA

FMECA has become a favored tool across various industries for its thorough evaluation of hardware failures in complex technical systems, as well as for providing detailed insights into the criticality of each failure. The outcomes from an FMECA serve as valuable inputs for guiding further analyses in reliability, risk, and barrier analyses. It should be noted that the critical failure modes identified in the analysis often require additional focus, prompting further investigations and allocating necessary resources toward the needed design modifications and maintenance actions to enhance system resilience and safety.

However, the deployment of FMECA also presents notable challenges. It is a highly time-intensive process and can impose significant financial burdens on an organization. While FMECA is excellent in analyzing and evaluating the failures of individual components within a system based on its structured breakdown, it has limitations in addressing complex interactions and dependencies between failures. This makes FMECA less effective for analyzing systemic hazards where failure interdependencies play a critical role.

### 4.6.2 Failure Mode, Effect, and Diagnostic Analysis

#### 4.6.2.1 Introduction of FMEDA

FMEDA is a refined technique dedicated to identifying and analyzing the failure modes, failure rates, and diagnostic capabilities of technical systems, subsystems, and components. This methodology differentiates itself from FMECA by incorporating component-specific failure rates and distributions, and by factoring in the system ability of detecting failures through automatic diagnostics. FMEDA was developed by William M. Goble and other engineers at the American company Exida (Goble, 1992; Goble & Brombacher, 1999).

FMEDA specifically targets the failure analysis of technical safety barriers or components used in safety-critical applications, rather than identifying safety barriers for broader technical systems or processes. It assesses specific functional failure modes, categorized as safe and dangerous, as well as detected and undetected. We previously introduced these failure types and their characteristics in Chapter 3, with detailed computational formulas to be discussed in Chapter 8. Here, it is needed to mention that FMEDA employs failure rates to quantify the occurrence frequency of different failure modes:

- $\lambda_{SD}$: Safe detected failure rate;
- $\lambda_{SU}$: Safe undetected failure rate;
- $\lambda_{DD}$: Dangerous detected failure rate;
- $\lambda_{DU}$: Dangerous undetected failure rate;

Such metrics are essential for evaluating the integrity of technical safety barriers, including the fraction of safe failures and the diagnostic coverage, which represents the proportion of detected failures in FMEDA.

As a thorough documentation of quantitative analysis for safety barriers, FMEDA relies heavily on robust data sources. This includes manufacturer data and specialized databases such as OREDA or those developed by Exida, which provide detailed information on various failure modes and the residual useful life of components.

#### 4.6.2.2 Main Steps of FMEDA

An example of FMEDA applied to a safety valve is illustrated in Figure 4.8. The main steps involved in an FMEDA project are outlined below:

**Step 1: Preparing the project.**

As in FMECA, the preparation phase is the groundwork for the FMEDA process. Details of this step have been outlined previously and are similar in nature, focusing on assembling the right team and defining the scope and objectives of the analysis.

**Step 2: Identifying failures.**

This step is to identify all potential failure modes of the technical safety barrier or safety-critical system, along with their causes and effects. In FMEDA, the effects are specifically categorized as safe or dangerous. Additionally, it is essential to determine whether each failure mode can be detected by the automatic diagnostic modules equipped on the safety-critical system.

**Step 3: Determining the rates of different failure modes.**

Failure rates are obtained based on the historical data, manufacturer's data, or specific reliability databases. In the context of the example provided in Figure 4.8, the analysis is limited to the four types of failure rates previously outlined.

**Step 4: Calculating the integrity measures.**

As demonstrated in Figure 4.8, this step involves calculating metrics such as the *Safety Failure Fraction* (SFF) and *Diagnostic Coverage* (DC). The methodologies for these calculations are detailed in Chapter 7 of this book.

Study object: Oil transmission in a pipeline | Date:

Study node: Safety valve | Compiled by: Yiliu Liu

Reference drawing: XXX | Approved by:

| Study node/Block | | Description of failure | | | | Failure rate ($10^{-6}$/h) | | | | Safety failure fraction | Diagnostic coverage | Comments |
|---|---|---|---|---|---|---|---|---|---|---|---|---|
| Ref. | Function | Failure mode | Failure causes | Effect | Diagnostic | $\lambda_{SD}$ | $\lambda_{SU}$ | $\lambda_{DD}$ | $\lambda_{DU}$ | | | |
| (1) | (2) | (3) | (4) | (5) | (6) | (7) | (8) | (6) | (10) | (11) | (12) | (13) |
| 1 | Close flow in case of danger | Failure to move upon a demand | Spring broken | Dangerous | Undetected | | | | 2.2 | | | |
| | | Failure to open after closure | Loss of power supply | Safe | Undetected | | 33 | | | 60% | 0% | |
| | | Leakage across the valve | Corrosion rupture | Dangerous | Detected | | | 19.8 | | | 90% | |

**FIGURE 4.8** An illustrative example of FMEDA worksheet.

#### 4.6.2.3 Advantages and Limitations of FMEDA

FMEDA is specifically designed for evaluating technical safety barriers, offering a detailed understanding of potential failures and enabling the development of effective safety strategies. By implementing FMEDA during the design and development phases, potential vulnerabilities can be identified and rectified early, thereby preventing major issues later on. This proactive, quantitative approach facilitates prioritized allocation of resources to maintain the integrity of safety barriers.

On the other hand, the effectiveness of a FMEDA project heavily relies on accurate and reliable data regarding failure mode, and diagnostic capabilities. However, acquiring such is often challenging, which may impact the quality of the analysis. Furthermore, FMEDA involves various assumptions in its quantitative assessments, the implications of which will be discussed further in subsequent chapters. Dependence on these assumptions can introduce uncertainties into the analysis, potentially limiting its applicability in certain scenarios.

## 4.7 FAULT TREE ANALYSIS

### 4.7.1 Introduction of Fault Tree Model

FTA is a systematic, failure-oriented analytical approach represented by a tree-like graphical model to describe how an accidental event occurs due to different causes. Originally developed by Bell Telephone Laboratories in 1962, FTA has become one of the widely utilized tools for safety and reliability analysis. It is now standardized under the international standard IEC61025 (2006).

The main objectives of FTA include the following:

- Identifying all potential events and combinations of events that could lead to an accident.
- Determining the causal relationships between these events and the resultant accident.
- Analyzing the likelihood of an accident by examining the causal events and their interrelationships.
- Identifying vulnerabilities within a system or process and devising improvement strategies, such as implementing safety barriers.

A fault tree model is essentially a top-down logic diagram, as depicted in the simplified example shown in Figure 4.9. The topmost node, referred to as the "*TOP event*", represents a system-level failure from a reliability engineering perspective, or a pivotal event or an accident from a risk assessment perspective, and is illustrated using a rectangle symbol, which is commonly used to denote general events in fault tree models.

Starting from the TOP event, a tree is constructed by tracing back to identify all contributing causes of the accident. These causes can include individual component failures, combinations of component failures, human errors, or procedural errors, among others. These causes are represented as lower-level or input events to the TOP event. For example, in Figure 4.9, the TOP event is influenced by two input events,

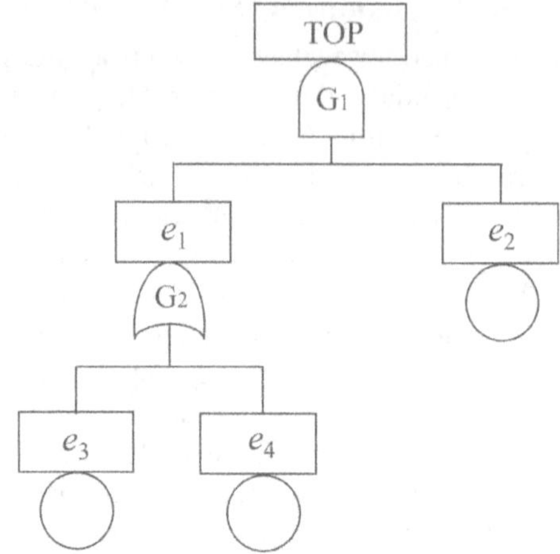


**FIGURE 4.9** A simple fault tree model example.

labeled $e_1$ and $e_2$. Conversely, the TOP event can be regarded to serve as the *upper-level event* or *output event* of $e_1$ and $e_2$. In this example, event $e_1$ also can be further decomposed into its lower-level events, $e_3$ and $e_4$. Through this hierarchical structure, the fault tree model visualizes the pathways leading to a critical failure or accident.

On the other hand, event $e_2$ is identified as a *basic event*, illustrated by a rectangle with a circle below it. This graphical representation indicates that $e_2$ is a terminal element in the fault tree, one that requires no further development. Events $e_3$ and $e_4$ are also basic events. Basic events are the leaf nodes of a fault tree model, and they can be root causes of failures or represent failures whose causes are either unknown or not practical to investigate further. IEC61025 (2006) uses another term of "primary event" for basic event, which means the event at the bottom or lowest level of a fault tree.

An event that is not classified as either the TOP event or a basic event is designated as an *intermediate event* (IEC61025, 2006), like $e_1$ in Figure 4.9. Descriptions of an event can be given in the associated rectangle. To efficiently use space in large fault tree diagrams, intermediate events are described within rectangles, while basic events may be denoted simply with circles.

Fault tree analysis utilizes logic gates to connect events, facilitating the representation of conditions under which upper-level events occur based on the status of lower-level events. The main two types of logic gates used in fault trees are the following:

- *AND-gate (as $G_1$ in* Figure 4.9*)*: The output event of the gate occurs only if all the input events occur. Namely, it symbolizes a conjunction of conditions.
- *OR-gate (as $G_2$ in* Figure 4.9*)*: The output event of the gate will occur if any of the input events occurs. Namely, it symbolizes a disjunction of conditions.

In the fault tree model example of Figure 4.9, the structure allows us to deduce that the TOP event will occur if event $e_1$ happens (which itself happens if either event $e_3$ or

$e_4$ occurs), and event $e_2$ also occurs simultaneously. In other words, this leads to the identification of two combinations of basic events that, if all occur, would directly lead to the occurrence of the TOP event: $\{e_2, e_3\}$ and $\{e_2, e_4\}$. In FTA, such a set of basic events leading to the TOP events are called as *cut set*, which can be defined as:

- *Cut set*: A set of basic events that, if all occur, would cause the occurrence of the TOP event.

A cut set is termed a *minimal cut set* if it cannot be further reduced without affecting its capability to cause the TOP event.

By comparing the fault tree in Figure 4.9 with the safety-barrier diagram in Figure 4.4, we can observe that the fault tree models the occurrence of "Consequence 2" without considering any intervening safety barriers. This methodical exclusion of safety barriers in the fault tree provides a view of the potential failure paths and their impacts, facilitating a focused analysis on inherent system vulnerabilities that need addressing.

### 4.7.2 Risk Analysis with a Fault Tree

Typically, FTA serves as a critical tool used to analyze the likelihood of a hazardous event. This hazardous event may be either the pivotal event within the Bow-Tie diagram or an end event with a severe consequence. In the context of the fault tree, this hazardous event is referred to as the TOP event.

A general procedure of using a fault tree in risk analysis include the following main steps:

**Step1: Determining the scope and the TOP event to be studied.**

The first step in conducting an FTA is to define the scope of the analysis and identify the critical event, or TOP event, to be studied. This involves a thorough understanding of the system or process, including which components or sequences are involved and the external threats that may impact the system. Identifying the TOP event is the most critical task and can be supported by insights gained from other risk assessment tools like HAZOP or FMECA, which help pinpoint the most significant risks associated with the system.

**Step 2: Constructing the fault tree model.**

Once the TOP event is determined, the next step is to construct the fault tree. Starting with the TOP event, the analysis works backward to identify all contributing causes and factors. These are represented as lower-level events within the fault tree. The logical relationships of these events to the TOP event are defined using logic gates, namely AND-gates or OR-gates in the basic method, to model how combinations of events lead to the occurrence of the TOP event. Events within the fault tree are categorized as either basic or intermediate events, with intermediate events further broken down until all paths are fully developed. This work continues until no further intermediate events remain undeveloped.

There is no definitive answer to where the appropriate level in the hierarchical structure of a system is. However, it is beneficial to consider that the level of resolution within a fault tree should align with the level of detail provided by the available information. A fault tree that is too detailed may introduce uncertainties or inaccuracies if it surpasses the detail of the data available. Conversely, an overly simplified fault tree may ignore critical factors or events, potentially undermining the integrity of the analysis.

It is essential to recognize that basic FTA operates on a binary premise, treating all events as either occurring or not occurring. This binary perspective extends to the components themselves, which are viewed as either fully "functioning" or fully "failed". States that represent intermediate conditions, such as "degraded" or "partially functioning", are generally not suitable for the traditional binary framework of FTA. This simplification helps maintain clarity in the analysis but can sometimes limit the understanding on the systems where gradual degradation is critical (see Chapter 9).

**Step 3: Qualitative analysis of the fault tree.**

The primary goal of qualitative analysis in FTA is to identify all minimal cut sets. For simpler fault trees, the MOCUS algorithm (Method for Obtaining Cut Sets) is commonly employed to determine these minimal cut sets. As described by Rausand et al. (2020), let us consider a small fault tree illustrated in Figure 4.10 as an example.

MOCUS starts at the TOP event. The logic gate linked to the TOP event is an AND-gate. According to Rule 1 in MOCUS:

- *Rule 1*: Input events of an AND-gate should be listed in separate columns.

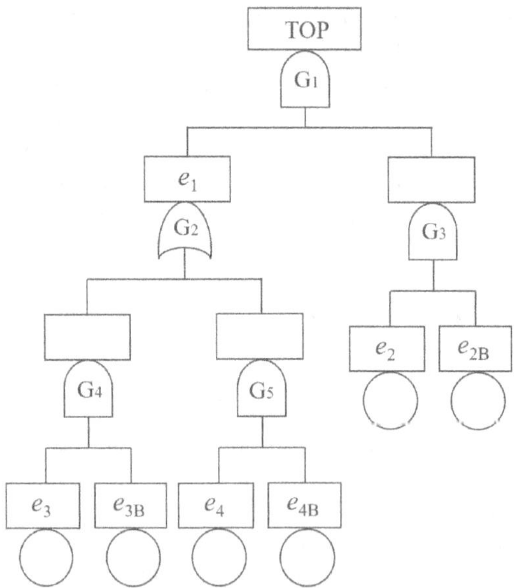


**FIGURE 4.10** A fault tree developed from Figure 4.9 with barriers.

Therefore, we represent this as

$$G_2\ (e_1), e_2$$

Here, $G_2$ as the connecting gate of $e_1$ denotes this intermediate event, thus, it is simplified to

$$G_2, e_2$$

Proceeding with the analysis, we replace all logic gates with basic events. For $G_2$, since it is an OR-gate, we can use the Rule 2 in MOCUS:

- *Rule 2*: Input events of an OR-gate should be listed in separate rows.

Thus, we have

$$e_3, e_2$$

$$e_4, e_2$$

Now, no more logic gates are left to replace, so we conclude the process. Each row now represents a minimal cut set, with the sets from the fault tree in Figure 4.10 being:

$$\{e_2, e_3\} \text{ and } \{e_2, e_4\}$$

Upon evaluation, both sets are minimal because removing any event from these sets results in the inability of the remaining event alone to cause the TOP event. Therefore, these are confirmed as minimal cut sets.

**Step 4: Quantitative analysis of the fault tree.**

The aim of this step is to estimate the likelihood of occurrence of the TOP event. Quantitative analysis methods, which will be explored in detail in Section 5.4, typically involve calculating the probability of the minimal cut sets, thereby deriving the probability of the TOP event.

**Step 5: Vulnerability analysis and improvements.**

Building upon the qualitative and quantitative analyses, this step focuses on identifying the specific basic events that significantly contribute to the occurrence of the TOP event. Such insights are important for revealing the potential weaknesses or vulnerabilities within the system. Based on these findings, protection strategies can be developed, and necessary measures (e.g., implementing safety barriers) can be taken to mitigate the risks associated with the identified vulnerabilities. Further discussion and methodology regarding this aspect will be provided in the subsequent subsection.

### 4.7.3 Barrier Identification and Analysis with a Fault Tree

In FTA, when modeling accidents in a system or process, the basic events represent hazards that can disrupt normal operations. Initially, these fault trees do not incorporate safety barriers or account for potential failures of these barriers.

After completing the qualitative and quantitative analysis steps in FTA, the next task is to develop a barrier strategy to enhance system safety. The most straightforward strategy is to introduce a safety barrier for each identified basic event. This approach can modify the fault tree by replacing each basic event, which serves as a leaf in the tree, with a new branch. This new branch comprises:

- *Original event*: Represents the hazard or potential failure point.
- *Safety barrier event*: A new event introduced to control the impact of the original event.
- *AND-gate*: Connects the original event and the safety barrier, indicating that for the system to be affected by the basic event, the safety barrier must fail.

The core idea of this approach is that the system remains unaffected by a basic event as long as the associated safety barrier functions effectively.

This barrier strategy can be visually represented in a fault tree diagram, for example, the updated version of Figure 4.10. In this revised diagram, safety barriers are added to each basic event previously identified in the original fault tree. Each safety barrier is linked to its corresponding basic event through an AND-gate. The addition of these barriers transforms the fault tree, showing how each layer of protection contributes to the overall safety of the system.

In the model of Figure 4.10, $e_{2\text{B}}$, $e_{3\text{B}}$, and $e_{4\text{B}}$ represent the failures of the associated safety barriers of events $e_2$, $e_3$, and $e_5$, respectively. Using the MOCUS algorithm, we identify the minimal cut sets in this enhanced model, which include the following:

$$\{e_2, e_{2\text{B}}, e_3, e_{3\text{B}}\} \text{ and } \{e_2, e_{2\text{B}}, e_4, e_{4\text{B}}\}$$

These sets demonstrate the scenarios in which multiple barrier failures, in combination with their corresponding basic events, could lead to the TOP event.

Given that the original fault tree in Figure 4.9 represents an accident scenario leading to Consequence 2 without any barriers, as depicted by the safety-barrier diagram in Figure 4.4, we can extend this model to include a comprehensive safety barrier for each basic event and logic gate. This expansion results in a new fault tree, illustrated in Figure 4.11, where a new TOP event is introduced and Barrier 1, Barrier 2, and Barrier 3 are in place.

The new TOP event is controlled by a reactive barrier, specifically designed to mitigate the damage that the original TOP event would cause. Occurrence of the new TOP event, indicating Consequence 2, only occurs in case of the failure of Barrier 6 ($e_{\text{TOPB}}$), as represented in Figure 4.4. In addition, $e_{1\text{B}}$ represents the failure of Barrier 4 in the safety-barrier diagram. It is important to note that this model does not consider Consequence 1, thereby excluding Barrier 5 from consideration.

The minimal cut sets for this revised fault tree are more complex, reflecting multiple layers of safety barriers:

$$\{e_{1\text{B}}, e_2, e_{2\text{B}}, e_3, e_{3\text{B}}, e_{\text{TOPB}}\} \text{ and } \{e_{1\text{B}}, e_2, e_{2\text{B}}, e_4, e_{4\text{B}}, e_{\text{TOPB}}\}$$

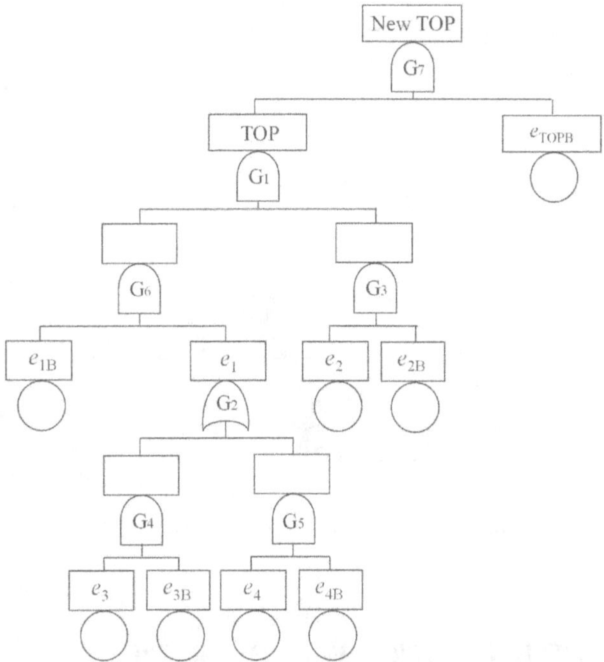


**FIGURE 4.11** A fault tree involving more barriers.

The strategy illustrated in Figure 4.11, developed through the fault tree model, presents a robust approach to accident prevention by minimizing risk through numerous safety measures. However, such a strategy may be too cautious and financially burdensome due to the extensive deployment of safety barriers.

To reach a balance between safety and cost-effectiveness, an alternative barrier strategy can involve installing safety barriers primarily at the output events of OR-gates. These gates generally signal a higher probability of occurrence for the output event, making them critical points for intervention. This targeted approach allows for a more economical allocation of safety resources without compromising critical safety needs.

With this revised strategy in mind, an alternative fault tree model is presented in Figure 4.12, which builds upon the structure initially laid out in Figure 4.9. The minimal cut sets for this new model would be

$$\{e_{1B}, e_2, e_3\} \text{ and } \{e_{1B}, e_2, e_4\}$$

With the quantitative analysis methods to be introduced in Chapter 5, we can further investigate these fault tree models in Figures 4.9–4.12. Our analysis can focus on calculating the occurrence probability of the TOP event and assessing the effectiveness and adequacy of various barrier strategies. This investigation will help us understand how different approaches impact the overall safety and efficiency of the system.

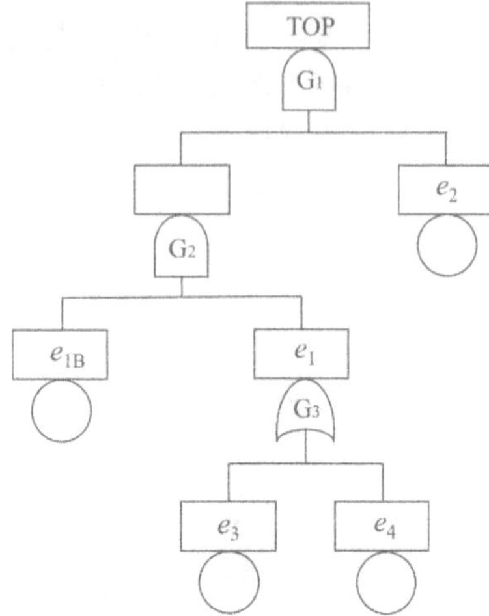


**FIGURE 4.12** A fault tree developed with a barrier only for the OR-gate.

## 4.8 MANAGEMENT OVERVIEW AND RISK TREE

The Management Oversight and Risk Tree (MORT) is a logical expression of the functions and a structured methodology for managing organizational risk, which was first documented by W.G. Johnson in 1973 (Kingston et al., 2009). MORT is more often used in business management for ensuring organizational safety, but it is also useful in safety analysis for technical systems.

A key component of MORT is the energy trace and barrier analysis (ETBA), which is used to identify hazards and their corresponding barriers. ETBA shares similarities with the EFBA discussed in Section 4.3. The tool focuses on how barriers influence energy flows, such as how a properly functioning valve can adjust an energy flow from excessive to normal levels.

A basic ETBA framework only includes three columns: energy flow, target (asset), and barrier. However, in Figure 4.13, we provide a more comprehensive ETBA worksheet, based on the table provided by Lewis and Haug (2009), serving as a counterpart to the EFBA worksheet depicted in Figure 4.2.

According to the International Crisis Management Association (ICMA, 2006), an ETBA includes the following steps:

**Step 1: Identifying energies in a system.**

The first step is to identify the various types of energies present within the system, along with their attributes and sources. It is important to recognize that energies can originate from within the system or be introduced from external sources. This information is recorded in Columns 1 and 2 of the ETBA worksheet, as depicted in Figure 4.13.

Study object: Hydrogen refueling station | Date:

Reference: XXX | Compiled by: Yiliu Liu

| No. | Energy origin and amount | Energy flow route | Energy release | Affected asset | Barriers | Barrier evaluation | Risk | Recommended action | Comments |
|---|---|---|---|---|---|---|---|---|---|
| (1) | (2) | (3) | (4) | (5) | (6) | (7) | (8) | (9) | (10) |
| 1 | 700-bar compressed gaseous hydrogen in tank (pressure, flammable and explosive gas, 149 L) | Flow in pipeline from tank to dispensers | Leakage (flow by wind) | Facilities and people in the station | 1.1 Gas detector<br>1.2 Alarm<br>1.3 Shut-down valve | SIL 2<br>SIL 3<br>SIL 2 | Low | Regular tests on the barrier equipment | |

**FIGURE 4.13** An illustrative example of ETBA worksheet.

**Step 2: Tracing energy flow.**

This step requires tracing the flow paths of each type of energy throughout the system, documenting any energy exchanges between components if applicable. It is also essential to track how energy exits the system. The findings from this analysis are recorded in Column 3 of the worksheet.

**Step 3: Recognizing the energy exchange.**

Here, the focus is on identifying both the energy exchanges, particularly undesired releases of energy, and the assets that can be impacted by these exchanges. This includes identifying any individuals or equipment that may be in the path of an unintended energy discharge. Relevant information is documented in Columns 4 and 5 of the ETBA worksheet shown in Figure 4.13.

**Step 4: Barrier analysis**

This step involves identifying and documenting the safety barriers that are in place to manage energy releases, recording them in Column 6. An analysis is then conducted on understanding how these barriers modify the energy amount and flow, what are the potential failures of these barriers, and their integrity and availability. In the provided example, the safety integrity levels (SILs) of the barriers are provided in Column 7. The explanations on the values of SIL can be found in Section 6.4.

**Step 5: Overall risk control assessment.**

The final step assesses the overall risk control strategies implemented across the system. If the identified risks are considered unacceptable, proposals for safety improvements are necessary. The outcomes of this analysis are listed in Columns 8 and 9 of the worksheets, with further comments provided in Column 10. This step aims to ensure a comprehensive and effective risk management strategy is in place for the system as a whole.

Another important tool in MORT is MORT tree. This tree model is constructed based on fault tree models with input information derived from ETBA. However, in comparison to basic fault trees, MORT trees have relatively fixed structures. For instance, they must include S-branches for specific control factors and

M-branches for management system factors. MORT trees incorporate a broader range of symbols, such as triangles that indicate transfers to or from the MORT tree. Moreover, MORT trees employ strict terminology to provide guidance to analysts when investigating latent technical or organizational anomalies within an organization.

Since MORT tree is mainly used for exploring and addressing potential weaknesses or vulnerabilities in an organization, not for barrier analysis and engineering. For this reason, further discussion on MORT trees is not provided here. Readers interested in exploring this method are encouraged to refer to the EU MORT Manual by Kingston et al. (2009), which provides comprehensive information and guidance on implementing MORT trees effectively.

## 4.9 SYSTEMS-THEORETIC PROCESS ANALYSIS

STPA is a component of the System-Theoretic Accident Model and Processes (STAMP) framework, providing a unique perspective on risk analysis and management, as briefly introduced in Chapter 2. First proposed by Nancy Leveson (2004), STPA has obtained attentions across various industries over recent decades.

Unlike traditional hazard identification methods that often focus on hardware component failures, STPA is predicated on the assumption that risks or accidents arise from inadequate controls over processes or systems. Thus, it emphasizes identifying potential deficiencies in process control and the interactions between various elements within a system.

An STPA study involves several main steps and requires a dedicated team of domain experts, along with a team leader who is proficient in control theory. The steps to execute a STPA project typically include the following:

**Step 1: Identifying system-level accidents.**

The first task of STPA is to identify critical accidents or unacceptable losses at the system level. The determination of what constitutes an unacceptable loss varies and depends on factors such as legal requirements, regulations, societal norms, business objectives, and rational judgment.

Once the critical accident to be studied is determined, it is essential to clarify the scope of the STPA study. This includes defining the boundaries of the system or process under investigation and identifying the hazardous events that could lead to the accident or unacceptable loss. It should be noted that Leveson (2012) uses "hazards" to describe the term of "hazardous event", but we maintain a distinction in this book to keep terminology consistent: hazards may evolve into hazardous events, which in turn can lead to accidents. Historical HAZOP and FMECA studies can be utilized to inform the identification of these hazardous events in STPA.

Additionally, this step involves identifying system-level requirements that specify behaviors and conditions necessary to avert accidents. These requirements are referred to as *safety constraints* in STPA.

### Example 4.2:

#### *Scope, Hazardous Events, and Safety Constraints in STPA*

Consider the scenario of barrier management designed to prevent a collision between a ship and an iceberg, as explored in the master thesis by Nesse (2019) under the supervision of the author of this book at NTNU. In this example, the STPA system boundary encompasses a floating production, storage, and offloading (FPSO) platform used in the offshore oil and gas industry, ice management vessels (IMVs), and sensor platforms for ice monitoring. The unacceptable losses are defined as injuries or deaths of personnel, and damage to equipment resulting from a collision between the FPSO and an iceberg. External factors such as weather and sea conditions also influence the likelihood of such an accident.

A few hazardous events are linked to the accident, but we will focus on two in this case study:

- *H-1*: The FPSO fails to disconnect and evade when an iceberg drifts toward it.
- *H-2*: The IMVs employ water cannons or towing during dangerous weather conditions, which is hazardous as the iceberg might collide with the IMVs, or falling ice may injure crew members.

The safety constraints identified in response to these hazardous events include the following:

- *SC-1*: The IMVs must prevent dangerous icebergs from drifting toward the FPSO (response to H-1).
- *SC-2*: The FPSO must disconnect and evade in emergency situations (response to H-1).
- *SC-3*: The IMVs must not use towing or water cannons in dangerous weather conditions (response to H-2). ■

**Step 2: Constructing a control structure.**

The next task is to establish a control structure for the study object. Any system or process is regarded as a looped control structure in STPA, with the basic elements as illustrated in Figure 4.14.

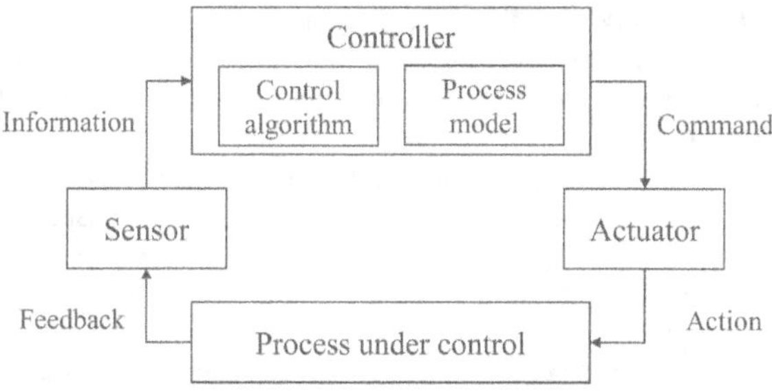


**FIGURE 4.14** Basic control structure in STPA.

In STPA, we use the terminology similar to that for technical safety barriers, since these barriers also incorporate a defined control structure. Within this structure, a *controller* is tasked with making decisions based on a *control algorithm* that utilizes collected information and a process model to influence a *process* (or a *process under control, controlled process*). The controller gathers this information primarily through *feedback* obtained via a *sensor* and executes control actions using an *actuator.*

It is significant to identify all the associated elements involved in this control loop. The controller, sensor, actuator, and even the process under control can comprise humans, hardware, software systems, or combinations thereof. The structure depicted in Figure 4.14 represents a basic or high-level abstraction of a control structure, serving as a foundational model from which a hierarchical control structure can be developed.

For instance, both the controller and the process under control can be decomposed into subsystems, with control models for these subsystems crafted to include more detailed information. However, it is important to carefully consider the appropriate level of abstraction for STPA. A hierarchical structure inherently involves complex interactions both within and between different levels, which can complicate the analysis and management of the system. This requires a thoughtful approach to determine the most effective structure for identifying and mitigating risks in the system.

## Example 4.3:

### *The Basic Elements in the Control Structure*

In continuation from Example 4.2, the key elements of the control structure can be outlined as follows:

- *Controller*: This role is fulfilled by the ice management team on the FPSO and the captain, who are responsible for decision-making.
- *Process under control*: The primary process being managed is the safe operation of the FPSO.
- *Sensor*: Monitoring is conducted through human observation and dedicated sensor platforms that track ice movements.
- *Actuator*: IMVs and the FPSO connectors.
- *Feedback information from process to sensor*: This includes updates on the status of water cannons and towing, the safety status of personnel and equipment, and the location of the FPSO.
- *Information from sensor to controller*: Sensors provide vital feedback to the controller, including data on iceberg characteristics (size, speed, and trajectory), current weather conditions (waves, wind speed and direction, and any precipitation), and the distance between the FPSO and the approaching iceberg.
- *Control action by actuators*: Potential actions taken by the actuators involves using water cannons or towing by IMVs to alter the path of the iceberg or initiating the FPSO to disconnect from its moorings to evade the iceberg.

- *Command*: Commands are issued to direct the FPSO and IMVs to execute specific actions based on the assessed risks and operational requirements. ■

**Step 3: Defining process model for each controller.**

In STPA, the "process model" refers to how the controller understands the operational mechanism and behavior of the process under control. For example, in the context of ice management, the process model may include the understanding that using water cannons and towing by IMVs can effectively drive away icebergs. The timing of these operations is critical, and initiating IMV actions earlier could result in diverting the iceberg sooner. In this scenario, one key *process variable* is the timing of IMV operations, which can be measured by the distance between the FPSO and the iceberg. Controllers make decisions based on the observed values of these process variables, and their decisions can also alter the values of certain variables.

Process variables can serve various objectives. In Example 4.3, another critical process variable is the trajectory of the iceberg. If the iceberg is on a direct collision course with the FPSO, immediate action is necessary. Otherwise, the action taken by the IMVs may be unnecessary. The trajectory is influenced by several factors, including wind direction, which introduces uncertainties. The further away the iceberg, the higher the trajectory uncertainty, which diminishes as it draws nearer. To minimize uncertainty and avoid superfluous actions, it is generally advisable to initiate IMV operations as late as possible while ensuring safety.

**Step 4: Identifying unsafe control actions.**

Given the process model defined in Step 3, the next task is to ascertain the required actions to control the process as expected and then identify any *Unsafe Control Actions* (*UCAs*) that may lead to a hazardous event under certain conditions. Some control actions are typically safe but can become hazardous in specific situations. For instance, IMV actions are safe under good weather conditions but can become hazardous during heavy rain accompanied by strong winds.

STPA provides guidelines for pinpointing UCAs. According to Leveson (2012), there are four generic types of UCAs:

- A control action is not provided when it is necessary.
- A control action is provided leading to a hazardous event.
- A control action is provided in incorrect time, meaning too soon, too late or in a wrong order.
- A control action is provided with too short or too long time.

**Example 4.4:**

*UCAs in Ice Management*

In this example, we analyze a specific control action within the ice management context: The decision to direct IMVs to use water cannons or towing to move an iceberg that poses a threat, assuming that weather conditions are favorable.

**TABLE 4.2**
**Some Unsafe Control Actions in the Ice Management Example**

| Guideword | Not Provide | Provide | Provide Early/Late | Provide Short/Long |
|---|---|---|---|---|
| UCA | UCA-1: Water cannon or towing is not provided when an iceberg is approaching. | UCA-2: Water cannon or towing is provided in dangerous weather conditions. | UCA-3: Water cannon or towing is provided too late, and the iceberg collides with FPSO. | UCA-4: Water cannon or towing is stopped too soon, and the iceberg collides with FPSO. |

Based on the STPA guidance, the following UCAs can be identified and are detailed in Table 4.2. ■

**Step 5: Identifying the causes of unsafe control actions.**

The subsequent step is to determine why a control action may fail or be considered unsafe. In STPA, the sources of unsafety are not limited to failures of hardware components, and actually they often come from the interactions and information exchanges among the elements within the control structure. The STPA Handbook (Leveson & Thomas, 2018) outlines generic causes of UCAs. Some of these causes are elaborated in Table 4.3.

### Example 4.5:

#### *Causes of UCAs in Ice Management*

For the UCAs provided in Table 4.3, we can identify their causes following the guidance in Table 4.4. ■

**TABLE 4.3**
**Generic Causes of Unsafe Control Actions**

- Failure of the controller
- Incorrect process model (due to design or changes)
- Incorrect implementation of the process model
- Interruption of control action from controller to actuator
- Failure or unexpected behavior of the actuator
- No feedback from the process under control
- Failure of sensor
- No (wrong) information transmission from sensor to controller

*Source:* Adapted from Leveson and Thomas (2018).

**TABLE 4.4**
**Some Causes of UCAs in Ice Management**

| UCA | Causes |
|---|---|
| UCA-1 | Failure of water cannons. |
| | Failure of the ice monitoring system. |
| UCA-2 | The management team does not notice the alarm. |
| | The management team does not realize the severity of bad weather. |
| | Short-term weather forecast is wrong. |
| UCA-3 | Preparing water cannon and towering equipment takes too much time. |
| | The management team does not notice the alarm on time. |
| | The management team waits for good weather for actions. |
| UCA-4 | Failure of water cannons. |
| | The management team thinks the distance between iceberg and FPSO has been enough large. |

**Step 6: Identifying safety constraints.**

The concept of "safety constraint" in STPA has the similar meaning with the terms of "risk reduction measure" or "safeguard". However, it is typically expressed as a design or operational requirement. Within the STPA framework, safety barriers are considered a specialized form of safety constraints. The primary goal of this step is to identify constraints that can effectively prevent UCAs, or at least, reduce their likelihood.

The process of identifying safety constraints can provide insights about the underlying causes of UCAs. It involves an assessment of both the existing constraints that are already in place and potential new constraints that could serve as improvement measures. These additional constraints may need to be implemented based on the findings of the STPA study to enhance the overall safety and integrity of the system.

### Example 4.6:

#### *Safety Constraints in Ice Management*

For the case to avoid collision of FPSO with an iceberg, the existing safety constraints can be identified finally, as listed in Table 4.5. ■

**TABLE 4.5**
**Some Existing Safety Constraints in Ice Management**

| UCA | Existing Safety Constraints |
|---|---|
| UCA-1 | Water cannons and relevant equipment must be tested and inspected regularly. |
| | Ice monitoring system must be verified and updated promptly. |
| | Alarms must be highly noticeable. |

*(Continued)*

**TABLE 4.5** *(Continued)*
**Some Existing Safety Constraints in Ice Management**

| UCA | Existing Safety Constraints |
|---|---|
| UCA-2 | Safety training is mandatory for all FPSO crews. |
| | IMVs must be equipped with an adequate amount of evacuation equipment. |
| UCA-3 | Specific training and examination are mandatory for relevant crews before they are onboard. |
| | Alarms must be highly noticeable. |
| | Short-term and mid-term weather forecasts are necessary. |
| | Water cannons and relevant equipment must be tested and inspected regularly. |
| UCA-4 | Specific training and examination are mandatory for relevant crews. |

As illustrated in Example 4.6, the current safety constraints may not always be sufficient to prevent UCAs. For instance, even though the ice management team has received specialized trainings, their lack of experience in real emergency situations could lead to errors. It is therefore essential to evaluate whether the existing safety constraints adequately maintain an acceptable level of risk for UCAs. If found lacking, it is advisable to implement additional measures that strengthen the safety constraints. For example, including a few experienced members in the management team can help minimize errors committed by less experienced personnel.

Compared to traditional hazard identification methods like FMECA, STPA introduces a novel approach by focusing on the control structure and advocating for systematic thinking. STPA uniquely considers human factors and the critical role of human operators in ensuring system safety, thereby identifying a wider array of safety issues linked to human interactions within a system or process. Additionally, STPA considers the dynamic behavior of systems, which involves time-dependent and feedback-driven processes, facilitating the identification of potential hazards and failures that arise from dynamic interactions.

While STPA presents these significant advantages, it is important to recognize that it is still a relatively novel methodology without a widely recognized international standard. Research integrating STPA with quantitative analysis is very limited. Furthermore, applying STPA can be time-intensive and demands a high level of expertise for effective execution. STPA is not intended to completely replace traditional methods like HAZOP or FMECA, as each technique has its unique strengths. Instead, integrating STPA with these methods can provide a more thorough and robust approach to system safety and barrier analysis.

## NOTE

1. HAZOP Manager is a trademark of Lihoutech Technical & Software Services.

## REFERENCES

Crawley, F., & Brian, T. (2015). *HAZOP: Guide to best Practice*. Elsevier.
DOE, U. S. (1996). *Hazard and Barrier Analysis Guidance Document*.

Duijm, N. J. (2009). Safety-barrier diagrams as a safety management tool. *Reliability Engineering & System Safety, 94*(2), 332–341.
Duijm, N. J., & Markert, F. (2009). Safety-barrier diagrams as a tool for modelling safety of hydrogen applications. *International Journal of Hydrogen Energy, 34*(14), 5862–5868.
Goble, W. M. (1992). *Evaluating Control Systems Reliability-Techniques and Applications.* ISA.
Goble, W. M., & Brombacher, A. C. (1999). Using a failure modes, effects and diagnostic analysis (FMEDA) to measure diagnostic coverage in programmable electronic systems. *Reliability Engineering & System Safety, 66*(2), 145–148.
ICMA. (2006). *The Energy Trace and Barrier Analysis.* International Crisis Management Association.
IEC61025. (2006). *Fault tree analysis (FTA).* International Electrotechnical Commission.
IEC61882. (2016). *Hazard and operability studies (HAZOP studies) – Application guide.* International Electrotechnical Commission.
Kingston, J., Koornneef, F., Ruit, J., Frei, v. d., & Schallier, P. (2009). *NRI MORT User's Manual: For Use with the Management Oversight & Risk Tree Analytical Logic Diagram (Second Edition).*
Leveson, N. (2004). A new accident model for engineering safer systems. *Safety Science, 42*(4), 237–270.
Leveson, N. (2012). *Engineering a Safer World: Systems Thinking Applied to Safety.* The MIT Press.
Leveson, N., & Thomas, J. P. (2018). *STPA Handbook.* Massachusetts Institute of Technology.
Lewis, C. L., & Haug, H. (2009). *The System Safety Handbook.*
MIL-STD-1629A. (1980). *Procedures for performing a failure mode, effect and criticality analysis.* Washington, DC: U.S. Department of Defense.
MIL-STD-882E. (2012). *Standard practice for system safety.* Washington, DC: U.S. Department of Defense.
Nesse, C. V. (2019). *Ice Management with System Theoretic Process Analysis (STPA) for Marine Operations in Arctic Environments.* Norwegian University of Science and Technology.
Rausand, M., Barros, A., & Hoyland, A. (2020). *System Reliability Theory: Models, Statistical Methods, and Applications* (3rd ed.). Wiley-Blackwell.
Rausand, M., & Haugen, S. (2020). *Risk Assessment: Theory, Methods and Applications.* Wiley.
Xie, L. (2022). *Safety Barriers in Complex Systems Dependent Failures: Modeling and Assessment Approaches.* Norwegian University of Science and Technology.

# 5 Probabilistic Modeling Approaches for Barrier Engineering

## 5.1 INTRODUCTION

The qualitative methods discussed in Chapter 4 provide a comprehensive understanding on the functions of barriers. These methods are highly effective in illustrating how barriers can contribute to asset protection and risk management. However, to confirm the performance of a barrier component or system with more robust evidence, qualitative methods alone is insufficient. Therefore, quantitative assessment, which can offer more detailed information, becomes necessary essential after the qualitative analysis.

Quantitative models in barrier engineering can be generally divided into two categories, each distinguished by its approach to understanding failures, as well as required information required and modeling resolution:

- *Physical models or physics-based models*: These models rely on fundamental principles of physics and engineering to explain barrier failures through specific physical phenomena or particularly physical processes. Such kind of models often require detailed information about design specifications, material characteristics, geometry, and operating conditions. They are particularly useful when a thorough understanding of the physical behavior and failure mechanisms of barriers is crucial for precise analysis and informed decision-making. For example, the following two models are useful in barrier analysis:
  - *Deterioration model*, which views failure as a function of a specific load. For example, when the load is increasing, a failure is more likely to occur.
  - *Mechanic model*, which regards that failure occurs when the working load exceeds the strength of the component.
- *Actuarial models or statistic-based models*: These models employ a statistical perspective to analyze barrier failures and regard failures as probabilistic events. Relying on historical data and failure statistics, the actuarial models can reflect the statistical patterns (e.g., probability distributions) and trends of barrier failures. They do not need much detailed information about the physical attributes of a barrier, and thus, they are more cost-effective when the detailed physical behavior is has little influence on the analysis outcomes.

In this book, most of the modeling methods, such as reliability block diagrams, fault tree analysis, and Markov models, belong to the actuarial models. The main reason is that for performance analysis of complex systems with numerous components,


DOI: 10.1201/9781003245636-5

it is often impractical or unnecessary to investigate the physical details of each component. Additionally, the outputs of actuarial models, such as time-to-failure or failure frequency, facilitate communication across diverse disciplinary backgrounds.

It is also noteworthy that in some situations, a hybrid approach combining both physical and actuarial models can provide a more holistic understanding of barrier performance. For example, integrating probability with a mechanical model can be useful when considering that failure may occur with a higher likelihood under an increasing load.

In the subsequent sections of this chapter, we start from introducing probability theories and distributions utilized in barrier engineering. Then, various modeling approaches and their quantitative analysis methods for system failure/reliability assessment are presented. These approaches include reliability block diagrams, fault trees, Bayesian networks, Markov models, and Petri nets. After that, we introduce the event tree model for consequence analysis and the Monte Carlo simulation for validating models for system analysis.

## 5.2 PROBABILITY THEORIES FOR BARRIER ENGINEERING

The introduction of probability theory in this section is very brief. It only selects several aspects that are crucial for understanding reliability, availability, and safety analysis in the remainders of this book. As a complimentary, we provide references for most methods discussed to enable further exploration. For readers interested in learning more about probability theory and its applications in reliability engineering, we recommend the textbooks by Rausand et al. (2020).

### 5.2.1 General Probability Metrics

#### 5.2.1.1 Time-to-Failure

The state of an item at time $t$ may be described by *state variable*:

$$X(t) = \begin{cases} 1 & \text{if the item is the functioning state at time } t \\ 0 & \text{if the item is in the failed state at time } t \end{cases} \tag{5.1}$$

The *time-to-failure* ($T$) of an item is defined as the duration from the initiation of operation ($t = 0$) to its failure. The relation between the time-to-failure and the state variable is shown in Figure 5.1. In this book, we use $T$ to denote a variable, while $t$ to represent the observed value of time. For a safety barrier, the starting point of $t = 0$ also can be the moment that a proof test is completed, assuming the barrier is restored to an "as-good-as-new" state.

The concept of "time" can be quantified here in several ways:

- Calendar time, for example, 8,760 hours, 365 days, or 1 year since an item was bought,
- Operational time, for example, 1,760 working hours in 1 year (meaning 8 hours per workday),

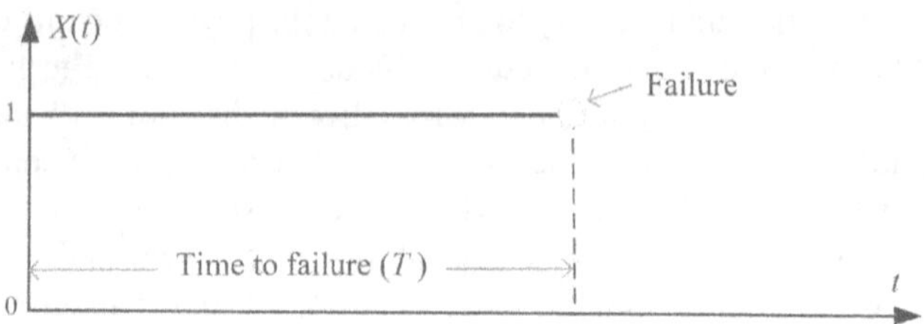


**FIGURE 5.1** State variable and time-to-failure.

- Other indirect time concepts, for example, the number of kilometers driven by a car, the number of uses of a camera shutter, or the number of rotations of a bearing.

For the indirect time concepts, these measures are often discrete variables but also can be approximated using continuous variables to facilitate analysis. In this book, we assume that time-to-failure as a continuous variable, governed by a *probability distribution function* $F(t)$, and a *probability density function* (PDF), $f(t)$. The relationship of the two functions can be described as:

$$F(t) = \Pr(T \leq t) = \int_0^t f(u)du \text{ for } t > 0 \tag{5.2}$$

The transition of an item from the functioning state to the failed state is regarded as a *failure event* in this book. We always assume that the duration of an event is zero or ignorable. When a failure event occurs by time $t$, it means that $T \leq t$ is true, and thus, $F(t)$ is the failure probability of the item in the interval of $(0, t]$.

Based on Equation (5.2), we can know that $f(t)$ is the derivative of $F(t)$ as:

$$f(t) = \frac{d}{dt}F(t) = \lim_{\Delta t \to 0} \frac{F(t+\Delta t) - F(t)}{\Delta t} = \lim_{\Delta t \to 0} \frac{\Pr(t < T \leq t + \Delta t)}{\Delta t} \tag{5.3}$$

Thus, when $\Delta t \to 0$, we have

$$\Pr\left(t < T \leq t + \Delta t\right) \approx f(t)\Delta t \tag{5.4}$$

This means that when examining a very small interval $(t, t+\Delta t]$ from the starting point $t = 0$, the failure probability of the item within this period is equal to the product of its failure density and the interval duration. Additionally, $F(t)$ is a probability measure with a value in the range of $[0,1]$ for $t \geq 0$, and $F(+\infty) = \int_0^{+\infty} f(t)dt = 1$, implying that a failure must occur if the timeframe extends indefinitely.

The probability that a failure occurs in an interval $(t_1, t_2]$ is:

$$\Pr(t_1 < T \leq t_2) = F(t_2) - F(t_1) = \int_{t_1}^{t_2} f(u)du \tag{5.5}$$

Consider a simple example in which the PDF of an item is given by $f(t) = 0.01t$, and then we can calculate its probability distribution function as $F(t) = \int_0^t 0.01u du = 0.005t^2$. Assuming the time unit is hours, the failure probability of this item in the interval $***(1,10]$ is 0.495, meaning that there is a 49.5% chance the item will fail between 1 hour and 10 hours. Then, by 14.142 hours, the item will almost fail because $F(14.142) \approx 1$.

Probability density function is simplified as PDF. To avoid confusion with this term, probability distribution function is also called as *cumulative distribution function* in some literature, or CDF in short.

#### 5.2.1.2 Survivor Function and Reliability

The *survivor function* of an item is defined by:

$$R(t) = 1 - F(t) = \Pr(T > t) = 1 - \int_0^t f(u)du = \int_t^\infty f(u)du \tag{5.6}$$

$R(t)$ represents the probability that the item does not fail within the time interval $(0, t]$ and remains functioning at time $t$. Thus, $R(t)$ is also called as the *survival probability.* If we regard *reliability* as the probability of an item still functioning by a specific time, $R(t)$ can be considered as its *reliability function.* We will further explore the definition of reliability and its relevant concepts in Chapter 6.

For the example provided earlier, where PDF of an item is $f(t) = 0.01t$, $R(10)$ can be easily calculated as 0.5, indicating that there is a 50% chance the item will still be functioning after it has been in operation for 10 hours. In other words, its reliability by 10 hours is 0.5.

#### 5.2.1.3 Failure Rate Function

Now we consider that an item has survived by time $t$, and then the probability that it fails in the interval $(t, t + \Delta t]$ is a conditional probability:

$$\Pr(t < T \le t + \Delta t | T > t) = \frac{\Pr(t < T \le t + \Delta t)}{\Pr(T > t)} = \frac{F(t + \Delta t) - F(t)}{R(t)} \tag{5.7}$$

When this time interval is very small, namely $\Delta t \to 0$, we can introduce the *failure rate function* $z(t)$ as

$$\lambda(t) = \lim_{\Delta t \to 0} \frac{\Pr(t < T \le t + \Delta t | T > t)}{\Delta t} = \lim_{\Delta t \to 0} \frac{F(t + \Delta t) - F(t)}{\Delta t} \cdot \frac{1}{R(t)} = \frac{f(t)}{R(t)} \tag{5.8}$$

We also can see:

$$\Pr(t < T \le t + \Delta t | T > t) \approx \lambda(t)\Delta t \tag{5.9}$$

If we stand at time $t = 0$ and ask, "What is the probability that a new item will fail in the interval $(t, t + \Delta t]$?", the answer is approximately as $f(t)\Delta t$. In this case, the estimation is *unconditional*, meaning that we do not consider what have happened before $t$. In fact, the item may have failed or not. On the other hand, if we stand at time $t$, knowing that the item is still functioning and ask, "What is the probability that this survivor item will fail in the interval $(t,t+\Delta t]$?", and the answer is approximately as $z(t)\Delta t$. This estimation is *conditional*, based on the fact that the item has not failed in the interval $(0,t]$. Thus, it can be found that failure density and failure rate represent two distinct concepts.

Then, for failure rate, we can have:

$$\lambda(t) = \frac{f(t)}{R(t)} = -\frac{R'(t)}{R(t)} = -\frac{d}{dt} logR(t) \tag{5.10}$$

Namely

$$\log R(t) = -\int_0^t \lambda(u)\,du \tag{5.11}$$

or

$$R(t) = e^{-\int_0^t \lambda(u)du} \tag{5.12}$$

Sometimes, it is meaningful to calculate the average failure rate in an interval $(t_1,t_2]$:

$$\bar{\lambda}(t_1,t_2) = \frac{1}{t_2 - t_1}\int_{t_1}^{t_2} \lambda(u)\,du$$

#### 5.2.1.4 Mean-Time-to-Failure

Suppose that we have a group of items ($n$ in total), and we can observe the time-to-failure ($t_i$) of any of these items. It is natural to calculate the average value ($\bar{t}$) of all the observed times-to-failures as:

$$\bar{t} = \frac{\sum_{i=1}^n t_i}{n}$$

In statistics, we have the *law of large numbers*, where given that $n$ is approaching to infinite, the empirical mean, $\bar{t}$, will be stable around a constant value that does not depend on $n$ anymore. We can call such a value the *expected value*, or *mean value* of variable. Given that the variable is time-to-failure, $T$, this mean value is *mean time-to-failure* (MTTF). Thus, MTTF is the expected value $E(T)$ of variable $T$ (Rausand et al., 2020):

$$MTTF = E(T) = \lim_{n\to+\infty} \frac{\sum_{i=1}^n t_i}{n} \tag{5.13}$$

When $n$ is approaching to infinite, each possible failure time $t$ is multiplied by its probability of occurrence $f(t)dt$, and the summation can be replaced by an integral:

$$\text{MTTF} = E(T) = \int_0^{\infty} tf(t)\,dt = \int_0^{\infty} tR'(t)\,dt \tag{5.14}$$

By taking partial integration, we have:

$$\text{MTTF} = -\left[tR(t)\right]_0^{\infty} + \int_0^{\infty} R(t)\,dt$$

Since MTTF is always a finite number, it can be shown that $\left[tR(t)\right]_0^{+\infty} = 0$, then:

$$\text{MTTF} = \int_0^{\infty} R(t)\,dt \tag{5.15}$$

In the formulas above, we regard a failure as an event, and MTTF actually represents the average time to the occurrence of this event. Similarly, if we consider repair or maintenance as a process, take the starting point of this process as $t = 0$, and regard the completeness of repair as the event of interest, it is also possible to calculate the *mean-time-to-repair* (MTTR).

In some cases, the focus is on the *mean downtime* (MDT), which starts from the moment when the item fails and ends when it is restored to be functioning. In the calculation of MDT, the *restoration rate function* (always denoted as $\mu(t)$ in this book) is relevant. It is always challenging to accurately estimate such a rate, but even an approximate value can be useful given that some historical maintenance data is available.

#### 5.2.1.5 Mean Residual Lifetime

Suppose an item that is put into operation at time $t = 0$, and it is functioning at time $x$. Then, the residual lifetime of this item considering its random failure time $T$ is $T - x$, and the *mean residual lifetime* (MRL) is the expected time from $x$ to $T$ with the condition $T > x$:

$$\text{MRL}(x) = E\left(T - x \middle| T > x\right) \tag{5.16}$$

MRL($x$) can be regarded as the MTTF of an item that is put into operation at $t = x$ given the condition that it is functioning at the moment, thus:

$$\text{MRL}(x) = \int_x^{\infty} R(t|x)\,dt = \frac{1}{R(x)}\int_x^{+\infty} R(t)\,dt \tag{5.17}$$

Here, we do not consider any history of the item before $x$. The *Remaining useful lifetime* (RUL) is a similar concept that we will also use in the subsequent sections of this book, but the difference is that RUL($x$) is always specific to a particular item for which performance and maintenance data is available in the period (0, $x$) (Rausand et al., 2020).

#### 5.2.1.6 Variance and Standard Deviation

*Variance* is a measure of the dispersion of values around their mean. In an experiment involving $n$ items, if we can observe their lifetimes, the empirical variance of these lifetimes is given by:

$$s^2 = \frac{1}{n-1}\sum_{i=1}^{n}(t_i - \overline{t})^2 \tag{5.18}$$

where $\overline{t}$ is the mean value of all the observed samples. In this context, we have a finite number ($n$) of observed values as discrete samples, although the lifetime of each individual item is a continuous variable. The standard deviation, which is the square root of the variance, is defined as:

$$s = \sqrt{\frac{1}{n-1}\sum_{i=1}^{n}(t_i - \overline{t})^2} \tag{5.19}$$

When $n$ approaches infinity, the empirical variance converges to a constant, known as the variance of the lifetime ($T$) of these items, which can be calculated as:

$$var(T) = \int_0^{\infty}[t - E(T)]^2 f(t)dt = E(T^2) - [E(T)]^2 \tag{5.20}$$

Then, the associated standard deviation is:

$$SD(T) = \sqrt{\text{var}(T)} \tag{5.21}$$

Both variance and standard deviation are extensively used in data analysis. Within the field of barrier engineering, they can provide insights for diverse performance measures.

### 5.2.2 Continuous Probabilistic Distributions

Probabilistic distributions are the statistic patterns of variables or observed data. Many probabilistic distributions have been proposed and widely applied in reliability and safety engineering. In general, there are two types of distributions: *continuous probability distribution* and *discrete probability distribution*. The previous one applies to continuous variables that can take on any value within a range, while the latter one applies to discrete variables that take on a countable number of distinct values.

In these two subsections, we will only introduce several continuous and discrete distributions that are needed in the rest parts of this book.

#### 5.2.2.1 Exponential Distribution

Firstly, we consider the simplest and most widely used continuous distribution, where an item is put into operation at time $t = 0$, and its time-to-failure $T$ has the probability density function as:

$$f(t) = \begin{cases} \lambda e^{-\lambda t} & \lambda > 0, t > 0 \\ & \text{otherwise} \end{cases} \tag{5.22}$$

Such a distribution is called the *exponential distribution* with parameter λ, and it can be written as $T \sim \exp(\lambda)$.

The probability distribution function $F(t)$ with the exponential distribution is:

$$F(t) = \Pr(T \leq t) = \int_0^t \lambda e^{-\lambda u}\, du = 1 - e^{-\lambda t} \text{ for } t > 0 \tag{5.23}$$

And its survivor function is thus:

$$R(t) = 1 - F(t) = e^{-\lambda t} \text{ for } t > 0 \tag{5.24}$$

The failure rate function can be calculated as:

$$\lambda(t) = \frac{f(t)}{R(t)} = \frac{\lambda e^{-\lambda t}}{e^{-\lambda t}} = \lambda \tag{5.25}$$

This result is significant for reliability analysis, indicating that an item characterized by an exponential distribution of time-to-failure has a constant failure rate, independent of time. Conversely, we also can conclude that an item with a constant failure rate follows an exponential distribution for its time-to-failure.

Regarding the MTTF of the item with exponential distribution:

$$\text{MTTF} = \int_0^{+\infty} R(t)\,dt = \int_0^{+\infty} e^{-\lambda t}\, dt = \frac{1}{\lambda} \tag{5.26}$$

It indicates that for an item with an exponential distribution, its MTTF is the inverse of the failure rate. Then, the probability that an item survives until the MTTF can be calculated as:

$$R(\text{MTTF}) = R\left(\frac{1}{\lambda}\right) = e^{-\lambda \cdot \frac{1}{\lambda}} = e^{-1} \approx 0.3679$$

This probability remains constant for all values of failure rates.

Moreover, the conditional probability that an item will continue to be functioning for an additional time $t$ given that it is functioning at time $x$ is:

$$R(t + x|x) = \frac{\Pr(T > t + x)}{\Pr(T > x)} = \frac{e^{-\lambda(t+x)}}{e^{-\lambda x}} = e^{-\lambda t} = R(t) \tag{5.27}$$

Equation (5.27) shows that the survivor probability of an item over a period of $t$, whether new or used, remains unchanged. In other words, the reliability of an item with the exponential distribution remains unchanged over any time interval. When calculating the MRL of an item under these conditions, we can find:

$$\text{MRL}(x) = \int_{x}^{+\infty} R(t|x)\,dt = \int_{0}^{+\infty} R(t)\,dt = \text{MTTF} \tag{5.28}$$

The MRL of such an item is equal to its MTTF, regardless its current age, $x$. This attribute implies that the item can always be considered in an "*as-good-as-new*" (AGAN) state, and it is also referred to as the "memoryless" property of the exponential distribution.

For many electronic components and those complex systems involving a variety of hardware and software elements, it is reasonable to assume that their failure times follow an exponential distribution. Many reliability databases, such as OREDA (2009) and MIL-HDBK-217F (1991), adopt this assumption to present data in a format of a constant failure rate.

#### 5.2.2.2 Gamma Distribution

The time-to-failure $T$ of an item is considered to follow a *Gamma distribution* when its PDF is described by:

$$f(t) = \frac{\lambda}{\Gamma(\alpha)}(\lambda t)^{\alpha-1} e^{-\lambda t} \text{ for } t > 0 \tag{5.29}$$

where $\Gamma(\cdot)$ is called as the *Gamma function*. The Gamma distribution is written as $T \sim Gamma(\alpha, \lambda)$, indicating that it has two parameters, *shape parameter* $\alpha$ and *inverse scale parameter* $\lambda$ ($\alpha > 0$ and $\lambda > 0$). Sometimes, $\lambda$ is called as *rate parameter*, and thus *scale parameter* $\theta = 1/\lambda$.

In this book, the Gamma distribution is utilized to model various stages of degradation or partial failures of an item. For a positive real number $\alpha$, the Gamma function is derived by Swiss mathematician and physicist Daniel Bernoulli as:

$$\Gamma(\alpha) = \int_{0}^{\infty} x^{\alpha-1} e^{-x}\,dx \tag{5.30}$$

If $\alpha$ is a positive integer, $\Gamma(\alpha) = (\alpha - 1)!$, and in such situations, the Gamma distribution is also called as an *Erlang distribution*. The CDF of the Gamma distribution is:

$$F(t) = \frac{1}{\Gamma(\alpha)}\gamma(\alpha, \lambda t) \text{ for } t > 0 \tag{5.31}$$

where $\gamma(\alpha, \lambda t)$ is called as the *lower incomplete Gamma function*, defined as

$$\gamma(\alpha, \lambda t) = \int_{0}^{\lambda t} u^{\alpha-1} e^{-u}\,du \tag{5.32}$$

Then we can obtain the failure rate function $\lambda(t)$ based on Equation (5.8).

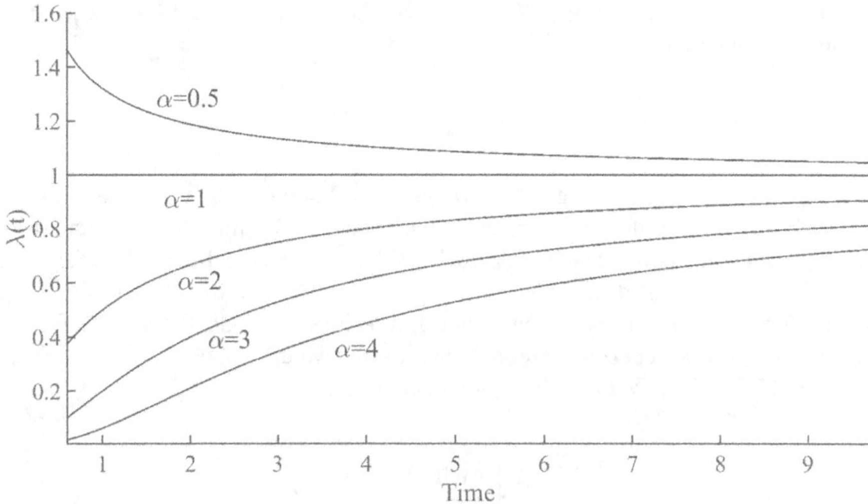


**FIGURE 5.2** Failure rate functions of Gamma distributions with selected values of $\alpha$, when $\lambda = 1$.

Figure 5.2 displays the failure rate functions of the Gamma distribution for various values of the shape parameter $\alpha$ ($\alpha = 0.5$, 1, 2, 3, and 4), when $\lambda$ is set as 1. It can be found that when $\alpha = 1$, the failure rate is a constant. This means that under such a condition, the Gamma distribution simplifies to an exponential distribution. In other words, the exponential distribution is a specific instance of the Gamma distribution where $\alpha = 1$. When $\alpha > 1$, the failure rate of an item increases with time, reflecting that the item is more likely to fail as it ages. When $\alpha < 1$, the failure rate decreases with time, indicating that the item becomes less likely to fail as time progresses. When we consider a degradation process of a mechanical component, it can be assumed that the failure rate of this item increases, and so the Gamma distribution with $\alpha > 1$ is applicable for this situation.

#### 5.2.2.3 Weibull Distribution

The time-to-failure $T$ of an item is a two-parameter *Weibull distributed* if its PDF is described by:

$$f(t) = \begin{cases} \alpha\lambda^{\alpha}t^{\alpha-1}e^{-(\lambda t)^{\alpha}} \text{ for } t > 0 \\ 0 \text{ otherwise} \end{cases} \tag{5.33}$$

Such a distribution has a shape parameter $\alpha$ ($\alpha > 0$) and a rate parameter $\lambda$ ($\lambda > 0$), and it can be written as $T \sim Weibull(\alpha, \lambda)$, with the CDF as:

$$F(t) = \begin{cases} 1 - e^{-(\lambda t)^{\alpha}} \text{ for } t > 0 \\ 0 \text{ otherwise} \end{cases} \tag{5.34}$$

It can be found that when $\alpha = 1$, the Weibull distribution becomes to an exponential distribution.

The survivor function of the Weibull distribution is $R(t)=e^{-(\lambda t)^{\alpha}}$ for $t>0$, and its failure rate function is:

$$\lambda(t)=\frac{f(t)}{R(t)}=\frac{\alpha\lambda^{\alpha}t^{\alpha-1}e^{-(\lambda t)^{\alpha}}}{e^{-(\lambda t)^{\alpha}}}=\alpha\lambda^{\alpha}t^{\alpha-1} \text{ for } t>0 \tag{5.35}$$

When $\alpha=1$, the failure rate of a Weibull distribution is equal to $\lambda$, a constant value. When $\alpha>1$, failure rate is increasing. When $\alpha<1$, failure rate is decreasing function of time, as illustrated in Figure 5.3. When comparing these failure function figures of the Weibull distribution and those of the Gamma distribution shown in Figure 5.2, it is evident that despite some similarities in trends, the profiles for the same value of $\alpha$ (except $\alpha=1$) are different for the two distributions.

The MTTF of the Weibull distribution is:

$$\text{MTTF}=\int_{0}^{+\infty}R(t)\,dt=\frac{1}{\lambda}\Gamma\left(1+\frac{1}{\alpha}\right) \tag{5.36}$$

The Weibull distribution was originally developed by Swedish engineer and mathematician Waloddi Weibull (1887–1979). This distribution is very flexible, and by setting parameters, it can model different types of failure behaviors. The Weibull distribution can also be extended to include a third parameter, a threshold parameter $\xi$, leading to the following probability distribution function:

$$F(t)=1-e^{-[\lambda(t-\xi)]^{\alpha}} \text{ for } t>0 \tag{5.37}$$

In this book, we only use the two-parameter model.

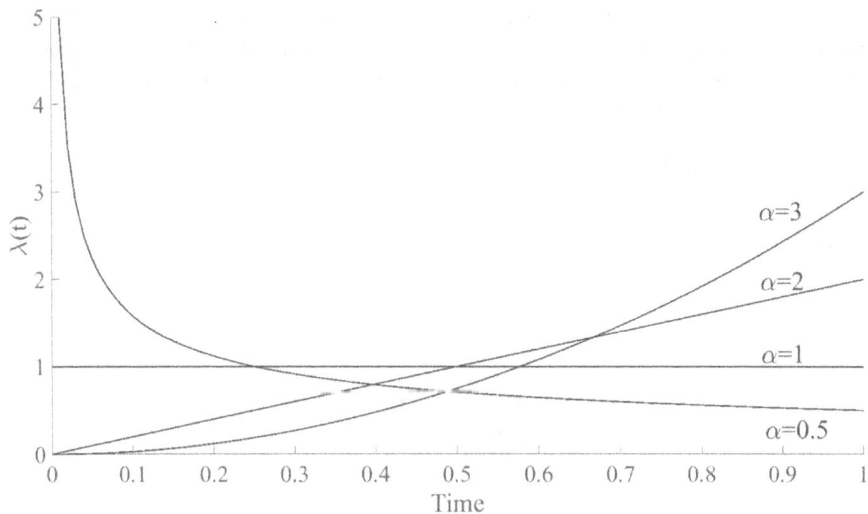


**FIGURE 5.3** Failure rate functions of the Weibull distribution with selected values of $\alpha$, when $\lambda=1$.

#### 5.2.2.4 Other Continuous Distributions

In this subsection, we will introduce three types of probabilistic distributions that describe continuous variables. Although these distributions are not commonly used for modeling time-to-failure, they are valuable for other analytical tasks related to safety barriers.

##### 5.2.2.4.1 *Uniform Distribution*

Consider a random variable, $X$, which may represent a quantity like time, diameter, or length. This variable follows a *uniform distribution* across an interval $[a,b]$ if its PDF is given by:

$$f(x)=\begin{cases}\frac{1}{b-a} \text{ for } a<x<b \\ 0 \text{ otherwise}\end{cases} \tag{5.38}$$

The uniform distribution can be denoted as $X$~unif$(a,b)$, and its expected value of is:

$$E(X)=\int_a^b xf(x)dx=\frac{a+b}{2} \tag{5.39}$$

This means that the occurrence of an event is evenly distributed in the interval. In the analysis of occurrences of hidden failures on safety barriers in Chapter 6, we will use this distribution.

##### 5.2.2.4.2 *Normal Distribution*

The *normal distribution*, also called as *Gaussian distribution* (named after the German mathematician Johann Carl Friedrich Gauss (1777–1855)), is often regarded as the most widely applied distribution in statistics and everyday applications. A random variable $X$ is normally distributed, given that it has the following PDF:

$$f(x)=\frac{1}{\sqrt{2\pi}\sigma}e^{-\frac{(x-\mu)^2}{2\sigma^2}} \text{ for } -\infty<x<+\infty \tag{5.40}$$

The normal distribution, denoted as $X \sim \mathcal{N}(\mu,\sigma)$, has two parameters: mean value $\mu$ and standard deviation $\sigma$. The CDF of $X \sim \mathcal{N}(\mu,\sigma)$ can be written as

$$F(x)=\Pr(X\le x)=\Phi\left(\frac{x-\mu}{\sigma}\right) \text{ for } -\infty<x<+\infty \tag{5.41}$$

The $\mathcal{N}(0,1)$ distribution is called as the *standard normal distribution*, and its distribution function is denoted by $\Phi(\cdot)$. The PDF of the standard normal distribution is:

$$f(x)=\frac{1}{\sqrt{2\pi}}e^{-\frac{x^2}{2}} \text{ for } -\infty<x<+\infty \tag{5.42}$$

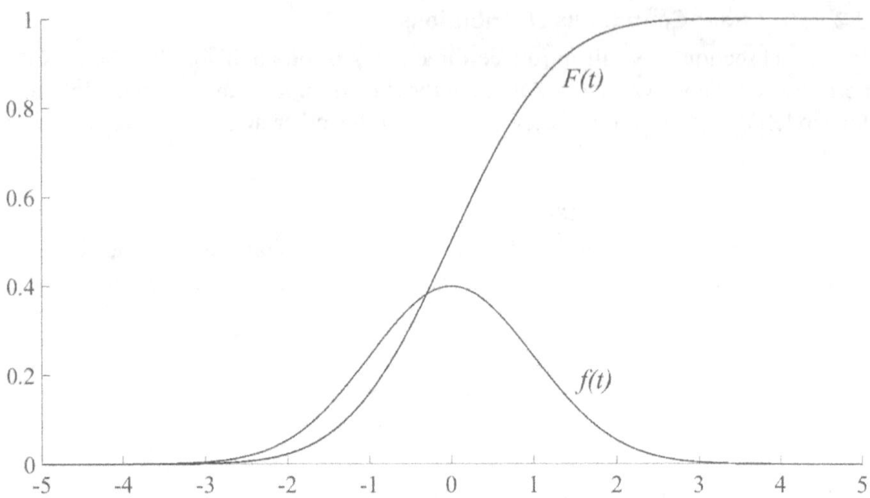


**FIGURE 5.4** Probability distribution function and probability density function of the standard normal distribution.

Figure 5.4 displays the CDF and PDF of the standard normal distribution, $\mathcal{N}(0,1)$, and the failure rate of the normal distribution for $x > 0$ is illustrated in Figure 5.5.

The normal distribution is utilized extensively in engineering, particularly in statistical process control. Although it is uncommon for the time-to-failure of a technical system or component to follow a normal distribution, it can be useful for studying the failures due to wear-out and fatigue in reliability analysis. More importantly, the normal distribution underpins various statistical methods employed in data-driven analysis.

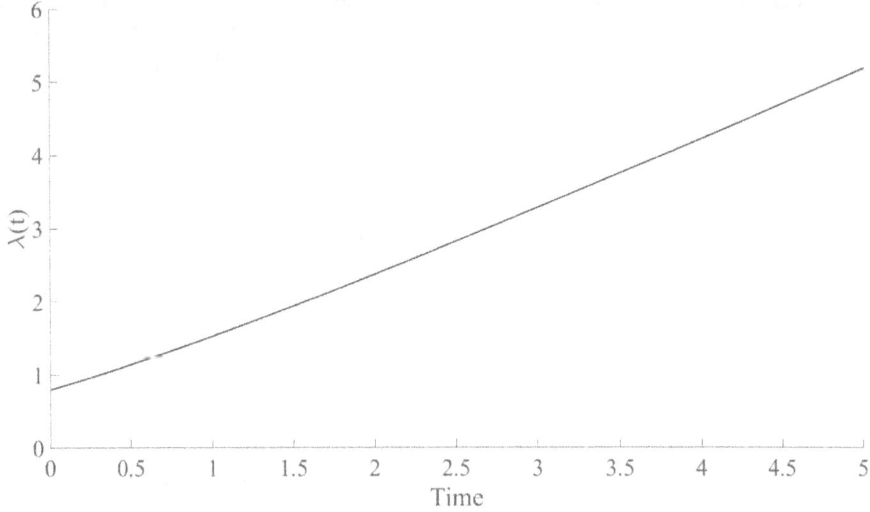


**FIGURE 5.5** Failure rate function of the standard normal distribution.

#### *5.2.2.4.3 Lognormal Distribution*

A random variable $X$ is *lognormally distributed* with two parameters, $\mu$ and $\sigma$, denoted as $X$~lognorm($\mu,\sigma$) when $Y = \ln X$ is normally distributed with the mean $\mu$ and the standard deviation $\sigma$ [$Y \sim \mathcal{N}(\mu,\sigma)$]. $X$~lognorm($\mu,\sigma$) has the following PDF:

$$f(x) = \begin{cases} \frac{1}{\sqrt{2\pi}\sigma x} e^{-\frac{(\ln x-\mu)^2}{2\sigma^2}} & \text{for } x > 0 \\ 0 \text{ othersiwe} \end{cases} \tag{5.43}$$

This distribution is introduced since it can be used to model the distribution of repair time, and the expected value of $X$ in this case is the *mean time to repair* (MTTR), which can be calculated as MTTR $= e^{\mu+\frac{\sigma^2}{2}}$.

A repair rate function $\nu(t)$ is defined here, analogous to the failure rate function $\lambda(t)$. We use $t$ to replace $x$ for the lognormal distribution, and the repair rate can be given by

$$\nu(t) = -\frac{d}{dt}\left[\log\Phi\left(\frac{\mu - \ln t}{\sigma}\right)\right] = \frac{\phi\left[\frac{\mu-\ln t}{\sigma}\right]}{\Phi\left[\frac{\mu-\ln t}{\sigma}\right]\sigma t} \tag{5.44}$$

where $\phi[\cdot]$ is the probability density function of the standard normal distribution.

Figure 5.6 illustrates the profile of the repair rate function, $\nu(t)$, with parameters $\mu = 5$ and $\sigma = 0.2$. In this example, the MTTR is calculated to be 151.41, and the time unit can be hour. The figure demonstrates that the repair rate initially increases, implying the rising probability of completing the repair as time progresses. However, some factors such as a lack of knowledge or absence of spare parts can significantly

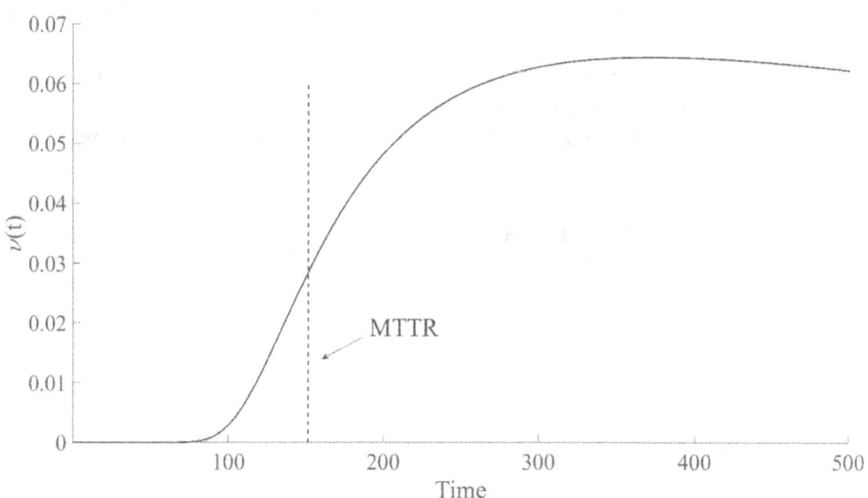


**FIGURE 5.6** Repair rate function of the lognormal distribution with $\mu = 5$ and $\sigma = 0.2$.

extend the repair time, potentially causing the repair rate to decrease after a certain period. More discussions on the lognormal distribution can be found in the report by Almog (1979).

### 5.2.3 Discrete Probabilistic Distributions

In discrete probabilistic distributions, random variable(s) are supposed to have a finite or countable number of distinct values. In this section, we will introduce a simple discrete distribution and relevant stochastic processes that are particularly useful for the analyses in the remainder of this book.

For discrete distributions, we utilize the probability mass function (PMF) rather than the PDF. The PMF describes the probability that a discrete random variable is exactly equal to a specific value.

Consider a discrete random variable $X$ whose possible values are enumerated as $\{x_1, x_2, \ldots, x_n\}$. The probability that the variable $X$ takes the value $x_i$ is denoted by $\Pr(X = x_i)$:

1. $\Pr(X = x_i) \geq 0$ for all $i = 1, 2, \ldots$
2. $\sum_{i=1}^{n} \Pr(X = x_i) = 1$

Then, the CDF of the variable $X$ is given by:

$$F(x) = \Pr(X \leq x) = \sum_{x_i \leq x} \Pr(X = x_i) \tag{5.45}$$

#### 5.2.3.1 Binominal Distribution

A random experiment can yield a countable set of results, or outcomes. Then, the situation follows the *binominal distribution* if it satisfies the following three criteria:

1. The experiment consists of $n$ independent trials, known as *Bernoulli trials.*
2. Each trial has two possible outcomes: $o$ and $o^*$.
3. The probability of obtaining the outcome $o$, denoted as $\Pr(o) = p$, is constant across all trials.

Let $X$ represent the number of trials that have outcome $o$, and $X$ is a discrete random variable described by the PMF:

$$\Pr(X = x) = \binom{n}{x} p^x (1-p)^{n-x} \text{ for } x = 0, 1, 2, \ldots n \tag{5.46}$$

where $\binom{n}{x}$ is called as the binominal coefficient:

$$\binom{n}{x} = \frac{n!}{x!(n-x)!}$$

The binominal distribution is denoted as $X \sim \text{bin}(n, p)$. It is useful to calculate the expected mean and variance $X \sim \text{bin}(n, p)$ as the follows:

$$E(X) = np \tag{5.47}$$

$$\text{var}(X) = np(1-p) \tag{5.48}$$

There are some other discrete distributions, such as the geometric distribution and the negative binomial distribution, that also apply in the binomial scenarios. More details for these distributions can be found in Rausand et al. (2020).

#### 5.2.3.2 Homogeneous Poisson Process

The *Homogeneous Poisson process* (HPP) is named after the French mathematician Siméon Denis Poisson (1781–1840), and it serves as a stochastic process for tracking the occurrences of a specific type of event $E$ within a given time interval. Such an event can be a system failure, a deviation in a process, the observation of a particular outcome, or the completion of a task.

When the following conditions can be satisfied, a stochastic process can be regarded as an HPP:

1. Event $E$ can occur at any time within the interval, and the probability of $E$ occurring in $(t, t+\Delta t]$ is independent of time $t$.
2. The probability of more than one event occurring in $(t, t+\Delta t]$ is $o(\Delta t)$, approaching to zero when $\Delta t$ is very short, namely $\lim_{\Delta t \to 0} \frac{o(\Delta t)}{\Delta t} = 0$.
3. Given that $(t_{11}, t_{12}]$, $(t_{11}, t_{12}]$, … are a sequence of disjoint intervals, the events "$E$ occurs in $(t_{j1}, t_{j2}]$" are independent for all $j$ values, $j = 1, 2, \ldots$.

Such a process starts at $t = 0$, and $N(t)$ denotes the number of occurrences of event $E$ within the time interval $(0, t]$. Then, the probability that $E$ occurs exactly $n$ times within the interval $(0, t]$ is given by:

$$\Pr[N(t) = n] = \frac{(\lambda t)^n}{n!} e^{-\lambda t} \quad \text{for } n = 0, 1, 2, \ldots \tag{5.49}$$

where $\lambda$ represents the occurrence rate of event $E$, and that the stochastic process $\{N(t), t > 0\}$ is an HPP with the rate of $\lambda$. The distribution of events within a time interval is called as the *Poisson distribution*, denoted as $N(t) \sim \text{Poisson}(\lambda t)$. It should be noted that the Poisson distribution, as a discrete distribution, has the random variable of $N(t)$, which is a count of events, rather than time-to-failure.

Based on the first assumption, the occurrence rate of events, $\lambda$, remains constant regardless of when the observation of the process begins. For example, if we start observing at time $x$, the probability that $E$ occurs exactly $n$ times in the interval $(x, x+t]$ is given by:

$$\Pr\left[N(x+t) - N((x) = n\right] = \frac{(\lambda t)^n}{n!} e^{-\lambda t} \quad \text{for } n = 0, 1, 2, \ldots \tag{5.50}$$

Comparing Equations (5.49) and (5.50), it can be found that the number of events only depends on the length $t$ of the observed interval, independent from when the interval starts.

Based on assumption (2), the mean number of events in $(0,t]$ can be derived using the Maclaurin series expansions of the exponential function:

$$
\begin{aligned}
E[N(t)] &= \sum_{n=0}^{+\infty} n\Pr[N(t)=n] = 0\cdot\Pr[N(t)=n] \\
&\quad +1\cdot\Pr[N(t)=1]+2\cdot\Pr[N(t)=2]+\cdots \\
&= e^{-\lambda t}\left[\frac{\lambda t}{1!}+2\cdot\frac{(\lambda t)^2}{2!}+\cdots+n\cdot\frac{(\lambda t)^n}{n!}\right] \\
&= e^{-\lambda t}\lambda t\left[1+\frac{\lambda t}{1!}+\frac{(\lambda t)^2}{2!}+\cdots+\frac{(\lambda t)^n}{n!}\right] \\
&= e^{-\lambda t}\cdot\lambda t\cdot e^{\lambda t} = \lambda t
\end{aligned}
\tag{5.51}
$$

Equation (5.51) implies that $\lambda$ can be calculated as $\lambda = E[N(t)]/t$, as the expected number of events per time unit. When the event $E$ is a component failure, $\lambda$ can be called as the *rate of occurrences of failures* (ROCOF) of the failure process of multiple components. It is noted that this concept is different from failure rate. As defined by OREDA (2009), ROCOF measures the mean number of failures per time unit and is specifically applicable to repairable items.

Let $T_1$ represent the time at which the event of interest (e.g., a failure) firstly occurs. The probability distribution function of this event can be expressed as:

$$
F(t) = \Pr(T_1 \le t) = 1-\Pr(T_1 > t) = 1-\Pr[N(t)=0] = 1-e^{-\lambda t} \text{ for } t>0 \tag{5.52}
$$

This formulation means that the time $T_1$ to the first event follows an exponential distribution with the parameter $\lambda$. In other words, if such events occur in a time interval following a HPP, the time to the first event is exponentially distributed. If we set the time from the first event to the second event as $T_2$, and the time from the $(i-1)$th event to the $i$th event is $T_i$, it also can be found that the times between the events $T_1$, $T_2$,... $T_i$,... are independent and exponentially distributed with the rate of $\lambda$, and thus $T_1$, $T_2$,... $T_i$,... can be called as the *interoccurrence times* of the process.

For a repairable item, if the repair time is negligible compared to its time-to-failure, the lifespan of this item can be modeled by an HPP, assuming a constant failure rate. In this book, we also apply the HPP to model the demands on safety barriers (see Chapter 8) since the external events or shocks also can be considered reoccurrences with a certain frequency.

#### 5.2.3.3 Renewal Process

In a more general way, a stochastic process $\{X(t),\ t \in \Theta\}$ can be regarded as a collection of random variables, indexed by $t$ (time). Here, $X(t)$ is referred to the state of this process at time $t$, and $\Theta$ is the set of indices. If $\Theta$ is countable, the process is a *discrete-time stochastic process*. Otherwise, if $\Theta$ is continuous and non-countable, the process is a *continuous-time stochastic process*. HPP belongs to the latter type. In this subsection, we will introduce another type of process used in barrier engineering, the *renewable process*, which can be regarded as a generalization of the HPP.

In this book, a renewal process is understood as a stochastic process where the interoccurrence times are independent and identically distributed with an arbitrary distribution. If we consider the event $E$ as a failure, the failed item will be replaced or restored to an "as-good-as-new" state, namely *renewed*.

We can let $t = 0$ as the starting point of such a process and let $N(t)$ as the number of occurrences of event within the time interval $(0, t]$, with the independent and identically distributed interoccurrence times $T_1, T_2, \ldots$. The CDF of a renewal process is:

$$F_T(t) = Pr(T_i \le t) \text{ } for \text{ } t > 0, i = 1, 2, \ldots, n \tag{5.53}$$

This means the probability that the $i$th event (failure) has occurred by time $t$.

In a renewal process, the distribution of times between events does not need to follow any specific distribution and can be modeled based on empirical data or theoretical needs. This brings flexibility to this model and allows it to be used in the cases where the exponential model does not suit the characteristics of the data.

The renewal function, $M(t)$, represents the expected number of renewals (or events) that have occurred by time $t$. It is given by:

$$M(t) = \sum_{n=1}^{\infty} F_T^{(n)}(t) \tag{5.54}$$

where $F_T^{(n)}(t)$ is the $n$-fold convolution of $F_T(t)$ with itself, as the probability that the $n$th renewal has occurred by time $t$.

The renewal process provides a framework for modeling situations where events recur independently, and the system is restored or renewed after each event. For example, in barrier engineering, the renewal process can be used to model the operation of a barrier system with periodic tests and replacements.

## 5.3 RELIABILITY BLOCK DIAGRAM

Based on the probability theory and probability distributions discussed above, from this section, we will introduce several methods for failure frequency analysis and quantitative system analysis to understand and estimate the performance of barrier systems.

### 5.3.1 Basic Elements of Reliability Block Diagram

One of the most important models in both quantitative system analysis and system reliability assessment, is the *reliability block diagram* (RBD). As illustrated in Figure 5.7, an RBD is a schematic representation that consists of two basic elements: functional blocks (represented by rectangles) and connections (represented by lines) between these blocks. RBD is success-oriented, meaning that if a block is functioning (indicated as "up"), the lines on either side of the block are connected,

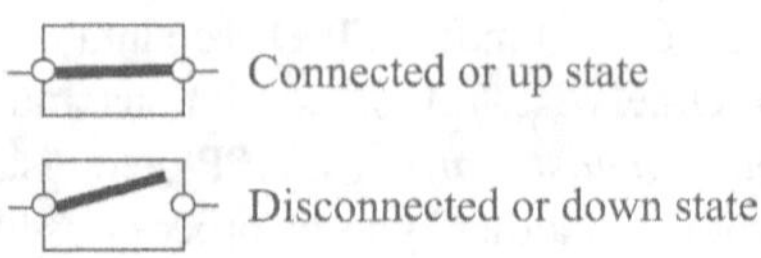


**FIGURE 5.7** Connection mechanism in a function block.

as depicted in the upper part of Figure 5.7. Otherwise, if a block fails (indicated as "down"), the lines on both sides of the block become disconnected, as shown in the lower part of Figure 5.7.

If at least one path exists that successfully connects the left end (or starting point, as denoted by point *a* in Figure 5.8) to the right end (or ending point, as denoted by point *b* in Figure 5.8) of the whole RBD, the system modeled by the model is considered as functioning.

It is important to note that an RBD describes how a specified system function is achieved through the integration of subfunctions. It represents the logical connections necessary for function, rather than the physical linkages between hardware components. Therefore, the same functional block may appear in several places within an RBD. However, in many practical applications, because one function is often fulfilled by one component (a one-to-one mapping), it is common to regard a block as a component, and the RBD as a representation of a system. Thus, the terminology is not used very strictly for RBD-based analysis, but it is crucial to recognize that an RBD is not the physical layout of a system.

For more details and terminology related to RBD and its analysis methods, readers are encouraged to check the latest version of IEC61078 (2016), which is an international standard for the method.

### 5.3.2 Qualitatively Structural Analysis

An RBD can exhibit two basic structures: parallel structure (part of $c_1$ and $c_2$ in Figure 5.8) and series structure (part of $c_3$ and $c_4$ in Figure 5.8). For a parallel structure, it is functioning if at least one of its $n$ components are functioning, because any connection in those parallel channels can maintain the linkage between the two ends of the structure. For a series structure, it is functioning if and only if all of its $n$ components are functioning, because any disconnection within a block can break the linkage between the two ends of the structure.

In barrier engineering, the model shown in Figure 5.8 can represent a barrier system equipped with two identical sensors positioned in different locations, a

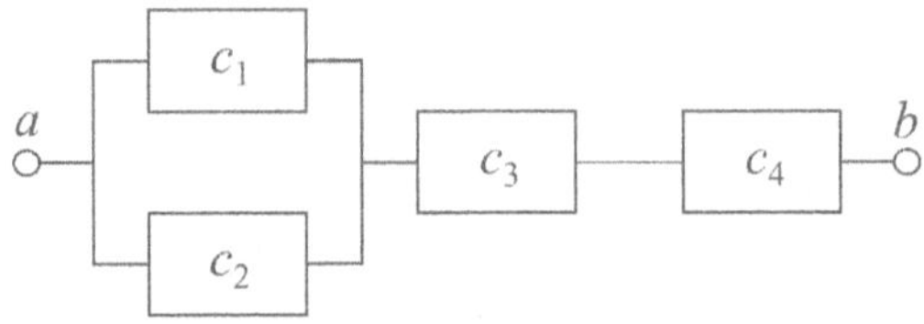


**FIGURE 5.8** A simple reliability block diagram.

programmable logic controller (PLC) as the analyzer, and a shutdown valve as the actuator. In this scenario, if either sensor detects a deviation, the valve can be activated to perform a barrier function.

The analysis of an RBD model typically begins with the identification of successful paths that can connect the two endpoints. These connected blocks, which ensure system functionality, are referred to as a *tie set* in IEC61078 (2016), or a *path set.* A *minimal tie set* is defined as a tie set that cannot be reduced any further without compromising its functionality. In Figure 5.8, we can find two tie sets $\{c_1\ c_3\ c_4\}$ and $\{c_2\ c_3\ c_4\}$. These are considered minimal because the failure of any single block in a set would disrupt the connection between points $a$ and $b$.

### 5.3.3 Structure Function and Probabilistic Analysis

The state of block (function) $i$ at time $t$ can be regarded as a random variable $X_i(t)$ with two possible values: 0 (disconnected or down) and 1 (connected or up). When the block is used to model the operation of a component, 0 is failed state, while 1 is the functioning state. For a system with $n$ components, its state vector can be written as $\boldsymbol{X}_S(t)$, and we can use a *structure function* to link $X_i(t)$ and $\boldsymbol{X}_S(t)$:

$$\boldsymbol{X}_S(t) = \varphi\left[X_1(t), X_2(t), \ldots, X_n(t)\right] \tag{5.55}$$

For a series structure with $n$ components, the structure function is:

$$\boldsymbol{X}_S(t) = \prod_{i=1}^{n} X_i(t) \tag{5.56}$$

This function shows that if any $X_i(t) = 0$ (i.e., any component fails), $\boldsymbol{X}_S(t)$ will be equal to 0. Only when all components are functioning $X_i(t) = 1$ for all $i$, $\boldsymbol{X}_S(t)$ can be equal to 1.

For a parallel structure with $n$ components, the logic becomes as $\boldsymbol{X}_S(t)$ is equal to 0 only when all components fail; while if at least one component is functioning, $\boldsymbol{X}_S(t)$ is equal to 1. Thus, the structure function for a parallel structure is:

$$\boldsymbol{X}_S(t) = 1 - \prod_{i=1}^{n} \left[1 - X_i(t)\right] \tag{5.57}$$

This implies that a system also has two states, and thus, a basic RBD is regarded as a *Boolean model.* In this textbook, our interest lies in determining the probability that a component or a system is functioning, expressed as:

$$\Pr\left[X_i(t) = 1\right] = R_i(t) \text{ for } i = 1, 2, \ldots, n \tag{5.58}$$

$$\Pr\left[\boldsymbol{X}_S(t) = 1\right] = R_S(t) \tag{5.59}$$

Here $R_i(t)$ is the reliability of component $i$ by time $t$, and $R_S(t)$ is the reliability of the system.

Since $X_i(t)$ for $i = 1,2,\ldots,n$ is a binary variable, its expected value can be easily calculated as:

$$E[X_i(t)] = 1\cdot \Pr[X_i(t)=1] + 0\cdot \Pr[X_i(t)=1] = \Pr[X_i(t)=1] = R_i(t) \quad (5.60)$$

Similarly, $E[X_S(t)] = R_S(t)$. When all components in a series structure are independent, the reliability function can be derived from the structure function as follows:

$$R_S(t) = E[X_S(t)] = E\left(\prod_{i=1}^{n} X_i(t)\right) = \prod_{i=1}^{n} E[X_i(t)] = \prod_{i=1}^{n} R_i(t) \quad (5.61)$$

Given that each $R_i(t) \le 1$, it follows that $R_S(t) \le \min[R_i(t)]$, indicating that the reliability of a series structure is constrained by its least reliable component.

For a parallel structure, its reliability function is derived as:

$$\begin{aligned} R_S(t) &= E[X_S(t)] = E\left(1 - \prod_{i=1}^{n}[1 - X_i(t)]\right) \\ &= 1 - \prod_{i=1}^{n}[1 - E[X_i(t)]] = 1 - \prod_{i=1}^{n}[1 - R_i(t)] \end{aligned} \quad (5.62)$$

Given that each $R_i(t) \le 1$, we can find that $R_S(t) \ge \max[R_i(t)]$, suggesting that the reliability of a parallel structure is at least as high as that of its most reliable component.

**Example 5.1:**

***Reliability Calculation of the Structure in Figure 5.8***

Consider the system depicted in Figure 5.8. Assume the reliability of each component by a specific time $t$ is as follows: $R_1 = 0.80$, $R_2 = 0.85$, $R_3 = 0.90$, and $R_4 = 0.95$. Utilizing Equations (5.61) and (5.62), we can calculate the system reliability as:

$$R_S = [1-(1-R_1)(1-R_2)]\cdot R_3 \cdot R_4 = [1-(1-0.80)(1-0.85)]\cdot 0.90 \cdot 0.95 = 0.82935$$

Based on the calculation result, the reliability of the parallel section of components $c_1$ and $c_2$ is 0.97, illustrating a significant improvement compared to the individual reliability of each of the two components. ■

A commonly used structure in barrier engineering is called as the *k*-out-of-*n* (*koon*) voting structure, which is functioning when at least *k* components out of the total *n* components are functioning. In many literatures, such a structure is called as *koon*: G structure, and G means "Good", which distinguishes itself from the *koon*: F structure. The latter fails when at least *k* components out of the *n* components fail, and F means "Fail". In the subsequent chapters, we omit G in expression, only

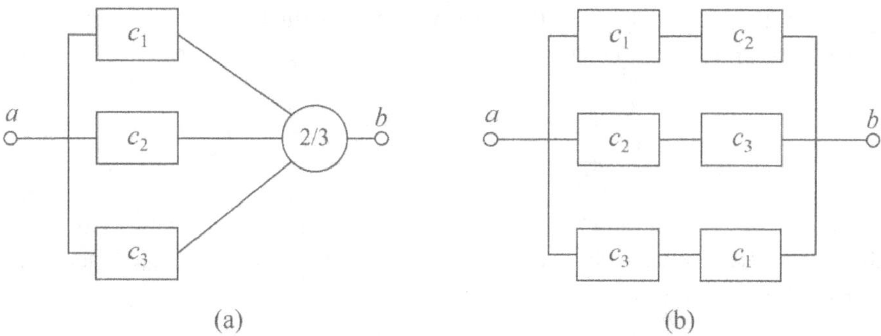


**FIGURE 5.9** Physical diagram (a) and RBD model (b) for a 2oo3 vote structure.

describe as *koon*. This is because we do not use the term of *koon*: F, but regarding *koon*: F equivalent to the $(n-k)$*oon*: G structure. This is called as a voting structure since the functionality of the system depends on the vote among the functioning and failed components.

Figure 5.9(a and b) illustrate the simplified RBD model (or physical diagram) for a 2oo3 voted structure and its real logic connection, respectively. The notation "2/3" in Figure 5.9(a) symbolizes the voting mechanism. Any of the three combinations $\{c_1\ c_2\}$, $\{c_1\ c_3\}$, and $\{c_2\ c_3\}$ can form a tie set. In other words, any one of three channels in Figure 5.9(b) is a path from end $a$ to end $b$, the entire system is functioning.

Voting structures are frequently utilized in barrier systems, particularly for the sensing subsystems. For example, consider gas detectors monitoring the concentration of a specific flammable gas. Installing only two detectors can lead to confusion when they provide conflicting signals, making it challenging to determine which one reflects the situation accurately and whether it is needed to activate the barrier function. When the third detector is introduced, given that the likelihood of two detectors failing simultaneously is significantly lower than a single failure, it is generally reasonable to trust the two agreeing readings, even if the other one is different.

Structure function of a *koon* voting structure is:

$$X_S(t) = \begin{cases} 1 \text{ for } \sum_{i=1}^{n} X_i(t) \geq k \\ 0 \text{ for } \sum_{i=1}^{n} X_i(t) < k \end{cases} \tag{5.63}$$

Suppose that all the $n$ components in the structure are identical, with the reliability $R_i(t) = R(t)$ for $i = 1,2,\ldots,n$, and failures of individual components are independent, the probability that exactly $k$ components functioning at time $t$ follows a binomial distribution:

$$Pr\left[\sum_{i=1}^{n} X_i(t) = k\right] = \binom{n}{k} R(t)^k \left[1 - R(t)\right]^{n-k} \tag{5.64}$$

The system is reliable in all the cases where the number of functioning components is no less than $k$, so that:

$$R_S(t) = \Pr\left[\sum_{i=1}^{n} X_i(t) \geq k\right] = \sum_{i=k}^{n} \binom{n}{i} R(t)^i \left[1 - R(t)\right]^{n-i} \tag{5.65}$$

For the 2oo3 structure in Figure 5.9, it can be regarded as a parallel structure with three channels, and each channel is a series of structure of two components. Thus, its structure function is:

$$\begin{aligned} X_S(t) &= 1 - \left[1 - X_1(t)X_2(t)\right]\left[1 - X_2(t)X_3(t)\right]\left[1 - X_3(t)X_1(t)\right] \\ &= X_1(t)X_2(t) + X_2(t)X_3(t) + X_3(t)X_1(t) - 2X_1(t)X_2(t)X_3(t) \end{aligned} \tag{5.66}$$

Using Equation (5.60), reliability of this 2oo3 structure is calculated as:

$$R_S(t) = R_1(t)R_2(t) + R_2(t)R_3(t) + R_3(t)R_1(t) - 2R_1(t)R_2(t)R_3(t) \tag{5.67}$$

According to Equation (5.64), we can obtain the same result when $n = 3$ and $k = 2$. If these three components are identical with the reliability $R_i(t) = R(t)$ for $i = 1,2,\ldots,n$, the reliability $R_S(t)$ can be calculated as:

$$R_S(t) = 3R(t)^2 - 2R(t)^3$$

For more comprehensive insights into RBD models and their quantitative analyses, the latest edition of IEC61078 (2016) is a valuable resource. Additionally, most reliability analysis software current in the market incorporates the RBD modeling and analysis functions, facilitating reliability assessments.

### 5.3.4 Importance Analysis

In this subsection, we explore *importance analysis* based on the RBD model. The objective of such an analysis is to identify the critical components that significantly influence the performance of a system, including aspects such as reliability, availability for specific functions, and overall safety. At first, we only consider the influence on reliability, and thus, the measures in the subsection are called *reliability importance measures*. The importance or criticality of a component typically depends on its inherent performance and its logical position within the system. We will introduce and discuss several reliability importance measures using RBD model and the structure function to provide a deeper understanding.

#### 5.3.4.1 Birnbaum's Measure

The Hungarian-American professor Z. W. Birnbaum (1969) proposed one of the earliest measures of reliability importance, defined as follows:

$$I^B(i|t) = \frac{\partial R_S(t)}{\partial R_i(t)} \quad \text{for } i = 1,2,\ldots,n \tag{5.68}$$

Here, $I^{\mathrm{B}}(i|t)$ is called as the *Birnbaum's measure* (B), which calculates the partial derivative of the system reliability with respect to the reliability of component $i$. This measure actually aligns with the principles of classical sensitivity analysis by assessing how the change in reliability of an individual component affects the overall system reliability.

In practice, when determining the importance of a component at a specific time $t$, the analysis can be simplified by omitting the explicit time dependency, considering $R_i(t) = R_i$ and then referring to the measures as $R_S$ and $I^{\mathrm{B}}(i)$, respectively.

For a series structure, such as the structure consisting of components $c_3$ and $c_4$ in Figure 5.8, Birnbaum's measure for component $i$ can be calculated by regarding the reliability of this component as the variable, while keeping the reliability of other components as constants. Therefore, we have:

$$I^{\mathrm{B}}(i) = \frac{\partial R_S}{\partial R_i} = \frac{\partial \prod_{i=1}^{n} R_i}{\partial R_i} = \frac{\partial \prod_{i=1}^{n} R_i}{\partial R_i} = \frac{\partial \left[ R_i \prod_{j \neq i} R_j \right]}{\partial R_i} = \prod_{j \neq i} R_j \tag{5.69}$$

This calculation shows that the sensitivity of the system reliability to changes in a component reliability within a series structure is directly related to the product of the reliabilities of the other components.

Similarly, for a parallel structure, such as the structure consisting of components $c_1$ and $c_2$ in Figure 5.8, the Birnbaum's measure of component $i$ can be calculated as

$$I^{\mathrm{B}}(i) = \frac{\partial R_S}{\partial R_i} = \frac{\partial \left[ 1 - \prod_{i=1}^{n} (1 - R_i) \right]}{\partial R_i} = \prod_{j \neq i} (1 - R_j) \tag{5.70}$$

In this scenario, the sensitivity of the system reliability to changes in a component reliability within a parallel structure is related to the probability that all the other components fail.

## Example 5.2:

### *Birnbaum's Importance of Components in Series and Parallel Structures*

Continuing with the data from Example 5.1, we first calculate the importance of components 1 and 2 within the parallel subsystem shown in Figure 5.8. The Birnbaum's measures $I^{\mathrm{B}}(1)$ and $I^{\mathrm{B}}(2)$ are calculated to be 0.15 and 0.2, respectively. For the series structure consisting of components $c_3$ and $c_4$, the importance of these two components to the subsystem is 0.90 and 0.95, respectively. ■

It can be found through Equations (5.69) and (5.70), as well as the findings from Example 5.2, we observe that in a series structure, the Birnbaum's measure indicates that the less reliable component is more critical to the system's overall reliability. Conversely, in a parallel structure, components with higher reliability are regarded more important.

Since we have assumed that each component only has two possible states: functioning or failed, system reliability can be expressed in the following way:

$$R_S(t) = R_i(t) \cdot R_S\left(t \middle| X_i(t) = 1\right) + \left(1 - R_i(t)\right) \cdot R_S\left(t \middle| X_i(t) = 0\right) \tag{5.71}$$

where $R_S(t|X_i(t) = 1)$ denotes the system reliability given that the component $i$ is functioning by time $t$. To calculate the Birnbaum's measure using the derivative of $R_i(t)$ in Equation (5.71), we have:

$$I^{\mathrm{B}}(i|t) = \frac{\partial R_S(t)}{\partial R_i(t)} = R_S\left(t \middle| X_i(t) = 1\right) - R_S\left(t \middle| X_i(t) = 0\right) \tag{5.72}$$

This equation implies that the Birnbaum's measure evaluates the impact of a component on the system reliability by comparing the system reliability when the component is functioning versus when it has failed.

### Example 5.3:

### *Birnbaum's Importance of Components in a Larger System*

Consider a larger system with more components, as illustrated in Figure 5.10. This system has a 2oo3 voting structure involving components 1–3, and a parallel structure with components 4–5. Assuming that all components have the same reliability of 0.90, we can evaluate the importance of components 1 and 5 within this system using the Birnbaum's measure.

According to Equations (5.68)–(5.70), the Birnbaum's measures of the two components can be calculated as follows:

$$I^{\mathrm{B}}(1) = \frac{\partial R_S}{\partial R_1} = \frac{\partial[(0.9 + 0.9)R_1 + 0.9 \cdot 0.9 - 2 \cdot 0.9 \cdot 0.9 R_1][1 - (1 - 0.9)(1 - 0.9)]}{\partial R_1} = 0.1782$$

$$I^{\mathrm{B}}(5) = \frac{\partial R_S}{\partial R_5} = \frac{\partial\left[3 \cdot 0.9^2 - 2 \cdot 0.9^3\right][1 - (1 - 0.9)(1 - R_5)]}{\partial R_5} = 0.0972$$

It is clear that the importance of components 2 and 3 is same with that of component 1, and similarly, the importance of components 4 and 5 is equal. Components within the 2oo3 voting structure are regarded more critical according

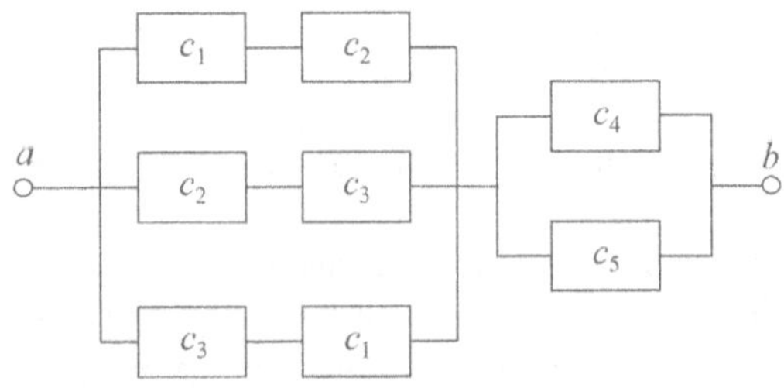


**FIGURE 5.10** A RBD example for a system with five components.

to the Birnbaum's measure. This is because a failure of any one component significantly impacts the system, reducing the potential paths within the structure from three to one, which is only 33% of the original paths. In contrast, within the parallel structure of two channels, a failure of component 4 or 5 reduces the potential paths from two to one, or 50% of the original paths.

The system illustrated in Figure 5.10 can be seen as a 2oo3 sensing subsystem and a 1oo2 activation subsystem within a barrier system. However, when the availability of a barrier system is critical in risk control, the calculation in importance analysis of the barrier components might differ. We will explore importance analysis of barrier components in Chapter 8. ■

In the above calculation, it can be found that the Birnbaum's measure of a component is only dependent on its location in the system structure and reliabilities of other components. $I^{B}(i|t)$ is actually independent from the reliability the component itself. Thus, with Birnbaum's measure, we do not provide any suggestions on the reliability design of components.

#### 5.3.4.2 Improvement Potential

The *improvement potential* (IP) is another importance measure. It is defined as the difference between the system reliability assuming component $i$ is perfect (always functioning by time $t$) and the actual system reliability with the current state of component $i$. Mathematically, this measure can be calculated as:

$$I^{\mathrm{IP}}(i|t) = R_S(t|X_i(t)=1) - R_S(t) \text{ for } i = 1,2,\ldots,n \tag{5.73}$$

The improvement potential measure indicates how much the system reliability can potentially increase if component $i$ were replaced by a perfect component that is always functioning.

Figure 5.11 illustrates the relationship between $I^{B}(i|t)$ and $I^{\mathrm{IP}}(i|t)$. In practice, when time $t$ is specified, we can simplify the notation to $I^{\mathrm{IP}}(i)$. In addition, it is unrealistic to expect that the reliability of a component can be improved to 1 or perfect,

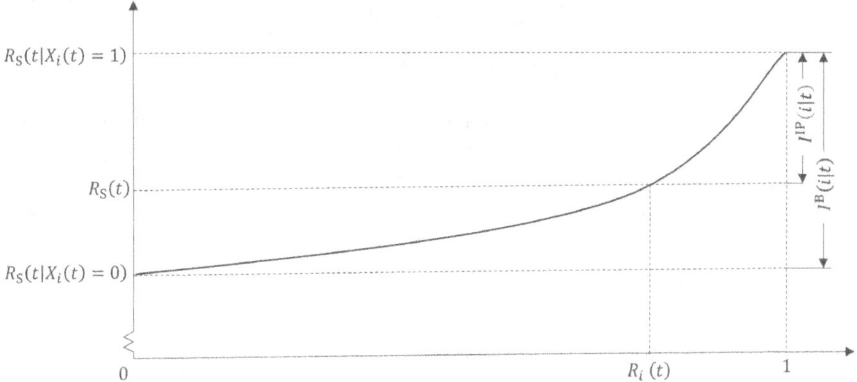


**FIGURE 5.11** Relationship between Birnbaum measure and improvement potential.

and therefore, a more realistic measure, called *credible improvement potential* (CIP), can be introduced as:

$$I^{\text{CIP}}\left(i|t\right) = R_S\left(t|R_i(t) = R_i^*\right) - R_S(t) \text{ for } i = 1,2,\ldots,n \tag{5.74}$$

where $R_i^*$ is the reliability target for component $i$.

**Example 5.4:**

***Improvement Potential of Components of Components in Series and Parallel Structures***

Continuing with the data in Example 5.1, we can calculate the improvement potential of components 1 and 2 within the parallel subsystem in Figure 5.8:

$$I^{\text{IP}}(1) = \left[1-(1-1)(1-0.85)\right] - \left[1-(1-0.80)(1-0.85)\right] = 0.03$$

$$I^{\text{IP}}(2) = \left[1-(1-0.80)(1-1)\right] - \left[1-(1-0.80)(1-0.85)\right] = 0.03$$

These calculation results indicate that both components are same important within the parallel structure. Since it is impractical to increase the reliability of a component to perfect, by setting a more realistic reliability target of 0.9 for these two components, we have:

$$I^{\text{CIP}}(1) = \left[1-(1-0.90)(1-0.85)\right] - \left[1-(1-0.80)(1-0.85)\right] = 0.015$$

$$I^{\text{CIP}}(2) = \left[1-(1-0.80)(1-0.90)\right] - \left[1-(1-0.80)(1-0.85)\right] = 0.010$$

When using the CIP measure, component 1 is regarded more important since it facilitates greater increase in system reliability when it is improved to the target, compared to component 2.

For the series structure consisting of components $c_3$ and $c_4$, the improvement importance is calculated as:

$$I^{\text{IP}}(3) = 1 \cdot 0.95 - 0.90 \cdot 0.95 = 0.095$$

$$I^{\text{IP}}(3) = 0.90 \cdot 1 - 0.90 \cdot 0.95 = 0.045$$

Based on the calculation results, the relatively unreliable component 3 is regarded more important in the series structure. When we use a more realistic target, e.g., 0.99 of reliability to the two components, the conclusion is same: The less reliable components in a series structure is more important for the overall system reliability. ■

#### 5.3.4.3 Risk Achievement Worth

The *Risk achievement worth* (RAW) of component $i$ by time $t$, as an importance measure, is defined as:

$$I^{\text{RAW}}\left(i|t\right) = \frac{1 - R_S\left(t|X_i(t) = 0\right)}{1 - R_S(t)} \tag{5.75}$$

In this equation, the denominator represents the unreliability or risk of the current system with component $i$ (functioning or failed with a probability), and the numerator denotes the unreliability or risk of the system with a definitely failed component $i$. The ratio $I^{\text{RAW}}(i|t)$ thus indicates the value of maintaining component $i$ at its current reliability level.

**Example 5.5:**

***Risk Achievement Worth of Components in Series and Parallel Structures***

Continuing using the data in Example 5.1, we calculate the RAWs of components 1 and 2 in the parallel structure as shown in Figure 5.8:

$$I^{\text{RAW}}(1) = \frac{1-[1-(1-0)(1-0.85)]}{1-[1-(1-0.80)(1-0.85)]} = 5$$

$$I^{\text{RAW}}(2) = \frac{1-[1-(1-0.80)(1-0)]}{1-[1-(1-0.80)(1-0.85)]} = 6.667$$

From the perspective of RAW, the more reliable component in a parallel structure is more important. This insight can guide priorities in maintenance and reliability enhancement strategies.

For the series structure containing components $c_3$ and $c_4$, the RAW calculations are:

$$I^{\text{RAW}}(3) = \frac{1-0\cdot 0.95}{1-0.90\cdot 0.95} = 6.897$$

$$I^{\text{RAW}}(4) = \frac{1-0.90\cdot 0}{1-0.90\cdot 0.95} = 6.897$$

It shows that both components are equally important in terms of RAW for the series structure, and any failure of these components will affect the reliability of the overall structure similarly. ■

#### 5.3.4.4 Risk Reduction Worth

The *Risk reduction worth* (RRW) of component $i$ by time $t$, as an importance measure, is defined as:

$$I^{\text{RRW}}(i|t) = \frac{1-R_S(t)}{1-R_S\left(t|X_i(t)=1\right)} \tag{5.76}$$

In this equation, the denominator represents the unreliability or risk of a system with a perfect component $i$ (always functioning), and the numerator denotes the risk of the current system with the actual component $i$ (functioning or failed with a probability). The ratio $I^{\text{RRW}}(i|t)$ thus represents how much ratio of risk can be reduced if component $i$ is replaced by a perfect one.

**Example 5.6:**

***Risk Achievement Worth of Components in Series and Parallel Structures***

Continuing using the data from Example 5.1, the RRWs of components 1 and 2 in the parallel structure shown in Figure 5.8 can be calculated as:

$$I^{RRW}(1) = \frac{1-[1-(1-0.80)(1-0.85)]}{1-[1-(1-1)(1-0.85)]} = \infty$$

$$I^{RRW}(2) = \frac{1-[1-(1-0.80)(1-0.85)]}{1-[1-(1-0.80)(1-1)]} = \infty$$

Such results indicate that according to RRW, both components in the parallel structure are same important or extremely important. If one of them becomes perfect, the structure will never fail.

For the series structure containing components $c_3$ and $c_4$, the RRW calculations are:

$$I^{RRW}(3) = \frac{1-0.90\cdot 0.95}{1-1\cdot 0.95} = 2.9$$

$$I^{RRW}(4) = \frac{1-0.90\cdot 0.95}{1-0.90\cdot 1} = 1.45$$

It shows that the less reliable component in the series structure is more important according to RRW. This suggests that efforts to improve reliability should prioritize the weaker components that have a higher potential to reduce the overall risk of the structure more significantly. ■

Due to the page limitation, we discuss only four basic importance measures in this section. However, there are numerous other importance measures used in system reliability analysis, including criticality importance and the Fussell-Vesely measure. For a more comprehensive understanding of these measures, readers can refer to the reviews by Kuo and Zhu (2012), Amrutkar and Kamalja (2017), and Si et al. (2020). Furthermore, applications of the above importance measures in barrier engineering will be explored in Chapter 8.

## 5.4 QUANTITATIVE FAULT TREE ANALYSIS

### 5.4.1 Comparison of Fault Tree and Reliability Block Diagram

In Section 4.7, we have explored the concept and visual representation of a fault tree, including the procedure of fault tree analysis (FTA) and some associated qualitative analysis methods. Here, we will shift our focus to quantitative analysis, specifically to Steps 4 and 5 in the FTA procedure as discussed.

First, we can compare fault tree with the RBD method introduced in the last section. Consider a system represented by an RBD model in Figure 5.8, where the

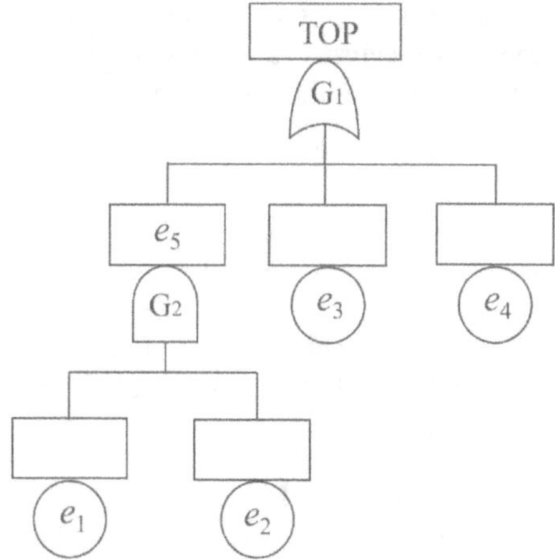


**FIGURE 5.12** A simple fault tree model with four basic events.

system failure can be regarded as the pivotal or TOP event of a fault tree. According to the RBD, system failure can be caused by the failure of either component $c_3$ or $c_4$, or a simultaneous failure of both components $c_1$ and $c_2$. Here, we can use $e_i$ to denote the event that component $i$ fails, and then based on this information, we can construct a fault tree model as shown in Figure 5.12.

When comparing Figure 5.12 with Figure 5.8, it can be found that the AND-gate in the fault tree has a similar role with a parallel structure in the RBD, whereas an OR-gate is similar to a series structure in an RBD. In fact, FTA and RBD models have similarities in foundational principles and functionalities, and the two models depicted in Figures 5.8 and 5.12 are equivalent.

Both RBD and FTA typically rely on a Boolean model, implying that each node in a fault tree, whether it represents the TOP event or a basic event, has only two states: occurred or not occurred. In the context where an event is considered as a failure of a component, these two states can correspond to the failed (disconnected) state and the functioning (connected) state of a block in a RBD.

### 5.4.2 Structural and Probabilistic Analysis

The main task of structural analysis for a fault tree is to identify the minimal cut sets, as we have defined in Section 4.7. It is natural to consider that a minimal cut set in FTA can correspond to the minimal tie set in an RBD, but the two concepts have distinct implications.

For the fault tree illustrated in Figure 5.12, the identified minimal cut sets include: $\{e_1, e_2\}$, $\{e_3\}$, and $\{e_4\}$. This means that the TOP event will occur if event $e_3$ occurs, event $e_4$ occurs, or if events $e_1$ and $e_2$ occur at the same time. For fault trees of smaller size, an approach known as MOCUS can be utilized to identify all minimal cut sets, as discussed in Section 4.7. For large-size fault trees, various software tools can be employed, such as Fault Tree Analyzer™ and Blocksim™ by Reliasoft®.

The occurrence of basic event $e_i$ by time $t$ is represented as a random variable $Y_i(t)$ with two possible outcomes: 1 (occurred) and 0 (not occurred). If the basic event $i$ is one of $n$ input events of an AND-gate, the occurrence of the upper/output event of the AND-gate can be expressed as:

$$Y_G(t) = \prod_{i=1}^{n} Y_i(t) \tag{5.77}$$

On the other hand, if the basic event $i$ is one of $n$ input events to an OR-gate, the occurrence of the upper event of the OR-gate is expressed as:

$$Y_G(t) = 1 - \prod_{i=1}^{n} [1 - Y_i(t)] \tag{5.78}$$

Let $q_i(t)$ denote the occurrence probability of event $i$ (or the failure probability of component $i$ by time $t$), denoted as $F_i(t) = \Pr[Y_i(t) = 1]$. The occurrence probability of the output event of an AND-gate is then:

$$F_G(t) = \prod_{i=1}^{n} F_i(t) \tag{5.79}$$

Similarly, the occurrence probability of the output event $q_j(t)$ of an OR-gate is:

$$F_G(t) = 1 - \prod_{i=1}^{n} (1 - F_i(t)) \tag{5.80}$$

For the TOP event in Figure 5.12, its occurrence probability can then be calculated as:

$$\Pr[Y_{TOP}(t) = 1] = F_S(t) = 1 - [1 - F_1(t)F_2(t)] \cdot [1 - F_3(t)] \cdot [1 - F_4(t)]$$

## Example 5.7:

### *System Failure Probability Calculation Using FTA*

Revisiting the system in Example 5.1, the failure probabilities of the components by the specific time $t$ are: $R_1 = R_2 = 0.15$, $R_3 = 0.05$, and $R_4 = 0.10$. The fault tree for system failure is illustrated in Figure 5.12. Using Equations (5.79) and (5.80), we can calculate the failure probability of the system:

$$F_S(t) = 1 - [1 - 0.15 \cdot 0.15] \cdot [1 - 0.05] \cdot [1 - 0.10] = 0.1642$$

This result corresponds to the overall unreliability of the system based on the calculation result in Example 5.1. ■

For a *koon* voting structure, it fails when at least $(n-k+1)$ components have failed. Assuming that all $n$ components in the structure are identical with the failure probability $F_i(t) = F(t)$ for $i = 1,2,\ldots,n$, and failures of these individual components are independent, the probability that exactly $(n-k+1)$ components have failed by time $t$ follows a binomial distribution:

$$\Pr\left[\sum_{i=1}^{n} Y_i(t) = n-k+1\right] = \begin{pmatrix} n \\ n-k+1 \end{pmatrix} F(t)^{n-k+1}\left[1-F(t)\right]^{k-1} \tag{5.81}$$

Equation (5.81) aligns with Equation (5.64) when considering $F(t) = 1 - R(t)$. The system is considered to be failed when the number of failed components exceeds $(n-k)$, thus:

$$R_S(t) = \Pr\left[\sum_{i=1}^{n} Y_i(t) \geq n-k+1\right] = \sum_{i=n-k+1}^{n} \begin{pmatrix} n \\ i \end{pmatrix} F(t)^{i}\left[1-F(t)\right]^{n-i} \tag{5.82}$$

For a 2oo3 structure, it fails when at least any two components have failed. The fault tree model for this configuration is illustrated in Figure 5.13.

In the model, the TOP event is connected to three intermediate events ($e_4$, $e_5$, $e_6$) via an OR-Gate ($G_1$), and each of the intermediate event links to two basic events via an AND-Gate. It is important to note that a same event can appear in multiple places within a fault tree, for example, $e_1$ is under both $G_2$ and $G_4$. The structure function for the TOP event in Figure 5.13 can be expressed as follows:

$$Y_{\text{TOP}}(t) = 1 - \left[1 - Y_1(t)Y_2(t)\right]\left[1 - Y_2(t)Y_3(t)\right]\left[1 - Y_3(t)Y_1(t)\right] \tag{5.83}$$

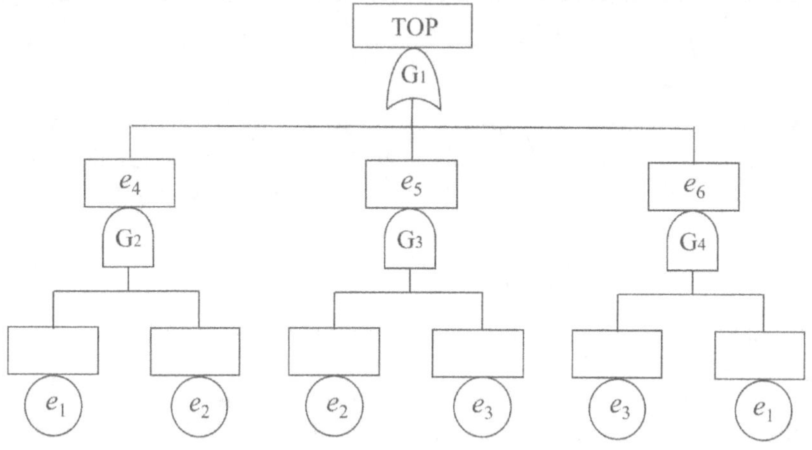


**FIGURE 5.13** Fault tree for a 2oo3 voted structure.

The occurrence probability of the TOP event is then calculated as:

$$F_S(t) = 1 - [1 - F_1(t)F_2(t)][1 - F_2(t)F_3(t)][1 - F_3(t)F_1(t)] \tag{5.84}$$

When the three components are identical with same failure probability $F(t)$, the above calculation can be simplified according to Equation (5.81) as:

$$F_S(t) = 3F(t)^2 - 2F(t)^3$$

We can also derive this with Equation (5.64) obtained using the RBD as:

$$\begin{aligned} F_S(t) &= 1 - R_S(t) = 1 - 3R(t)^2 + 2R(t)^3 = 1 - 3[1 - F(t)]^2 + 2[1 - F(t)]^3 \\ &= 3F(t)^2 - 2F(t)^3 \end{aligned}$$

Fault trees can also serve as the basis for importance analysis, where we consider the impact of a basic event on the occurrence of the TOP event. For example, the Birnbaum's measure of event $i$ can be calculated as:

$$I^B(i|t) = \frac{\partial F_S(t)}{\partial F_i(t)} \tag{5.85}$$

For a fault tree with an AND-gate with $n$ independent input basic events to the output TOP event, the Birnbaum's measure of event $i$ is:

$$I^B(i|t) = \frac{\partial F_S(t)}{\partial F_i(t)} = \frac{\partial \Pi_{j=1}^n F_j(t)}{\partial F_i(t)} = \prod_{j \neq i}^{n} F_j(t) \tag{5.86}$$

For a fault tree with an OR-gate connecting $n$ independent input basic events to the output TOP event, the Birnbaum's measure of event $i$ is:

$$I^B(i|t) = \frac{\partial F_S(t)}{\partial F_i(t)} = \frac{\partial \left[1 - \Pi_{j=1}^n (1 - F_j(t))\right]}{\partial F_i(t)} = \prod_{j \neq i}^{n} (1 - F_j(t)) \tag{5.87}$$

Another way to express the Birnbaum's measure is by noting that $I^B(i|t)$ is actually equal to the occurrence probability of the TOP event when event $i$ always occurs minus the probability of the TOP event when event $i$ never occurs, as:

$$I^B(i|t) = \frac{\partial F_S(t)}{\partial F_i(t)} = F_S(t|Y_i(t) = 1) - F_S(t|Y_i(t) = 0) \tag{5.88}$$

Readers can derive formulas by themselves for the other importance measures based on the fault tree model.

FTA is a widely used tool in the safety analysis of complex and large-scale technical systems, as well as in barrier analysis. Qualitative FTA (top-down method) often works together with FMECA (bottom-up method) to provide both deductive and inductive reasoning for a comprehensive analysis of system failure (TOP event), while quantitative FTA is useful for calculating the probabilities of intermediate and TOP events when the occurrence probabilities of basic events are known. Here, we have briefly introduced the simplest models and algorithms. More symbols and information about FTA, as well as development and evaluation procedures, can be found in the international standard IEC61025 (2006).

## 5.5 BAYESIAN BELIEF NETWORK

### 5.5.1 Model Description

As introduced in Section 5.4, a limitation of the basic FTA method lies in that it cannot analyze the dependency between basic events very well. For example, if the occurrence probability of event $e_1$ is influenced by the occurrence of $e_2$, the basic FTA does not provide a direct method for incorporating such dependencies into the modeling and analysis.

An effective alternative of FTA that allows for both the graphical representation of large-scale systems and the illustration of logical connections between various events (such as dependencies) is the *Bayesian belief network* (BBN), or *Bayesian network* (BN) in short. This model comprises a *directed acyclic graph* (DAG) characterized by a finite set of nodes connected by directed edges, along with conditional probability tables (CPTs) associated with each node, as depicted in Figure 5.14.

Within a BBN, each node can represent an event, designated as $e_i$ for node $i$. Therefore, a BN can be formalized as $N = \langle (E, V), P \rangle$, where $E$ the vector of nodes, $V$ is the vector of edges, $(E, V)$ represents the whole DAG, and $P$ is the vector containing the probability distributions of all nodes.

One of the functions of a BN is to utilize directed edges (arrows) between nodes to describe the causal relationships between causes and outcomes. As illustrated in Figure 5.14, the node TOP is connected to three input nodes via directed arcs 3, 4, and 5, which are termed its *parent nodes*. The state of node TOP, can be used to model the occurrence of the TOP event in FTA or system failure, is influenced directly by these parent nodes. Conversely, the node TOP is considered the *child node* of nodes 3, 4, and 5, indicating that an arc extends from a parent node to its child. Similarly, nodes 1 and 2 are the parent nodes of node 5. Nodes 1–4, without

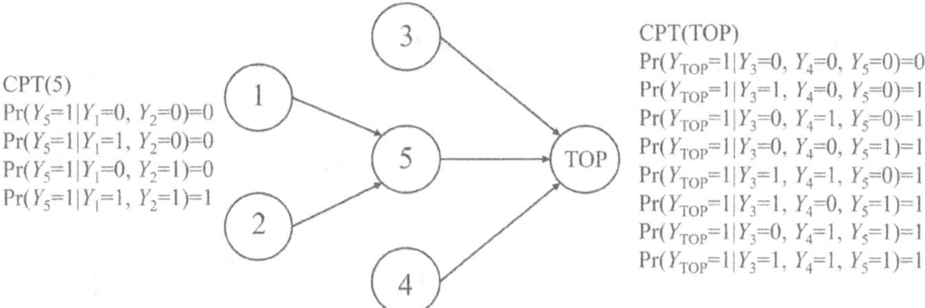


**FIGURE 5.14** Illustration of Bayesian network.

parent nodes, are regarded as *root nodes*, while the node TOP, having no child nodes, is referred to as a *leaf node* in this BN.

As depicted in Figure 5.14, a node can be a child of multiple parent nodes and a parent to several child nodes simultaneously. The CPT of a node illustrates the probability distribution of the associated event, conditional on the states of its parent nodes. For instance, the event associated with node $i$ is denoted by $e_i$, and the occurrence of $e_i$ at a specific time is represented by a random variable $Y_i$ with two possible values of 1 (occurred or present) and 0 (not occurred or not present). For node 5 in Figure 5.14, the occurrence of its associated event $e_5$ relies on of both input events $e_1$ and $e_2$ occur. Thus, event $e_5$ will only occurs if both $Y_1 = 1$ and $Y_2 = 1$. Otherwise, the probability that $Y_5 = 1$ is equal to 0. It can be found that such a CPT has an equivalent function with an AND-gate in a fault tree, and we can express the conditional dependence between nodes 1, 2, and 5 as $\text{Parent}(Y_5) = Y_1 \cap Y_2$. For the CPT of the TOP node, it specifies that when any of the events associated with the input nodes have occurred, namely $Y_i = 1$, $i = 3,4,5$, the TOP event can occur. It is equivalent with an OR-gate in a fault tree.

The comparison illustrates that the BN in Figure 5.14 can effectively correspond to the fault tree shown in Figure 5.12, meaning that a fault tree model can be accurately translated into a BN model without any loss of detail. Moreover, the algorithms used in FTA are also applicable to BNs.

The flexibility of BNs comes from their ability to accommodate various probability distributions within a CPT, as demonstrated in the subsequent example.

**Example 5.8:**

*Conditional Probability Table*

For the node 5 in Figure 5.14, its CPT can be updated as follows:

$$\Pr\left(Y_5 = 1 | Y_1 = 0,\ Y_2 = 0\right) = 0.1$$

$$\Pr\left(Y_5 = 1 | Y_1 = 1,\ Y_2 = 0\right) = 0.2$$

$$\Pr\left(Y_5 = 1 | Y_1 = 0,\ Y_2 = 1\right) = 0.4$$

$$\Pr\left(Y_5 = 1 | Y_1 = 1,\ Y_2 = 1\right) = 0.9$$

This revised CPT indicates that the occurrence probability of event $e_5$ is dependent on events $e_1$ and $e_2$. In other words, the occurring likelihood of $e_5$ increases significantly when the two causing events are present, but the occurrence is 100% guaranteed. Conversely, even if neither causing event occurs, there is still a low occurring likelihood of $e_5$. In FTA, it is very challenging to represent such kind of causality using the basic logical gates. ■

### 5.5.2 Bayesian Theorem and Quantitative Analysis

Quantitative analysis of BN is based on the Bayesian probability theory. In general, there are two perspectives of interpreting probability in academics: *Frequentist probability* and *Bayesian probability*. Researchers advocating the previous one think that the probability of an event $E$ can be unknown but is an existing number. With a series of independent and identical experiments, the frequency of observed occurrence of event $E$ approaches to a limiting value when the number of experiments increase, as the probability of this event $\Pr(E)$. While for those supporting the Bayesian perspective, probability is subjective as the measure how much a person believes that an event will occur, in other words, the degree of belief on an outcome. The degree of belief is dependent on the knowledge of the analyst on the event, available evidence and data, and personal biases, such as optimism or pessimism, and this degree of belief may change when obtaining new information.

It is apparent that the quantitative analysis of BN is based on the Bayesian perspective. Considering two events, $A$ and $B$, the probability of event $A$ occurring given that $B$ has occurred is the conditional probability $\Pr(A|B)$. In the Bayesian theory, $\Pr(A)$ is referred to as the *prior probability*, representing the initial belief in $A$ occurring without any conditions. After considering $B$, $\Pr(A|B)$ becomes the *posterior probability*, which is the updated likelihood of $A$ when considering $B$ as a condition. If $\Pr(B) \neq 0$, the relationship can be written as:

$$P\left(A|B\right) = \frac{P(A \cap B)}{P(B)} \tag{5.89}$$

Considering $\Pr(A \cap B) = \Pr(B|A)\Pr(A)$, the above equation can be written as:

$$\Pr\left(A|B\right) = \frac{\Pr\left(B|A\right)P(A)}{\Pr(B)} \tag{5.90}$$

This formula is the basic format of the *Bayes theorem*. Here, $B$ serves as evidence influencing the belief of the analyst in $A$.

Furthermore, considering a sample space $\Omega$ with mutually exclusive events $A_1$, $A_2,\ldots, A_m$, so that $\bigcup_{i=1}^{m} A_i = \Omega$, and $\sum_{i=1}^{m} \Pr(A_i) = 1$. Using the *law of total probability* to calculate $\Pr(B)$, we have:

$$\Pr(B) = \sum_{i=1}^{m} \Pr\left(B|A_i\right)P(A_i) \tag{5.91}$$

As a result, Equation (5.90) can be extended to calculate the conditional probability that event $A_j$ ($A_j \in \Omega$) occurs given that event B has occurred:

$$\Pr\left(A_j|B\right) = \frac{\Pr\left(B|A_j\right)\Pr(A_j)}{\Pr(B)} = \frac{\Pr\left(B|A_j\right)\Pr(A_j)}{\Sigma_{i=1}^{m}\Pr\left(B|A_i\right)\Pr(A_i)} \tag{5.92}$$

This is a more general form of Bayes theorem.

In a very basic BN model as illustrated in Figure 5.15, where node $B$ is a root node and node $A$ is a leaf node. The joint probability distribution of variables $Y_A$ and $Y_B$ at the possible states $a$ and $b$ respectively can be calculated as:

$$\Pr(Y_A = a \cap Y_B = b) = \Pr\left(Y_A = a|Y_B = b\right)\Pr\left(Y_B = b\right)$$

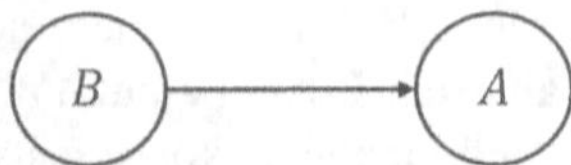


FIGURE 5.15 The simplest Bayesian network.

## Example 5.9:

### *Braking Failure Analysis Using Bayesian Network*

In Figure 5.15, node $A$ represents the performance of a braking system, and node $B$ corresponds to road conditions. We consider two possible states for each variable: $Y_A = 1$ if the braking system successfully stops the car within a safe distance, and $Y_A = 0$ if the braking system does not work as expected, causing the car to stop with a longer distance. Similarly, $Y_B = 1$ if the road is in the normal condition, and $Y_B = 0$ if the road is icy. In a cold region, it is assumed that the probability of encountering icy roads is 0.2, thus $\Pr(Y_B = 0) = 0.2$ and $\Pr(Y_B = 1) = 0.8$.

With the extensive driving experience in Norway, we can determine the conditional probabilities as follows:

| $b$ | $\Pr(Y_A = 1 \mid Y_B = b)$ | $\Pr(Y_A = 0 \mid Y_B = b)$ |
|---|---|---|
| 0 | 0.6 | 0.4 |
| 1 | 0.9 | 0.1 |

For instance, if the roads are icy ($Y_B = 0$), the probability that the braking system does not work ($Y_A = 0$) is $\Pr(Y_A = 0 \mid Y_B = 0) = 0.4$. Therefore, the overall probability that the braking is not within the expected distance can be calculated using the law of total probability:

$$\Pr(Y_A = 0) = \Pr(Y_A = 0 \mid Y_B = 0) + \Pr(Y_A = 0 \mid Y_B = 1) = 0.4 \cdot 0.2 + 0.1 \cdot 0.8 = 0.16$$

The distribution of a root node in a BN is called as a *marginal distribution*. For the example, the probabilities of $\Pr(Y_B = 0) = 0.2$ and $\Pr(Y_B = 1) = 0.8$ are the marginal distribution of node $B$ in the BN. ■

## Example 5.10:

### *Probability Calculation of Variable State in Bayesian Network*

Consider two parent nodes and one child node, such as nodes 1, 2, and 5 in Figure 5.14. Here, we use the DAG given in Figure 5.14 and the probabilities in Example 5.9. The marginal distribution is $\Pr(Y_1 = 0) = \Pr(Y_2 = 0) = 0.1$, and thus $\Pr(Y_1 = 1) = \Pr(Y_2 = 1) = 0.9$.

Then, we can calculate the probability of $Y_5 = 1$:

$$\begin{aligned}\Pr(Y_5 = 1) = {} & \Pr(Y_5 = 1 \mid Y_1 = 0, Y_2 = 0)\Pr(Y_1 = 0 \cap Y_2 = 0) + \Pr(Y_5 = 1 \mid Y_1 = \\ & 1, Y_2 = 0)\Pr(Y_1 = 1 \cap Y_2 = 0) + \Pr(Y_5 = 1 \mid Y_1 = 0, Y_2 = 1)\Pr(Y_1 = 0 \cap Y_2 = 1) + \\ & \Pr(Y_5 = 1 \mid Y_1 = 1, Y_2 = 1)\Pr(Y_1 = 1 \cap Y_2 = 1) = 0.1 \cdot 0.1 \cdot 0.1 + 0.2 \cdot 0.9 \cdot 0.1 + 0.4 \cdot \\ & 0.1 \cdot 0.9 + 0.9 \cdot 0.9 \cdot 0.9 = 0.784\end{aligned}$$

In such a BN, when new knowledge is introduced to change the values of $\Pr(Y_1 = 1)$, $\Pr(Y_2 = 1)$ or the CPT of node 5, the value of $\Pr(Y_5 = 1)$ will be updated. ■

Consider three events $A$, $B$, and $C$, when we know that event $C$ has occurred, if the following equation is satisfied:

$$\Pr\left(A \cap B \middle| C\right) = \Pr\left(A \middle| C\right) \Pr\left(B \middle| C\right) \tag{5.93}$$

Event $A$ and event $B$ are regarded *conditional independent.* If we generally consider a BN with $n$ nodes, and these nodes are conditional independent, then the probability that $Y_i = y_i$ (for $i = 1,2,\ldots,n$) is:

$$\Pr(Y_1 = a \cap Y_2 = b \cap \ldots \cap Y_n(t) = m) = \prod_{i=1}^{n} \Pr\left(Y_i = j \middle| \text{Parent}(Y_i)\right) \tag{5.94}$$

**Example 5.11:**

*Probability of the TOP Event*

For the node TOP in Figure 5.14, considering its CPTs, the probability of $Y_{TOP} = 0$ can be calculated as:

$$\Pr(Y_{TOP} = 0) = \Pr(Y_3 = 0 \cap Y_4 = 0 \cap Y_5 = 0) = \Pr(Y_3 = 0)\Pr(Y_4 = 0)\Pr\left(Y_5 = 0 \middle| \text{Parent}(Y_5)\right)$$

Then, based on the CPT of node 5 in Figure 5.14, we have:

$$\Pr\left(Y_5 = 0 \middle| \text{Parent}(Y_5)\right) = 1 - \Pr\left(Y_5 = 1 \middle| \text{Parent}(Y_5)\right) = 1 - \Pr(Y_1 = 1)\Pr(Y_2 = 1)$$

Thus,

$$\Pr(Y_{TOP} = 0) = \Pr(Y_3 = 0)\Pr(Y_4 = 0)\left[1 - \Pr(Y_1 = 1)\Pr(Y_2 = 1)\right]$$

If the data for the root nodes is available, it is possible to calculate the probability of the occurrence or non-occurrence of the TOP event. ■

BN is highly effective for analyzing and evaluating complex systems where dependencies among events exist. BNs offer more flexibility than traditional FTA methods and can be applied in various scenarios. Several studies, such as those by Cai et al. (2012, 2013) and Wu et al. (2016), have successfully utilized BBNs in the reliability and availability analysis of technical safety barriers. However, implementing BNs need to collect extensive data and the develop CPTs for all nodes. Even for relatively small systems, manual calculations can be extremely complex.

### 5.5.3 Procedure of Bayesian Network Analysis

A basic BN analysis typically includes the following main steps:

**Step 1: Hazard and impact factor identification**

The first step involves identifying potential hazardous events or failures that need to be avoided. In barrier analysis, such a step needs to identify the relevant accidental scenarios, such as root causes, barriers, and consequences, and clarify the causal relationships between the events. Impact factors, including human errors, software and hardware failures, and organizational issues, should be considered and roughly assessed for their criticality. Some tools, such like FMECA, can be helpful in this step for determining the factors for BN analysis.

**Step 2: Establishment of the directed acyclic graph**

When establishing a DAG for a BN, it is important to specify all the leaf nodes clearly and unambiguously. These nodes represent the end events in accidental scenarios. Subsequently, the impact factors or parent nodes of these leaf nodes should be identified, and directed arcs can be drawn to describe the dependencies among nodes. In the analysis of large systems equipped with safety barriers, a barrier is considered as a risk reduction factor, while in the availability analysis of the barrier itself, the root causes of failures and other operational and environmental factors are considered as nodes within the BN.

It is suggested to name the nodes in addition to assigning numbers to them. The names or captions of the intermediate and leaf nodes should reflect events and maintain consistency in their expression, using either all negative terms like "wrong operation" or "failure of control", or all positive terms such as "machine is functioning" and "maintenance is on time". Names or captions of the root nodes should describe a condition or status, such as "high temperature" or "lack of knowledge".

The scope of factors considered in a BN should align with the objectives of the analysis, data availability, and other resources such as computing capacity. In some studies, impact factors are categorized into different groups and represented at various layers within a BN. For example, hardware failures might be considered direct causes of an accident and placed closest to the leaf nodes, while human and operational errors, which influence hardware performance, are positioned in a second layer. This layered approach helps in structuring the BN to effectively reflect the complexity of the system and interdependencies among the elements within the system.

**Step 3: Definition of node states**

The state variable of a node in a BN can either be continuous or have several discrete levels. In most practical applications of BN analysis, discrete states can be considered. For classical reliability analysis, the state variable is binary, typically represented by values of 0 and 1. However, for more complex systems, such as those experiencing degradation, the number of states can be more than two.

The states defined for a parent node can influence its child node, meaning that changes in the state of the parent node will change the state of the child. It is not needed to identify numerous discrete states for a node if they only mean negligible or minor differences for the child node. Identifying an excessive number of states can significantly increase the workload in data collection and analysis, and more states also exponentially increase the complexity of the computational process.

**Step 4: Construction of conditional probability tables**

The relationships between nodes and their parent nodes in a BN are quantitatively defined using CPTs. These tables assign probabilities to the various states of a node.

In contrast to constructing a DAG, the development of CPTs starts from the root nodes. Once the states of the parent nodes are established, the CPTs for the child nodes are then specified. The values within these probability distributions are based on a priori knowledge, which may include data from industry databases, expert judgments, experimental analyses, or a combination of these sources. These probabilities can be updated when new evidence is introduced. The update is not only a simple change of values but includes a process known as *inference*, which means the computation of conditional probabilities and the adjustment of posterior distributions.

**Step 5: Quantitative analysis**

The final step involves calculating the probabilities of the leaf nodes based on the inputs from the root nodes, the CPTs, and the dependencies among nodes. Various software tools, like Netica,[1] have been developed to facilitate the quantitative analysis of BN, but most of these tools are not freely available.

Many textbooks on reliability analysis include sections dedicated to BN, such as Section 6.9 in Rausand et al. (2020), and several international standards regard BBN as their recommended methods for system risk and dependability analysis, such as IEC31010 (2019) – *Risk Assessment Techniques* and EN60300-3-15 (2009) – *Engineering of System Dependability*, but we did not recognize an international standard specific for BN. A recent review paper by Cai et al. (2019) provides more insights into BN applications in reliability evaluations, which can be interesting to readers.

It also should be noted that DAGs of BBNs in literature can be different from what we use in this book in terms of graphic representation. For example, in IEC31010 (2019), the nodes are drawn with rectangles including marginal distributions.

### 5.5.4 Dynamic Bayesian Network

An ordinary BN can be developed into a Dynamic Bayesian Network (DBN) by incorporating temporal evolution, which can effectively model the time-dependent behaviors of random variables (Khakzad, 2015). A DBN is structured as a series of temporal slices, each containing an ordinary or static BBN that describes the states of random variables at a specific time. The variable value evolves according to the probabilistic temporal dynamics. The CPT of each variable in a DBN can be then calculated accordingly.

Let $\Pr(Y_i(t)=j)$ denote the probability that the random variable $Y_i$ (associated with the node $i$ in the static BBN) is equal to a certain value $j$ at time $t$. In DBN analysis, two fundamental assumptions are typically made:

- $\Pr(Y_i(t+1)=k \mid Y_i(t)=j, Y_i(t-s)=i)=\Pr(Y_i(t+1)=k \mid Y_i(t)=j)$. This indicates that the probability of a variable transitioning from its current state to a future state only depends on its current state and not on its historical

states. This property characterizes the process as Markovian (see the next subsection), where the future state depends only on the present state and not on the sequence of preceding events.
- The transition probability of the variable $Y_i$ from value $j$ to value $k$ remains consistent across all time $t$.

Thus, we can extend Equation (5.94) by introducing time, as:

$$\Pr\left(Y_i(t+1)=k \middle| Y_i(t)=j\right)=\prod_{i=1}^{n}\Pr\left(Y_i(t+1)=k \middle| \text{Parent}\left(Y_i(t+1)\right)\right) \quad (5.95)$$

At any given time slice within a DBN, new information about model parameters, such as marginal distributions and CPTs, can be integrated into the current network. The values of the variables can then be recalculated using the Bayesian formula outlined in Equations (5.90)–(5.92).

A DBN facilitates both predictive or forward analysis and backward or diagnostic analysis. For example, DBN can be used for estimating the evolving probability of occurrence of the TOP node in the future in Figure 5.14, and the network also can calculate the posterior probability distribution of each root cause given that the TOP event is known occurred.

DBNs have been widely adopted in recent studies on safety barriers due to their effectiveness in analyzing system performance, particularly in contexts where time-dependent changes in events, conditions, and interrelationships between impact factors are influential. Readers interested in exploring DBNs further may refer to the works by Khakzad (2015) and Wu et al. (2016).

## 5.6 MARKOV METHOD

In assessing the performance of safety barriers, it is crucial to consider various dynamic behaviors such as the degradation of barriers, testing, preventive maintenance, and the restoration of barriers from a faulty state. FTA and RBD are highly effective for modeling large-scale systems, but they need the compliment of the state transition models, which provide more detailed insights into the operation of safety barriers. Among these, the *Markov method*, named after Russian mathematician Andrei Markov (1856–1922), is one of the most extensively utilized state transition models.

### 5.6.1 Markov Diagram

The term "Markov method" generally refers to a collection of approaches based on various types of *Markov models*, such as *semi-Markov chains* and *hidden Markov chains*. Initially, we can start from representing dynamic behaviors of an item graphically using a Markov diagram. Figure 5.13 illustrates a simple Markov diagram describing the transitions between two discrete states of an item. The diagram includes two circles representing the states of the item:

- 0: Operational or functioning state, and
- 1: Failed or faulty state.

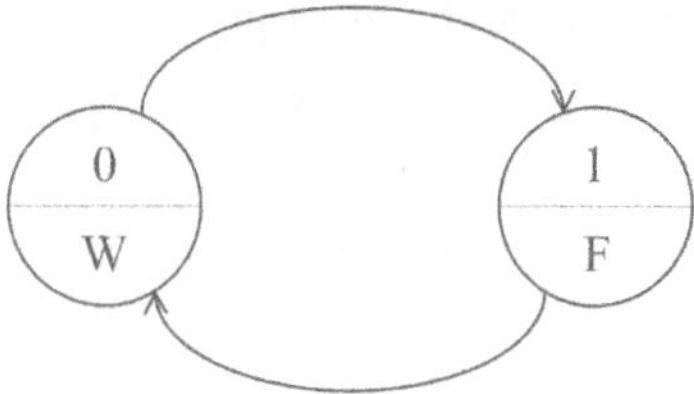


**FIGURE 5.16** A simple Markov diagram.

The directed arcs between the circles indicate the transitions or dynamic behaviors of the item: an arrow from 0 to 1 denotes a failure, and an arrow from 1 to 0 denotes a repair.

In this book, we recognize that the performance of an item is time-dependent. In such scenarios, the process underlying the Markov diagram is referred to as a *continuous-time Markov chain* (CTMC). The Markov diagram in Figure 5.16 is an simple example of a CTMC. In performance assessment, it is of interest to estimate the sojourn time or sojourn probability of the item in States 0 and 1. If the chain includes discrete time steps, it is classified as a *discrete-time Markov chain* (DTMC). For example, given that we only consider the moments when transitions occur, and these events are numbered as $n = 0,1,2,\ldots$, the process is a DTMC. DTMC also referred to as the *embedded Markov chain* of the continuous process.

### 5.6.2 Markov Property

In the Oxford dictionary, a Markov chain is defined as "*a stochastic model describing a sequence of possible events in which the probability of each event depends only on the state attained in the previous event*". This characteristic, where future events only depend on the current state and not on how the state was reached, is known as the *Markov property*, meaning that the future events of a Markov process are not influenced by how it arrived in the current state. Consequently, such a process is termed "*memoryless*" (Rausand, 2014).

In mathematics, the Markov property can be expressed as

$$\Pr\left[X(t+s)=j|X(s)=i,\ X(u)=k, 0\le u<s\right]=\Pr\left[X(t+s)=j|X(s)=i\right] \quad \text{for all possible } k \tag{5.96}$$

where $X(t)$ is the state variable of the process at time $t$, and $i$, $j$, and $k$ are possible values of the variable, or the states of the process. A stochastic process showing this property is also known as having stationary transition probabilities or being *time-homogeneous*, meaning that transition times in a Markov process are exponentially distributed.

When a stochastic process owns the Markov property, it can be quantitatively analyzed. Consider a Markov chain $\{X(t), t\ge 0\}$ with a state space $\Omega = \{0,1,2,\ldots,n\}$. The transition probability from state $i$ to state $j$ by time $t$ is defined as:

$$p_{ij}(t)=\Pr\left[X(t)=j|X(0)=i\right] \tag{5.97}$$

For an item with $n$ states, we can build a transition matrix as:

$$\mathcal{P}(t) = \begin{bmatrix} p_{00}(t) & p_{01}(t) & \cdots & p_{0n}(t) \\ p_{10}(t) & p_{11}(t) & & p_{1n}(t) \\ \vdots & & \ddots & \vdots \\ p_{n0}(t) & p_{n1}(t) & \cdots & p_{nn}(t) \end{bmatrix} \tag{5.98}$$

When a process is in state $i$ at its starting point $t_0 = 0$, it must either stay in the same state or transit to a different state by time $t$, Thus, the sum of transition probabilities from any state $i$ satisfies:

$$\sum_{J=0}^{N} p_{ij}(t) = 1 \text{ for all } i \in \Omega \tag{5.99}$$

For the element on the diagonal of the matrix in Equation (5.98), $p_{ii}(t)$, it is used to describe the probability that the process is in state $i$ at time 0 and remains in state $i$ at time $t$, regardless of any intermediate states between 0 and $t$.

Then, we consider the *sojourn time* ($ST$) of a Markov process in state $i$, from the moment it enters the state until it exits. The $ST$ is a random variable, and probability that $ST$ is longer than $t$ is given by $\Pr(ST_i > t)$. Since the Markov process is memoryless, this probability is only determined by the current state, not influenced by the duration already spent in state $i$, namely:

$$\Pr\left(ST_i > t + s \middle| ST_i > s\right) = \Pr\left(ST_i > t\right) \text{ for } t,\ s \geq 0$$

In a CTMC, the memoryless property also indicates that the sojourn time $ST_i$ is exponentially distributed, and the sojourn times in different states $ST_1$, $ST_2$... are independent. This means that when a Markov process enters state $i$, the duration it remains in this state before moving to another state follows an exponential distribution. To quantify this, we introduce the term *transition rate* or *leaving rate* ($\alpha_i$) to denote the reciprocal of the mean sojourn time of the process in a state $i$, namely $E(ST_i) = 1/\alpha_i$. For example, as illustrated in Figure 5.16, the transition rate or leaving rate from State 0 ($\alpha_0$) represents the failure rate or the reciprocal of mean time between failures of the system ($\lambda$). Because there is only one transition from State 0 in this case, $\alpha_0 = \alpha_{01} = \lambda$. Similarly, the transition rate from State 1 ($\alpha_1$) is equal to the rate from State 1 to State 0 as the restoration rate or the reciprocal of the mean downtime of the system ($\mu$), therefore, $\alpha_1 = \alpha_{10} = \mu$.

When $\alpha_i = \infty$, the process immediately exits state $i$ after entering the state, and state $i$ is called as an *instantaneous state.* When $\alpha_i = 0$, process will not leave state $i$ after it enters this state, and state $i$ is called as an *absorbing state.* If the transition from State 1 to State 0 in Figure 5.13 is removed, State 1 becomes an absorbing state. This can be used for modeling a non-repairable item where, once a failure occurs, the item stays in the faulty state forever.

### 5.6.3 Transition Rate Matrix

For a continuously stochastic process with an embedded Markov chain, let $\alpha_{ij} = \alpha_i p_{ij}$, where $\alpha_i$ denotes the leaving rate of the process from state $i$, and $p_{ij}$ is the probability that the process transits from state $i$ to state $j$ at a certain time. According to Equation (5.99), we define the total leaving rate from state $i$ as:

$$\alpha_i = \sum_{\substack{j=0 \\ j\neq i}}^{n} a_{ij} \tag{5.100}$$

The sojourn time in state $i$ is exponentially distributed with a constant rate $\alpha_i$, thus the probability that the process leaves state $i$ by time $t$ from the moment that we start observing is $\Pr(ST_i < t) = 1 - e^{-\alpha_i t}$.

Given that the process follows the Markov property, the transition probability from state $i$ to state $j$ remains constant over time, implying that $p_{ij}$ and $\alpha_{ij}$ are constants. Therefore, the probability that the process remains in state $i$ before transiting to state $j$ is:

$$\Pr\left(ST_{ij} < t\right) = 1 - e^{-\alpha_{ij} t} \tag{5.101}$$

Consider a very short time interval $\Delta t$, and utilizing the Taylor series expansion $e^x = \sum_{n=0}^{\infty} x^n / n!$, we have:

$$\lim_{\Delta t \to 0} \frac{p_{ij}(t)}{\Delta t} = \lim_{\Delta t \to 0} \frac{\Pr\left(ST_i < \Delta t\right)}{\Delta t} = \frac{1 - 1 + \alpha_{ij}\Delta t}{\Delta t} = \alpha_{ij} \text{ for } i \neq j \tag{5.102}$$

Thus, $\alpha_{ij}$ can be defined as the transition rate from state $i$ to state $j$, and it can be organized into a transition rate matrix $\mathcal{A}$ for all possible states of the Markov process:

$$\mathcal{A} = \begin{pmatrix} \alpha_{00} & \alpha_{01} & \cdots & \alpha_{0n} \\ \alpha_{10} & \alpha_{11} & & \alpha_{1n} \\ \vdots & & \ddots & \vdots \\ \alpha_{n0} & \alpha_{n1} & \cdots & \alpha_{nn} \end{pmatrix} \tag{5.103}$$

In this matrix, we can introduce the following notation for the diagonal elements (Rausand et al., 2020):

$$a_{ii} = -\alpha_i = -\sum_{\substack{j=0 \\ j\neq i}}^{n} a_{ij} \tag{5.104}$$

Thus, for the simplest case in Figure 5.13, the transition rate matrix of this Markov model is

$$\mathcal{A} = \begin{pmatrix} -\lambda & \lambda \\ \mu & -\mu \end{pmatrix}$$

**TABLE 5.1**
**Descriptions of States in Figure 5.17**

| State | Description |
|---|---|
| 0 | Both components are functioning (working, W) |
| 1 | One component is failed (F), and one is functioning |
| 2 | Both components are failed |

## Example 5.12:

### *Markov Models for a Parallel Structure with Two Components*

Consider a parallel structure comprising two identical components. Each component has a failure rate $\lambda$ and repair rate $\mu$ in case of failure. Three states of the whole structure can be identified as shown in Table 5.1.

A Markov model representing this structure is illustrated in Figure 5.17.

It is important to note that the transition rate from State 0 to State 1 is $2\lambda$ since each of the two components can fail independently at a rate of $\lambda$. Once one component has failed, the failure of the second leads to a transition from State 1 to State 2 at a rate of $\lambda$. On the other hand, the model assumes a sequential repair strategy, meaning that only one component is repaired at a time. As a result, the system transits from State 2 to State 1 with a repair rate of $\mu$, meaning the repair of one failed component. Then, a transition from State 1 to State 0 occurs when the other component is repaired.

If the repair strategy is different, for example, simultaneous repairs on two components are possible given that sufficient technicians and maintenance resources are available, the transition from State 2 to State 1 can be omitted, with introducing a direct transition from State 2 to State 0 at a rate of $\mu$.

A transition rate matrix for the Markov model in Figure 5.17 can be established based on Equation (5.103):

$$\mathcal{A} = \begin{bmatrix} -2\lambda & 2\lambda & 0 \\ \mu & -(\lambda+\mu) & \lambda \\ 0 & \mu & -\mu \end{bmatrix}$$

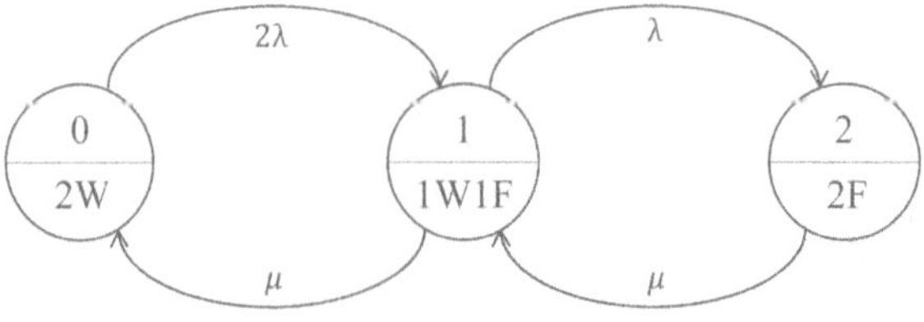


**FIGURE 5.17** Markov model for a parallel structure.

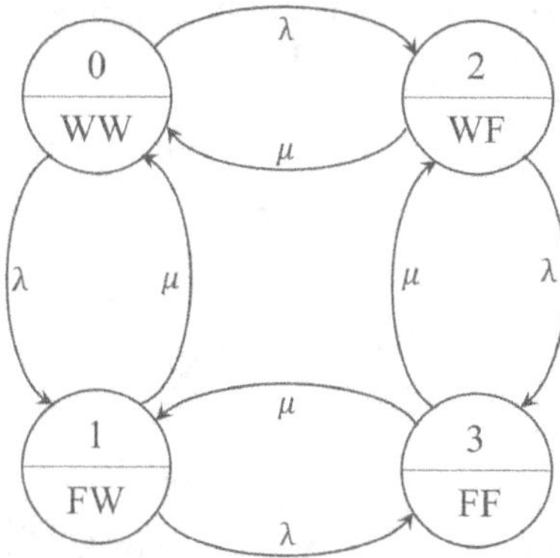


**FIGURE 5.18** An alternative Markov model for the parallel structure.

Different modeling approaches are possible with the Markov method. For example, an alternative Markov model with four states could be utilized for a parallel structure of two components, as depicted in Figure 5.18:

The descriptions of the states in Figure 5.18 are provided in Table 5.2.

The model in Figure 5.18 is more flexible in accommodating different transition rates, especially considering that the components are not identical, with distinct failure and repair rates. Unlike the simpler model in Figure 5.17, this four-state model does not presume the availability of only one maintenance crew. Instead, it allows for the possibility that both components may be repaired simultaneously, with the leaving rate from State 3 being $2\mu$. ■

### 5.6.4 Kolmogorov Equations

Following the last subsection, Equation (5.102), $\lim_{\Delta t \to 0} p_{ij}(t)/\Delta t = \alpha_{ij}$ can be written as

$$p_{ij}(t) = \alpha_{ij}\Delta t + o(\Delta t) \text{ for } i \neq j \tag{5.105}$$

**TABLE 5.2**
**Descriptions of States in Figure 5.18**

| State | Description |
|---|---|
| 0 | Components 1 and 2 are functioning |
| 1 | Component 1 is failed, component 2 is functioning |
| 2 | Component 2 is failed, component 1 is functioning |
| 3 | Components 1 and 2 are failed |

In Equation (5.105), $o(\Delta t)$ is a function that ensures $\lim_{\Delta t \to 0} o(\Delta t)/\Delta t = 0$, and the derivative of $p_{ij}(t)$ with respect to time $t$ is:

$$\dot{p}_{ij}(0) = \frac{d}{dt} p_{ij}(t)\Big|_{t=0} = \lim_{\Delta t \to 0} \frac{p_{ij}(0+\Delta t) - p_{ij}(0)}{\Delta t} = \alpha_{ij} \tag{5.106}$$

And thus

$$p_{ii}(0) = \alpha_{ii} = -\sum_{\substack{j=0 \\ j \neq i}}^{n} a_{ij}$$

Considering the transition rate matrix in Equation (5.103), now we can have:

$$\mathcal{A} = \begin{pmatrix} \alpha_{00} & \alpha_{01} & \cdots & \alpha_{0n} \\ \alpha_{10} & \alpha_{11} & & \alpha_{1n} \\ \vdots & & \ddots & \vdots \\ \alpha_{n0} & \alpha_{n1} & \cdots & \alpha_{nn} \end{pmatrix} = \dot{\mathcal{P}}(0) \tag{5.107}$$

For a Markov process, considering a scenario where the item is in state $i$ at time 0, and is in state $j$ at time $s$, the probability that the item will be in state $k$ at time $(s+t)$ can be calculated as:

$$p_{ik}(s+t) = \sum_{j=0}^{n} p_{ij}(s) p_{jk}(t) \tag{5.108}$$

Here, the transition probability $p_{jk}(t)$ is only dependent on the state $j$, consistent with the memoryless property of the Markov process. Equation (5.108) is known as the *Chapman-Kolmogorov equation*,[2] which can also be expressed in the matrix notation:

$$\mathcal{P}(s+t) = \mathcal{P}(s)\mathcal{P}(t) \tag{5.109}$$

By differentiating Equation (5.109) with respect to $t$ when $s = 0$, and integrating with Equation (5.106), we have:

$$\mathcal{P}(t)\mathcal{A} = \dot{\mathcal{P}}(t) \tag{5.110}$$

This matrix equation is called as the *Kolmogorov forward equation*, and its complete expression can be written as:

$$\left[p_0(t), p_1(t), \ldots, p_n(t)\right] \cdot \begin{pmatrix} \alpha_{00} & \alpha_{01} & \cdots & \alpha_{0n} \\ \alpha_{10} & \alpha_{11} & & \alpha_{1n} \\ \vdots & & \ddots & \vdots \\ \alpha_{n0} & \alpha_{n1} & \cdots & \alpha_{nn} \end{pmatrix} = \left[\dot{p}_0(t), \dot{p}_1(t), \ldots, \dot{p}_n(t)\right] \tag{5.111}$$

**Example 5.13:**

*State Equations of Markov Model*

Revisiting the Markov model depicted in Figure 5.14 and referring to Equation (5.111), we can derive the following state equations for this model, as:

$$\begin{bmatrix} p_{00}(t) & p_{01}(t) & p_{02}(t) \\ p_{10}(t) & p_{11}(t) & p_{12}(t) \\ p_{20}(t) & p_{21}(t) & p_{22}(t) \end{bmatrix} \cdot \begin{bmatrix} -2\lambda & 2\lambda & 0 \\ \mu & -(\lambda+\mu) & \lambda \\ 0 & \mu & -\mu \end{bmatrix} = \begin{bmatrix} \dot{p}_{00}(t) & \dot{p}_{01}(t) & \dot{p}_{02}(t) \\ \dot{p}_{10}(t) & \dot{p}_{11}(t) & \dot{p}_{12}(t) \\ \dot{p}_{20}(t) & \dot{p}_{21}(t) & \dot{p}_{22}(t) \end{bmatrix}$$

The item is assumed to be functioning at $t = 0$, namely $p_0(0) = 1$.

By multiplying the first row of the transition probability matrix by the elements in the transition rate matrix, we can obtain the following equations:

$$-2\lambda p_{00}(t) + \mu p_{01}(t) = \dot{p}_{00}(t)$$

$$2\lambda p_{00}(t) - (\lambda + \mu) p_{01}(t) + \mu p_{02}(t) = \dot{p}_{01}(t)$$

$$\lambda p_{01}(t) - \mu p_{02}(t) = \dot{p}_{02}(t)$$

At a certain moment $t$, since the item must be in one of the three states, there is $p_{00}(t) + p_{01}(t) + p_{02}(t) = 1$. Taking the derivative with respect to time, we have $\dot{p}_{00}(t) + \dot{p}_{01}(t) + \dot{p}_{02}(t) = 0$. It can be found that the sum of any two of the above equations is equivalent to the remaining equation. Therefore, it is sufficient to just analyze two of these equations. For example, we can keep the first and third equations. Then, with the equation $p_{00}(t) + p_{01}(t) + p_{02}(t) = 1$, there can be three equations with three variables. The subsequent subsection will discuss how to obtain the time-independent solutions. ■

### 5.6.5 Asymptotic Solution

The *steady-state* or *long-term* probability that an item remains in a particular state, denoted as $p_i(t)$ as $t \to \infty$, is of interests in many applications. It can be regarded as an average value in the long term and thus time-independent. For example, it is valuable to determine the average availability of an item depicted in Figure 5.13 by evaluating the long-term probability of the Markov process sojourning in State 0.

If a Markov process is *irreducible*, meaning that every state of the process is reachable from every other state, the long-term probability of being in any state can be calculated as:

$$\lim_{t \to \infty} p_i(t) = p_i(\infty) = p_i \quad \text{for } i = 1, 2, \ldots, n$$

Here, $p_i$ represents the *asymptotic value* or *limit* that $p_i(t)$ approaches when $t \to \infty$, and it is referred to the *steady-state probability*. The value of $p_i$ is independent of the initial state of the process due to the long time elapsed.

Given that $p_i$ is a constant value, we have:

$$\lim_{t\to\infty} \dot{p}_i(t) = 0 \text{ for } i = 1,2,\ldots,n$$

Thus, for the steady-state probabilities $\mathcal{P} = [p_0, p_1, \ldots, p_n]$, Equation (5.111) can be written as

$$[p_0, p_1, \ldots, p_n] \cdot \begin{pmatrix} \alpha_{00} & \alpha_{01} & \cdots & \alpha_{0n} \\ \alpha_{10} & \alpha_{11} & & \alpha_{1n} \\ \vdots & & \ddots & \vdots \\ \alpha_{n0} & \alpha_{n1} & \cdots & \alpha_{nn} \end{pmatrix} = [0,0,\ldots,0] \quad (5.112)$$

In the matrix format, this can be expressed in a compact way:

$$\mathcal{P} \cdot \mathcal{A} = 0 \quad (5.113)$$

Also based on Equation (5.99), we have:

$$\sum_{i=0}^{n} p_i = 1 \quad (5.114)$$

Thus, now there are $n$ linear algebraic equations based on Equation (5.112). We can use $(n-1)$ of them along with Equation (5.114) to calculate the steady state probabilities, $p_0, p_1, \ldots, p_n$. The values of such probabilities are independent with the initial state of the Markov process.

**Example 5.14:**

### *Solving the State Equations*

Revisiting the Markov model in Figure 5.17 and referring to Equation (5.112), we have the following equation for calculating steady-state probabilities:

$$[p_0, p_1, p_2] \cdot \begin{bmatrix} -2\lambda & 2\lambda & 0 \\ \mu & -(\lambda+\mu) & \lambda \\ 0 & \mu & -\mu \end{bmatrix} = [0,0,0]$$

This matrix multiplication results in the following linear equations:

$$-2\lambda p_0 + \mu p_1 = 0$$

$$2\lambda p_0 - (\lambda+\mu)p_1 + \mu p_2 = 0$$

$$\lambda p_1 - \mu p_2 = 0$$

Utilizing the first and third equations alongside the equation $p_0 + p_1 + p_2 = 1$, we can solve for the steady-state probabilities as:

$$p_0 = \frac{\mu^2}{2\lambda^2 + 2\lambda\mu + \mu^2} \approx \frac{\mu}{2\lambda + \mu}$$

$$p_1 = \frac{2\lambda\mu}{2\lambda^2 + 2\lambda\mu + \mu^2} \approx \frac{2\lambda}{2\lambda + \mu}$$

$$p_2 = \frac{2\lambda^2}{2\lambda^2 + 2\lambda\mu + \mu^2} \approx \frac{2\lambda^2}{2\lambda\mu + \mu^2}$$

It is noted that we have some approximations for the final results. The approximations are reasonable when the failure rate of a component is much lower than its repair rate. Assuming a failure rate $1\times10^{-4}$/hour (slightly lower than once per year), and repair rate is 0.125/hour (equivalent to the downtime of one working day), the steady-state probabilities are calculated as follows:

$$p_0 = 0.9984, p_1 = 1.5997\times10^{-3}, p_2 = 1.2759\times10^{-6}$$

Here, $p_2$ is the sojourn probability of this 1oo2 structure during the downtime or its average unavailability. ■

## Example 5.15:

### *State Sojourning Probabilities of the Markov Model in Figure 5.18*

Referring to the scenario in Figure 5.18 and based on Equation (5.112), we have the following equation for calculating steady-state probabilities:

$$[p_0, p_1, p_2, p_4]\cdot\begin{bmatrix} -2\lambda & \lambda & \lambda & 0 \\ \mu & -(\lambda+\mu) & 0 & \lambda \\ \mu & 0 & -(\lambda+\mu) & \lambda \\ 0 & \mu & \mu & -2\mu \end{bmatrix} = [0,0,0,0]$$

This results in the following set of linear equations:

$$-2\lambda p_0 + \mu p_1 + \mu p_2 = 0$$

$$\lambda p_0 - (\lambda+\mu)p_1 + \mu p_4 = 0$$

$$\lambda p_0 - (\lambda+\mu)p_2 + \mu p_4 = 0$$

$$p_0 + p_1 + p_2 + p_4 = 1$$

Solving this system gives us the steady-state probabilities:

$$p_0 = \frac{\mu^2}{\lambda^2 + 2\lambda\mu + \mu^2}$$

$$p_1 = \frac{\lambda\mu}{\lambda^2 + 2\lambda\mu + \mu^2}$$

$$p_2 = \frac{\lambda\mu}{\lambda^2 + 2\lambda\mu + \mu^2}$$

$$p_3 = \frac{\lambda^2}{\lambda^2 + 2\lambda\mu + \mu^2}$$

In most cases, given that the failure rate $\lambda$ is much lower than the repair rate $\mu$, the value of $\lambda^2 + 2\lambda\mu + \mu^2$ approximates the value $2\lambda\mu + \mu^2$, and then result of $p_0$ aligns with those in Example 5.14. In this case, $p_3$ denotes the system unavailability, as the half of that in previous example. This is because maintenance resource here is not limited, and thus, the leaving rate from the unavailable state is doubled. ■

It is important to note that the analysis discussed previously is only applicable to Markov models without absorbing states. If we modify the model by removing the transition from State 2 to State 1 in Figure 5.17, as depicted in Figure 5.19, we will have a new model where the process, once in State 2, cannot leave. In other words, if the system fails, it permanently stays in the faulty state without possible restoration. Thus, State 2 is an absorbing state.

According to Equation (5.111), the state equations of this model can be expressed as:

$$[p_0(t), p_1(t), p_2(t)] \cdot \begin{bmatrix} -2\lambda & 2\lambda & 0 \\ \mu & -(\lambda+\mu) & \lambda \\ 0 & 0 & 0 \end{bmatrix} = [\dot{p}_0(t), \dot{p}_1(t), \dot{p}_2(t)]$$

Since all elements in the third row of the transition rate matrix are equal to 0, the matrix equation can be reduced to:

$$[p_0(t), p_1(t)] \cdot \begin{pmatrix} -2\lambda & 2\lambda & 0 \\ \mu & -(\lambda+\mu) & \lambda \end{pmatrix} = [\dot{p}_0(t), \dot{p}_1(t)]$$

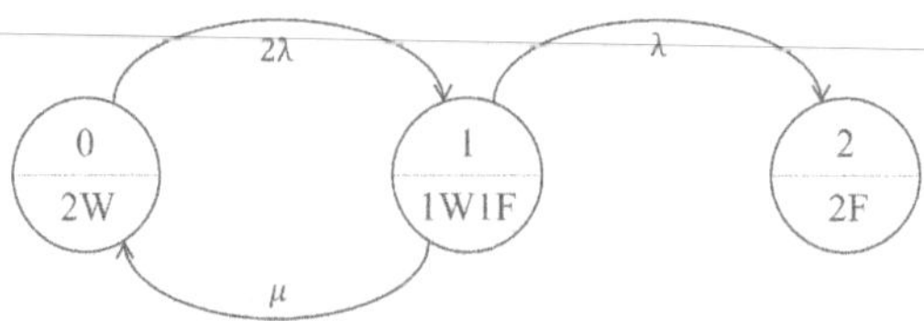


**FIGURE 5.19** Markov model of a parallel structure with an absorbing state.

However, it is impossible to calculate steady-state probabilities of this system for this system because the sojourn probability in the faulty State 2 changes along with time. Laplace transform and inverse Laplace transform techniques can be employed to estimate the expected time for the system to enter the faulty state, but we will not introduce these methods here, particularly when analyzing larger systems with more states. For readers interested in time-dependent solutions and Laplace transform-based calculations for Markov models, readers can refer to Ross (1996). Currently, the time-dependent solutions can typically be calculated using some computational tools like Matlab®. Monte Carlo simulation offers a practical alternative for deriving solutions.

### 5.6.6 General Procedures Using a Markov Model

A basic analysis using the Markov model involves the following steps:

**Step 1: State identification**

The first task is to identify all possible states of the study object, which are represented as circles in this book. Typically, for a component, only two states are considered in a simple Markov analysis: functioning and failed. Occasionally, a "degraded" state is also recognized as a meaningful state. It is suggested to limit the number of states in a model to avoid the exponential growth of the model size with more states. It is only necessary to involve those critical states in a Markov diagram while merging the similar states if possible. For example, if the reliabilities of two components in a parallel structure are nearly identical, a single state is able to denote the situation where one component is functioning and the other has failed.

These identified states should be listed in a table and numbered. The implication of each state should be clearly described. In this book, the fully functioning state is consistently denoted by 0, and higher numbers progressively indicate worse conditions.

**Step 2: Construction of state-transition diagram**

After all states are identified, it is essential to provide all possible transitions of the study object between these states. In this step, it is important (but sometimes difficult) to keep the layout of the states as simple as possible to make the diagram more understandable. To be simple, it is needed to minimize the length of arcs and avoid intersections. A good practice in drawing a Markov diagram is to position states with numerous incoming and outgoing transitions near the center. The computerized drawing programs can support this process, and most of the reliability analysis programs provide such functions.

**Step 3: Establishment of transition rate matrix**

The next task is to identify the rates of all transitions involved in the Markov diagram and to establish a transition rate matrix. When setting up the matrix, each transition rate $\alpha_{ij}$ $(i \neq j)$ for the transition from state $i$ to state $j$ should be placed at the intersection of the $i$th row and the $j$th column, while keeping the diagonal elements empty initially. These diagonal elements, $\alpha_{ii}$, should then be filled according to Equation (5.104) as the negative sum of other elements in the same row.

Once the transition rate matrix is constructed, it is necessary to compare with the state-transition diagram to ensure that the matrix and diagram are in alignment with each other.

**Step 4: Quantitative analysis**

Next step is to list the state equations and solve them, like what was demonstrated in Example 5.13, given that the aim of the Markov analysis to find steady-state solutions. Using the methods outlined above, a Markov model can be used to calculate the numerical values of several measures of a barrier system:

- Long-term or average sojourn probability of a barrier system in each state.
- Long-term or average availability and unavailability of a barrier system, for instance, as seen in Example 5.14, where the unavailability of the system is equal to the sojourn probability of the process in State 2, and its availability is equal to the sum of sojourn probabilities in States 0 and 1.
- The frequency of system failures, such as the visit frequency to State 2 in Figure 5.19.
- The expected number of system failures within a specific period.
- The mean downtime and uptime of a barrier system, which can be determined by multiplying the unavailability and availability by the total time under consideration.

The Markov model is favored by both researchers and practitioners for its clear description of the dynamics of a system. This method is versatile and effective in modeling the complex behaviors of smaller systems. The IEC61508 (2010) standard includes the Markov model in the analysis of barrier systems. However, applying the Markov model to larger systems is challenging, since the model size increases exponentially with the addition of more states.

The classical (continuous-time) Markov model has numerous extensions, and we will introduce several of them that can be useful in barrier engineering in the following subsections.

### 5.6.7 Semi-Markov Process

A strict requirement of a CTMC is that the time the process sojourns in a state is exponentially distributed. However, in barrier engineering, the real operations do not always follow such an assumption. For example, the repair times are found in many studies are log-normally distributed (Almog, 1979). Therefore, to utilize the strong modeling capability of the Markov diagram, it is meaningful to adjust the basic Markov models for more applications.

If the duration a process remains in a state follows a general distribution, the process is referred to as a *semi-Markov process*. More specifically, a semi-Markov process is defined as a stochastic process with a state space $\Omega = \{0, 1, \ldots, n\}$, where upon entering state $i$, it has a probability $p_{ij}$ to enter state $j$ in the next transition, and the duration ($ST_{ij}$) spent in state $i$ before transitioning to state $j$ follows a general distribution.

In a semi-Markov process, the memory is not only limited within the current state but also covers the duration the process has spent in that state. For example, when considering the completion of a repair, not only the current moment is the determinator, but the time elapsed since the repair began is also considered.

The distribution of the sojourn time $ST_i$ in state $i$ at time $t$ is given by:

$$ST_i(t) = \sum_{\substack{j=0 \\ j \neq i}}^{n} p_{ij} ST_{ij}(t) \tag{5.115}$$

We also have $\Sigma_{J=0}^{N} p_{ij}(t) = 1$, for all $j \in \Omega$. If $\Pr(ST_i < t) = 1 - e^{-\alpha_i t}$, the semi-Markov process simplifies as a basic Markov process.

This topic will not be explored further in this chapter, but readers can find much more details about semi-Markov process in the works of Korolyuk et al. (1975) and Sahner et al. (1996).

### 5.6.8 Multiphase Markov Process

Multiphase Markov processes are employed in some scenarios in barrier engineering, particularly when some internal or external events at predefined time points, such as proof tests and preventive maintenance, can change the progression of a Markov process. Here, the term of "phase" denotes the period between these changes.

Two types of changes can occur in the Markov model for reflecting the impacts of events:

- The transition rate matrix is changed after an event. For instance, preventive maintenance (before a failure) improves the component reliability, thereby reducing the transition rate from the functioning state to the failed state.
- The state in which the system resumes post-event is changed. For example, corrective maintenance (after a failure) restores the system from a failed state to a partially functioning state, rather than the as-good-as-new state as the system is put in operation.

With these changes, a classical Markov process becomes multiphase, in other words, a group of basic Markov processes reflecting different scenarios. Such an extension enables a more accurate analysis of systems under various conditions, adapting for more complex real-world situations. Readers can refer to the paper by Innal et al. (2016) for understanding more about using the multiphase Markov model in the integrity analysis of technical safety barriers.

### 5.6.9 Piecewise Deterministic Markov Process

The Piecewise Deterministic Markov Process (PDMP) represents a more generalized model than the Semi-Markov Process and the multiphase Markov Process. It is particularly useful when the distribution of the time to the next transition is unknown, or when changes occur at non-predefined time points. Therefore, PDMP can be a more flexible approach in analyzing the systems with more complex dynamics.

A PDMP can be defined as a hybrid Markov process $(X(t), \overline{y}(t), t \geq 0)$, which integrates both a discrete variable or vector $X(t)$ in a finite state space $\Omega$ and a vector $\overline{y}(t)$ in a continuous space $\mho$. This dual representation allows the PDMP to study the systems where state transitions are influenced by both sudden changes and gradual evolutions over time.

In barrier engineering, the PDMP is particularly effective in analyzing the performance of mechanical barrier components. Consider a scenario where a barrier component undergoes degradation due to both sudden shocks and continuous corrosion. The influence of sudden shocks can be modeled with the discrete variable $X(t)$ that can cause immediate changes of the system state. On the other hand, the continuous corrosion process can be modeled by the vector $\overline{y}(t)$ that evolves gradually until it reaches a critical threshold leading to a failure.

Considering the safety barriers that need to be exposed to corrosive environment in the normal operation and to withstand extremely high load when demands arise, the PDMP model is very useful for predicting their failure points, planning maintenance schedules, and improving the overall reliability and resilience of these systems.

Readers can refer to the PhD thesis of Torres (2022) for a comprehensive understanding on the PDMP model and its applications in maintenance of technical systems.

## 5.7 PETRI NET

### 5.7.1 Model Description

A Petri net (PN) is named after Carl Adam Petri (1926–2010), who first developed this model to describe chemical processes. His work documenting the method dates back to 1962. PN is an alternative to Markov models, and similar to Markov models, PN also provides a useful method for reflecting the state transitions of an item.

Figure 5.20 provides a very simple example of a basic Petri net, which is represented by a directed bipartite graph. The *static elements* of a basic PN include:

- *Places*: Represented as circles and denoted as $pa_i$ in Figure 5.20,
- *Transitions*: Represented as rectangles or bars, and denoted as $tr_j$, and
- *Directed arcs*: Represented as arrows connecting places and transitions.

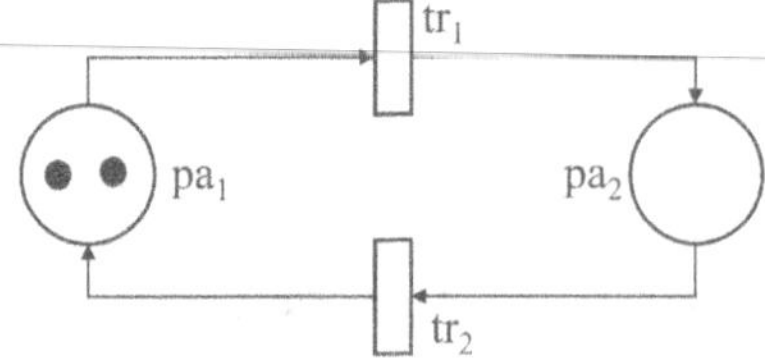


**FIGURE 5.20** An introductory example of Petri net.

These elements are termed static or structural components of the PN model because they keep unchanged regardless of which state the system under study is in. It is important to note in PN that a place can be directly connected to a transition, with a directed arc $\langle \text{pa}_i \rightarrow \text{tr}_j \rangle$, where $\text{pa}_i$ is the *input place* and $\text{tr}_j$ is the *output transition*, and a transition also can be directly connected to a place, with a directed arc $\langle \text{tr}_i \rightarrow \text{pa}_j \rangle$, where $\text{tr}_i$ is the *input transition* and $\text{pa}_j$ is the *output place*. However, within a PN, neither places nor transitions can directly connect to another node of the same type.

As the counterpart of static elements, the *dynamic elements* of a PN are:

- *Tokens*: Represented as solid points in places, as illustrated in Figure 5.16.

These tokens denote the movable resources within the system, and they are only stored in places in a PN model. The distribution of tokens in the places is referred to as the *marking* of the PN, which describes a specific state of the system. For example, the system state depicted in Figure 5.20 can be expressed as $X_0 = \{\text{pa}_1(2), \text{pa}_2(0)\}$, meaning that at State 0, two tokens are in place $\text{pa}_1$, and no tokens are in place $\text{pa}_2$.

Tokens in can be moved via directed arcs, either from a place to a transition or from a transition to a place. Each arc in a PN is assigned a *capacity*, often referred to as *multiplicity*, which defines the number of tokens it can transport at a single time. Although the multiplicity of an arc is always existing in the PN model, in many cases, this indication is omitted when the multiplicity (the arc can transport only one token at one time) is one for simplifying the diagram. In Figure 5.20, all the four directed arcs $\langle \text{pa}_1 \rightarrow \text{tr}_1 \rangle$, $\langle \text{tr}_1 \rightarrow \text{pa}_2 \rangle$, $\langle \text{pa}_2 \rightarrow \text{tr}_2 \rangle$, and $\langle \text{tr}_2 \rightarrow \text{pa}_1 \rangle$ have a multiplicity of one. Such multiplicity means that although two tokens are present in place $\text{pa}_1$in the current state, only one token can be delivered to $\text{tr}_1$in the subsequent state change.

It should be noted that an arc in a PN can be bidirectional, meaning that a place can simultaneously serve as both the input and the output of a transition. Such bidirectional arcs are used in the scenarios where the occurrence of an event does not change the quantity of a specific resource, and we will see examples later in this section.

Transitions in a PN do not keep any tokens, and therefore, when a token moves from a place to a transition, it is actually absorbed by the transition from the modeling perspective. On the other hand, a transition can generate or release a new token to then deliver it to the output place. In some situations, the absorbing and releasing actions occur simultaneously, allowing the PN to maintain a constant number of tokens while redistributing them in different places. However, it is very important to recognize that this redistribution actually involves two distinct actions: absorption and release. The token absorbed and the token released by the same transition may have different meanings or represent different types of resources. For instance, a chemical reaction described by the equation: $CH_4 + 2O_2 = 2H_2O + CO_2$, can be modeled using the PN depicted in Figure 5.21, where Figure 5.21(a) is the state/marking before the reaction and Figure 5.21(b) is the state after the reaction. In state (a), the token at $\text{pa}_1$ denotes a methane molecule ($CH_4$) and the two tokens at $\text{pa}_2$ represent two oxygen molecules ($O_2$). Whereas in state (b), the token in $\text{pa}_3$ is carbon oxygen ($CO_2$) and those in $\text{pa}_4$ are two water molecules ($H_2O$).

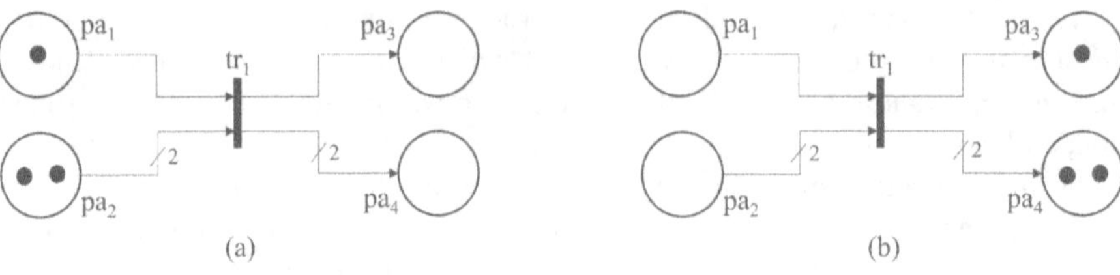


**FIGURE 5.21** Petri net model for a chemical reaction: (a) before reaction and (b) after reaction.

In some advanced PN models, such as colored PNs, different colors are used to signify various types of resources. This book does not investigate those models, and readers interested in exploring colored PNs can refer to the book by Jensen (1996).

When a transition either absorbs or releases tokens, or performs a combination of these actions, it is regarded as being *fired*. Firing a transition means a change of token distribution within a PN. Before a transition is fired, if all its input places contain a number of tokens at least equal to the multiplicities of the corresponding input arcs, this transition is considered as *enabled*, indicating that it is prepared for firing. For example, in Figure 5.14, transition $tr_1$ is enabled, while transition $tr_2$ is not enabled because its input place $pa_2$ has no token, more specifically, the number of tokens is below the multiplicity of the arc $\langle pa_2 \rightarrow tr_2 \rangle$.

From the moment when a transition becomes enabled to when it is actually fired, there may be a delay, referred to as the *firing time* of the transition. This results in what is called as a *timed transition*. Conversely, if there is no delay, meaning that the transition is fired immediately upon being enabled, it is called as an *untimed transition, immediate transition,* or *zero-time transition*. The system state before the activation of an immediate transition is a *transient state* since the system sojourns zero time in this state. In the situation where an immediate transition and a timed transition share the same input places, the immediate transition is given priority to fire.

The firing time of a transition can be modeled using various probability distributions. The international standard IEC62551 (2013) specifies symbols for transitions associated with different types of firing time distributions, as illustrated in Table 5.3.

**TABLE 5.3**
**Symbols of Transitions with Different Firing Times, Specified in IEC62551 (2013)**

| **Firing Time** | **Zero-Time** | **Exponential Distribution** | **Deterministic Delay** | **Other Distribution** |
|---|---|---|---|---|
| Symbol | | | | |

According to David and Alla (2005), if from any reachable state, it is possible to eventually fire a transition, the transition is considered *live*. Then, a PN is considered as live if all its transitions are live. A live PN can effectively model an item without absorbing states, such as a repairable system. We can find that the model in Figure 5.20 is a live PN, whereas the model in Figure 5.21 is not.

### 5.7.2 Stochastic Petri Net

If a PN has at least one timed transition, it is a *timed PN*. When the firing times of all transitions in a PN are stochastic, this model is a *stochastic Petri net* (SPN) (Malhotra & Trivedi, 1995).

**Example 5.16:**

***Petri Net Model of a Parallel Structure with Two Components***

The PN depicted in Figure 5.20 is an SPN, for modeling the failure and repair behaviors of a parallel structure including two identical components.

In this model, the presence of two tokens in place $pa_1$ indicates that both components are functioning, and the transition $tr_1$ being enabled indicates that either component is vulnerable to failure. As described by an empty rectangle in Table 5.1, transition $tr_1$ has an exponentially distributed firing time, reflecting the failure time distribution for the components. Since the directed arc $\langle pa_1 \rightarrow tr_1 \rangle$ has the multiplicity of one, only one component can fail at the same time.

When $tr_1$ is fired, one token in $pa_1$ is absorbed by $tr_1$, and a new token is released from $tr_1$ to place $pa_2$. The system transits from state $X_0 = \{pa_1(2), pa_2(0)\}$ to state $X_1 = \{pa_1(1), pa_2(1)\}$, where one component continues to be functioning, while the other within the parallel structure has failed.

Then, transition $tr_2$ becomes enabled, implying the potential for repairing the failed component. At the same time, transition $tr_1$ remains enabled, meaning that the functioning component also can fail. The firings of these transitions are independent, allowing $tr_2$ to potentially fire before or after the next firing of $tr_1$. If $tr_2$ fires first, the token in $pa_2$ is absorbed by $tr_2$, and a new token is released to $pa_1$, reflecting that the failed component is repaired and thus the system is restored to a fully functioning state. The repair time is also assumed to be exponentially distributed in Figure 5.20.

While $tr_1$ fires first, the rest token in $pa_1$ is removed, leading to two tokens sojourning in $pa_2$, and transiting the system into state $X_2 = \{pa_1(0), pa_2(2)\}$. The duration that both tokens stay in $pa_2$ reflects the downtime of the parallel structure.

Given that the arcs $\langle pa_2 \rightarrow tr_2 \rangle$ and $\langle tr_2 \rightarrow pa_1 \rangle$ each have a multiplicity of one, only one failed component can be repaired at a time, implying the limitation in maintenance resources.

Thus, it can be found that the PN model in Figure 5.20 can be used to express the same system with the Markov model in Figure 5.17. ■

If a PN includes both immediate and timed transitions, it is referred to as a *generalized stochastic Petri net* (GSPN) (Malhotra & Trivedi, 1995). A GSPN can distinguish between *vanishing markings* and *tangible markings*. In the situation where at least one immediate transition is enabled, the system will change states instantaneously, representing a vanishing marking. Conversely, a tangible marking occurs when no immediate transitions are enabled, causing the system to remain in the current state for a measurable duration.

When all timed transitions in a GSPN have an exponentially distributed firing time, the model is functionally equivalent to a continuous-time Markov process. Numerical analysis techniques applicable to Markov models are also suitable for such GSPNs.

Moreover, a GSPN can incorporate a specific type of arc known as an *inhibitor arc* (or simply *inhibitor*), which is included in the international standard IEC62551 (2013) as a directed arc from a place to a transition, terminating in a small circle, as shown in Figure 5.22. The role of an inhibitor is to prevent the activation of its target transition when the input place holds a specified number of tokens. For example, in Figure 5.22(a), the capacity of the inhibitor is two, and when its input place $pa_2$ has two or more tokens, the output transition $tr_1$ is disabled or blocked. When there is only one or no token in place as shown in the figure, $tr_1$ is enabled. It also should be noted that when $tr_1$ fires, the token in place $pa_2$ is **not** removed since it is an inhibitor arc linking the place and the transition.

The inhibitor is useful in modeling scenarios where the availability of resources is influential on an event, but the event can only occur if the resource levels are below a specified threshold, and the occurrence (or non-occurrence) of the event does not affect these resources. For example, Figure 5.22 can be used to model how alarm signals function for two trains traveling in opposite directions on a shared track (represented by 2 tokens in $pa_1$). When two signaling systems on both ends of the section are functioning (2 tokens in $pa_2$), alarms alert both train drivers, telling them the situation and thus they can sequentially pass through the section. If only one driver receives the alarm (1 token in $pa_2$), the situation is hazardous since the other train proceeds unaware, and there is a risk of collision (firing of $tr_1$). However, the number of functioning signaling systems is independent of whether a collision occurs (firing of $tr_1$ does not change the number of tokens in $pa_2$).

GSPN can provide greater flexibility and expressiveness compared to Markov models in many applications. The Markov approach needs to determine a known set of possible system states, while GSPN analysis does not require predefining potential states before modeling. This flexibility allows for a deeper understanding of complex systems (how many states they really have) by observing the dynamics of tokens.

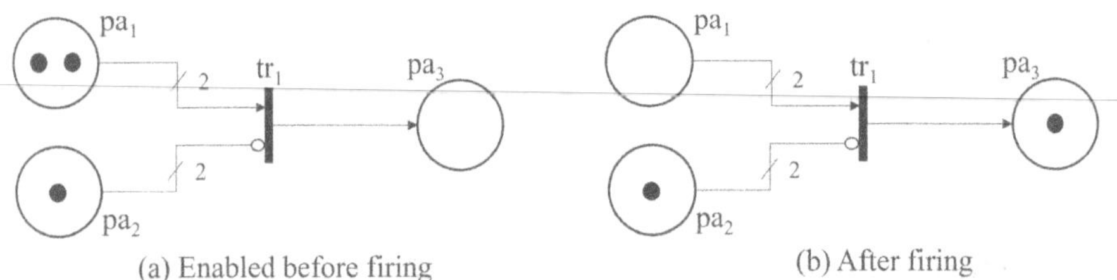


**FIGURE 5.22** A Petri net model with an inhibitor (a) Transition is enabled, (b) Transition fires.

However, SPN is not always equivalent with a basic Markov process. When deterministic transitions or other non-exponential transitions are involved, the model is called as a Deterministic and Stochastic Petri Net (DSPN). Analyzing a DSPN typically requires Monte Carlo simulation. For more comprehensive details on basic Petri net models, refer to IEC62551 (2013).

### Example 5.17:

#### *Petri Net Model of a 1oo2 Barrier System*

Consider a 1oo2 barrier subsystem, which operates as follows:

- Two independent barrier components (A and B) are functioning in the normal situation, and their time-to-DU-failure follows the exponential distribution;
- Periodical proof tests are simultaneously conducted on both components for detecting DU-failures;
- If a DU-failure in any of the two components is detected, it can be repaired, with the repair time following a lognormal distribution;
- The subsystem is unavailable when neither of the components is functioning (due to undetected failure or ongoing repair).

We can build a DSPN for this system as illustrated in Figure 5.23.

The model functions as follows:

The tokens in $pa_1$ and $pa_3$ represent two functioning barrier components, A and B, respectively. The firings of $tr_1$ and $tr_3$ are exponentially distributed, reflecting the failure of these components. On the right side of Figure 5.23, firing of transition $tr_5$ with the deterministic delay indicates a proof test, depositing a token in $pa_6$.

If no DU-failures are detected, no token in $pa_2$ and $pa_4$, and one token in $pa_6$, the enabling condition of immediate transition $tr_6$ is met. Firing this transition to release a new token in $pa_5$, initiating a new proof test period.

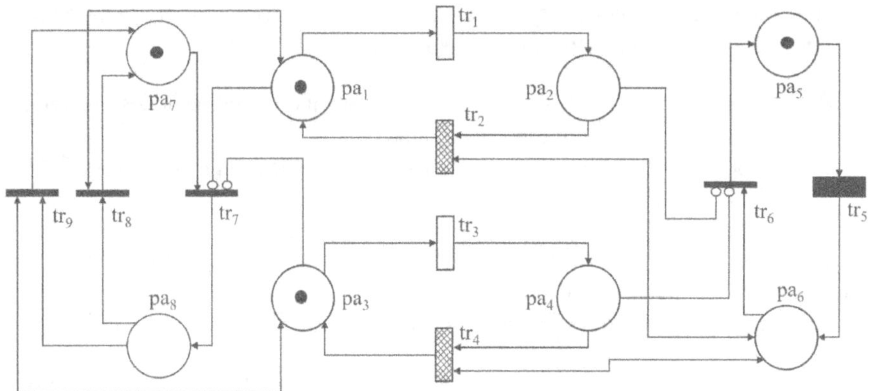


**FIGURE 5.23** Petri net model for a 1oo2 barrier subsystem with periodical proof tests.

When a token is in $pa_2$, it indicates that component A has failed. Repair cannot start until the failure is detected during a proof test, and thus, transition $tr_2$ remains disabled without a token in $pa_6$. However, when tokens are present in both $pa_6$ and $pa_2$, $tr_2$ is enabled and then is fired following a lognormal distribution. It is noted that the arcs between $pa_6$ and $tr_2$ are bidirected, meaning that the firing of $tr_2$ does not change the number of tokens in $pa_6$. The token in $pa_6$ is then removed when no token is in $pa_2$ (indicating repair completion). The process for component B mirrors the above, utilizing $tr_4$ and $pa_4$ instead of $tr_2$ and $pa_2$.

The left part of Figure 5.23 captures the overall system state. A token in $pa_7$ indicates that the system is functioning. The system fails when neither component is functioning. With the control of inhibitors from $pa_1$ and $pa_3$ to $tr_7$, transition $tr_7$ is only enabled when no token is in $pa_1$ or $pa_3$, and then it is fired in no time. The system can be restored either by repairing component A or B. For example, depositing a token to $pa_1$ enables and immediately fires $tr_8$. The bidirectional arcs between $pa_1$ and $tr_8$, ensuring that the firing of $tr_8$ does not affect $pa_1$. The mechanism for system restoration through the repair of component B is same. ■

The DSPN model depicted in Figure 5.23 enhances traditional Markov methods by relaxing several underlying assumptions, thereby increasing flexibility to capture more details of system operations. For example, this DSPN enables modeling of sequential tests on the two components (e.g., testing component A first, followed by component B) and facilitates the evaluation of various repair strategies (e.g., add an inhibitor from $pa_2$ to $tr_4$ to prioritize component A in repair).

However, the introduction of numerous arcs, which represent different logics and interactions, can the model difficult to be understood. Therefore, it is important to develop methods that improve the readability of PN models with complex logics for larger systems with multiple components.

### 5.7.3 Reliability Block Diagram Driven Petri Net

Petri net has numerous extensions tailored for various applications, such as the colored Petri net mentioned in the previous subsection, which are often referred to as high-level PNs. In IEC61508 (2010), a reliability block-diagram-driven Petri net (RBD-PN) is recommended for functional safety and barrier system analysis.

The RBD-PN enhances the basic SPN mainly in two ways:

- It integrates PN models of individual components within the nodes of a reliability block diagram of the system; and
- It employs control formulas to simplify the graphical representation of PN models.

The purpose of utilizing RBD-PN is to achieve a balance between analytical capability and model readability.

Signoret et al. (2013) have defined two general formulas in the RBD-PN models:

- Predicate (described by "?" or "??"): Control the enabling condition of a transition. It is like "If" in most coding languages, and for example, "??*a*==*b*" can return a message of "True" when *a* is equal to *b*;

- Assertion (described by "!" or "!!"): Update one variable when the transition fires. It is like "Then" in some codes, and for example, "!!*a*=*a*+1" updates the value of *a* by adding 1.

These formulas can simplify the PN model by replacing or eliminating some arcs that could otherwise complicate the diagram. In subsequent RBD-PN illustrations within this book, we use short messages to simplify the above expressions:

- "?Mes" signifies "??Mes==true"
- "?-Mes" signifies "??Mes==false"
- "!Mes" signifies "!!Mes=true"
- "!-Mes" signifies "!!Mes=false"

This modeling approach is both reasonable and beneficial, particularly when constructing a PN for multiple interacting items, as demonstrated in Figure 5.23. When some arcs are used to describe relatively straightforward logic, they can be efficiently simplified. We continue to explore the RBD-PN using the 1oo2 barrier system introduced in Example 5.17.

### Example 5.18:

#### *Reliability Block Diagram Driven Petri Net Model for the 1oo2 Barrier System*

Revisiting the 1oo2 barrier system discussed in Example 5.17, the DSPN model depicted in Figure 5.23 can be segmented into several distinct modules ($M_i$):

- $M_1$: Including $pa_1$, $pa_2$, $tr_1$, and $tr_2$, for modeling the failures and repairs of component A;
- $M_2$: Including $pa_3$, $pa_4$, $tr_3$, and $tr_4$, for modeling the failures and repairs of component B;
- $M_3$: Including $pa_5$, $pa_6$, $tr_5$, and $tr_6$, for modeling proof tests;
- $M_4$: Including $pa_7$, $pa_8$, $tr_7$, $tr_8$, and $tr_9$, for capturing the system state.

Based on these modules, we can redraw the above PN model in Figure 5.24, where each module is delineated with dotted lines to define boundaries. Connections between modules are replaced with small boxes linked to or from transitions, indicating relationships between modules. A box directed into a transition represents a control demand from another module, with the text within the box denoting the relevant place. On the other hand, a box directed from a transition (no such box in Figure 5.24) indicates which place needs the updated information.

Next, these boxes are replaced with control formulas of predicates and assertions and so that less symbols are needed for those repeated elements to simplify the illustration. For example, transitions $tr_8$ and $tr_9$ have rather similar functions and are represented by a single transition with common control formulas. A revised PN using predicates to replace all inter-module arcs is shown in Figure 5.25. In this model, we do not use assertions since the basic PN elements can indicate how the numbers of tokens change after firings.

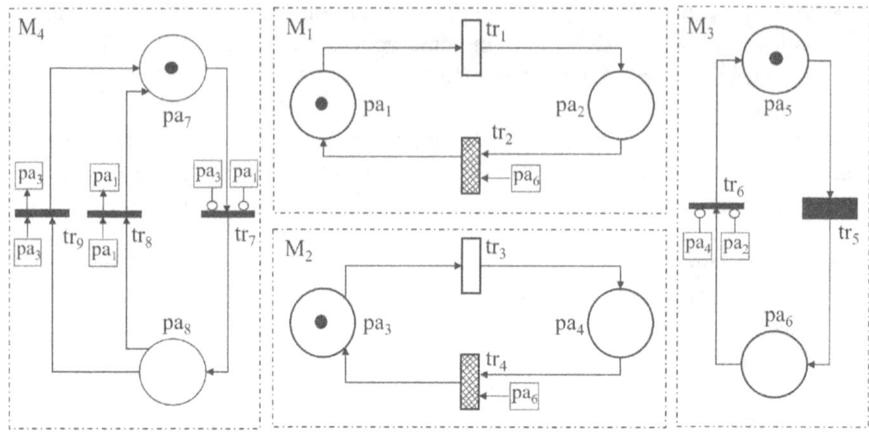


FIGURE 5.24 Modular redrawn Petri net model for a 1oo2 barrier subsystem.

In module $M_1$, a predicate "$??n_{pa_6} == 1$" indicates that for transition $tr_2$ to be enabled, place $pa_6$ must hold exactly one token in addition to the required token at the input place $pa_2$ of this transition. When $tr_2$ fires, a token is deposited to $pa_1$ without affecting $M_3$. Similarly, two predicates in $M_4$ can control this module: "$??n_{pa_1} + n_{pa_3} == 1$" is the predicate of $tr_7$ indicates the additional requirement of enabling the transition is that neither $pa_1$ nor $pa_3$ holds a token. "$??npa_1 + npa_3 > 0$" is the predicate of $tr_8$, meaning that if any of $pa_1$ and $pa_3$ has been deposited a token, the transition can be enabled.

$M_1$ and $M_2$ are independent with each other, but both have mutual controls with $M_3$, and at the same time, $M_4$ is controlled by $M_1$ and $M_2$. We adopt $X_i$ to denote the state variable of component $i$, and $X_S$ to denote the state variable of the system. A reliability block diagram is used to express the system structure and

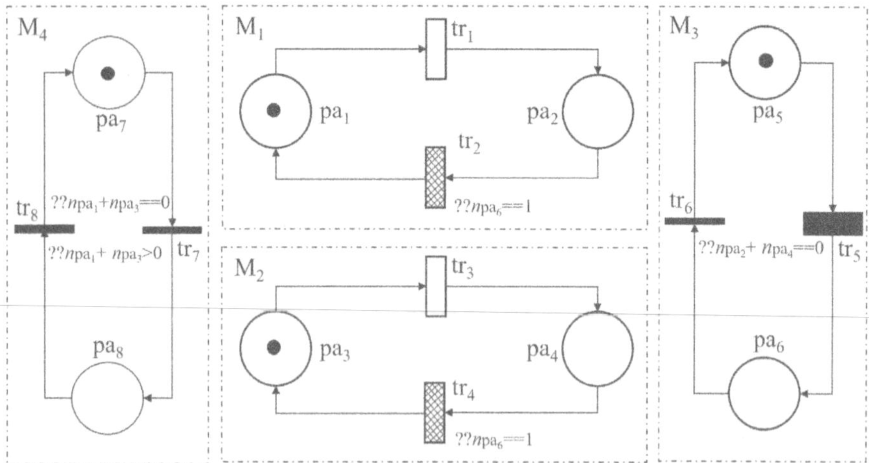


FIGURE 5.25 Re-drawn Petri net model with predicates.

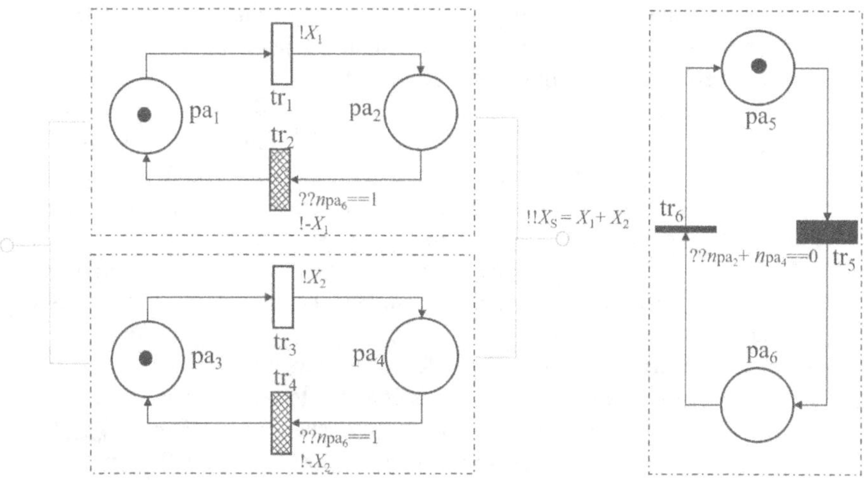


**FIGURE 5.26** An RBD-PN model for the 1oo2 barrier subsystem.

a *global assertion* is introduced to express the module $M_4$. Since $X_i$ only has two values in reliability analysis (1 or true, and 0 or false), the firing of $tr_1$ updates the values of $X_1$ as true, asserted as "!$X_1$", while the firing of $tr_2$ updates the values of $X_1$ as false, asserted as "!–$X_1$".

According to Signoret et al. (2013), symbols (+) for logical OR and (•) for logical AND are used in the control formulas of an RBD-PN. Consequently, operation of the 1oo2 barrier system is modeled using an RBD for its structure and a PN for capturing its dynamic behaviors, as illustrated in Figure 5.26. The global assertion in the model is "!!$X_s = X_1 + X_2$", meaning that when $X_1$ or $X_2$ is true (at least one barrier component is functioning), the value of the system state variable $X_s$ is equal to 1. ■

The software suite GRIF-Workshop® includes a module specifically designed for Petri net analysis, which is based on the RBD-PN and Monte Carlo simulation for quantitative analysis.[3] While GRIF-Workshop® is commercial platform, it offers a free trial version that allows students to conduct analyses on small systems. The symbols and terminology used in the software are similar to those presented in this book, but readers are highly recommended to check the software manual for detailed guidance.

### 5.7.4 Procedure of Petri Net Analysis in Barrier Engineering

In barrier engineering, Petri net analysis includes the following steps:

**Step 1: Understanding of the study object**

Petri net analysis does not require identifying all states of the study object at first. However, it still needs a clear understanding of the system performance indicators of interest, the resources impacting performance, and the relationships between various system elements. For example, in barrier operational analysis, the focus is often on different operational modes of barrier components, their potential failures, and the influence of testing and maintenance/repair resources on system availability.

**Step 2: Establishment of the model structure**

With insights into how resources change in terms of quantity and location, we can define the structure of the PN model. This includes specifying places, transitions, and the initial token distribution. The network diagram should be designed to simplify the subsequent analyses and ensure the readability for potential users. To realize this, the number of arcs should be minimized, and any arc intersections should be avoided. Depending on the system properties and scale, appropriate Petri net models such as SPN and RBD-PN can be selected.

**Step 3: Qualitative analysis and model validation**

After constructing a PN model, it is essential to verify whether the model accurately reflects reality and is suitable for further quantitative analysis. For smaller systems, it is possible to manually explore all potential token movements within the network to identify all possible states and construct a *reachability graph* for the PN. The diagram shown in Figure 5.27 represents the reachability graph for the PN in Example 5.18, where six states have been identified. These states are represented by vectors indicating the number of tokens in different places. Since the tokens in $pa_7$ and $pa_8$ are totally controlled by other modules, these two places are excluded from this reachability graph. A reachability graph is similar with the Markov state transition diagram of the same system but also includes transient states, such as state VI depicted in Figure 5.27. The transient states can be removed if necessary, transforming the reachability graph into a state transition diagram that can even support Markov analysis.

Furthermore, it is necessary to assess the *boundary conditions* and *liveness* of the constructed PN model. For example, if a token cannot exit a place, it is important to examine if the real system possesses an absorbing

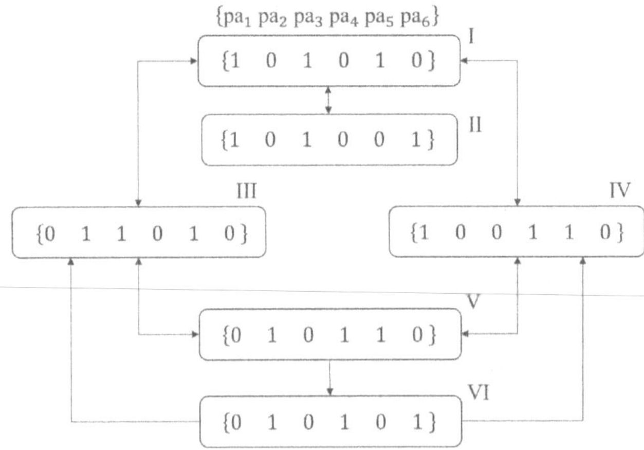


**FIGURE 5.27** Reachability graph of the PN model in Example 5.18.

state. Sometimes, a place may accumulate tokens, meaning that tokens are deposited, but the enabling condition of the output transition of this place is not easy to be satisfied. Validation can be conducted manually or by utilizing specialized software, such as GRIF-Workshop®.

**Step 4: Quantitative validation and analysis**

For GSPN models, numerical solutions for the sojourn probability of states are derived from Markov analysis based on the reachability graph. Otherwise, Monte Carlo simulation (see Section 5.10) is needed for quantitative analysis as suggested in IEC62551 (2013).

Even for the models that cannot be directly converted to continuous time Markov processes, approximate calculations using the Markov method are still valuable for validating the Petri net. Discrepancies between the approximated results and those obtained from Monte Carlo simulations must be analyzed to determine if they arise from approximation errors or modeling mistakes.

Petri nets, including GSPN and RBD-PN, offer robust tools for modeling complex behaviors and parallel events under varying conditions. Despite their effectiveness, the adoption of Petri nets in industry remains limited, mainly because specialized training is needed for the engineers who are not familiar with this method.

## 5.8 COMPARISON OF MODELS

This chapter has introduced several probabilistic models for analyzing barrier performance or likelihood of hazardous events. Here, we compare these models very briefly and provide some suggestions for their appropriate applications, as depicted in Figure 5.28:

- When the study object is a single item, such as a barrier component, and there is no need to break down the item into smaller units or consider the effects of various operational modes on performance, emphasis should be placed on hazard identification. Statistical methods are employed to estimate reliability parameters using experimental and field data.
- For systems where performance is closely tied to structure, particularly those with multiple components, static Boolean models such as FT and RBD are recommended. These methods typically assume binary states (functioning or failed) for components and aim to link component performance directly to overall system performance. These models are straightforward to construct and understand, especially useful for modeling large-scale systems.
- When dependencies between components need consideration, BN is a more effective alternative for modeling such more complex systems. BN can enable an insightful assessment of interdependencies beyond what basic fault trees and reliability block diagrams can provide. BN is also suitable for modeling large-scale systems, but manual calculation is very difficult for BN.
- For scenarios where the dynamic behaviors of systems and their components are critical, Markov methods and Petri nets are applicable. These

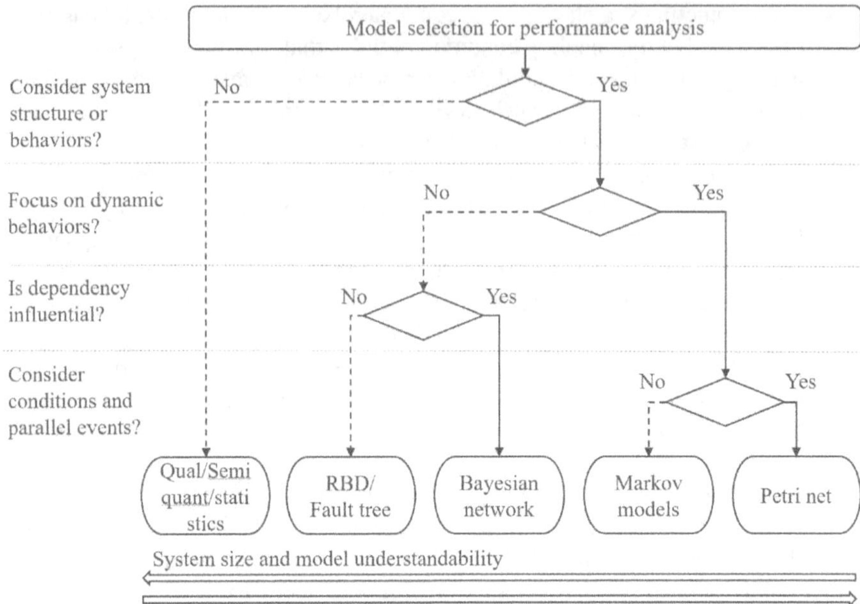


**FIGURE 5.28** Model selection for performance analysis of safety barriers.

approaches are capable of modeling complex activities during system operations, but both of them are sensitive to the number of elements in modeling. The complexity of these models (especially the Markov model) increase exponentially with the addition of more components in a system.

- Petri net is more flexible than the Markov methods in modeling variety of operational conditions and parallel events, even though some extensions of Markov model, such as hidden Markov process and semi-Markov process, have been generalized the original model in more applications. However, Petri net can be challenging for those users who have not received the specific trainings.

For a more exhaustive discussion on modeling methods applicable in reliability analysis (though not specifically tailored for barrier systems), the PhD thesis by Zhang (2018) provides comprehensive insights.

## 5.9 EVENT TREE ANALYSIS

The methods described previously are mainly used for likelihood analysis, focusing on quantifying the probability of hazardous events, particularly system failures, considering specific system structures, operational modes, and environmental conditions. This section shifts the focus toward analyzing development of hazardous/pivotal events in barrier engineering, discussing the methods that assess the impact and outcomes of such events.

Event Tree Analysis (ETA) is a probabilistic method extensively applied in barrier analysis, specifically tailored for the consequence analysis following a hazardous event, supporting the decision-making process for mitigation efforts. It is noted that the analysis for a safety-barrier diagram (see Section 4.4) relies on the principles of ETA.

As suggested by its name, ETA employs a graphical tree-like model to depict possible outcomes following an event. Each node (denoted as A, B, ..., E in Figure 5.29) on this tree, known as *partition point* or *division point*, represents a potential subsequent event or a decision point, with branches illustrating the possible developments or consequences arising from each event. The paths traced from the initiating event to various endpoints (labeled 1, 2, ..., 5 in Figure 5.29) represent different accident scenarios. This graphical representation is not only useful in tracing the sequence of events along different branches but also able to support evaluating the probabilities of varying scenarios and understanding the overall impacts of the initial pivotal event.

In a Bow-Tie diagram, a fault tree is used on the left side to describe the logical interrelationships among events leading up to the pivotal event, while an event tree is located on the right side to illustrate how this pivotal event may develop into a variety of consequences. An event tree typically depicts the chronological progression of events.

Here we use the event tree in Figure 5.29 as the example to introduce the main steps of ETA, which include:

**Step 1: Determining the pivotal event to be studied**

This step is conducted in parallel with the initial phase of a Bow-Tie analysis using FTA, and thus, we do not repeat the details here. However, it is important to note that the pivotal event in many ETA applications should

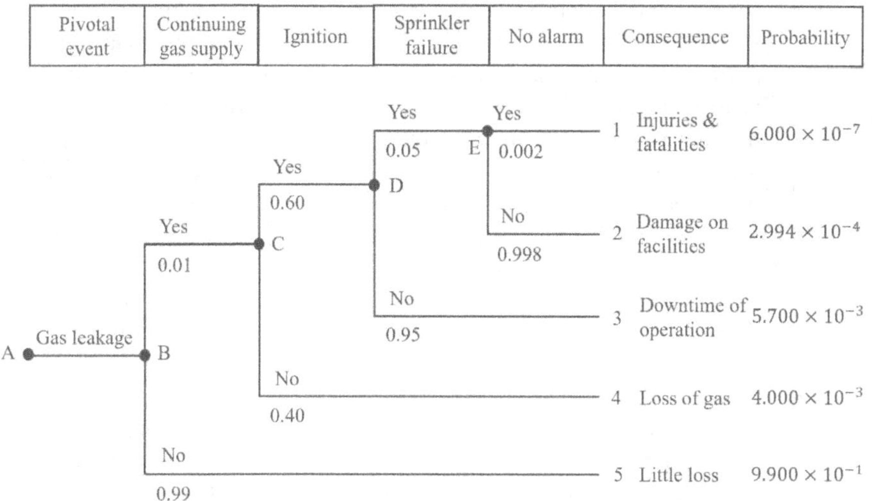


**FIGURE 5.29** An illustrative example of event tree.

be elaborated with additional details, including the type of event, its occurrence time, and physical location.

**Step 2: Identifying the unwanted consequences**

This step is to ascertain the most severe consequence that can reasonably result from the pivotal event. For example, in the event tree depicted in Figure 5.29, the worst consequence identified is the potential of injuries or fatalities due to a fire caused by a gas leakage. In practice, other potential consequences, ranging from minor to catastrophic, should also be cataloged and detailed in a tabular format for clarity and risk assessment. It is important to ensure that the consequences identified are consistent with the level and scope of the risk analysis being conducted. In the given example, it may be unnecessary to consider the impact of the gas leakage on climate change.

**Step 3: Identifying the events and decisions on accidental scenarios**

After determining the most severe consequence, the objective of this step is to identify the key events and decision points that can either prevent this consequence or potentially escalate the situation from the pivotal event to the severe outcome. It is also important to consider how interventions can effectively stop or alter the progression of these events.

In the example depicted in Figure 5.29, the following potential events that can alter the course of the accidental scenario are identified:

- Prompt activation of a safety valve stops the gas flow in the pipeline, effectively preventing gas leakage and minimizing potential losses.
- Avoidance of gas ignition due to a low gas concentration, given strong winds. The leakage continues for a period but results in limited economic impact before it is detected and stopped.
- Effective operation of a water sprinkler system, or another type of fire-fighting system, suppresses a pool fire ignited by the leaked gas. This action causes a significant disruption to the normal operations of the process plant but prevents severe damage to equipment and facilities.
- If the fire-fighting system fails, the fire spreads to additional areas. A fire alarm can be activated, leading to the prompt evacuation of all nearby operators, thereby ensuring their safety. Although some facilities may be damaged, no lives are lost.
- The absence of an alarm results in a delayed response to the fire, resulting in the injuries and even death of several individuals.

These identified events should be sequentially arranged and clearly depicted at the top of the tree model, as shown in Figure 5.29. Each event should be consistently described, demonstrating how it contributes to the progression of the accident scenario toward the most severe consequence or alter its direction to less severe ones.

**Step 4: Constructing the event tree**

With a sequence of potential events defined, an event tree can be constructed, originating from the pivotal event and dividing each event into two branches, representing possible outcomes. In the example provided in Figure 5.29, the initial event following a gas leakage concerns the continuation of the gas supply. This can be determined in terms of whether the gas supply continues or is stopped if the safety valve operates effectively. It is important to note each event in a way that clearly defines the potential to alter the progression of the accident scenario.

Each division on the event tree represents a dichotomy, for example:

- *Yes*: Gas supply continues
- *No*: Gas supply is stopped

When developing an event tree, it is essential that the branches coming from each event are mutually exclusive, meaning only one outcome can occur at a time.

While this example illustrates each event with two outcomes, it is possible, and sometimes necessary, to consider multiple outcomes for a single event. For example, with gas leakage, factors like wind direction could significantly affect the dispersion of the gas, leading to varied scenarios such as spreading "towards a residential area", "towards an expressway", or "towards an unoccupied area".

The construction of the event tree is completed once all relevant events and their respective outcomes have been fully included and recorded. It is important to review these consequences against those identified in Step 2, ensuring that no potential events or consequences are omitted in the construction process.

**Step 5: Assigning probabilities to outcomes**

Each branch stemming from a decision or event node in the event tree should be assigned a probability. These probabilities are conditional, reflecting the likelihood of progressing along each branch given that the preceding event has occurred. It is necessary to emphasize that the sum of the probabilities for all branches emerging from a single node must be equal to 1, regardless of the number of branches, ensuring a balanced probabilistic model.

To determine the probability of a specific outcome, probabilities along the path leading to that outcome are multiplied together. For instance, in Figure 5.29, considering a gas leakage (event A) as the pivotal event, the probabilities for five potential outcomes are calculated as follows:

$$p_1 = p_B p_C p_D p_E = 0.01 \times 0.60 \times 0.05 \times 0.002 = 6.000 \times 10^{-7}$$

$$p_2 = p_B p_C p_D p_E^* = 0.01 \times 0.60 \times 0.05 \times 0.998 = 2.994 \times 10^{-4}$$

$$p_3 = p_B p_C p_D^* = 0.01 \times 0.60 \times 0.95 = 5.700 \times 10^{-4}$$

$$p_4 = p_B p_C^* = 0.01 \times 0.40 = 4.000 \times 10^{-3}$$

$$p_5 = p_B^* = 0.99 = 9.900 \times 10^{-1}$$

where $p_i^*$ represents the probability that event $i$ does not occur, corresponding to the "No" branch.

In the event tree depicted in Figure 5.29, these probabilities are noted alongside each branch. In some cases, the likelihood of each consequence may also be expressed in terms of frequency instead of probability.

Using the probabilities derived from the ETA, we can assess whether the likelihood of severe consequences is acceptably low and determine strategies to further mitigate these risks if needed. This may involve adding additional safety barriers or enhancing existing measures to better protect against identified risks.

ETA can be seamlessly integrated with Bow-Tie analysis, providing a clear visualization of cause-effect relationships of consequential events following the pivotal one. The straightforward nature of probability assessment in ETA makes it easily supported by various software programs, as a practical tool in barrier engineering. Specifically, ETA supports identifying reactive safety barriers and assessing the potential impacts of their absence, thereby effectively supporting the risk-informed decision-making process.

However, it is important to acknowledge that ETA is inherently subjective, as it is the analyst who is responsible for identifying all the paths within the analysis. This method is not particularly effective in analyzing dependencies among parallel events. Additionally, ETA typically focuses on a single pivotal event, which may limit its effectiveness in complex systems that require multi-event analysis.

Given these advantages and limitations, employing ETA along with other physical consequence modeling approaches is highly beneficial. Integrating ETA with methods such as fire and explosion load models, gas cloud dispersion models, and computational fluid dynamics (CFD) analysis can enrich the understanding of the consequences of hazardous events and the efficacy of safety barriers.

NASA (2011) has developed a similar approach with ETA, known as *event sequence diagram*. Readers who have interests can find more introductions about ETA in the literature by Andrews and Dunnett (2000); CCPS (2008); Rausand and Haugen (2020).

## 5.10 MONTE CARLO SIMULATION

We have referenced *Monte Carlo simulation* several times in the previous sections, particularly in the discussion of Petri nets. Monte Carlo simulation can be applied directly in barrier analysis or used to verify results obtained from the models previously mentioned. Here, we outline some foundational principles of Monte Carlo simulation and its application in programming.

Named after the renowned casino city of Monte Carlo in Monaco, the Monte Carlo simulation, or method, is a computational algorithm that uses repeated random sampling to obtain numerical results. This technique is extensively utilized to risk and uncertainty analysis, and since it requires a vast number of samplings that are impractical to perform manually, it is necessary to use computerized programs.

Monte Carlo simulation generally comprises three primary steps:

**Step 1: Building a probabilistic model**

The first step of a Monte Carlo simulation is to define a stochastic problem and develop an appropriate probabilistic model, using the probability distributions introduced in this chapter.

**Step 2: Generating random numbers**

Monte Carlo simulation depends on generating a series of independent random numbers that conform to a specified distribution with a finite mean and variance (Bonate, 2001). Most programming languages offer a random number generator. For example, MATLAB uses the "`rand()`" to generate numbers uniformly distributed between 0 and 1.

### Example 5.19:

#### *Generating Time-to-Failure for an Exponential Distribution*

If we want to generate a sequence of random time-to-failure ($t_1, t_2, \ldots, t_n$) of component whose failure follows the exponential distribution with the failure rate $\lambda = 1$. This time can be from 0 to infinite. If we use the uniform distribution to generate random numbers, we only can obtain a number within (0, 1). For example, using "`rand()`" in MATLAB, we obtain the following five numbers:

0.2191, 0.1800, 0.8272, 0.5100, 0.4170

According to Equation (5.24), we can have:

$$t = -\frac{\ln[1 - F(t)]}{\lambda}$$

Since $F(t)$ can be any value within (0, 1), it can be reflected by the random numbers. Then, using the equation above, we can obtain the following five random time-to-failure (given that $\lambda = 1$, in what time the failure probability can approach to the respective random values) as:

0.2473, 0.1985, 1.7555, 0.6952, 0.5396 ■

In MATLAB, the function "`exprnd()`" can directly generate random numbers following the exponential distribution, but it follows the same consideration in Example 5.19.

While theoretically random numbers are fully random and independent for all practical purposes, in reality, their generation relies on softwarc-specific algorithms, and thus these numbers are known as "*pseudo-random numbers*".

**Step 3: Estimating the statistic**

This step uses the random number sampling to estimate a relevant statistical measure, such as failure rate or MTTF. Generally, the accuracy of the estimation can become higher with the number of samplings and iterations in the simulation.

### Example 5.20:

#### *PDF of an Exponential Distribution Using Monte Carlo Simulation*

In this example, we employ Monte Carlo simulation to approximate the probability density function (PDF) of an exponential distribution, specifically modeling time-to-failure. We set the failure rate $\lambda = 1$, and simulate 1,000 samples from

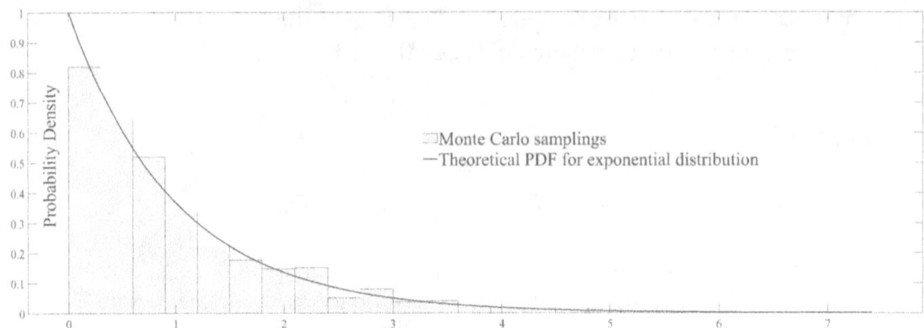


**FIGURE 5.30** Comparison of Monte Carlo simulation results and theoretical PDF curve.

this distribution. The pseudocode for a MATLAB program with explanations is as follows:

```
Lambda = 1;              % Rate of the exponential distribution
Samplesize = 1000;      % Number of samples to generate
Unisamples = rand(Samplesize, 1); % Generate random numbers with a
uniform distribution within (0,1)
Expsamples = -log(1 - Unisamples)/lambda; % Transform the uniform
samples to exponential distribution
```

The transformation of uniform samples to the exponential distribution can be simplified using MATLAB built-in function:

```
Expsamples = exprnd(1/lambda, Samplesize, 1);
```

We then visualize the distribution of these samples using a histogram with 50 bins, as depicted in Figure 5.30. Along with the histogram, we also plot a theoretical PDF curve to compare the empirical data with theoretical expectations. Furthermore, the failure rate can be estimated with the following MATLAB code:

```
Estrate = Samplesize/sum(Expsamples);
```

The data shown in Figure 5.30 suggests an estimated failure rate of 0.9630. When the number of samples is increased to 10,000, the new estimate is 0.9943, which is closer to the theoretical value of 1.0000. ■

## Example 5.21:

### *Monte Carlo Simulation for a Simple Petri Net*

This example explores the use of Monte Carlo simulation to analyze the performance of a barrier component modeled as a simple Petri net. Initially, the Petri net has one token in Place 1, indicating that the component is functioning. Firing of the output transition from Place 1, modeling component failure, follows an exponential distribution with a rate of $10^{-5}$/hour. When firing the transition, the token in Place 1 is removed, and one token is released to Place 2, it means that the component fails. The output transition of Place 2 has a fixed firing time of 720 hours, which corresponds to one regular test per month. When Transition 2 fires, one token is removed from Place 2, and one token is released back to Place 1.

To determine the average unavailability of the component, we implement a Monte Carlo simulation with 10,000 iterations. The pseudocode for the simulation is outlined below:

```
SojournTime2 =0; % Initial variable
TotalSojournTime = 0; % Initial variable as the accumulated time in
Place 2
Trate1 = 0.00001; % Firing rate of output transition from Place 1
for i = 1:10000
      SojournTime1 = exprnd(1/Trate1); % Firing time
      if SojournTime1 < 720 % Failure occurs before a test
            SojournTime2 = 720 - SojournTime1; % Downtime to next
test
            TotalSojournTime = TotalSojournTime + SojournTime2;
      end
      Probability = TotalSojournTime / (720*10000);
      % Calculate the average sojourn probability
end
```

The simulation estimates the average unavailability of this barrier component as 0.0036, consistent with the approximated calculation result. The approximation algorithm for such a failure that is only detected in a regular test can be found in Chapter 8. ■

These two simple examples just illustrate how Monte Carlo simulation can be used in barrier engineering. For further reading on Monte Carlo simulation and its applications in reliability engineering, the book by Zio (2013) is recommended.

## NOTES

1. Developed by Norsys Software Corp (https://www.norsys.com/).
2. It is named after the British mathematician Sydney Chapman (1888–1970) and the Russian mathematician Andrey Kolmogorov (1903–1987).
3. GRIF-Workshop is developed by TotalEnergies. The website of this program is: http://grif-workshop.com/. GRIF means GRaphical Interface for reliability Forecasting.

## REFERENCES


Almog, R. (1979). *A Study of the Application of the Lognormal Distribution to Corrective Maintenance Repair Time*. U.S. Naval Postgraduate School.

Amrutkar, K., & Kamalja, P. (2017). An overview of various importance measures of reliability system. *International Journal of Mathematical, Engineering and Management Sciences*, *2*(3), 150–171.

Andrews, J. D., & Dunnett, S. J. (2000). Event-tree analysis using binary decision diagrams. *IEEE Transactions on Reliability*, *49*(2), 230–238.

Birnbaum, Z. W. (1969). On the importance of different components in a multicomponent system. In P. R. Krishnaiah (Ed.), *Multivariate Analysis II* (pp. 581–592). Academic Press.

Bonate, P. L. (2001). A brief introduction to Monte Carlo simulation. *Clinical Pharmacokinetics, 40*, 15–22.

Cai, B., Kong, X., Liu, Y., Lin, J., Yuan, X., Xu, H., & Ji, R. (2019). Application of Bayesian networks in reliability evaluation. *IEEE Transactions on Industrial Informatics, 15*(4), 2146–2157.

Cai, B., Liu, Y., Liu, Z., Tian, X., Dong, X., & Yu, S. (2012). Using Bayesian networks in reliability evaluation for subsea blowout preventer control system. *Reliability Engineering & System Safety, 108*, 32–41.

Cai, B., Liu, Y., Liu, Z., Tian, X., Zhang, Y., & Ji, R. (2013). Application of Bayesian networks in quantitative risk assessment of subsea blowout preventer operations. *Risk Analysis, 33*(7), 1293–1311.

CCPS. (2008). *Guidelines for Hazard Evaluation Procedures.* Wiley and Center for Chemical Process Safety.

David, R., & Alla, H. (2005). *Discrete, Continuous, and Hybrid Petri Nets.* Springer-Verlag.

EN60300-3-15. (2009). Dependability management – Part 3–15: Application guide – engineering of system dependability. Brussels, Belgium: European Committee for Electrotechnical Standardization.

IEC31010. (2019). Risk management – Risk assessment techniques. Geneva, Switzerland: International Electrotechnical Commission.

IEC61025. (2006). Fault tree analysis (FTA). Geneva, Switzerland: International Electrotechnical Commission.

IEC61078. (2016). Reliability block diagram. Geneva, Switzerland: International Electrotechnical Commission.

IEC61508. (2010). Functional safety of electrical/electronic/programmable electronic safety-related systems. Geneva, Switzerland: International Electrotechnical Commission.

IEC62551. (2013). Analysis techniques for dependability – Petri net techniques. Geneva, Switzerland: International Electrotechnical Commission.

Innal, F., Lundteigen, M. A., Liu, Y., & Barros, A. (2016). PFDavg generalized formulas for SIS subject to partial and full periodic tests based on multi-phase Markov models. *Reliability Engineering & System Safety, 150*, 160–170. https://doi.org/10.1016/j.ress.2016.01.022

Jensen, K. (1996). *Coloured Petri Nets* (2nd ed.). Heidelberg. U. S. National Aeronautics and Space Administration.

Khakzad, N. (2015). Application of dynamic Bayesian network to risk analysis of domino effects in chemical infrastructures. *Reliability Engineering & System Safety, 138*, 263–272.

Korolyuk, V., Brodi, S., & Turbin, A. (1975). Semi-Markov processes and their applications. *Journal of Soviet Mathematics, 4*, 244–280.

Kuo, W., & Zhu, X. (2012). Some recent advances on importance measures in reliability. *IEEE Transactions on Reliability, 61*(2), 344–360.

Malhotra, M., & Trivedi, K. S. (1995). Dependability modeling using petri-nets. *IEEE Transactions on Reliability, 44*(3), 428–440.

MIL-HDBK-217F. (1991). Reliability prediction of electronic equipment. Washington DC, USA: U.S. Department of Defense.

NASA. (2011). *Probabilitic Risk Assessment Procedures Guide for NASA Managers and Practitioners. Guide NASA/SP-2011-3421.* Det Norske Veritas.

OREDA. (2009). *Offshore Reliability Data* (5th ed.). O. Participants. Det Norske Veritas.

Rausand, M. (2014). *Reliability of Safety-Critical Systems: Theory and Applications.* John Wiley & Sons.

Rausand, M., Barros, A., & Hoyland, A. (2020). *System Reliability Theory: Models, Statistical Methods, and Applications* (3rd ed.). Wiley-Blackwell.

Rausand, M., & Haugen, S. (2020). *Risk Assessment: Theory, Methods and Applications.* Wiley.

Ross, S. M. (1996). *Stochastic Processes.* Wiley.

Sahner, R., Trivedi, K. S., & Puliafito, A. (1996). Semi-Markov chains. In *Performance and Reliability Analysis of Computer Systems.* Springer.

Si, S., Zhao, J., Cai, Z., & Dui, H. (2020). Recent advances in system reliability optimization driven by importance measures. *Frontiers of Engineering Management, 7*, 335–358.

Signoret, J. P., Dutuit, Y., Cacheux, P. J., Folleau, C., Collas, S., & Thomas, P. (2013). Make your Petri nets understandable: Reliability block diagrams driven Petri nets. *Reliability Engineering & System Safety, 113*, 61–75.

Torres, R. J. A. (2022). Piecewise deterministic Markov processes for condition-based maintenance modelling: Applications to critical infrastructures. Norwegian University of Science and Technology.

Wu, S., Zhang, L., Zheng, W., Liu, Y., & Lundteigen, M. A. (2016). A DBN-based risk assessment model for prediction and diagnosis of offshore drilling incidents. *Journal of Natural Gas Science and Engineering, 34*, 139–158.

Zhang, J. (2018). Contribution to reliability and availability analysis of novel subsea technologies – Methods and approaches to apply in the early design phase. Trondheim: Norwegian University of Science and Technology.

Zio, E. (2013). *The Monte Carlo Simulation Method for System Reliability and Risk Analysis.* Springer.

# 6 Performance Measures

## 6.1 INTRODUCTION

### 6.1.1 Relevant Concepts

Performance measures are essential in assessing safety barriers to determine if they adequately control risks and mitigate potential hazards. In systems engineering, performance of an item is always linked to a specific function, which defines the intended operation and capability. With this relationship, *performance* can be defined as the degree to which the item fulfills the requirements on the specific function. In barrier engineering, the term "*function*" specifically refers to the reduction of the likelihood of a hazardous event and/or mitigation of its impact (see those in Example 6.1) Therefore, performance of a barrier is about how effectively it can manage a hazardous event when it is implemented and operated.

**Example 6.1:**

*Functional Requirements of Emergency Shutdown System*

ISO13702 (2015) specifies the functional requirements of an emergency shutdown (ESD) system, including as follows:

- Preventing the release of hydrocarbon inventories within pipelines and reservoirs;
- Limiting the quantity of released materials;
- Controlling subsurface safety valve(s); and
- Depressurizing hydrocarbon materials.

It is essential to verify whether an installed ESD system realize all its expected functions. The verification process includes comparing the actual performance of a function against its expected performance, using defined indicators or measures. If the system fails to perform any of the above functions adequately, there can be safety issues. These failures may arise from a malfunction within the ESD system or its limited capacity. ■

In this context, *performance measures* refer to the metrics that gauge the extent to which safety barriers fulfill their intended function. The performance of a barrier is often reflected by a vector of variables, each representing a quantifiable aspect of the barrier system or its subsystems and components. In some literature, researchers also use the terms such like *performance indicators* or *performance variables* as alternatives to performance measures.



DOI: 10.1201/9781003245636-6

### 6.1.2 Commonly Used Measures in Literature

Prashanth et al. (2017) have conducted an exploratory analysis based on interviews with cross-industry experts and built a list of 25 performance variables of safety barriers. These include reliability/repeatability, response time, robustness, trigger event, capacity, efficiency, ability to withstand loads, integrity, implementation delay, adequacy, safety-critical tasks applicability, human dependence, availability, validity, completeness, maintainability, lagging indicators, effectiveness, and level of confidence during operation. Johansen and Rausand (2015) have identified several key requirements on the performance of safety barriers, such as specificity, functionality, reliability, response time, capacity, durability, robustness, auditability, and independence.

Not all the aforementioned performance variables and requirements are quantifiable measures, and some of them actually can be considered design principles and operational strategies. For instance, as discussed in Chapter 3, specificity, functionality, auditability, independence, and human dependence belong to these categories. Additionally, overlaps exist among some of the measures, for example, capacity and load resistance are defined in a very similar way in the literature.

In Table 6.1, we present a list of selected performance measures based on the study of Prashanth et al. (2017), along with the explanations from the literature reviews. In

**TABLE 6.1**
**Selected Performance Measures of Safety Barriers**

| Performance Measures | Definitions/Explanations According to Literature |
|---|---|
| Availability (average) | Percentage of time that a barrier is ready to perform its function in a certain period. |
| Availability (instantaneous) | The probability that a barrier is ready to perform its function at a certain moment. |
| Capacity | Maximum limit to which a barrier can perform its function without being destroyed. |
| Cost | The associated cost or investment in the whole life cycle of a barrier. |
| Cost-effectiveness | Ratio of the benefit of installing a safety barrier (risk reduction) to the cost. |
| Detectability | Easiness to detect the degradation and failures in a safety barrier. |
| Effectiveness | Extent to which a barrier is able to prevent accidents. |
| Maintainability | Easiness, rapidity, and cost to maintain the performance of safety in repairs and preventive maintenances. |
| Reliability | The capability of a safety barrier to be functioning as required in a certain period under a certain environmental and operational condition. |
| Response time | Length of time that a barrier reacts to a certain event/demand. |
| Robustness | Ability of a barrier to perform as expected when the underlying assumptions and operational environment change. |
| Safety integrity | Ability of a barrier to perform its required function in a certain period. |
| Spurious trip frequency | Frequency that a barrier activates when it is not needed. |
| Sustainability | Ratio of risks reduced by a barrier to the needed resources. |

*Source:* Based on literature review (e.g., Prashanth et al., 2017).

subsequent sections, we will further explore these measures, offering more precise definitions for those listed in Table 6.1 as well as others not previously mentioned. Additionally, we will organize these measures into several distinct categories for clearer analysis and discussion.

### 6.1.3 Classification of Measures

In ISO13702 (2015), the commonly used performance measures of safety barriers are categorized into three groups: functionality, integrity, and vulnerability measures. However, based on the discussions in previous chapters, we consider performance measures of safety barriers from two main perspectives: risk management, and barrier design and operation.

- *Measures in risk management*: These measures are employed in the operational management and generic risk management of a process, a project, or an asset. They aim to evaluate the capability of a safety barrier for reducing the risk to the specific asset, as well as its role within the overall risk management framework. Typical measures in this category include effectiveness and cost, and they can be interested by corporate management, organizational-level risk analysts, and policymakers.
- *Measures in barrier design and operation*: These measures are utilized in the specification, verification, and operational evaluation of safety barriers, making them of interest to designers, engineers, and operators. They are used to evaluate whether the design options and operational strategies of barriers are sufficient to fulfill the intended functions of barriers within a given context. This type of measures can be further categorized into several groups based on their applications:
  - *Capacity measures*: This group of measures quantifies the theoretical maximum risk reduction a safety barrier is expected to achieve or its utmost design capability in impeding the development of associated accidental scenarios, according to the design specifications. These measures are particularly relevant to the intended functionalities of safety barriers and are of interest to barrier designers and operators.
  - *Responsiveness measures*: This group of measures evaluate whether or how effectively safety barriers can respond to a hazardous event, ensuring that they perform their barrier function as required. In most international standards addressing technical safety barriers, availability and safety integrity are the primary performance measures within this group. This category also includes measures related to incorrect responses.
  - *Resiliency measures*: This group of measures assesses the performance of safety barriers following an abnormality or failure. This category includes the measures related to how fast and how effectively safety barriers can recover their capacity and responsiveness.

It should be noted that the classifications are not rigid or strictly partitioned. Some measures may still have overlaps with others in terms of definitions and applications,

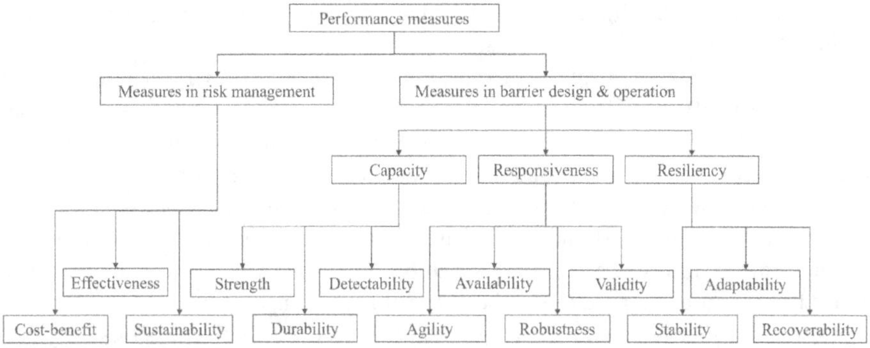


**FIGURE 6.1** Categorization of performance measures of safety barriers.

and it is possible to reassign a measure from one group to another. The primary objective of classification is to facilitate understanding of the meanings and utilities of these measures. Figure 6.1 provides a summary of the classification of performance measures.

## 6.2 BARRIER PERFORMANCE MEASURES IN RISK MANAGEMENT

### 6.2.1 Effectiveness

One of the key performance metrics for barriers in risk management is their effectiveness, which is also known as *sufficiency* or *adequacy*. In this book, we define the effectiveness of a safety barrier as follows:

- *Effectiveness (1)*: The extent to which a safety barrier can reduce the risk associated with specific hazards to the expected level.

According to this definition, effectiveness evaluates whether a barrier can reduce risk sufficiently (Kang et al., 2016) and the amount of risk reduction it achieves given that it is capable. Khakzad et al. (2017), describe effectiveness as the degree to which damage is mitigated, while ISO13702 (2015) uses the term "capacity" for describing effectiveness. The term "*risk-reduction factor*" is also equivalently used as effectiveness in some literature.

It is also important to note that we use "expected level" in the above definition rather than "acceptable level" as in the ALARP principle, suggesting that effectiveness is a relative measure of the actual performance of a barrier compared to the required risk reduction. A barrier is considered fully effective here when it reduces risk to the acceptable level while performing its intended functions as expected. For example, a shutdown valve that stops all gas flow in a pipeline or a braking system that halts a vehicle traveling at 30 km/h within 4.5 m exemplify full effectiveness. On the other hand, a barrier is also regarded effective if it can at least reduce the risk to a required lower level, even this level remains higher than what is considered acceptable. For example, a gas detection system that triggers an alarm is effective, despite risk still exists that operators may not be able to evacuate quickly enough.

The effectiveness of a safety barrier is influenced by the operational context. A barrier that is effective in reducing risk associated with a specific hazard in one context might not be as effective in another. For example, a braking system that performs effectively on a dry road may not stop a vehicle safely on an icy surface.

In many cases, effectiveness is considered independent of time, meaning that once a barrier is installed and operational in a particular setting, its effectiveness is determined. However, effectiveness can become time-dependent if the barrier is subject to potential degradation or the external load changes over time.

One of the foundational metrics for quantifying the effectiveness of safety barriers is the *effectiveness coefficient* (EC). When physical parameters are accessible, the EC of a barrier system can be empirically assessed. For example, Landucci et al. (2015) have introduced a formula to calculate the intensity reduction factor of a water deluge system, which quantifies the effectiveness of this barrier system in mitigating the effects of radiant heat. The EC is calculated in the context as follows:

$$\text{EC} = \frac{\text{Reduced heat load}}{\text{Heat radiation}}$$

More generally, EC can be defined as the ratio of the damage or risk on the asset after installing a safety barrier to the risk prior to installation:

$$\text{EC} = \frac{\text{Risk}(fr,\, cq) - \text{Risk}\left(fr^*,\, cq^*\right)}{\text{Risk}(fr,\, cq)} \tag{6.1}$$

where Risk ($fr$, $cq$) represents the risk estimated according to the frequency ($fr$) and consequence of a hazardous event, and Risk ($fr^*$, $cq^*$) refers to the modified risk with the potentially reduced frequency ($fr^*$) and reduced consequence ($cq^*$) of the same event when a barrier is in place.

Typically, $fr^* \leq fr$, and $cq^* \leq cq$, and thus EC is in the range of [0, 1]. An EC value of 0 indicates that the barrier has no impact on the risk of the studied hazardous event (fully ineffective), whereas a value of 1 signifies that the barrier reduces the risk to 0 or the acceptable level as specified by applicable regulations and standards (fully effective). In practices, since we have considered effectiveness of a barrier is evaluate its capability to reduce the risk to expected level, EC of an effective barrier can be less than 1. In such a situation, multiple barriers are needed to reduce the risk to the acceptable level.

To this point, effectiveness is defined deterministically, which is not aligned with the stochastic nature of risk. Therefore, it is also meaningful to define effectiveness as a probabilistic measure (Landucci et al., 2015):

- *Effectiveness (2)*: The probability that a safety barrier will either stop the progression of an accidental scenario or reduce the risk to an asset of interest to an acceptable level.

Accordingly, we introduce a measure of *effectiveness in probability* (EP), defined as:

$$\begin{aligned} \mathrm{EP} = \Pr\big(&\text{Risk is acceptable with barrier}\,\big|\,\text{Risk is unacceptable} \\ &\text{without barrier}\big) \; or \; \mathrm{EP} = \Pr\big(\text{Risk reduction requirement is met}\big) \end{aligned} \tag{6.2}$$

This definition uses the term "acceptable level", and it implies that the barrier is supposed available and successfully activated at the required moment. However, it acknowledges that a barrier may not be able to fully stop an accidental scenario due to limitations in its functionality. The probability values of EP can thus be incorporated into risk assessment methodologies such as fault tree analysis or event tree analysis. It can be found that this definition uses the reduction of probability to evaluate effectiveness, in contrast to the reduction of damage amount as stated in the first definition.

In cases where there is a specified minimum risk reduction requirement for a safety barrier (denoted as $\vartheta$), Equation (6.1) can be reformulated to define EP as the probability that this requirement is met:

$$\mathrm{EP} = \Pr(\mathrm{EC} \geq \vartheta) = \Pr\left[\frac{\mathrm{Risk}(fr, cq) - \mathrm{Risk}(fr^*, cq^*)}{\mathrm{Risk}(fr, cq)} \geq \vartheta\right] \tag{6.3}$$

Effectiveness of a barrier may vary over time due to potential degradation and changes in external conditions. Consequently, EP as a function of time (denoted as $t$) can be expressed using Equation (6.3) as:

$$\mathrm{EP}(t) = \Pr\left[\frac{\mathrm{Risk}(fr, cq, t) - \mathrm{Risk}(fr^*, cq^*, t)}{\mathrm{Risk}(fr, cq, t)} \geq \vartheta\right] \tag{6.4}$$

When dealing with multiple types of hazards in a process where a single barrier is expected to mitigate the risks associated with not just one hazard, the concept of *coverage* becomes relevant. Coverage evaluates the breadth of accidental scenarios that a safety barrier is capable of impeding and is considered a specific dimension of effectiveness.

### 6.2.2 Cost and Cost-Benefit

Cost serves as a critical performance measure for evaluating the financial implications of both the safety barrier itself and its associated activities. From a lifecycle perspective, the costs related to a safety barrier include the following:

- *Initial cost or initial capital expenditure (CAPEX)*: Capital expenditure relates to the financial resources allocated during the procurement and installation phases of a safety barrier. This typically includes the expenditure incurred in the activities for deploying the barrier, such as feasibility studies, risk assessments, barrier design, and the development of detailed technical specifications in design and development phases. Additionally, CAPEX covers the costs of materials and manufacturing, procurement of necessary components or equipment, and the labor and logistics costs incurred during the installation of barriers.

- *Operational cost or expenditure (OPEX)*: Operational expenditure includes all the costs associated with the ongoing activities necessary to maintain the required integrity and effectiveness of the safety barrier or barrier system after installation. OPEX includes expenses related to the training of personnel to ensure proper operation and responsiveness in emergency scenarios, system monitoring for the continuous assessment of the barrier performance, regular maintenance and timely upgrades for preventing system obsolescence and failure, as well as periodic tests and inspections mandatory to comply with safety regulations and standards.

When designing and specifying barriers, it is crucial to consider the *present value* (PV) of the future operational cost. The general relationship between the *future value* (FV) and the PV of these costs can be represented mathematically as:

$$\text{FV} = \text{PV} \cdot (1 + r)^n \tag{6.5}$$

where $r$ is the *inflation rate*, and $n$ denotes the number of years in operation. To convert future values back to present values, a *discount rate* is used:

$$\text{PV} = \text{FV} \cdot (1 - d)^n \tag{6.6}$$

In this discussion, we will not conduct a detailed decomposition of these costs or present specific cost calculation algorithms for safety barriers, due to the significant difference in cost elements associated with various types of barriers.

In most applications, *cost-benefit* analysis provides a more meaningful evaluation of a safety barrier by comparing the financial investment in the barrier to the potential benefits related to safety improvement or risk reduction. A higher *cost-benefit ratio* indicates that the investment in the barrier yields substantial risk reduction relative to the expenditure. This relationship can be expressed mathematically as follows:

$$\text{Cost}_{\text{B}} = \text{CAPEX}_{\text{B}} + \text{OPEX}_{\text{B}} < \text{Benefit}_{\text{B}} = \text{Cost}_{\text{Risk}} = \Pr(\text{E}) \cdot \text{Cost}_{\text{E}} \tag{6.7}$$

where, $\text{Cost}_{\text{B}}$ denotes the cost of barrier, and $\text{Cost}_{\text{Risk}}$ is the potential cost avoided due to risk reduction for the asset. This potential cost is calculated by considering the probability of a hazardous event (E) occurring and $\text{Cost}_{\text{E}}$, which is the cost or equivalent economic loss as the consequence of the hazardous event.

Cost-benefit analysis should be conducted together with other evaluations, including those addressing regulatory integrity requirements. Given that failures in safety barriers can lead to catastrophic outcomes that cannot be adequately quantified in monetary terms, such as permanent disabilities, human fatalities, and severe environmental damage, integrity requirements always serve as non-negotiable constraints in the analysis. Therefore, while it is interesting to balance the costs against the risk reduction benefits, any scenario failing to meet required integrity is regarded unacceptable.

In certain situations, such as when establishing insurance schemes for employees, it becomes unavoidable to quantify human lives by determining the amount of compensation to be paid in insurance claims. However, when using probabilities and present values of potential costs occurring far into the future, it can lead to significantly underestimated costs and, consequently, unreasonable insurance strategies. This is mainly because of the inherently low probability of a severe accident.

**Example 6.2:**

*Financing Horizon*

Consider an example where the frequency of a severe accident resulting in ten fatalities is once per millennium, and the discount rate is 1% per year. Calculating the present value of the potential cost, assuming a $1.0 million compensation per fatality, yields:

$$PV = (1-0.01)^{1000} \times 10 \times 10^6 \approx \$432$$

This estimate appears unreasonably low given the high severity of potential consequences.

As a response, Lind (2007) introduces the concept of *"financial horizon"*, which defines the maximum timeframe meaningful for financing a project, process, or individual. For equipment or a facility, the financing horizon typically coincides with the end of its useful service life. For individuals, financing horizon may correspond to their expected remaining lifespan. For example, if the average life expectancy in a country is 82 years, and the average age of workers is 42 years, the financing horizon would be $82-42=40$ years. Using this approach for the above scenario, the present value of the potential cost calculates as follows:

$$PV = (1-0.01)^{40} \times 10 \times 10^6 \approx \$6.69 \text{ million}$$

This revised estimation provides a more realistic reflection of the financial impact of a potential fatal accident. ■

### 6.2.3 Sustainability

In the global context of climate change, recognizing the importance of sustainability as a performance criterion for safety barriers is very important. This requires assessing not only their immediate impact on safety of the asset but also considering their broader environmental, social, and economic implications. Currently, the integration of sustainability principles into the analysis of safety barriers is still very limited.

In Chapter 14, we will explore this topic in greater depth, examining the applicability of sustainability principles within the domain of safety barrier engineering. By doing so, our aim is to provide a holistic framework that not only enhances the role of safety barriers in mitigating risks but also aligns them with sustainable development goals, ensuring that these systems contribute positively to both safety outcomes and broader environmental and societal objectives.

## 6.3 CAPACITY MEASURES OF BARRIERS

*Capacity* is a critical concept in safety barrier design and operation, and it is typically reflected in the design specifications as essential parameters. Capacity can also be determined through direct observation or analysis of the design features and operational strategies of a barrier. This section will introduce some key measures of capacity, including strength, detectability, and durability.

### 6.3.1 Strength

In this book, we define the *strength* of a safety barrier as

- *Strength*: The maximum load that a barrier can withstand while performing its intended function without failure.

This load could be any significant physical force or stressor, including barometric pressure, extreme temperature, or shock waves from explosions. The strength of a barrier ensures that it remains functioning under the specified conditions. If the applied load remains within this predefined maximum threshold, the barrier is assumed to be performing effectively. On the other hand, exceeding this threshold indicates a failure of the barrier to maintain its function or at least a high possibility of failure.

The concept of capacity in this context is inherently associated with a specific physical parameter and is determined during the design and manufacturing process of a barrier. It typically serves as a deterministic measure for the performance of a safety barrier. For example, the capacity of a blast-resistant wall is established based on its material properties, wall thickness, reinforcement, and anchoring mechanisms. If the capacity of a wall has been determined to be 5000 bars, it means that the wall should be able to withstand any blast overpressure below this threshold, thereby safeguarding the facility. However, when the blast overpressure exceeds 5000 bars, any deformation, rupture, or complete failure of the wall should be regarded reasonable.

### 6.3.2 Durability

Within the framework of safety barrier analysis, *durability* is defined as:

- *Durability*: The capability of a safety barrier to withstand the load of a demand for the required duration.

Strength assesses the barrier capacity of withstanding the maximum working load, while durability concerns the barrier capacity from the perspective of maximum working time. More specifically, durability is about the endurance of a barrier under a specified load or upon demand (e.g., fire) for a certain period before experiencing a failure or disruption. The required *withstanding time* or *duration* ($t_D$) is typically documented in the specifications of barriers, as the indicator of durability.

However, the actual duration it withstands the demand, denoted as $T_{WS}$, is a stochastic variable. If $T_{WS}$ exceeds the required duration of a specific hazard, such as fire for firewall and flood for dam, the barrier successfully performs its function as expected. Otherwise, if $T_{WS}$ is shorter than the required duration, the barrier fails to meet its required function. Therefore, durability also can be quantitatively assessed as a *durable probability* (DR):

$$\mathrm{DR}(t) = \Pr(T_{WS} \geq t_D) \tag{6.8}$$

In the evaluation of physical passive barriers, durability or withstanding time can be a useful metric. For example, Xie et al. (2022) examine the performance degradation and durability of safety barriers in withstanding thermal propagation within a battery pack, which can illustrate whether the provided safety methods are sufficient.

It is important to distinguish between the withstanding time and the time-to-failure of a barrier system. The time-to-failure refers to the duration from the moment the barrier is put into operation to the occurrence of a failure that hinders the required barrier function. When we use time-to-failure as a measure, we consider a failure in a generic way, for example the failure that can impede the response of a safety barrier, or the failure during the demand if it is not ignorable short. However, we must note that these two failures can be different, in terms of failure modes, failure causes, failure frequency. Consider a pressure relief valve (PRV), the first type of failure can be failure to open since the PRV becomes blocked due to debris, while the second type of failure can be failure to relieve pressure adequately due to partial obstruction under the high-pressure. In practice, it is meaningful to consider the impacts of such differences in performance analysis. We will see such an example in Figure 6.5 in Section 6.4.4.

In the following section, we will discuss the integrity measures commonly used to evaluate the responsiveness of a safety barrier upon a demand. Durability serves as a complementary measure to these integrity metrics. Furthermore, durability has been found valuable in risk assessment scenarios, as it allows for comparison between the duration a safety barrier can withstand a hazard and the actual duration of the hazardous event, thereby determining the associated risk of an asset.

### 6.3.3 Detectability

Within the framework of safety barrier analysis, *detectability* is defined as:

- *Detectability*: The capability of a safety barrier to gather information in order to identify potential hazards to an asset.

Based on this definition, the measure can be quantified as a detectability probability (DT) as:

$$\begin{aligned} \mathrm{DT} &= \Pr(\text{Hazard event is detected} \mid \text{Hazard event occurs}) \\ \text{or } \mathrm{DT} &= \Pr(\text{Hazard is detected} \mid \text{Hazard exists}) \end{aligned} \tag{6.9}$$

Detectability is composed of two elements: *detection coverage* and *detection sensitivity*.

Detection coverage refers to the breadth or comprehensiveness with which the sensing elements of a safety barrier can identify all potential hazards to a specific asset. The sensing elements can be humans or machines, and the detection can be conducted online or offline. High detection coverage reduces the risk of unobserved hazards and is influenced not only by the design and placement of sensors but also by environmental conditions where sensors are used. For example, the effectiveness of catalytic bead sensors and electrochemical sensors in detecting gas leaks depends crucially on their placement, as these sensors must come into direct contact with the gas to detect it. Environmental factors such as wind speed can significantly affect their performance. In cases where the leaked gas disperses rapidly due to strong winds, these sensors may fail to detect the leakage.

Detectability or detection coverage in this subsection is not equivalent with the *diagnosis coverage* (DC), which is used in IEC61508 (2010) and some other standards, measuring the extent to which failures within a safety barrier can be identified by its automatic diagnostic module. DC is actually an indicator of automatic detection coverage for technical safety barriers.

Detection sensitivity, on the other hand, considers the precision and necessity of the information collection capabilities of a safety barrier. A high detection sensitivity of a barrier system increases the probability of revealing hazards, thereby enhancing safety. However, increased sensitivity can also result in a higher incidence of spurious alerts. For example, a catalytic bead sensor utilizes a heated wire coated with a catalyst that reacts with specific gases like methane. This reaction generates heat, causing a change in resistance. While a highly sensitive sensor detects even trace amounts of methane, allowing for early intervention, it may also react to non-target gases or environmental interferences such as certain chemicals, aerosols, or high humidity, leading to false alarms. Alarm management can be used to find an optimal level of detection sensitivity, which has been briefly discussed in Section 3.2.2.

## 6.4 RESPONSIVENESS MEASURES OF BARRIERS

Responsiveness measures assess the capability of a safety barrier to effectively perform its intended function when encountering a hazardous event. This category includes several sub-types of measures, including agility, reliability, availability, accuracy, and robustness. We will introduce these measures but particularly focus on reliability and availability, given their critical relevance across various industries and their incorporation into numerous international technical safety standards.

### 6.4.1 Agility

In this book, *agility* of a safety barrier is defined as:

- *Agility*: The speed with which a safety barrier responds to a demand.

In practical applications, agility is often quantified by the variable of *expected response time* (denoted as $T_{RP}$), representing the duration from the reception of a demand to the activation of the barrier system. In fact, agility can be considered as the reciprocal of $T_{RP}$. This metric is important for evaluating the performance of active safety barriers.

Agility can reflect the efficiency and timing of the process including detection, decision-making, and action. Essentially, $T_{RP}$ includes the following key elements, comprises several critical elements, each determined by the system's design and operational strategies:

- *Detection time*: The interval from the occurrence of a hazardous event to being detected by the sensing subsystem of a barrier.
- *Activation delay*: The duration from detecting the event to initiating a protective action.
- *Action execution time*: The duration required to perform the protective action fully.

For example, in a case study by Yuan et al. (2022), the response time of an ammonia gas detection and alarm system is documented as 60 seconds from when ammonia concentration reaches 17 ppm at the detection point. Then, the response time of an ESD as a barrier includes this detection time plus the time to isolate the gas container. However, if we regard the whole evacuation program as a barrier, its response time of includes the duration from gas leakage to gas dispersion to the detection system, the aforementioned gas detection and alarm system response time, and the time taken for individuals to evacuate from the hazard area.

A shorter response time is generally preferable in barrier engineering. In the study of Landucci et al. (2015), the authors use the term of "time-to-mitigation" as the maximum time required to initiate the pre-planned actions ($t^*_{RP}$). The probability that the actual response time does not exceed this required response time (RPV) can be used to describe responsiveness as:

$$\text{RPV} = \Pr\left(T_{RP} \leq t^*_{RP}\right) \tag{6.10}$$

Given the assumption that a quicker response (higher response rate) can mitigate more risks is valid and suitable for a certain context, response time also can be used to serve as a physical parameter for assessing the effectiveness of a barrier. In these cases, the EC can be defined based on response time as:

$$\text{EC} = \frac{1/T_{RP}}{1/t^*_{RP}} = \frac{t^*_{RP}}{T_{RP}} \tag{6.11}$$

### 6.4.2 Reliability and Availability

Reliability is one of most widely used measures for system performance evaluation. We have seen the definition of reliability in Section 3.3 and in Table 6.1: Reliability in barrier analysis refers to the capability of an item to perform its required barrier function without failure for a stated period of time under the specific environmental and operational conditions.

Based on such a definition, it becomes evident that reliability is influenced by time and operational conditions. From a mechanical perspective, an item is considered reliable if its strength ($S$, see Section 6.3.1) is no less than the load ($L$) that it needs to withstand. Thus, the reliability of an item (e.g., a barrier component) can be expressed as:

$$R(t) = \Pr\left[S(t) \geq L(t)\right] \tag{6.12}$$

We can interpret $S(t) \geq L(t)$ as an event or a condition, the difference between strength and load as the *reliability margin*. This formulation is the foundation of the *physical approach* in reliability analysis. The strength of the item is established by design and operational strategies, but it may decrease over time due to deterioration mechanisms or increase due to improvement efforts. On the other hand, the load is related to the operational condition and can also fluctuate over time. With different load functions, or different environmental and operational conditions, reliability of the item may change.

The *time to failure*, $T$, is defined as the minimum time until the event $S(t) < L(t)$ occurs:

$$T = \min\{t; S(t) < L(t)\} \tag{6.13}$$

In many applications, modeling $S(t)$ and $L(t)$ for complex systems with diverse components can be challenging due to varying metrics and methods for measuring strengths and loads. Amount of information and knowledge in the relevant domain is needed. Consequently, system reliability analysis often employs an *actuarial approach*, where explicit models of strengths and loads are eschewed in favor of describing reliability as a probability distribution function of time to failure. This method allows the reliability of any type of item to be characterized consistently.

In Chapter 5, we introduced several metrics of reliability, such as the mean time to failure (MTTF), failure rate, and residual useful lifetime (RUL). However, throughout the remainder of this book, the term "reliability" specifically denotes the ability to perform a barrier function prior to the first failure or to operate without failure within a specified timeframe. In this context, $R(t)$ represents the probability that the barrier remains failure-free up to time $t$.

As discussed in Chapter 4, barriers can experience both safe and dangerous failures, either of which can compromise reliability. Based on such a consideration, reliability can also be viewed as the capability of an item to avoid unexpected repairs.

Since reliability considers only the conditions before a failure, we need to introduce another metric, *availability*, to evaluate the overall performance of repairable items before and after a failure. This concept is defined in Section 3.3 and detailed in Table 6.1. The symbol $A(t)$ denotes the instantaneous availability of an item at time $t$, and the average availability in a period of $[t_1, t_2)$ is expressed as:

$$A_{\text{avg}} = \int_{t_1}^{t_2} A(s)\,ds \tag{6.14}$$

Typically, the average availability of an item over a period is a more meaningful metric for barrier analysis, and it is influenced by the reliability of the item and the time taken for repairs. For instance, over the long term, the average availability can be calculated using the formula:

$$A_{\text{avg}} = \frac{\text{MTTF}}{\text{MTTF} + \text{MTTR}} \tag{6.15}$$

In this equation, MTTF is the reliability indicator, while MTTR, the mean time to repair, serves as an the indicator of maintainability and efficiency of maintenance. MTTR is determined by design characteristics and operational strategies and is affected by factors such as maintenance resources, spare part inventory, and logistics. It can be found that an increase in reliability, while maintainability remains constant, leads to higher availability. Conversely, if reliability is maintained unchanged but maintainability improves (resulting in a reduced MTTR), availability also increases.

However, the calculation for the availability of a safety barrier cannot simply follow the above general formula, considering that the DU failure of a technical safety barrier is only revealed in a proof test with the interval of $\tau$. Thus, the average availability of a barrier within a proof test interval can be calculated following Equation (6.14) as:

$$A_{\text{avg}} = \int_0^{\tau} A(t)dt \tag{6.16}$$

In practice, *unavailability* is more frequently used than availability for evaluating the performance of technical safety barriers because of their failure probability is very low. Using availability, percentages like 99.99% and 99.9% might appear very similar, but they actually represent a tenfold difference in unavailability. In many industrial practices, *safety integrity* is considered the main performance indicator for safety barrier systems, which is obtained based on unavailability.

Several standards, such as IEC61508 (2010) and IEC61511 (2016) use four discrete *Safety Integrity Levels (SILs)* as performance requirements for technical safety barriers, while ISO13702 (2015) implements the terminology of safety integrity for all barriers. SILs are designed to be straightforward, for facilitating communication and management since they are understandable even for those without an extensive engineering background. SIL of a barrier is determined using the calculated results of barrier unavailability. In the subsequent sections, we will demonstrate how SIL is determined with simple examples.

### 6.4.3 Probability of Failure on Demand

#### 6.4.3.1 Concepts and Assumptions

In Section 3.3, we have discussed the demand modes of safety barriers. In this subsection, we will introduce the performance evaluation measures of low-demand safety barriers and their relevance to SIL. Then, we will examine the measures for high-demand and continuous modes of barriers in the next subsection.

*The Probability of Failure on Demand (PFD)* is the widely utilized metric for assessing the reliability and availability of technical safety barriers in the low-demand mode. PFD is a time-dependent function that quantifies the likelihood of a barrier failing to fulfill its intended function when a demand occurs at a specific time, denoted as $t$. It essentially represents the instantaneous unavailability of the barrier and can be expressed as PFD($t$):

$$\text{PFD}(t) = 1 - A(t) \tag{6.17}$$

In Chapter 3, we have learned about two types of failures: DD (Dangerous Detected) failure and DU (Dangerous Undetected) failure. As technical safety barriers are routinely inspected at regular intervals, it becomes valuable to examine the average value of PFD($t$) in a specific period, which we refer to as average PFD or $\text{PFD}_{\text{avg}}$. We can define $\text{PFD}_{\text{avg}}$ as follows:

- **The average probability of failure on demand, $\text{PFD}_{\text{avg}}$,** represents the average probability over a defined period that a barrier (no matter it is a barrier system, subsystem, or component) is unable to perform its designated barrier function when a demand occurs.

$\text{PFD}_{\text{avg}}$ assesses the ability of a barrier to respond as needed. Accordingly, $\text{PFD}_{\text{avg}}$ is actually equal to the average proportion of downtime within the period of interest, which can be a proof test interval or longer-term. Therefore, $\text{PFD}_{\text{avg}}$ can be expressed as:

$$\text{PFD}_{\text{avg}} = \frac{E[D(t)]}{t} \tag{6.18}$$

where $E[D(t)]$ is the expected downtime. We will discuss various methods for calculating $\text{PFD}_{\text{avg}}$, focusing on a single barrier component in this section, and expanding to more complex configurations in Chapter 8.

Some assumptions and statements are needed for the calculations in this section:

- Proof tests are conducted on a technical barrier at a constant interval, denoted as $\tau$.
- Following each proof test, the barrier is restored to an as-good-as-new state.
- An automatic diagnostic function is integrated into the barrier, enabling the detection and rapid correction of DD failures, which can be fixed in a timeframe significantly shorter than the proof test interval.
- The time to DD failures and the time to DU failures follow the exponential distribution, with the rates of $\lambda_{\text{DD}}$ and $\lambda_{\text{DU}}$, respectively.
- No more than one failure occurs simultaneously, and no additional failures occur while the barrier is in a failed state.

Following the above assumption, the proportion of the DD failure rate relative to the total rate of dangerous (D) failures is termed *diagnostic coverage* (DC), denoted by $\alpha_{\mathrm{DD}}$ in this book, namely

$$\alpha_{\mathrm{DD}} = \frac{\lambda_{\mathrm{DD}}}{\lambda_{\mathrm{D}}} \tag{6.19}$$

And

$$\lambda_{\mathrm{DU}} = (1 - \alpha_{\mathrm{DD}})\lambda_{\mathrm{D}} \tag{6.20}$$

### 6.4.3.2 Approximation Formulas

Considering DD failures can be immediately identified and quickly repaired, their impact on the unavailability of a barrier is relatively minor. Therefore, in our initial analysis, we can disregard the influence of DD failures and provide approximate formulas for calculating $\mathrm{PFD}_{\mathrm{avg}}$ only considering DU failures.

Within a proof test interval, the barrier component is restored to a fully functioning state at the beginning, which we consider as time zero. PFD($t$) is the failure probability of this item, equivalent with $F(t)$. Then, for a single barrier component, to achieve the average probability $\mathrm{PFD}_{\mathrm{avg}}$, the calculation is as following:

$$\mathrm{PFD}_{\mathrm{avg}} = \frac{1}{\tau}\int_0^{\tau} \mathrm{PFD}(t)\,dt = \frac{1}{\tau}\int_0^{\tau} F(t)\,dt = 1 - \frac{1}{\tau}\int_0^{\tau} R(t)\,dt \tag{6.21}$$

Based on Equation (6.21), the relationship between PFD($t$) and $\mathrm{PFD}_{\mathrm{avg}}$ can be plotted, as illustrated in Figure 6.2.

Following the exponential distribution assumption, Equation (6.21) can be expressed as:

$$\mathrm{PFD}_{\mathrm{avg}} = 1 - \frac{1}{\tau}\int_0^{\tau} e^{-\lambda_{\mathrm{DU}} t}\,dt = 1 - \frac{1}{\lambda_{\mathrm{DU}}\tau}\left(1 - e^{-\lambda_{\mathrm{DU}}\tau}\right) \tag{6.22}$$

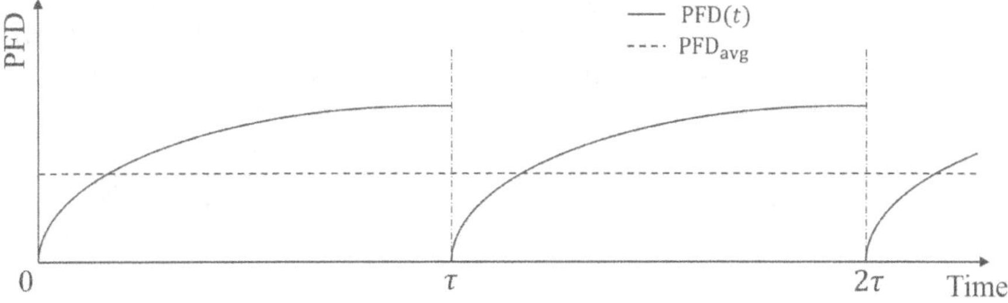


**FIGURE 6.2** Relationship between PFD($t$) and $\mathrm{PFD}_{\mathrm{avg}}$.

Next, we can replace $e^{-\lambda_{DU}\tau}$ by its Taylor series (see Note 6.1), so that

$$\text{PFD}_{\text{avg}} = 1 - \frac{1}{\lambda_{\text{DU}}\tau}\left(\lambda_{\text{DU}}\tau - \frac{(\lambda_{\text{DU}}\tau)^2}{2} + \frac{(\lambda_{\text{DU}}\tau)^3}{3!} - \cdots\right) = \frac{\lambda_{\text{DU}}\tau}{2} - \frac{(\lambda_{\text{DU}}\tau)^2}{6} + \cdots$$

In most applications, the DU failure rate of a barrier is very low, typically lower than $10^{-5}$ per hour, the proof test interval is always no longer than 1 year (8760 hours),

### NOTE 6.1

The Taylor series is a method for expressing a function as an infinite sum of terms calculated from the values of its derivatives at a single point. It is named after the English mathematician Brook Taylor (1685–1731). When the point at which the derivatives are evaluated is zero, the series is specifically known as a Maclaurin series, named after the Scottish mathematician Colin Maclaurin (1698–1746). The general format of the Taylor series of function $f(x)$ is given by:

$$f(x) = f(a) + \frac{f'(a)}{1!}(x-a) + \frac{f''(a)}{2!}(x-a)^2 + \frac{f'''(a)}{3!}(x-a)^3 + \cdots$$

In the compact format

$$f(x) = \sum_{n=0}^{\infty} \frac{f^{(n)}(a)}{n!}(x-a)^n$$

where $f^{(n)}(a)$ is the $n$th derivative of $f(x)$ at the point $a$. For the function $f(t) = e^{-\lambda t}$ around the point $a = 0$, its Taylor series or Maclaurin series can be written as:

$$e^{-\lambda t} = \sum_{n=0}^{\infty} \frac{(-\lambda t)^n}{n!} = 1 - \frac{\lambda t}{1!} + \frac{(\lambda t)^2}{2!} - \frac{(\lambda t)^3}{3!} + \frac{(\lambda t)^4}{4!} \cdots$$

We can assess the difference between the precise value of $1 - e^{-\lambda t}$ and its approximated value when keeping different numbers of terms in the series expansion, as listed in the table below:

| $\lambda t$ | $1-e^{-\lambda t}$ | Diff. (%) | $\lambda t-(\lambda t)^2/2!$ | Diff. (%) | $\lambda t-(\lambda t)^2/2!+(\lambda t)^3/3!$ | Diff. (%) |
|---|---|---|---|---|---|---|
| 0.01 | 0.00995 | 0.50 | 0.00995 | $\approx 0$ | 0.00995 | $\approx 0$ |
| 0.05 | 0.04877 | 2.52 | 0.04875 | 0.04 | 0.04877 | $\approx 0\%$ |
| 0.10 | 0.09516 | 5.09 | 0.09500 | 0.17 | 0.09517 | 0.01 |
| 0.20 | 0.18127 | 10.3 | 0.18000 | 0.70 | 0.18133 | 0.03 |

thus $\lambda_{DU}\tau$ is a small number ($<0.1$). Thus, we can ignore all the very small items since $(\lambda_{DU}\tau)^2/6$, then we have the following approximation:

$$\text{PFD}_{\text{avg}} \approx \frac{\lambda_{DU}\tau}{2} \tag{6.23}$$

Another method to calculate $\text{PFD}_{\text{avg}}$ considers it as the mean fraction of downtime due to a DU failure over the proof test interval, expressed as:

$$\text{PFD}_{\text{avg}} = \frac{E\left[D(0,\,\tau)\right]}{\tau} \tag{6.24}$$

It is important to note that the considered period includes the moment when the next proof test begins, thus the actual interval is $(0,\,\tau]$, but in the formula we write $D(0,\,\tau)$ as it is commonly referenced.

If we test the barrier at time $\tau$ and find that it is in a failed state, we need to determine the conditional mean downtime (MDT) within the $(0,\,\tau]$. First, we calculate the MDT. Given that the barrier operates in only two states: functioning (up) and failed (down), the expected value of downtime is determined by:

$$\begin{aligned} E\left[D(0,\,\tau)\right] &= E\left(D \middle| X(\tau)=0\right)\cdot \Pr\left(X(\tau)=0\right) + E\left(D \middle| X(\tau)=1\right)\cdot \Pr\left(X(\tau)=1\right) \\ &= E\left(D \middle| X(\tau)=0\right)\cdot F(\tau) \end{aligned} \tag{6.25}$$

Thus,

$$\begin{aligned} E\left(D \middle| X(\tau)=0\right) &= \frac{E\left[D(0,\,\tau)\right]}{F(\tau)} = \frac{1}{F(\tau)}\int_0^\tau F(t)\,dt \\ &= \frac{\tau}{F(\tau)}\cdot\frac{1}{\tau}\int_0^\tau F(t)\,dt = \frac{\tau}{F(\tau)}\cdot \text{PFD}_{\text{avg}} \end{aligned} \tag{6.26}$$

It can be found that when $\lambda t < 0.1$, the error due to approximation based on Taylor series deployment is very small.

When a proof test indicates that a barrier component has failed due to a DU failure, the conditional MDT in the proof test interval $(0,\,\tau]$ can be approximated by replacing $e^{-\lambda_{DU}\tau}$ with its Maclaurin series. The equation then simplifies as follows:

$$E\left(D \middle| X(\tau)=0\right) = \frac{\tau}{F(\tau)}\cdot \text{PFD}_{\text{avg}} \approx \frac{1}{1-e^{-\lambda_{DU}t}}\cdot\frac{\lambda_{DU}\tau}{2} \approx \frac{\tau}{2} \tag{6.27}$$

Equation (6.27) implies that when the barrier with constant DU failure rate is failed by time $\tau$, the time of failure is uniformly distributed over $(0,\,\tau]$. The MDT in $(0,\,\tau]$ is therefore $\tau/2$.

Furthermore, $PFD_{avg}$ can be explained as the average unavailability of the barrier due to a DU failure across a proof test interval:

$$PFD_{avg} = \bar{A} = \frac{MDT}{MDT + MTTF_{DU}} \tag{6.28}$$

For a single reliable barrier component, $MTTF_{DU} \gg MDT$, and then we can obtain the same approximation result as:

$$PFD_{avg} = \bar{A} \approx \frac{MDT}{MTTF_{DU}} = \frac{\frac{\tau}{2}}{\frac{1}{\lambda_{DU}}} = \frac{\lambda_{DU}\tau}{2} \tag{6.29}$$

Given the exponential distribution assumption for DU failures, $PFD_{avg}$ remains consistent across all test intervals, as outlined in Equation (6.28). Therefore, $PFD_{avg}$ is equal to the long-term average unavailability of the barrier.

**Example 6.3:**

*Approximation of Average PFD for a Single Barrier Component*

Consider an emergency shutdown (ESD) ball valve. According to OREDA (2009) data book, we can find that the rate of failure to close on demand of such a valve is $3.0 \times 10^{-6}$ per hour. Suppose the proof test interval for such a topside equipment is six months (4380 hours), and its $PFD_{avg}$ can be calculated as:

$$PFD_{avg} = \frac{\lambda_{DU}\tau}{2} = \frac{3.0 \times 10^{-6} \times 4380}{2} = 6.57 \times 10^{-3} \ \blacksquare$$

#### 6.4.3.3 Analysis with Markov Method

We also can calculate $PFD_{avg}$ using a Markov model. For a barrier component with DD and DU failures, it has three possible states, as listed in Table 6.2.

Given the identified states, a Markov transition diagram of this barrier component can be established as shown in Figure 6.3.

**TABLE 6.2**
**States of a Barrier Component with DD and DU Failures**

| State | Description |
|---|---|
| 0 | The barrier is functioning |
| 1 | The barrier is failed due to a DD failure |
| 2 | The barrier is failed due to a DU failure |

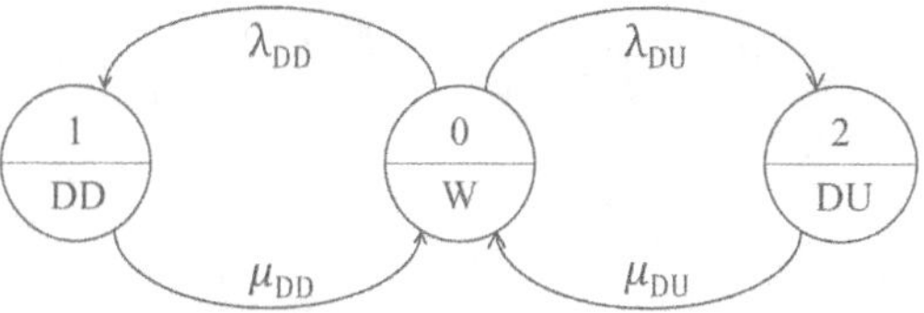


**FIGURE 6.3** Markov model for a barrier with DD and DU failures.

Then, it is not difficult to build a transition matrix according to Figure 7.1 as:

$$\boldsymbol{A} = \begin{pmatrix} -(\lambda_{DD} + \lambda_{DU}) & \lambda_{DD} & \lambda_{DU} \\ \mu_{DD} & -\mu_{DD} & 0 \\ \mu_{DU} & 0 & -\mu_{DU} \end{pmatrix}$$

Here $\mu_{DD}$ and $\mu_{DU}$ represent the restoration rates of the barrier from a failed state. If we assume that the repair time after a DD failure follows an exponential distribution, $\mu_{DD}$ equals the reciprocal of the MTTR. While for $\mu_{DU}$, the scenario is more complex. It encompasses the time from the occurrence of a DU failure to when the barrier is restored, including both the downtime before the failure is detected and the duration of the repair process. Thus, the *mean time to restoration* (MTR) is the sum of the *MDT* and the *mean repair time* (MRT):

$$\text{MTR} = \text{MDT} + \text{MRT} = \frac{\tau}{2} + \text{MRT}$$

Here, we roughly suppose that the total unavailable time of the barrier due to a DU failure also follows an exponential distribution, and $\mu_{DU}$ is thus given as:

$$\mu_{DU} = \frac{1}{\text{MTR}} = \frac{1}{\frac{\tau}{2} + \text{MRT}} \tag{6.30}$$

Now we can calculate the sojourn probability of the system in different state of the Markov model by resolving the following equations:

$$\lambda_{DD} P_0 - \mu_{DD} P_1 = 0$$

$$\lambda_{DU} P_0 - \mu_{DU} P_2 = 0$$

$$P_0 + P_1 + P_2 = 1$$

The solutions are

$$P_0 = \frac{\mu_{DD}\mu_{DU}}{\lambda_{DD}\mu_{DU} + \lambda_{DU}\mu_{DD} + \mu_{DD}\mu_{DU}}$$

$$P_1 = \frac{\lambda_{DD}\mu_{DU}}{\lambda_{DD}\mu_{DU} + \lambda_{DU}\mu_{DD} + \mu_{DD}\mu_{DU}}$$

$$P_2 = \frac{\lambda_{DU}\mu_{DD}}{\lambda_{DD}\mu_{DU} + \lambda_{DU}\mu_{DD} + \mu_{DD}\mu_{DU}}$$

Then, $PFD_{avg}$ of this system can be calculated as the sum of sojourn probabilities in states 1 and 2, as

$$PFD_{avg} = \frac{\lambda_{DD}\mu_{DU} + \lambda_{DU}\mu_{DD}}{\lambda_{DD}\mu_{DU} + \lambda_{DU}\mu_{DD} + \mu_{DD}\mu_{DU}}$$

**Example 6.4:**

*PFD Calculation Using the Markov Model*

Revisiting the ESD ball valve with a DU failure rate of $3.0\times10^{-6}$ per hour. OREDA (2009) data book has collected several types of DD failures of ESD, such as external leakages to process and utility, and abnormal instrument reading. The combined failure rates for these DD failures are estimated as $1.5\times10^{-5}$ per hour. Suppose the proof test interval for such a topside equipment is six months (4,380 hours), MTTR of a DD failure is 6 hours, and MRT of a DU failure is also 6 hours. We aim to calculate $PFD_{avg}$ based on the above Markov model.

First, we calculate $\mu_{DU}$ as follows:

$$\mu_{DU} = \frac{1}{\frac{\tau}{2} + MRT} = \frac{1}{\frac{4380}{2} + 6} = 4.55\times10^{-4}$$

Then,

$$PFD_{avg} = \frac{1.5\times10^{-5}\times4.55\times10^{-4} + 3.0\times10^{-6}\times\frac{1}{6}}{1.5\times10^{-5}\times4.55\times10^{-4} + 3.0\times10^{-6}\times\frac{1}{6} + \frac{1}{6}\times4.55\times10^{-4}} = 6.64\times10^{-3}$$

This result is a bit higher than that obtained using the approximation formula since the Markov method considers DD failures. The contributions of DD and DU failures can be differentiated by calculating the sojourn probabilities in states $P_1$ and $P_2$. $PFD_{avg}$ due to DD failures is $8.94\times10^{-5}$, while $PFD_{avg}$ due to DU failures is $6.55\times10^{-3}$. The latter value is very close to the $PFD_{avg}$ result by the approximation formula in Example 7.3. We find that the contribution of DD failures accounts for only 1.35% of the overall $PFD_{avg}$, explaining why the approximation in Section 7.3.1 is reasonable. ■

It is essential to recognize that all analyses in this section are only for a single barrier component, which might also be considered as a subsystem of a barrier without redundancy. In practical applications, technical safety barriers typically comprise sensing (SE), analyzing (AY), and actuating/executing (EX) subsystems. To determine the $PFD_{avg}$ for the entire barrier system, it is necessary to aggregate the PFD calculation results of all these subsystems.

Let $E_i$ denote the event that barrier subsystem $i$ fails, for $i = 1, 2, 3$. The barrier function is unavailable if anyone of the subsystems fails. Thus, the overall $PFD_{avg}$ of the barrier system can be calculated as:

$$\begin{aligned} PFD_{avg} &= \Pr(E_1 \cup E_2 \cup E_3) = \Pr(E_1) + \Pr(E_2) + \Pr(E_3) \\ &= \Pr(E_1 \cap E_2) - \Pr(E_2 \cap E_3) - \Pr(E_1 \cap E_3) + \Pr(E_1 \cap E_2 \cap E_3) \end{aligned}$$

Assuming that the three subsystems are independent, and each has high reliability (meaning $\Pr(E_i)$ is small), we can simplify the above equation to:

$$PFD_{avg} = \Pr(E_1) + \Pr(E_2) + \Pr(E_3) = PFD_{SE} + PFD_{AY} + PFD_{EX}$$

In the subsequent chapters, we will continue to discuss PFD calculation for more complex systems with different configurations.

#### 6.4.3.4 Safety Integrity Level for Low-Demand Barriers

In IEC61508 (2010) standard, the requirement for availability of technical barriers is categorized into four safety integrity levels known as SILs: SIL 1, SIL 2, SIL 3, and SIL 4. Among these levels, SIL 4 represents the strictest availability requirements, while SIL 1 denotes the least strict requirement for barrier availability.

It is important to understand that an SIL is actually attributed to a specific barrier function and is not directly associated with a barrier system or component. Nevertheless, it is common practice, both in this book and other literature, to use SIL for an entitative barrier when discussing a single barrier function, although this can be somewhat imprecise.

The $PFD_{avg}$ for a safety barrier or function is linked to its corresponding SIL, as outlined in Table 6.3. This table emphasizes the significance of the barrier function in terms of integrity level. For example, a barrier function rated at SIL 3 suggests that a failure of this barrier can increase the risk to the asset of interest by at least 1000 times compared to when the barrier is well functioning.

**TABLE 6.3**
**SIL Target Specified by $PFD_{avg}$ Value**

| SIL | $PFD_{avg}$ |
|---|---|
| SIL 4 | $10^{-5}$ to $10^{-4}$ |
| SIL 3 | $10^{-4}$ to $10^{-3}$ |
| SIL 2 | $10^{-3}$ to $10^{-2}$ |
| SIL 1 | $10^{-2}$ to $10^{-1}$ |

**TABLE 6.4**
**Minimum SIL Requirements from OLF070 (2022) for Some Barrier Functions**

| Barrier Function | SIL Requirement |
|---|---|
| Closing one ESD valve | SIL 1 |
| Opening one blowdown valve | SIL 1 |
| Gas detection with a detector | SIL 2 |
| Fire detection with a detector | SIL 2 |
| Isolating production bore upon high pressure | SIL 2 |
| Isolating annulus in one topside gas lift well from the gas injection manifold | SIL 3 |

In the case of the barrier component described in Example 6.4, its $PFD_{avg}$ is calculated as $6.64 \times 10^{-3}$, corresponding to SIL 2.

It is important to recognize that different industries have specific requirements for SILs. For example, Table 6.4 outlines the minimum SIL requirements for certain safety barrier systems in the offshore oil and gas industry, as provided in OLF070 (2022).

### 6.4.4 Frequency of a Dangerous Failure

For the barrier systems operated in the high-demand or continuous mode, the current version of IEC61508 (2010) suggests using the *average frequency of dangerous failures per hour* to be the performance measure. This measure is commonly referred to as PFH, although the abbreviation does not precisely match the full name. In fact, PFH originated from the term "Probability of dangerous Failure per Hour" used in the 1997 version of IEC 61508. While the calculation has been updated to reflect an average frequency, the abbreviation PFH has been kept, because the term PFH effectively conveys the nature of this measure. Although PFH can be expressed as a time-dependent function, denoted as PFH($t$), it is typically evaluated based on its average value over a specific period.

To calculate PFH, we start by introducing a fundamental reliability metric known as *the rate of occurrence of failures* (*ROCOF*). ROCOF is used to denote the unconditional failure rate of an item at any given time and is represented by $\omega(t)$. Let $N(t)$ r represent the cumulative number of failures that have occurred up to time $t$, with its expected value denoted as $E[N(t)]$, and the derivative of $E[N(t)]$ as $\omega(t)$. In mathematics, the relationship can be expressed as follows:

$$\omega(t) = \frac{d}{dt} E\left[N(t)\right] = \lim_{\Delta t \to 0} \frac{E\left[N(t+\Delta t) - N(t)\right]}{\Delta t} \tag{6.31}$$

Given that $\Delta t$ is a very short interval, it is reasonable to assume that at most one failure occurs within this interval. Consequently, $\omega(t)$ is actually approximately equal to the probability that the item has a (dangerous) failure in the interval of length $\Delta t$:

$$\Pr\left[\text{Failure in } (t, t+\Delta t)\right] \approx \omega(t)\Delta t \tag{6.32}$$

ROCOF is applicable to repairable items such as barriers, which differs from the failure rate $\lambda(t)$ discussed in Chapter 5. For more theoretical insights into ROCOF, readers can refer to the article by Yeh (1997).

Then, within a certain time interval $(0, t_1)$, for a barrier component operated in a high-demand mode, its average frequency (rate, $\bar{\omega}$) of dangerous failures is given by:

$$\text{PFH} = \bar{\omega}(0, t_1) = \frac{E[N(0, t_1)]}{t_1} = \frac{1}{t_1}\int_0^{t_1} \omega(t)dt \tag{6.33}$$

In this context, $t_1$ can be replaced with $\tau$ if proof tests on the barrier are conducted with the interval of $\tau$. This leads to the following expression:

$$\bar{\omega}(0,\tau) = \frac{E[N(\tau)]}{\tau} \tag{6.34}$$

Given that the expected number of failures $E[N(\tau)] = 0 \cdot \Pr[N(t) = 0] + 1 \cdot \Pr[N(t) = 1] + 2 \cdot \Pr[N(t) = 2] + \cdots$, and considering barrier components typically exhibit high reliability, we can reasonably assume that no more than one failure occurs within a proof test interval. This assumption implies that $\Pr[N(t) = k] = 0$ for $k > 1$, and thus we derive that

$$\bar{\omega}(0,\tau) = \frac{E[N(\tau)]}{\tau} = \frac{\Pr[N(t) = 1]}{\tau} = \frac{F(\tau)}{\tau} \tag{6.35}$$

Since $F(\tau) = \int_0^\tau f(t)dt$, this equation can be written as:

$$\bar{\omega}(0, \tau) = \frac{F(\tau)}{\tau} = \frac{1}{\tau}\int_0^\tau f(t)dt = \bar{f}(0,\tau) \tag{6.36}$$

where $\bar{f}(0, \tau)$ is the average failure probability density in the interval $(0, \tau)$.

For a single barrier component whose failures follow the exponential distribution, we use $\lambda_D$ to denote the rate of dangerous failure, which can be the sum of $\lambda_{DD}$ and $\lambda_{DU}$. Since the dangerous failure probability of the item can be obtained as $F(\tau) = 1 - e^{-\lambda_D \tau}$, integrating with Equations (6.33) and (6.35), and considering $1 - e^{-\lambda_D \tau} \approx \lambda_D \tau$ when $\lambda_D \tau$ is small, we derive the following approximation for PFH:

$$\text{PFH} = \frac{F(\tau)}{\tau} = \frac{1 - e^{-\lambda_D \tau}}{\tau} \approx \frac{\lambda_D \tau}{\tau} = \lambda_D \tag{6.37}$$

This implies that given that barrier components are highly reliable, the probability that more than one failure occur within one proof test interval is very low, ROCOF or PFH of the barrier is equal to its dangerous failure rate in terms of values.

Requirements of SIL can also be linked to PFH values, as depicted in Table 6.5.

**TABLE 6.5**
**SIL Target Specified by PFH Value**

| SIL | PFH |
|---|---|
| SIL 4 | $10^{-9}$ to $10^{-8}$ |
| SIL 3 | $10^{-8}$ to $10^{-7}$ |
| SIL 2 | $10^{-7}$ to $10^{-6}$ |
| SIL 1 | $10^{-6}$ to $10^{-5}$ |

**Example 6.5:**

*PFH of Anti-Lock Braking System*

Examples 6.3 and 6.4 featuring the ESD system may not perfectly illustrate the calculation of PFH, since these barriers primarily operate in the low-demand mode. A more relevant example for PFH might involve barrier systems in vehicles, such as the anti-lock braking system (ABS). The ABS serves as a barrier by preventing wheel lock-up during braking, thereby maintaining traction with the road surface. It can be viewed as both a proactive barrier against loss of control and a reactive barrier against lack of control.

In the study of Khorasani-Zavareh et al. (2013), the authors discover that the average rate of fatal road traffic crashes related to brake failures in ABS-equipped vehicles was significantly lower at 30.0 per 1000 vehicle-years, compared to vehicles without ABS. By assuming that 1% of these accidents are due to ABS failures and considering that one vehicle-year translates to approximately 500 hours of operation, we can roughly estimate the failure rate, denoted as $\lambda_D$ or PFH, of ABS at about $6.0 \times 10^{-7}$ per hour. This PFH corresponds to SIL 2. ■

As demand frequency increases, the operational mode of barrier systems begins to resemble that of regular production systems, thus making PFH comparable to normal reliability measures such as ROCOF. In the remainder of this book, the focus will predominantly be on the low-demand mode of safety barriers, discussing how to calculate $PFD_{avg}$ in most instances. However, Section 8.4 will further explore the applicability of PFD and PFH, investigating situations that can be categorized as either low-demand or high-demand modes for safety barriers.

### 6.4.5 Common Integrity Measures for Low- and High-Demands

Several researchers have introduced common integrity measures for safety barriers that function under both low- and high-demand conditions. For instance, Jin et al. (2011) proposed the measure known as *hazardous event frequency* (HEF), which represents the average frequency of hazardous events (on the asset) per hour resulting from the unavailability of a barrier. HEF accounts for the impact of demand frequencies and does not consider the unavailability of a barrier system as the worst-case scenario. Mathematically, for a low-demand barrier, HEF is calculated as follows:

$$\text{HEF} = \lambda_{\text{DE}} \text{PFD}_{\text{avg}} \tag{6.38}$$

where, $\lambda_{\text{DE}}$ represents the demand rate.

In the scenarios where demands are frequent, namely $\lambda_{\text{DE}} \gg \mu_{\text{DU}}$, if the barrier enters an unavailable state due to a DD or DU failure, there is a high likelihood that demands will come during the downtime of the barrier, leading to a hazardous event on the asset. As demand frequency increases, the value of HEF approaches that of PFH. Finally, in the continuous mode, the asset faces an accident immediately after the safety barrier is unavailable, thus HEF ≈ PFH.

Sobral and Soares (2019) have proposed a measure based on the similar consideration but named it as *probability of a hazardous situation* (PHS). In their framework, when $\text{PFD}_{\text{avg}}$ is calculated, probability of occurrence of the initiating event (equivalent with $\lambda_{\text{DE}}$) can be associated with SIL, taking into account the risk to the asset of interest. In the study of Kang et al. (2016), such a measure is named *level of confidence* (LC).

Further discussion on the impacts of demand frequencies and the HEF measure will be presented in Section 8.5.

### 6.4.6 Robustness

A safety barrier is considered robust if it can endure high workloads or unforeseen events that were not originally anticipated during its design phase (Liu, 2020). The robustness (definition in Table 6.1) of a barrier can be assessed by measuring the consistency in its effectiveness or performance under diverse operational conditions. A barrier that is highly robust will exhibit minimal deviations in its performance, maintaining its effectiveness regardless of changes in the operating environment.

Sensitivity analysis is an effective method for evaluating the robustness of a barrier. This approach involves systematically varying parameters such as workload, environmental factors, or system configurations and observing how the performance of a barrier changes in response to these changes. A barrier with low sensitivity in such an analysis indicates a high level of robustness.

In the manufacturing industries, the Taguchi method (Wu & Hamada, 2021) is widely used for improving product quality with providing higher robustness. This method also can be applied in the sensitivity analysis and robustness in barriers. The Taguchi method quantifies performance loss (or quality loss in the Taguchi terminology) occurring as soon as there is a deviation from *the target value* of a performance measure, such as response time. The Taguchi method differentiates between three types of target characteristics for performance measures:

- *Smaller-the-better*: Desirable characteristics such as failure rate, cost, and response time should be minimized. Ideally, the target value for these characteristics is zero.
- *Larger-the-better*: Characteristics that are preferable when higher, such as time to failure and barrier strength. For these, the inverse of the target value is zero.

- *Nominal-the-best*: Characteristics with a specific target value that must be achieved, such as the thickness and length of a component. These do not aim for minimal or maximal values but rather a precise target.

For the situation of nominal-the-best, the Taguchi's loss function ($L$) can be expressed as

$$L = k(y - m)^2 \tag{6.39}$$

where $y$ is the actual performance in terms of a characteristics, $m$ is the target value, and $k$ is a proportionality constant. When evaluating multiple items or stress levels, the average of $(y_i - m)^2$ is called the *mean-squared deviation* (MSD):

$$\text{MSD} = \frac{1}{n}\sum_{i=1}^{n}(y_i - m)^2 = (\bar{y} - m)^2 + \sigma^2 \tag{6.40}$$

Robustness is assessed by calculating the MSD. The expected performance loss can then be expressed as:

$$L = k\left[(\bar{y} - m)^2 + \sigma^2\right] \tag{6.41}$$

It can be found that there are two ways to reduce the performance loss: (1) reducing the variation around the average, denoted by $|\bar{y} - m|$; and (2) decreasing the deviation of the average from the target, $|\sigma|$.

In the smaller-the-better scenario, where the target is zero, the quality loss function is

$$L = ky^2 \tag{6.42}$$

And

$$\text{MSD} = \frac{1}{n}\sum_{i=1}^{n} y_i^2 = \bar{y}^2 + \sigma^2 \tag{6.43}$$

The main approach to reduce the performance loss here is to decrease the deviation.

For the larger-the-better scenario, where the target is effectively infinity and the inverse of the target is zero, the performance loss function can be described as:

$$L = \frac{k}{y^2} \tag{6.44}$$

$$\text{MSD} = \frac{1}{n}\sum_{i=1}^{n}\frac{1}{y_i^2} \tag{6.45}$$

The main strategy to reduce performance loss in this case is to enhance overall performance. Using robustness as a measure is particularly meaningful in this context.

For further details and calculations using the Taguchi method, readers are encouraged to refer to specialized textbooks, such as Wu and Hamada (2021).

### 6.4.7 Validity

In Section 3.6, we have explored the concept of safe failures in barrier systems, focusing notably on spurious activations, where a barrier unnecessarily activates in the absence of a genuine demand. To evaluate this aspect, we introduce the measure of *validity*, which is defined as:

- *Validity*: The ability of a safety barrier to activate correctly to an actual demand while avoiding unnecessary actions.

The concept of validity is further elaborated in IEC61511 (2016) through the specification of the *spurious trip rate* (*STR*). This metric represents the frequency at which spurious activations or trips occur, referring to situations where a technical safety barrier performs a complete action in implementing its intended function without any actual demand. A higher STR thus means lower validity of the barrier. In some literature, such as Lundteigen and Rausand (2008), a spurious trip means the whole barrier function is performed while a spurious activation means any wrong response as a part of the whole barrier function, but we do not distinguish these two terms in this book.

A company RISKNOWLOGY® has proposed the concept of spurious trip level (STL),[1] analogous to SIL. STLs range from 1 to 6, defined based on the mean costs resulting from spurious trips, where STL 6 indicates the most severe failures leading to the highest cost losses.

A higher STR/STL or a lower validity always implies more unexpected stops of normal production and greater production loss. Therefore, the maximum STR represents the highest allowable frequency of spurious trips. Given that spurious activations are often considered relatively safe, the STR for a barrier component is denoted by the rate of safe failure, $\lambda_{\mathrm{ST}}$. Detailed formulas for calculating STR across different barrier structures will be presented in Section 8.6.

When considering alarms as safety barriers, the concept of validity is expressed through the *false alarm rate*. This rate quantifies the probability of false positives, reflecting how often alarms incorrectly detect hazards and trigger alerts in the absence of actual danger. The methodology for calculating the false alarm rate parallels that used for the STR.

Within the framework of IEC61508 (2010), the *safe failure fraction* (SFF) is recognized as a performance metric of technical barriers. The SFF evaluates not only validity but also integrates considerations of availability. Specifically, SFF is calculated as the ratio of the sum of the average rates of safe failures (including spurious activations) and DD failures to the total failure rate, as:

$$\mathrm{SFF} = \frac{\lambda_{\mathrm{ST}} + \lambda_{\mathrm{DD}}}{\lambda_{\mathrm{ST}} + \lambda_{\mathrm{DD}} + \lambda_{\mathrm{DU}}} \tag{6.46}$$

It is important to note that the value of SFF does not directly indicate the level of validity of barrier system. If the total failure rate remains constant, typically implies a lower proportion of DU failure, and in turn a lower PFD. Hence, in IEC61508 (2010), the value of SFF is also used to determine the SIL of a barrier system. For example, a safety barrier at SIL 3 or SIL 4 is expected to have SFF > 99%. However, since it is not meaningful to consider the total failure rate as constant in system design, Lundteigen and Rausand (2009) have argued that SFF is an unnecessary concept in SIL evaluation.

## 6.5 RESILIENCY MEASURES OF BARRIERS

In this section, we will explore three measures pertaining to the resilience of a safety barrier: *survivability*, *recoverability*, and *adaptability*. Figure 6.4 illustrates the functions of these measures following the response of a barrier to a demand. Survivability evaluates the ability of the barrier in the initial performance declining phase (Phase I) due to natural aging or after a demand, recoverability assesses the ability of the barrier in Phase II when the performance regains from the lowest point, and adaptability measures the performance once it returns to a new steady state in Phase III.

### 6.5.1 Survivability

It is natural that the performance of an entitative barrier degrades with time, and we introduce the term of survivability with the following definition:

- *Survivability*: The ability of a barrier to sustain its performance over time, even in the face of gradual deterioration.

The concept of survivability is important in evaluating how well a barrier resists degradation. Barrier components may deteriorate under stress, occurring in both dormant and active states. For instance, as illustrated in Figure 6.4, barrier performance may decline rapidly or gracefully following a demand. Survivability can be

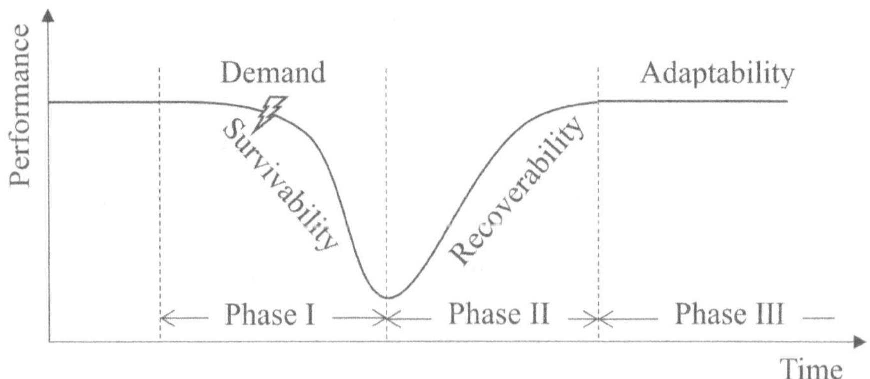


**FIGURE 6.4** Measures of a safety barrier after a demand.

reflected by how long the barrier can sustain essential functions despite this degradation. A prolonged duration in Phase I indicates greater survivability, reflecting the capability of a barrier to maintain its function over time. In this context, survivability is closely linked to durability.

However, durability only considers the performance before the demand completes, while survivability can take into account both the degradation due to a demand (during and after) and the natural aging process of the barrier. This dual consideration can lead to a more holistic understanding of how various factors contribute to the degradation of a barrier.

Survivability also can be quantified by the degradation rate or degradation speed, no matter in the demanding state or in the dormant state. Degradation rate is typically based on physical performance metrics, such as strength degradation rate or agility degradation rate. For instance, if the closing time of a shutdown valve at the onset of operation is 10 seconds or it responds at a rate of 0.1 cycles per second, and after five years the closing time has increased to 15 seconds or the response rate has decreased to 0.067 cycles per second, then the degradation speed would be calculated $(15-10)/5=1$ second per year, or $(0.1-0.067)/5=0.0066$ cycles per second per year.

Furthermore, it is important to recognize that the scope of survivability extends beyond simply measuring how long a barrier can last. Survivability is not only about the duration a barrier can withstand but also concerns the depth of the lowest performance level it reaches. For example, in Figure 6.4, performance level at the lowest point of the curve also reflects the survivability of a barrier. Reconsider the closing time of a valve, we can use the longest closing time or lowest closing rate to denote the lowest point.

Survivability can be quantified in a probabilistic way, where $\mathrm{SU}(t)$ represents the probability that the degraded actual performance (AP) remains above the required minimum performance (RP, or threshold value) by time $t$:

$$\mathrm{SU}(t)=\Pr\left(\mathrm{AP}(t)>\mathrm{RP}\right) \tag{6.47}$$

Figure 6.5 illustrates the concepts related to degradation and survivability. Since the degradation of a barrier component can be detected through continuous

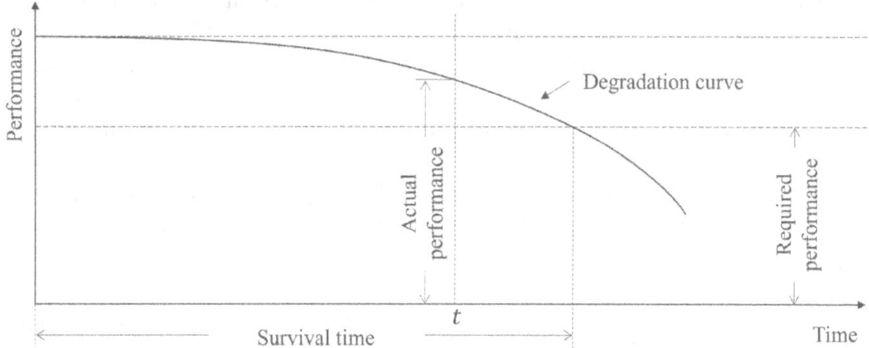


**FIGURE 6.5** Illustrations of concepts related to degradation and survivability.

monitoring or regular inspections (proof tests), specific actions can be implemented to slow or interrupt the degradation process. Consequently, the survivability of a barrier is significantly influenced by its testing and preventive maintenance strategy. These issues will be further explored in Chapter 9.

### 6.5.2 Recoverability

Recoverability, while related to maintainability within the context of safety barriers, serves as a distinct performance measure. Maintainability refers to the easiness and cost associated with keeping a barrier at a satisfactory performance level, and it is considered a design property and is not treated as a performance measure in this book. Conversely, recoverability focuses on the performance of a barrier following its response to a high-stress demand. It is defined as:

- *Recoverability*: The capability of a safety barrier to return to a functioning state after being subjected to a demand and be prepared for subsequent demands.

In this book, recoverability refers to the ability of bounce-back in many applications. The time it takes for a barrier to regain its functionality is a critical indicator of recoverability. As illustrated in Figure 6.4, a shorter duration in Phase II indicates a higher level of recoverability.

In practice, it may not be possible to fully restore the performance of a safety barrier to its original level. Nevertheless, if the performance in the new steady state is above the required threshold, this state is still considered functioning. The time taken from the point of lowest performance to reach this new steady state is utilized to assess recoverability. When quantified probabilistically, RC($t$) can be used to denote the recoverability of an item after a specific duration $t$ from the moment when its performance reaches its lowest level ($t = 0$), represented as:

$$\mathrm{RC}(t) = \Pr\left[\mathrm{AP}(t) > \mathrm{RP}\right] \tag{6.48}$$

RC($t$) is the probability that the actual performance of the barrier has surpassed the minimum required performance. It should be noted that the meaning of $t$ in Equation (6.48) is different with those in Equation (6.47) and all the previous ones.

A further consideration is that performance might be restored during a repair to a new steady state that satisfies the requirements. Therefore, it might be more practical to consider recoverability as the probability that the actual performance of the barrier has stabilized in a new state where the requirements are met, expressed as:

$$\mathrm{RC}(t) = \Pr\left[\mathrm{AP}(t) = \mathrm{AP}(t+1) \middle| \mathrm{AP}(t) > \mathrm{RP}\right] \tag{6.49}$$

This formula is illustrative, as specific time units are not provided. It highlights the meaning of recoverability in reaching a new stability of barrier functionality.

### 6.5.3 Adaptability

For applications involving safety barriers, adaptability can be defined as:

- *Adaptability*: The capability of a barrier to be restored, adjusted, or modified to respond to subsequent demands.

Adaptability is introduced in this chapter to assess the effectiveness of restoration activities on a safety barrier. As depicted in Figure 6.4, adaptability comes into play during Phase III. In a simplified explanation, adaptability can be indicated by the performance level of the new steady state achieved after repair and restoration, reflecting the readiness of a barrier to handle future demands.

Moreover, given that the actual demanding situations and operational conditions of a safety barrier may deviate from initial assumptions, adjustments, and modifications to barrier structures and configurations, operational and activation strategies, and testing and maintenance approaches are sometimes necessary. These changes are informed by insights gained from previous demands, making the barrier compatible with new conditions and continues to meet desired performance criteria. Therefore, adaptability also includes the capacity of the organization managing safety barriers to learn from past experiences and apply those lessons to enhance future barrier effectiveness.

From a risk management perspective, adaptability in a safety barrier implies that its effectiveness in handling a new hazardous event, as well as its integrity in response to the demand, can be improved compared to previous levels. This means that an adaptable barrier is not only able to meet but to exceed its prior performance standards when faced with subsequent challenges.

To quantify adaptability in the probabilistic way, we can model it as the likelihood that the barrier performance ($U$) after experiencing a demand and following adjustments will exceed its previous benchmarks under similar conditions. A potential formula to represent adaptability (AD) at time $t$ is

$$AD(t) = \Pr\left[U_{\text{new}}(t) > U_{\text{old}}(t) \middle| \text{Demand, adjustment}\right] \tag{6.50}$$

Sometimes, we can use the term of *restorability* as the overall measure of recoverability and adaptability, which also can be regarded as the counterpart to the measure of survivability if the moment with the lowest performance is taken as the key reference.

More discussion on resilience issues and the extension of barrier analysis methods in the resilience engineering will be in Chapter 14.

## NOTE

1. http://www.risknowlogy.com

## REFERENCES

IEC61508. (2010). Functional Safety of Electrical/Electronic/Programmable Electronic Safety-related Systems. In. Geneva, Switzerland: International Electrotechnical Commission.

IEC61511. (2016). Functional Safety – Safety Instrumented Systems for the Process Industry Sector. In. Geneva, Switzerland: International Electrotechnical Commission.

ISO13702. (2015). Petroleum and natural gas industries – Control and mitigation of fires and explosions on offshore production installations – Requirements and guidelines. In. Geneva, Switzerland: International Organization for Standardization.

Jin, H., Lundteigen, M. A., & Rausand, M. (2011). Reliability performance of safety instrumented systems: A common approach for both low- and high-demand mode of operation. *Reliability Engineering & System Safety*, *96*(3), 365–373.

Johansen, I. L., & Rausand, M. (2015). Barrier management in the offshore oil and gas industry. *Journal of Loss Prevention in the Process Industries*, *34*, 49–55.

Kang, J., Zhang, J., & Gao, J. (2016). Analysis of the safety barrier function: Accidents caused by the failure of safety barriers and quantitative evaluation of their performance. *Journal of Loss Prevention in the Process Industries*, *43*, 361–371.

Khakzad, N., Landucci, G., & Reniers, G. (2017). Application of graph theory to cost-effective fire protection of chemical plants during Domino effects. *Risk Analysis*, *37*(9), 1652–1667.

Khorasani-Zavareh, D., Shoar, S., & Saadat, S. (2013). Antilock braking system effectiveness in prevention of road traffic crashes in Iran. *BMC Public Health*, *13*, 439.

Landucci, G., Argenti, F., Tugnoli, A., & Cozzani, V. (2015). Quantitative assessment of safety barrier performance in the prevention of domino scenarios triggered by fire. *Reliability Engineering & System Safety*, *143*, 30–43.

Lind, N. (2007). Discounting risks in the far future. *Reliability Engineering & System Safety*, *92*(10), 1328–1332.

Liu, Y. (2020). Safety barriers: Research advances and new thoughts on theory, engineering and management. *Journal of Loss Prevention in the Process Industries*, *67*, Article 104260. https://doi.org/10.1016/j.jlp.2020.104260

Lundteigen, M., & Rausand, M. (2008). Spurious activation of safety instrumented systems in the oil and gas industry: Basic concepts and formulas. *Reliability Engineering & System Safety*, *93*(8), 1208–1217.

Lundteigen, M. A., & Rausand, M. (2009). Architectural constraints in IEC 61508: Do they have the intended effect? *Reliability Engineering & System Safety*, *94*(2), 520–525.

OLF070. (2022). Offshore Norge Recommended Guidelines for Application of IEC 61508 and IEC 61511 in the Norwegian Petroleum Industry (Recommended SIL requirements) version 5. In. Stavanger, Norway: Offshore Norge.

OREDA. (2009). *Offshore Reliability Data* (5th ed.). OREDA participants.

Prashanth, I., Fernandez, G. J., Sunder, R. G., & Boardman, B. (2017). Factors influencing safety barrier performance for onshore gas drilling operations. *Journal of Loss Prevention in the Process Industries*, *49*, 291–298. https://doi.org/10.1016/j.jlp.2017.07.009

Sobral, J., & Soares, C. G. (2019). Assessment of the adequacy of safety barriers to hazards. *Safety Science*, *114*, 40–48.

Wu, C. F. J., & Hamada, M. (2021). *Experiments: Planning, Analysis, and Optimization*. John Wiley & Sons.

Xie, L., Ustolin, F., Lundteigen, M. A., Li, T., & Liu, Y. (2022). Performance analysis of safety barriers against cascading failures in a battery pack. *Reliability Engineering & System Safety*, *228*, 108804.

Yeh, L. (1997). The rate of occurrence of failures. *Journal of Applied Probability*, *34*(1), 234–247.

Yuan, S., Cai, J., Reniers, G., Yang, M., Chen, C., & Wu, J. (2022). Safety barrier performance assessment by integrating computational fluid dynamics and evacuation modeling for toxic gas leakage scenarios. *Reliability Engineering & System Safety*, *226*, 108719.

# 7 Data Analysis in Barrier Engineering

## 7.1 DATA AND MODEL

Data analysis is a very broad topic, and in this chapter, we only briefly discuss how we utilize data in the modeling and analysis of barrier engineering and briefly introduce several relevant methods. In Chapter 5, we have discussed several probabilistic and stochastic models that are used to simulate and analyze the structures and behaviors of barrier systems. Data serves as the fundamental element for model development and as the driving force for using models. Conversely, models provide mechanisms to refine and interpret the raw data collected.

The dynamic interaction between models and data in barrier engineering is illustrated in the flowchart of Figure 7.1. On the one hand, data is gathered from various sources as indicated in the upper right of the diagram. This data serves as the foundational input for the model selection and the following analysis processes. On the other hand, after identifying hazards and barriers and conducting a qualitative analysis, the models potentially useful for barrier analysis can be evaluated based on

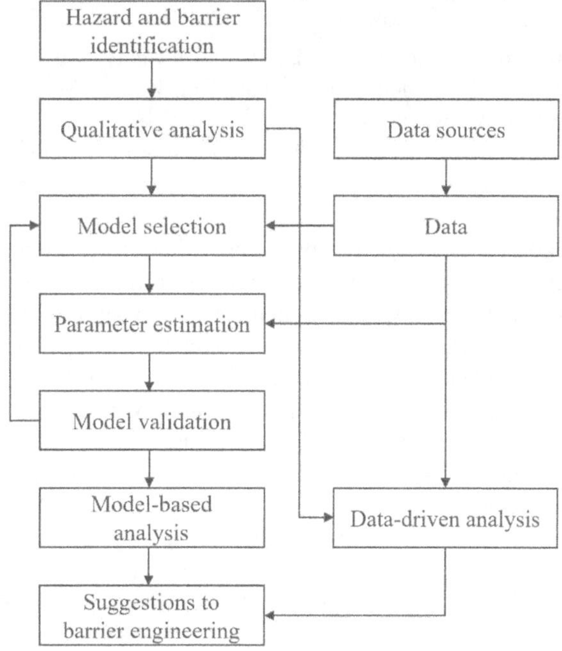


**FIGURE 7.1** Relationship of model and data in quantitative barrier analysis.

DOI: 10.1201/9781003245636-7

the available data. This suggests that the choice of model is influenced by the nature of model, as well as the quality and amount of the collected data. Once a model is selected, the parameters of the model are estimated using data to fit the observed reality as closely as possible. The next step in the process is model validation for ensuring that the model performs well in simulating the reality or predicting the outcomes related to the identified hazards and barriers.

In addition, barrier analysis is not limited to model-based approaches alone, and it also includes data-driven analysis, which operates independently of traditional probabilistic or physical models. The insights gained from the model-based and data-driven analyses are used to make suggestions for improving barrier effectiveness, responsiveness, and resilience. This can be changes to existing barriers or the development of new ones.

In this chapter, we will explore the application of data in both model-based and data-driven analysis methods.

## 7.2 TYPES OF DATA USED IN BARRIER ENGINEERING

Several types of data are useful for modeling the structure and behavior of barrier systems and for analyzing their performance. These include the following:

- *Incident and failure data*: This type of data is collected from past incidents or accidents involving similar processes or safety barriers. It can help identify failure patterns and determine the potential risks of the studied process/asset and vulnerabilities of the relevant barriers. Sources of incident and failure data include the following:
  - Detailed incident reports and root cause analysis documents within the organization, such as those based on Failure Mode, Effects, and Diagnostic Analysis (FMEDA).
  - Industry-wide databases that track accident reports and failure analyses in the relevant sector, often maintained by regulatory bodies or safety organizations.
  - Scholarly articles and case studies published in academic journals and conferences, providing in-depth analyses of notable incidents.
  - Media reports or news articles that cover relevant incidents.
- *Incident response data*: This type of data is related to the actions taken as the responses to incidents, including emergency response times, the effectiveness of emergency procedures, and the role of safety barriers in mitigating incident consequences. Sources of incident response data include the following:
  - Debriefing reports and emergency response evaluations from the organization.
  - Data from simulation drills and emergency response training sessions.
  - Feedback and reports from first responders and emergency services.
  - After-action reports or expert reviews conducted following the emergency procedures.
- *Technical specification data*: This type of data includes design and other technical specifications of the studied process, relevant safety barriers,

and assets, such as materials used, dimensions, strength, and installation details. Technical data is critical for evaluating the inherent reliability and safety characteristics of the asset and the integrity of the barriers. Sources of technical specification data include the following:
- Reliability testing reports, simulations, or other relevant experiments conducted in plant.
- Manufacturer specifications and documentation provided by suppliers.
- Construction or installation records.
- Technical databases managed by the third parties.
- Standards and guidelines from professional engineering bodies and safety regulators.

- *Operational and maintenance records*: Data from the actual operational context is necessary for system modeling and performance estimation. This type of data can reveal wears and tears, potential damages, abnormal conditions, hazardous environmental factors, or inadequate maintenance. The operational data collected within a company can better reflect specific operational situations, but the statistical confidence in the data may be low if based on only a few installations. Sources of operational and maintenance data include the following:
  - Real-time or historical operational data collected by field sensing and monitoring systems.
  - Work orders and records in the asset management systems, documenting the condition and functionality of equipment and barriers.
  - Inspection reports and maintenance logs, both scheduled and ad-hoc, by maintenance personnel.
  - Environmental data in operation period from weather stations or climate databases.
- *Human factor data*: This type of data is related to human factors and errors, such as those reports regarding the perceptions, understanding, and behaviors of individuals interacting with the safety barriers. Sources of human factor data include the following:
  - Surveys and interviews with employees and system users.
  - Field observation reports on how people engage with safety protocols and barriers.
  - Incident and near-miss reports that specifically note human error or behavioral factors.
- *Regulatory compliance data*: This type of data includes information about applicable regulatory requirements and standards, as well as historical records of compliance. Sources of regulatory compliance data include the following:
  - Compliance audit reports from both internal auditors and external regulatory inspectors.
  - Certification documents that verify adherence to national and international safety standards.
  - Regulatory updates and bulletins that inform organizations of changes to safety requirements and standards.

In the remaining sections of this chapter, we will discuss the analysis techniques for several data types, with a particular emphasis on those assessing the performance, such as reliability and availability, of technical safety barriers.

## 7.3 REFERENCE DATA AND PERFORMANCE PREDICTION OF BARRIERS

### 7.3.1 Generic Data and Reference Data

It is widely recognized that performance, especially the reliability, of a barrier element can be effectively predicted using the *generic data* provided in industrial and sectorial databases. The prediction results derived from this data are instrumental in setting requirements and objectives for subsequent stages such as reliability design, testing, and growth programs. In this chapter, according to ISO14224 (2016), generic data refers to the following:

- *Generic data*: Data collected based on operational experiences from multiple installations and comparable equipment types, which represents the average performance of the equipment under study.

Generic data is valuable in the early phases of a system lifecycle when specific information about the equipment is limited. In the development of many new products/systems, relevant commercial databases can be utilized, which record generic reliability data for equipment and components. For example,

- *Component reliability database* (CRD), which is used for FMEDA and managed by company called as Exida. It encompasses performance data of safety-critical devices, such as failure rates, sourced from various industries, including chemical, oil and gas, metals, mining, pharmaceutical, pulp and paper, automotive, and machinery (https://www.exida.com/).

This type of data is important as a reference for predicting reliability and overall performance of equipment and devices during their developmental stages.

Moreover, various industries have taken their initiatives for developing tailored databases. Notable examples in the oil and gas sector include the following:

- *Offshore and Onshore Reliability Data* (OREDA), which is a collaborative project involving several oil and gas companies. This project has been running for over 40 years, documenting each failure with detailed parameters such as item name, date of failure, impact, mode, and cause (OREDA, 2009).
- *Reliability Data for Safety Instrumented Systems* (PDS) is based on a cooperation between SINTEF (an independent Scandinavian-based research institution) and various stakeholders in the oil and gas industry, including operators, drilling contractors, engineering companies, consultants, and manufacturers, with a focus on safety instrumented systems (SISs) (https://pds-forum.com/) (Hauge et al., 2013).

Given that certain components are employed across multiple applications, the data within industry databases can serve as the *reference data* for performance prediction of the items for which reliability information is unknown. Sometimes, reference data also can be sourced directly from manufacturers. It is important to note that reference data is typically gathered under specific operating and environmental conditions, or it represents industrial average values. Once reference data is available for an item either newly designed or applied in a new context, the next step involves predicting the performance of this item.

In the field of barrier engineering, *reliability prediction* refers to the following:

- *Reliability prediction*: The process for forecasting the values of reliability metrics (such as failure rates) of an item, including converting the reference values of reliability metrics, into values that account for various risk influence factors (RIF) that may influence reliability.

RIFs include the factors in barrier system design and operation, such as novel design elements, structural attributes, materials used, installation methods, and operational circumstances. In the calculation process, RIFs should be quantifiable, allowing for their measurement and subsequent integration into reliability predictions. Ideally, an RIF is constant, like the strength of a structure, though it may also exhibit time-dependent variations, such as changes in temperature and humidity. For simplicity, the time dependence of RIFs is disregarded in the formulas introduced later in this chapter.

In the remainder of this section, we will introduce several commonly used reliability prediction methods.

### 7.3.2 Proportional Hazards Model for Failure Rate Prediction

Predictions of failure rates can be based on either physical or statistical models. Physical models consider the effects of physical laws on reliability calculations by incorporating parameters such as temperature, stress, and voltage. One of such methods is based on the Cox model (Cox, 1972) or the *proportional hazards model* in MIL-HDBK-217F (1991).

Let $\varphi_j$ denote value of the covariate/factor $j$ that influences the failure rate of the item, $\omega_j$ as the coefficient of $j$, and the Cox model is expressed as follows:

$$\lambda(t|\Phi) = \lambda_{\text{ref}}(t)\exp\left(\sum_{j=1}^{n}\omega_j\varphi_j\right) = \lambda_{\text{ref}}(t)\exp(\omega\cdot\Phi) \tag{7.1}$$

where

- $\lambda(t)$ denotes the failure rate of the item at time $t$;
- $\lambda_{\text{ref}}(t)$ denotes a *baseline failure rate* or *reference failure rate* at time $t$ obtained from a known item; and
- $\exp(\omega\cdot\Phi)$ is called the *scaling factor*, describing the multiplicative effect of RIFs, or the difference between the baseline failure rate and the predicted failure rate.

Then, in simplified terms, the failure rate of a new item can be expressed as

$$\lambda(t) = \lambda_{\text{ref}}(t)\Phi_{\text{rif}} \tag{7.2}$$

- $\Phi_{\text{rif}}$ is a vector related to RIFs that influence failure rate.

An illustrative example from MIL-HDBK-217F (1991) is for an electronic system, such as an electric relay used in safety control, comprising $n$ components in a series structure. The failure rate of the new system $\lambda_S(t)$ is calculated as

$$\lambda_S(t) = \sum_{i=1}^{n}\left[\lambda_i^{\text{ref}}(t) \times \varphi_V \times \varphi_I \times \varphi_T\right] \tag{7.3}$$

where

- $\lambda_i^{\text{ref}}(t)$ denotes the reference failure rate at time $t$ obtained from a known component $i$;
- $\varphi_V, \varphi_I, \varphi_T$ are coefficients indicating the influences of voltage, current, and temperature.

The proportional hazards model provides a simple, flexible but robust framework for understanding how different factors influence the failure rate of an item, and it acts as the foundation for other prediction methods.

### 7.3.3 Weight-Based Reliability Prediction

Brissaud et al. (2010) have developed the above proportional hazards model by generalizing the RIFs to the mechanisms with limited detailed knowledge and considering the different contributions (weights) of RIFs. The proposed method can thus be referred to as a *weight-based reliability prediction method.* This approach relies on statistical data, historical records, or expert judgment to assess the impacts of RIFs on reliability. For a new item, its failure rate is calculated using the following formula:

$$\lambda(t) = \lambda_{\text{ref}}(t)\sum_{j=1}^{n}\omega_j\varphi_j \tag{7.4}$$

where

- $\varphi_j$ is the value assigned to the $\text{RIF}_j$;
- $\omega_j$ is the weight of $\text{RIF}_j$ among all $n$ factors.

The first step in the weight-based reliability prediction is to enumerate all RIFs for both the reference and the new items and then define the nominal values of RIFs for the reference item. The next step is to compare the impacts of these RIFs on the reference item and the new item to determine the values of $\sigma_j$. In general:

- If the impacts of $\text{RIF}_j$ on the two items are similar, $\varphi_j = 1$;

- If $\text{RIF}_j$ is more benign to the new item than to the reference item, $\varphi_j < 1$;
- If $\text{RIF}_j$ is more hostile to the new item than to the reference item, $\varphi_j > 1$.

Subsequently, the value of $\omega_j$ for each RIF is determined. It should be noted that the sum of all weights in Equation (7.4) must be equal to 1. Both the values of RIFs and the weights of RIFs are derived from statistics, historical data, or expert assessments.

Using Equation (7.4), the failure rate of the new item can be calculated. Typically, the result of such a prediction should be conservative because this calculation is approximated.

### Example 7.1:

### *Failure Rate Prediction for a Safety Valve Using the Weight-Based Reliability Prediction Method*

This example is for the development of a new safety valve designed for use between an electrolyzer and a storage tank within an offshore hydrogen production system. It is critical to compare this new design with its existing onshore counterparts, given that similar valves are used in onshore facilities. A key part of this analysis involves addressing the unique challenges posed by the offshore operational environment.

For this simplified case study, our focus is on four key RIFs: ambient temperature, humidity levels in the operational environment, technical maturity of the valve, and the frequency of tests and maintenances. The values for these RIFs, both for the reference (existing onshore valve) and the new (offshore valve) items, are summarized in Table 7.1.

In this example, we roughly determine the values of $\sigma_j$:

- When the impacts of $\text{RIF}_j$ on the two items are similar, $\sigma_j = 1$;
- When $\text{RIF}_j$ is more benign to the new item, $\sigma_j = 0.6$;
- When $\text{RIF}_j$ is more hostile to the new item, $\sigma_j = 3.0$.

From the analysis, we find that for RIFs 2–4, their new values are more hostile to reliability, while $\text{RIF}_1$ of both items are similar. Assuming that $\lambda_{\text{ref}} = 2.0 \times 10^{-5}$ per hour, the predicted failure rate ($\lambda$) for the new valve can be calculated as follows:

$$\lambda = \lambda_{\text{ref}} \sum_{i=1}^{k} \omega_i \sigma_j = 2.0 \cdot 10^{-5} \cdot [0.25 \cdot 1 + 0.25 \cdot 3 + 0.25 \cdot 3 + 0.25 \cdot 3] = 5.0 \cdot 10^{-5} \text{per hour.}$$

■

**TABLE 7.1**
**Reference and New Values of RIFs**

| No. | RIF | Reference Value | New Value | Weight (%) |
|---|---|---|---|---|
| 1 | Temperature | −10°C to 30°C | −10°C to 25°C | 25 |
| 2 | Humidity | 30%–60% | 50%–90% | 25 |
| 3 | Technical maturity | High | Medium | 25 |
| 4 | Test frequency | Once per month | Once per year | 25 |

### 7.3.4 Similarity-Difference-Based Reliability Prediction

The weight-based reliability prediction model assumes that all RIFs are completely independent and that the same RIFs apply to both old and new items. Rahimi and Rausand (2013) propose a different approach, suggesting that the failure rate of an item is dependent on the rates of its potential failure modes, wherein each failure mode is linked to several failure causes susceptible to the influences of different RIFs. In light of this perspective, these researchers evaluate the similarities and differences between the new items and reference items in terms of failure modes and the associated failure causes and RIFs and propose a *similarity-difference-based reliability prediction* (SDRP) method.

The following procedure of SDRP is based on the study of Rahimi and Rausand (2013) with some adjustments in calculation.

**Step 1: Identifying failure modes and failure mechanisms**

The first step of this method is to identify failure modes, failure causes, and the associated RIFs for both the reference item and the new item. In general, this includes considering failure mode $i$ ($i = 1, \ldots, n_{\text{REF}}$), failure cause $k$ ($k = 1,\ldots,l_{\text{REF}}$), and RIF$_j$ ($j = 1,., m_{\text{REF}}$). The total failure rate of the reference item is calculated as follows:

$$\lambda_{\text{Total}}^{\text{REF}} = \sum_{i=1}^{n_{\text{REF}}} \lambda_{\text{i}}^{\text{REF}} \tag{7.5}$$

Subsequently, a structured factor-contribution diagram is formulated to describe the above elements influencing the total failure rate and their intricate interconnections, as the illustrative example in Figure 7.2. Such a diagram is developed based on the combination of historical data analysis, comprehensive literature reviews, and collaborative brainstorming sessions involving domain experts.

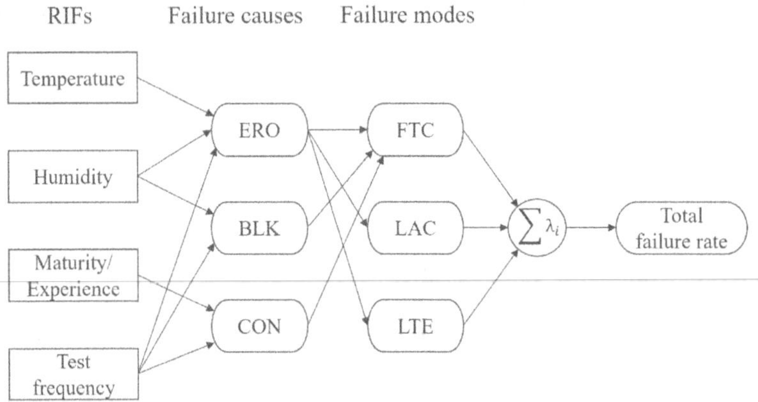


**FIGURE 7.2** Factor-contribution to the total failure rate of existing equipment.

**Example 7.2:**

*Building a Factor-Contribution Diagram for the Reference Item*

We can build a factor-contribution diagram for describing the interrelationships between failure modes, failure causes, and RIFs that contribute to the failure rate of the existing safety valve installed in an onshore factory. The values for different failure modes of this reference item are sourced from existing databases.

The total failure rate of the safety valve is calculated by summing the rates of three failure modes: failure to close on demand (FTC), leakage at the closure position (LAC), and leakage into the environment while in the open position (LTE).

Within this simplified model, we consider that these failure modes are influenced by three main failure mechanisms: erosion/corrosion (ERO), blockage/obstruction (BLK), and deficiencies in control system (CON). All the three mechanisms are related to FTC, while only ERO is related to LAC and LTE within this model.

Then, four key RIFs are identified with direct impacts on the aforementioned failure mechanisms: temperature, humidity, technical maturity or technological maturity or experience with the technology, and the frequency of testing. It should be noted that each of these RIFs corresponds specifically to distinct failure mechanisms, as depicted in Figure 7.2. ■

This factor-contribution diagram allows stakeholders to visualize and analyze the multiple factors contributing to the failure of the item. In Example 7.2, the RIFs are selected randomly, but they also can be identified more systematically, for example, within three categories: design and manufacturing, operation and maintenance, and environment. Rahimi and Rausand (2013) have provided a checklist for identifying the relevant RIFs.

**Step 2: Comparing the reference item and the new item**

The next step is a detailed comparison between the failure modes, failure causes, and RIFs associated with both the reference item and the new item. Since the two items have similarities in terms of structure, material, and operational strategies, some of failure modes, failure causes, and RIFs are common for both. However, there are also differences in operational conditions and the technologies employed which necessitate a thorough analysis to identify new factors that contribute to the total failure rate of the new item and to reveal which factors are no longer relevant.

For the new item, similar to the reference item, the total failure rate is calculated as follows:

$$\lambda_{\text{Total}}^{\text{NEW}} = \sum_{i=1}^{n_{\text{NEW}}} \lambda_i^{\text{NEW}} \tag{7.6}$$

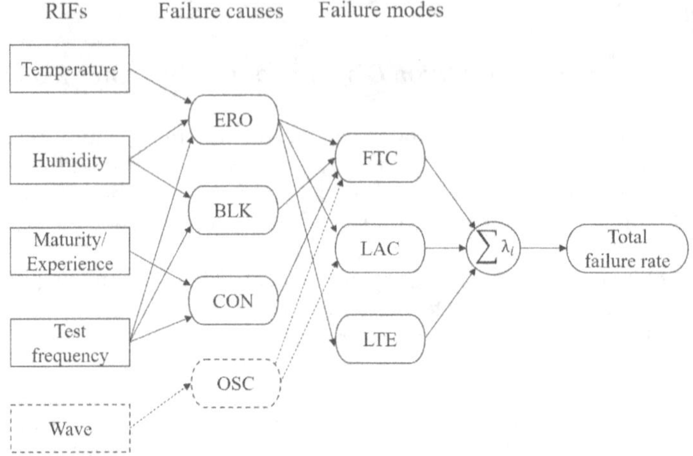


**FIGURE 7.3** Updated factor-contribution with new elements of new equipment.

## Example 7.3:

### *Building a Factor-Contributing Diagram for the New Item*

As demonstrated in Example 7.2, we can now construct a factor-contribution diagram for a new safety valve that is expected to be installed on an offshore platform. Our analysis continues to focus on the three critical failure modes: FTC, LAC, and LTE. Then, a new failure mechanism is considered in this scenario: the oscillation/shaking of the platform due to waves (OSC), which can result in FTC and LAC. Subsequently, we integrate the influence of wave motions, serving as the RIF of shaking. The updated diagram is presented in Figure 7.3. The new elements are depicted with dotted blocks, and dotted arrows are used to show the influence of these new elements. ■

**Step 3: Scoring the RIFs**

The selected RIFs should then be evaluated in pairs, based on their importance for each specific failure cause. For example, considering a pair of RIFs, denoted as $\text{RIF}_1$ and $\text{RIF}_2$, both associated with the failure cause $k$, distinctly weight values, namely $\omega_{1k}$ and $\omega_{2k}$, are allocated to these RIFs. These weights are assigned by evaluating the relative contributions of the two factors.

The weight $\omega_{jk}$ represents the importance of $\text{RIF}_j$ to the failure cause $k$. If an RIF is not relevant to a particular failure cause, it is assigned a weight of 0. It is essential to ensure that the cumulative weights of all contributing RIFs related to any given failure cause sum to 1. It is also important to note that the weight of an RIF in the context of the reference item may differ from its weight in the context of the new item.

Rahimi and Rausand (2013) have also proposed a practical approach employing an *influence score*, $\eta_{jk}$, for evaluating the different impacts of the same $\mathrm{RIF}_j$ on the same failure cause $k$ for the reference item and the new item. The value of $\eta_{jk}$ can be provided by domain experts. In this study, a scale ranging from –3 to +3 is used to limit the extreme values of $\eta_{jk}$. Within this framework, a higher numerical value of $\eta_{jk}$ indicates a higher likelihood that the corresponding RIF will lead to the occurrence/existence of the specific failure cause within the context of the new item. For example, increased humidity in offshore environments, which makes equipment more susceptible to erosion, can result in a higher $\eta_{jk}$ value of the new item. The value of 0 indicates no difference in the impact of the RIF in the two contexts. When a new RIF is identified for the new item, it is assigned an influence score of 3. Conversely, if an RIF from the reference item has no effect on the new item, its influence score is designated as –3. If an RIF is irrelevant with a failure cause for both reference item and new item, it is not meaningful to assign an influence score to it.

For a specific failure cause $k$, its weighted average influence factor of RIFs, $\overline{\eta_k}$, for the new item can be calculated as follows:

$$\overline{\eta_k} = \sum_{j=1}^{m} \omega_{jk}^{\mathrm{NEW}} \left( \frac{\eta_{jk} + 3}{3} \right) \tag{7.7}$$

where $\omega_{jk}^{\mathrm{NEW}}$ represent the importance of $\mathrm{RIF}_j$ to the failure cause $k$ of the new items.

Equation (7.7) employs a denominator of 3 due to the scale ranging from –3 to +3, while the numerator incorporates $\eta_{jk} + 3$, accounting for the distance $\eta_{jk}$ from the extreme situation (–3). This calculation result $\overline{\eta_k}$ illustrates how the exposure level of failure cause $k$ changes in response to all relevant RIFs, offering a detailed metric for assessing risk in the new operational context. It should be noted that the range of (–3, 3) is subjectively determined, and there is no specific physical meaning.

**Example 7.4:**

### *Scoring the RIFs*

Based on the process outlined in Step 3, we can construct a table for scoring the RIFs for the failure causes of the reference item and the new item. It is important to clarify that the weights provided in Table 7.2 are approximate and are used primarily for illustrative purposes.

In this case, the weighted averages of RIF scores for failure cause ERO, BLK, CON, and OSC are calculated as 1.33, 1.50, 1.50, and 2.00, respectively. ■

**Step 4: Weighing failure causes for failure modes**

In this step, weights are assigned to various failure causes in relation to their corresponding failure modes. Typically, the weights for the failure causes associated with the reference item are derived from existing

TABLE 7.2
Scoring RIFs for the Failure Causes

| RIFs | Item | Value | Weight for Failure Cause | | | |
|---|---|---|---|---|---|---|
| | | | ERO | BLK | CON | OSC |
| Temperature | Ref | $-10°C$ to $30°C$ | 0.40 | 0 | 0 | 0 |
| | New | $-10°C$ to $25°C$ | 0.30 | 0 | 0 | 0 |
| | $\eta_{jk}$ | | 0 | – | – | – |
| Humidity | Ref | 30%–60% | 0.40 | 0.60 | 0 | 0 |
| | New | 50%–90% | 0.40 | 0.50 | 0 | 0 |
| | $\eta_{jk}$ | | 1 | 1 | – | – |
| Maturity | Ref | High | 0 | 0 | 0.50 | 0 |
| | New | Medium | 0 | 0 | 0.50 | 0 |
| | $\eta_{jk}$ | | – | – | 2 | – |
| Test frequency | Ref | Once per month | 0.20 | 0.40 | 0.50 | 0 |
| | New | Once per year | 0.30 | 0.50 | 0.50 | 0 |
| | $\eta_{jk}$ | | 2 | 2 | 1 | – |
| Wave | Ref | No | 0 | 0 | 0 | 0 |
| | New | Exist with variations | 0 | 0 | 0 | 1.00 |
| | $\eta_{jk}$ | | – | – | – | 3 |

databases. For the new item, these weights need to be determined through expert judgment.

The weight $\omega_{ki}$ of a failure cause $k$ for the failure mode $i$ is defined as follows:

$$\omega_{ki} = \Pr\left(\text{The failure is caused by } k \middle| \text{Failure mode } i \text{ has occurred}\right) \quad (7.8)$$

This definition reflects a conditional relationship, emphasizing the direct influence of a specific failure cause on a particular failure mode. It is also important to ensure that the sum of the weights for all failure causes associated with a particular failure mode equals 1.

## Example 7.5:

### *Scoring the Failure Causes*

We can build the Table 7.3 for scoring the contributions of failure causes to the failure modes of the reference item and the new item. It should be noted that the weights are roughly given in this example only for illustration. ■

**Step 5: Calculating failure rate of the new item**

Once the failure rates of the reference item have been established using databases, the rate for a specific failure mode $i$ of the new item can be

**TABLE 7.3**
**Weights of Failure Causes for Each Failure Mode**

| | | Weight for Failure Mode | | |
|---|---|---|---|---|
| **Failure Cause** | **Item** | **FTC** | **LAC** | **LTE** |
| ERO | Ref | 0.40 | 1.00 | 1.00 |
| | New | 0.35 | 0.80 | 1.00 |
| BLK | Ref | 0.40 | 0 | 0 |
| | New | 0.35 | 0 | 0 |
| CON | Ref | 0.20 | 0 | 0 |
| | New | 0.15 | 0 | 0 |
| OSC | Ref | 0 | 0 | 0 |
| | New | 0.15 | 0.20 | 0 |

derived based on the rate of the corresponding failure mode of the reference item. This relationship is expressed as follows:

$$\lambda_i^{\text{NEW}} = \delta_i \lambda_i^{\text{REF}} \tag{7.9}$$

where $\delta_i$ is a *scaling factor* that accounts for the differences in how various failure causes affect failure mode $i$ of the new item compared with the reference item. The scaling factor is calculated using the following equation:

$$\delta_i = \frac{\sum_{\substack{k=1 \\ \omega_{ki}^{\text{REF}} \neq 0}}^{l} \frac{\omega_{ki}^{\text{NEW}} \overline{\eta_k}}{\omega_{ki}^{\text{REF}}}}{n_{\text{REF}}} + \sum_{\substack{k=1 \\ \omega_{ki}^{\text{REF}} \neq 0}}^{l} \omega_{ki}^{\text{REF}} \overline{\eta_k} \tag{7.10}$$

where $n_{\text{REF},i}$ is the number of failure causes of failure mode $i$, and $l$ is the total number of failure causes considered for the failure mode. The first item on the right side of Equation (7.10) reflects the contributions from the existing failure causes in the context of the reference item, while the second item is for the contributions of the newly identified failure causes. $n_{\text{REF},i}$ helps distribute the impact proportionally across all relevant causes.

## Example 7.6:

### *Determining the Failure Rate of the New Valve*

Suppose that the rates of FTS, LAC, and LTE of the reference safety valve used onshore are $0.4 \times 10^{-5}$, $0.6 \times 10^{-5}$, and $1.0 \times 10^{-5}$ per hour, and therefore, the total failure rate is $2.0 \times 10^{-5}$ per hour, same with that in Example 7.1.

Based on Equation (7.10), the scaling factor for the first failure mode FTS is ($c_{ki} = 1.5$ for the new failure cause OSC) as follows:

$$\delta_1 = \left(\frac{0.35\times1.33}{0.40} + \frac{0.35\times1.50}{0.40} + \frac{0.15\times1.50}{0.20}\right) / 3 + 0.15\times2.00 \approx 1.500$$

Similarly, we can calculate $\delta_2$ and $\delta_3$ for LAC and LTE as 1.464 and 1.333, respectively.

The total failure rate is then calculated as follows:

$$\lambda_{\text{Total}}^{\text{NEW}} = \sum_{i=1}^{n} \lambda_i^{\text{NEW}} = \sum_{i=1}^{n} \delta_i \lambda_i^{\text{REF}} = 1.500\times0.4 + 1.464\times0.6 + 1.333\times1.0$$

$$= 2.811\times10^{-5} \text{ per hour}$$

■

### 7.3.5 Bayesian Network for Reliability Prediction

A Bayesian network (BN) can be effectively used to model the impacts of RIFs and failure causes on the failure rates of an item. Zhang et al. (2016) employed this methodology to predict reliability of subsea high integrity pressure protection system (HIPPS), which is a safety-critical system in subsea production.

In the context of BN, the occurrence of a failure mode can be regarded as the "TOP event", and the BN structure is designed based on the causal relationships between RIFs, failure causes, and the failure modes. This network facilitates the updating of prior probabilities (as discussed in Chapter 5), which can be set based on reference data, experiences, and/or expert judgment.

As the system operates and new data becomes available, the evidence about the impacts of new factors and changes in conditions is collected. This information is used to update the system model by calculating the posterior probabilities of the nodes within the BN, utilizing the Bayesian theorem.

The procedure of implementing BN-based failure rate prediction will be provided with Example 7.7.

**Example 7.7:**

***BN-Based Failure Rate Prediction***

Following the example in Section 7.3.5, here we consider predicting the rate of failure mode FTC of the new valve. All failure modes, failure causes, and RIFs of the reference and new items keep unchanged.

**Step 1: Construct a BN**

A BN can be constructed following the guidance given in Section 5.5. The BN should include all RIFs and failure causes for the new item, as illustrated in Figure 7.4. For clarity, each node in the network is numbered (from 1 to 10 in this example). It can be found that the causal relationships here are consistent with those associated with the FTC in Figure 7.3.

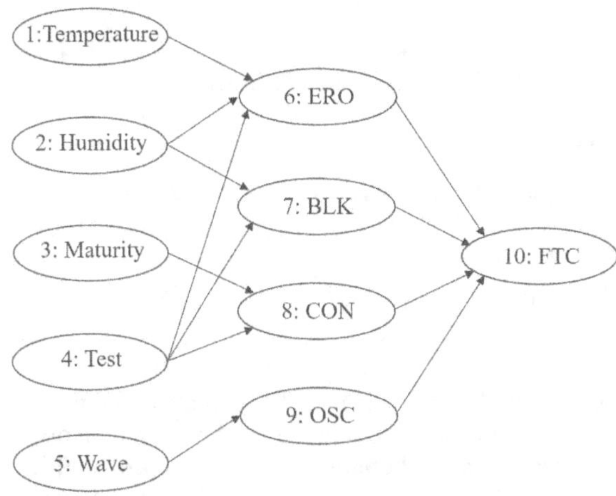


**FIGURE 7.4** Bayesian network diagram for predicting rate of FTC.

**Step 2: Determine the states of the nodes**

In this case, we assume that each RIF node and failure cause node have three states:

- *State 0*: Lower impact on the new item than on the reference item;
- *State 1*: Same impacts on the two items; and
- *State 2*: Higher impact on the new item than on the reference item.

For Node 10: FTC, it can have several states with different rates, for example,

- $8.0 \times 10^{-6}$, lower than the rate of the onshore item;
- $5.0 \times 10^{-6}$, similar with the rate of the onshore item; and
- $3.5 \times 10^{-6}$, higher than the rate of the onshore item.

Admittedly, this setting is very subjective, dependent on the knowledge and experience of the analysts involved.

**Step 3: Assign prior probabilities to RIFs**

The prior probabilities describe the initial estimated impacts of the RIFs and failure causes on the failure mode, based on reference data and expert judgment. For nodes without parent nodes (e.g., Node 2: Humidity), probabilities are assigned to reflect uncertainties: It can have 0.1 of probability in State 0, 0.7 of probability in State 1, and 0.2 of probability in State 1.

For nodes representing new factors (e.g., Node 5: Wave), which are irrelevant in the context of the reference item, they may be assigned a State 2 with a prior probability of 1.

**Step 4: Build conditional probability tables**

Conditional probability tables (CPTs) are developed for nodes associated with failure causes, presenting the conditional probabilities based on interactions with other nodes. For example, the CPT for Node 7 (BLK) might be in Table 7.4.

CPTs are established for all nodes with parent nodes in the BN. The conditional probabilities are set with reference data and expert judgment. It is then possible to calculate the probabilities that Node 10

**TABLE 7.4**
**Conditional Probability Table for Node 7: BLK**

| | $Pr(Y_7 = 0)$ | | | $Pr(Y_7 = 1)$ | | | $Pr(Y_7 = 2)$ | | |
|---|---|---|---|---|---|---|---|---|---|
| | ***$Y_2 = 0$ | $Y_2 = 1$ | $Y_2 = 2$ | $Y_2 = 0$ | $Y_2 = 1$ | $Y_2 = 2$ | $Y_2 = 0$ | $Y_2 = 1$ | $Y_2 = 2$ |
| $Y_4 = 0$ | 0.90 | 0.70 | 0.30 | 0.15 | 0.40 | 0.80 | 0.05 | 0.30 | 0.70 |
| $Y_4 = 1$ | 0.70 | 0.40 | 0.20 | 0.50 | 0.95 | 0.40 | 0.30 | 0.40 | 0.80 |
| $Y_4 = 2$ | 0.60 | 0.10 | 0.05 | 0.80 | 0.35 | 0.10 | 0.50 | 0.85 | 0.95 |

sojourns at a certain state. The calculation procedures are skipped here, and readers can go back to Chapter 5 for the relevant information.

**Step 5: Collect evidence and update the posterior probabilities**

As new operational evidence, or new technical information becomes available, the initial (unconditional) probabilities are updated, followed by the conditional probabilities. This continuous updating process improves the modeling accuracy over time. ■

For further reading and a more detailed exploration of BN-based failure rate prediction calculations, the master thesis of Kumar (2016) supervised by the author is recommended.

## 7.4 TESTING DATA AND CLASSIC STATISTIC METHODS FOR RELIABILITY ESTIMATION

### 7.4.1 Laboratory Test and Reliability Estimation

Manufacturers conduct laboratory tests on barrier components to observe their performance in a controlled environment. These tests are important for obtaining accurate reliability and failure information, and they can be categorized based on their objectives: *failure identification tests*, *reliability estimation tests*, and *environmental stress tests*.

In this section, we will focus on *reliability estimation tests*, which are designed to provide empirical evidence regarding the current reliability level of the tested components. These tests help to evaluate whether reliability goals have been achieved and allow for the comparison of design options in terms of reliability. Furthermore, reliability estimation tests can explore the potential relationships between reliability parameters and various stresses.

Although the reliability metrics measured in reliability estimation tests, such as mean time-to-failure (MTTF) and failure rate, are not directly applied in most cases for barrier performance analysis, they serve as intermediate variables that influence the values of performance measures introduced in Chapter 6, such as probability of failure on demand (PFD).

It is important to note that since barrier components are designed to be highly reliable, estimating their long-term performance and observing failures within limited timeframes pose significant challenges. To overcome these limitations, tests are often conducted at higher-than-normal levels of stress, such as accelerated thermal cycles, increased voltage, and more frequent vibration, to expedite failure occurrences. These are known as *acceleration tests*, which enable quicker collection of relevant data. The discussion of techniques and models for acceleration tests will not be covered extensively in this book. However, data and examples from both normal and acceleration tests will be used in this section.

The volume of data collected in laboratory tests can vary significantly, depending on the objectives of the tests and the precision required for data analysis. Nevertheless, the quality of testing data is generally high, thanks to deliberate and controllable collection methods. Two complementary methods commonly used in the data analysis of laboratory tests are *non-parametric* and *parametric* methods. Each method has its own purposes and offers unique advantages to reliability analysis.

### 7.4.2 Overview of Non-Parametric Methods

Non-parametric methods are statistical techniques used to estimate the reliability of a system or component without assuming a specific underlying probability distribution for time-to-failure. These methods are especially valuable in the early design and development phases, where limited historical data may be available.

Some classic non-parametric methods in reliability estimation include the following:

- *The total time on test (TTT) method*: This intuitive approach calculates the unreliability of items under test based on the ratio of the cumulative operating time of functioning items to the total time, including those that have failed. It provides a straightforward measure of the overall reliability of the test items at various time points.
- *Kaplan-Meier Estimator*: The Kaplan-Meier estimator is used to estimate the survival function (reliability) of an item over time. It is particularly useful when handling both complete data (where a failure is observed) and censored/incomplete data (where an item is removed from testing without failure). The estimator calculates reliability at different time points based on observed failure times and the number of failures.
- *Nelson-Aalen Estimator*: Similar to the Kaplan-Meier estimator, the Nelson-Aalen estimator provides a measure of the cumulative hazard rate, or the cumulative number of failures at a particular time divided by the time in test. This estimator is derived from the Kaplan-Meier estimator and offers an alternative perspective on reliability.

These non-parametric methods are preliminary tools for further reliability analysis. For instance, using the TTT method or the Nelson-Aalen estimator, we can approximately plot the curves that depict how the cumulative failure rate changes over time. This can aid in identifying the probable distribution of time-to-failure for items under test, preparing for the application of parametric methods.

Further exploration of these non-parametric methods is not included in this book since the calculations in subsequent chapters do not rely on them. Readers interested in more details and numerical examples are encouraged to refer to the textbook by Rausand et al. (2020).

### 7.4.3 Overview of Parametric Methods

Parametric methods can estimate the reliability of an item based on specific mathematical distributions. These methods are widely used in reliability and barrier engineering, offering a structured approach to derive reliability metrics.

Parametric methods depend on the assumption that the time-to-failure of an item follows a specific probability distribution, such as those introduced in Chapter 5. This assumption allows for the application of mathematical models to estimate key reliability metrics such as failure rate, MTTF, and survival probability. The choice of distribution can be influenced by the nature of the failure mechanism, historical data, or theoretical considerations.

The generic steps in using parametric methods in reliability estimation include the following:

**Step 1: Data collection and preprocessing**

Data collected from laboratory tests typically includes the time of failures and associated stress levels. In some instances, preprocessing of data is required to identify missing or erroneous data points and ensure all data is in a suitable format for analysis.

**Step 2: Selection of probability distribution**

The non-parametric methods mentioned above can be introduced in this step to help identify and choose an appropriate probability distribution that best fits the time-to-failure data. Still, this task is based on the knowledge of analysts and industrial experience.

**Step 3: Parameter estimation**

This step involves estimating the parameters of the chosen distribution. Commonly, the *maximum likelihood estimation* (MLE) method is utilized, which will be elaborated on in the following subsection.

**Step 4: Goodness-of-fit analysis**

It is necessary to determine whether the observed data aligns with the expected values under the selected probability distribution. *Goodness-of-fit* can be assessed through visual inspections, such as plotting the data against the fitted distribution, or through statistical tests like the *Chi-squared* or *Kolmogorov-Smirnov tests*. If the goodness-of-fit is unsatisfactory, it may be necessary to consider alternative distributions. Further details on goodness-of-fit analysis are beyond the scope of this book but can be found in statistical textbooks such as those by Walpole et al. (2016).

**Step 5: Reliability estimation**

Once the appropriateness of the selected distribution has been verified, and its parameters have been determined, the next task is to calculate reliability metrics. If an acceleration test is employed, wherein a higher stress level is applied compared to regular operational conditions, it is also necessary to investigate the relationship between stress level and reliability metrics. This work can be performed by conducting tests at various stress levels. A brief introduction to regression models for this purpose will be provided in Section 7.4.4.

**Step 6: Uncertainty analysis**

In some situations, it is meaningful to analyze the uncertainty associated with reliability estimation results. This involves calculating the confidence intervals for reliability metrics or performing sensitivity analyses to evaluate the impact of parameter uncertainty. Although this book primarily uses point estimation values in modeling analysis, it is helpful to understand interval estimation since it can offer a more comprehensive view and enhance the assessment of the probability distribution.

**Step 7: Interpretation and reporting**

Finally, the results of the reliability estimation should be interpreted to communicate effectively with relevant stakeholders, such as barrier design engineers, maintenance engineers, and product managers. Recommendations should be made regarding whether the reliability meets the required standards and how improvements could be implemented if necessary.

In the subsequent subsections, we will briefly introduce two mostly used parameter methods in reliability estimation.

### 7.4.4 Maximum Likelihood Estimation

One of the most important statistical techniques used to estimate the parameters of a probability distribution is the MLE. The fundamental principle of MLE is to identify the parameter values within the model that yield the highest possible value for the *likelihood function*. The likelihood function measures the possibility of the observed data given a specific setting of parameter values. By optimizing this function, MLE seeks to find the parameter values that make the observed data most probable.

The following are the main steps of MLE:

**Step 1: Defining the probabilistic model**

The first step in MLE is to select an appropriate probabilistic distribution that best describes the observed data. Common choices for time-to-failure data include the exponential and Weibull distributions, which are well-suited for modeling the lifetimes of barrier components or systems. Once a distribution is chosen, it becomes clear that one or more parameters of this distribution will need to be estimated.

**Step 2: Defining the likelihood function**

Consider a vector of parameter as $\Theta$, and then we estimate $\Theta$ by the value which is most likely given the observed data. The likelihood function can be defined as the probability density of observing the given data $x_1, x_2, \ldots x_n$, typically denoted as $L(\Theta|x_1, x_2, \ldots x_n)$. In reliability estimation test, $x_i$ can be replaced with $t_i$, as time-to-failure of barrier component $i$ in test. Assuming that failures of different components are independent and identically distributed, the likelihood function can be expressed as the product of the individual probability densities:

$$f(x_1|\Theta)f(x_2|\Theta)\ldots f(x_n|\Theta)=\prod_{i=1}^{n}f(x_i|\Theta)=L(\Theta|x_1,\ x_2,\ \ldots\ x_n) \quad (7.11)$$

To facilitate the calculation, the logarithm of the likelihood function (log-likelihood) is often used as follows:

$$\ln L(\Theta|x_1,\ x_2,\ \ldots\ x_n)=\sum_{i=1}^{n}\ln f(x_i|\Theta) \quad (7.12)$$

**Step 3: Maximizing the likelihood function**

After the likelihood function is established, the next step is to determine the parameter vector $\Theta$ that can maximize the likelihood function. Typically, this is accomplished by computing the derivative of the likelihood function concerning the parameters, equating it to zero, and then solving for the parameter values:

$$\frac{\partial L}{\partial \Theta}=0 \quad (7.13)$$

MLE is known for providing consistent and asymptotically efficient estimations. This means that as the sample size increases, the estimates converge to the real parameter values and exhibit the lowest possible variance among all consistent estimators.

## Example 7.8:

### *Estimate λ in the Exponential Distribution*

In reliability engineering, the exponential distribution is frequently used due to its simplicity and the constant failure rate it assumes. The PDF for a barrier component that follows an exponential distribution is given by the following:

$$f(t)=\lambda e^{-\lambda t}$$

In this MLE, $x_i$ is the observed time-to-failure of item $i$, denoted as $t_i$. Therefore, the likelihood function for a sample of $n$ failures can be expressed as follows:

$$L(\Theta|x_1,\ x_2,\ \ldots\ x_n)=\prod_{i=1}^{n}f(x_i|\Theta)=\lambda e^{-\lambda t_1}\cdot\lambda e^{-\lambda t_2}\cdot\ldots\cdot\lambda e^{-\lambda t_n}=\lambda^n e^{-\lambda\Sigma_{i=1}^{n}t_i}$$

To facilitate the calculation, the logarithm of the likelihood function is taken:

$$\ln L(\Theta|x_1,\ x_2,\ \ldots\ x_n)=n\ln\lambda-\lambda\sum_{i=1}^{n}t_i$$

Differentiating this log-likelihood function with respect to $\lambda$ and setting it to zero for maximization, we have the following:

$$\frac{\partial L}{\partial \lambda} = \frac{n}{\lambda} - \sum_{i=1}^{n} t_i = 0$$

Solving for $\lambda$ yields the estimator

$$\hat{\lambda} = \frac{n}{\sum_{i=1}^{n} t_i}$$

Consider an example where $n = 10$ independent lifetimes (given in hours) have been observed during a reliability test conducted under harsh environmental conditions:

113.8 137.0 154.1 190.3 212.2 270.8 284.0 305.2 309.3 335.7

Applying the exponential distribution assumption and the MLE method, we can calculate the following:

$$\hat{\lambda} = \frac{10}{\sum_{i=1}^{n} t_i} = 4.325 \times 10^{-3}$$

Thus, the MTTF for these items, under the specified testing conditions, is approximately 231.24 hours. ■

### 7.4.5 Regression Analysis

Parametric methods can also support the exploration of statistical relationships between various influencing factors and reliability metrics. These factors include design and operational parameters such as weight, diameter, and usage frequency, as well as environmental conditions like temperature and humidity. Since reliability tests cannot cover all possible operational conditions, understanding the relationship between operational parameter and performance is important for estimating performance in the untested scenarios.

To identify the relationship between a certain design or condition parameter as a variable ($x$) and the reliability metrics as a response ($y$), one of the simplest statistical techniques is *linear regression.* For the two variables, the basic linear model can be described as follows:

$$y = ax + b + \varepsilon \tag{7.14}$$

where $\varepsilon$ is a random error, assumed following the normal distribution, with the mean of $E(\varepsilon) = 0$ and variance as $\sigma^2$, which is known as the *error variance* or *residual variance.* Thus, the expected value of response $y$ is $E(y) = ax + b$, and the variance of $y$ is $\text{Var(y)} = \sigma^2$. $E(y) = ax + b$ can be called the *regression line.*

Suppose there are $n$ observations in a reliability test, and the value of the $i$th response of $y_i$ is

$$y_i = ax_i + b + \varepsilon_i \tag{7.15}$$

In such a case, $\varepsilon_i$ is the error of the $i$th actual response $y_i$ to its expected/estimated value on the regression line $(\widehat{y_i})$. To build a solid relationship between the variable ($x$) and response ($y$), it is needed to minimize the variance $\sigma^2$, namely, to find a relationship with minimized error variance. This is idea behind the method of *least square estimation*, where we can calculate the sum of squares of error $\varepsilon_i$ ($SS_E$), as a function of $a$ and $b$:

$$SS_E = \sum_{i=1}^{n}\left[y_i - (ax_i + b)\right]^2 \tag{7.16}$$

Take partial derivatives with respect to $a$ and $b$, set them to zero, and solve the following:

$$\frac{\partial(SS_E)}{\partial a} = -2\sum_{i=1}^{n} x_i\left(y_i - ax_i - b\right) = 0 \tag{7.17}$$

$$\frac{\partial(SS_E)}{\partial b} = -2\sum_{i=1}^{n}\left(y_i - ax_i - b\right) = 0 \tag{7.18}$$

The parameters of $a$ and $b$ in the linear relationship can be then obtained as follows:

$$\hat{a} = \frac{\sum_{i=1}^{n}(x_i - \overline{x})(y_i - \overline{y})}{\sum_{i=1}^{n}(x_i - \overline{x})^2} \tag{7.19}$$

$$\hat{b} = \frac{\sum_{i=1}^{n} y_i - \hat{a}\sum_{i=1}^{n} x_i}{n} = \overline{y} - \hat{a}\overline{x} \tag{7.20}$$

where $\overline{x} = \frac{1}{n}\sum_{i=1}^{n} x_i$ and $\overline{y} = \frac{1}{n}\sum_{i=1}^{n} y_i$.

**Example 7.9:**

***Estimate Parameters of the Weibull Distribution***

The Weibull distribution is a flexible model to represent various types of failure rates. The PDF of the Weibull distribution is defined as follows:

$$F(t) = 1 - e^{-(\lambda t)^{\alpha}} \quad \text{for } t > 0$$

Considering $t_i$ as the observed time-to-failure of item $i$, we first transform this equation by taking the natural logarithm on both sides:

$$\ln\left[1-F(t_i)\right]=\ln\left[e^{-(\lambda t_i)^\alpha}\right]=-(\lambda t_i)^\alpha$$

Taking the logarithm again, we have the following:

$$\ln\left\{-\ln\left[1-F(t_i)\right]\right\}=\alpha\ln t_i+\alpha\ln\lambda$$

Let $y=\ln\{-\ln[1-F(t_i)]\}$, $x=\ln t_i$, and $\beta=\alpha\ln\lambda$. This transformation can yield a linear relationship between $x$ and $y$.

To proceed with the analysis, we reconsider the data from Example 7.8 For computational feasibility, only the first nine failures are used because including the tenth failure would result in $F(t_{10}=335.7)=1$, making $\ln[1-F(t_i)]$ incalculable.

The following table shows the data to be used in the regression analysis:

| $t_i$ | 113.8 | 137.0 | 154.1 | 190.3 | 212.2 | 270.8 | 284.0 | 305.2 | 309.3 |
|---|---|---|---|---|---|---|---|---|---|
| $F(t_i)$ | 0.100 | 0.200 | 0.300 | 0.400 | 0.500 | 0.600 | 0.700 | 0.800 | 0.900 |
| $x_i$ | 4.734 | 4.920 | 5.038 | 5.249 | 5.358 | 5.601 | 5.649 | 5.721 | 5.734 |
| $y_i$ | −2.250 | −1.500 | −1.031 | −0.672 | −0.367 | −0.087 | 0.186 | 0.476 | 0.834 |

Using the least square estimation in Equations (7.19) and (7.20), the parameters $\alpha$ and $\beta$ in the linear relationship are estimated as 2.5979 and −14.3466, respectively. Then, the estimate of $\lambda$ is calculated from the estimated value of $\beta$ as 0.003996.

It is important to note that parameter estimates using least squares analysis can be sensitive to approximations, with deviations potentially exceeding 10%. Therefore, it is essential in practices to conduct a goodness-of-fit test to verify if the assumed distribution adequately models the observed data, as stated in Step 4 of Section 7.4.2. Due to the page limitation, we do not further discuss these methods here. ■

### Example 7.10:

#### *Estimate Mean Time-to-Failure under Different Conditions*

Assume that the reliability of the barrier components in Example 7.2 is susceptible to temperature, and the Arrhenius equation can be introduced for describing the relationship between temperature and time-to-failure (Vassiliou & Mettas, 2001).

The Arrhenius equation is commonly used in chemical kinetics and reliability engineering for modeling the effect of temperature on reaction rates and failure times:

$$\text{MTTF}=A\exp\left(\frac{E_a}{kT}\right)$$

- $A$ is a constant determined by test for the barrier components involved;
- $E_a$ is activation energy, as a unique value for a specific failure mechanism;

- $k$ is the Boltzmann's constant = $8.62 \times 10^{-5}$ eV/°C; and
- $T$ is the absolute temperature in Kelvin (the Celsius temperature + 273.15).

By taking the natural logarithm of both sides, the function can be linearized as follows:

$$\ln(\text{MTTF}) = \frac{a}{T} + b$$

where $b = \ln(A)$ and $a = E_a/k$. In the equation above, the variable is $1/T$, and the response is ln(MTTF).

Assuming the data from Example 7.8 was collected under a temperature of 80°C, where we have calculated the MTTF is 231.24 hour. Additional tests conducted at temperatures of 70, 60, 50, and 40°C yielded MTTF values of 311.55, 398.92, 501.30, and 654.64 hours, respectively. Based on these data, the following table summarizes the values for $x$ and $y$.

| | | | | | |
|---|---|---|---|---|---|
| $x = 1/T$ | 0.002832 | 0.002914 | 0.003002 | 0.003095 | 0.003193 |
| $y = \ln(\text{MTTF})$ | 5.4435 | 5.7416 | 5.9888 | 6.2172 | 6.4841 |

Using Equations (7.19) and (7.20), the parameters $a$ and $b$ can be estimated as −2.5213 and 2,825. Back to the Arrhenius equation, it can be written with calculated values of $A$ and $E_a$ based on $\hat{a}$ and $\hat{b}$:

$$\widehat{\text{MTTF}} = 0.0804\exp\left(\frac{2{,}825}{T}\right)$$

Given that 20°C is the normal operational condition, the MTTF at this temperature is estimated to be approximately 1,223.45 hours.

It is important to note that these estimates assume no new failure mechanisms are introduced as temperatures vary from 20°C to 80°C. When comparing MTTF values under various conditions, the importance of acceleration testing becomes evident.

For readers interested in a deeper exploration of life-stress models used in accelerated reliability tests, further reading is recommended on the works like that by Yang (2007). ■

The previous example demonstrates a simple linear regression analysis involving a single factor, but the real-world scenarios often require the consideration of multiple influencing factors. This complexity can be addressed through *multiple linear regression* (MLR).

MLR extends the concept of simple linear regression to accommodate multiple explanatory variables. Considering that $m$ variables are evaluated in a test, the $i$th response $y_i$ in MLR is represented as follows:

$$y_i = \sum_{j=1}^{m} a_j x_{ij} + b + \varepsilon_i \tag{7.21}$$

where $a_j$ is the parameter to be estimated related to the $j$th factor, and $x_{ij}$ is the $i$th value of variable $j$, and $\varepsilon$ is also a random error following the normal distribution, with $E(\varepsilon) = 0$ and variance as $\sigma^2$.

In practices, the assumptions of linear relationships between variables and response, as well as the normal distribution of variance, are always challenged. To overcome these problems, *generalized linear models* (GLMs) can be introduced. A GLM includes three components:

- **Systematic component** specifies the linear combination of the variables, e.g., $\Sigma_{j=1}^{m} a_j x_{ij}$, as we can find in the linear regression.
- **Random component** specifies the probability distribution of the variance $\varepsilon$. For example, it can be a binomial distribution or Poisson distribution.
- **Link function** connects the systematic component and the random component. For example, if we consider MTTF as the response in Example 7.8, ln(MTTF) is a link function.

The quality and appropriateness of regression models, whether MLR or GLMs, can be assessed using the analysis of variance (ANOVA) approach. ANOVA tests the significance of the model's parameters and helps in understanding the contribution of each factor to the variability in the response variable.

Due to the page limitation, we do not have detailed discussions on MLR, GLM, and ANOVA. Readers interested in a deeper understanding of these statistical techniques are encouraged to refer to comprehensive statistics textbooks such as that by Walpole et al. (2016).

## 7.5 FIELD DATA-BASED RELIABILITY ESTIMATION

When safety barriers are put into operation, a significantly larger amount of field data can be collected. This includes information about barrier statuses, operational behaviors, and environmental factors. Despite the relatively low incidence of barrier failures compared to many other technical systems due to their high reliability, the volume of operational data is still abundant compared to the testing data.

In this section, we aim to provide an overview of the widely used data-driven techniques for revealing the relationships between operational and environmental factors and estimating the reliability metrics of safety barriers. This section is based on a research for estimating the failure rate of shutdown valves based on the field data of the oil and gas industry, conducted by the author of this book and his colleagues (Xie et al., 2019).

### 7.5.1 Framework of Reliability Estimation

The general framework for reliability estimation based on field data has similarities to that for reliability prediction using reference data. There are three main steps:

**Step 1: Data collection and preprocessing**

The quality of field data can vary and is sometimes uncontrollable, so that preprocessing and cleaning data is necessary to ensure its utility for reliability estimation. Preprocessing involves interpreting, classifying, and cleaning the data by removing low-quality entries to ensure sufficient quality for reliable information extraction. For instance, when estimating the failure rate of equipment commonly used in the oil and gas industry, failures due to specific conditions such as icing problems can be excluded. Data classification is important, such as dividing dangerous undetected (DU) failure data of barriers into groups based on failure modes, causes, and detection approaches. It is also essential in this step to focus on the data related to the most critical safety barriers.

**Step 2: Identification of critical risk influence factors**

Each potential RIF is treated as a variable in reliability estimation. For technical safety barriers, DU failures are regarded as the main threats, and thus only those resulting in DU failures are considered critical. However, DU failures are not immediately observable, and their frequency and subsequent downtime can be influenced by various factors, including technical and temporal resources for inspection and maintenance, as well as the physical properties of safety equipment. Such factors should not be ignored.

This step also explores the correlations among potential RIFs and identifies the critical ones with substantial impacts on the performance of safety barriers. Due to the complex interactions of multiple factors and resulting collinearity issues (some variables/factors are correlated or have similar influences), even the GLM mentioned in the previous section may not effectively describe these correlations.

In this section, we provide an overview of widely used data-driven techniques for handling the complex relationships among factors and between factors and reliability. These methods include various multivariate statistical approaches, with a particular emphasis on *principal component analysis* (PCA) for extracting vital information of RIFs, and *principal component regression* (PCR) and *partial least squares regression* (PLSR) for establishing correlations between the critical RIFs and the reliability of safety barriers. We will discuss these methods subsequently. Machine learning (ML) approaches (see Section 7.5.4) can also be used to cluster RIFs and reveal the significant ones.

**Step 3: Reliability estimation**

Based on the identified significant RIFs, failure rates can be estimated by determining the weights of the RIFs similarly to reliability prediction. The DU failure rate $\lambda_{DU}$ of a barrier system can be estimated according to collected failure data, and this rate is the sum of the rates of different failure modes. Methods from Section 7.3 can be employed in this step, especially if new operational conditions are considered.

### 7.5.2 Principal Component Analysis

PCA is a widely used method for reducing the dimensionality of multivariate data while retaining the most critical information by identifying the most significant features in a dataset. In barrier engineering, PCA can be utilized to identify the critical

RIFs affecting the failure modes of a barrier system/component based on various field data. The main steps of PCA are as follows:

**Step 1: Establishment of covariance matrix**

Given a dataset with $n$ observations or samples (e.g., barriers in operation) and $m$ variables or factors (e.g., RIFs), we can set up an $n \times m$ matrix $\boldsymbol{X}$. Each variable can be regarded as a *dimension* in a multi-coordinate system, and the observations can be projected on the dimension. For the cases with many variables, the main idea behind PCA is to find the main directions in the dataset that can explain the maximum variance, namely the *principal components* (PCs). In general, the number of PCs is same with that of variables. Those PCs that do not explain much variance can be skipped in the following analysis, and we can call this process as *dimensionality reduction*.

The elements in $\boldsymbol{X}$ describe the observed situation relating to various influencing factors and the states of DU failures, where $X_{ij} = 1$ means that a DU failure is observed on barrier $i$ when RIF$_j$ is present, while $X_{ij} = 0$ represents that no DU failure is observed on barrier $i$ when RIF$_j$ is present. In some cases, data should be scaled for PCA, but here due to the binomial situation, no scaling is needed.

$\boldsymbol{X}$ is then decomposed into a *score matrix* $\boldsymbol{S}$ and a *loading matrix* $\boldsymbol{L}$:

$$\boldsymbol{X} = \boldsymbol{S}\boldsymbol{L}^T \tag{7.22}$$

The loading matrix $\boldsymbol{L}$ reflects the weight for each original RIF when calculating the PCs and the score matrix $\boldsymbol{S}$ illustrates how the observations project along the PCs. Considering the potential unmodeled part of the data space, a residual matrix ($\boldsymbol{E}$) can be added in Equation (7.22) as follows:

$$\boldsymbol{X} = \boldsymbol{S}\boldsymbol{L}^T + \boldsymbol{E} \tag{7.23}$$

The *covariance matrix* $\boldsymbol{C}$ of $n$ observations can be expressed by

$$\boldsymbol{C} = \frac{1}{n-1}\boldsymbol{X}^T\boldsymbol{X} \tag{7.24}$$

This matrix $\boldsymbol{C}$ describes the pairwise correlations between all RIFs in the dataset. The diagonal elements of this matrix stand for the variance of each RIF, while the other elements indicate the covariances between different pairs of RIFs.

**Step 2: Eigen decomposition**

Based on the covariance matrix $\boldsymbol{C}$, *eigenvalues* and *eigenvectors* can be calculated. The eigen decomposition is performed on $\boldsymbol{C}$ to obtain loading matrix $\boldsymbol{L}$. The vector of eigenvalues $\boldsymbol{V}$ is denoted as follows:

$$\boldsymbol{V} = [v_1, v_2 \ldots v_m] \tag{7.25}$$

The eigenvector $V$ reflects the directions of the PCs. The $i$th eigenvalue $v_i$ relates to the $i$th element of the score matrix $\boldsymbol{S}$:

$$v_i = \frac{1}{n-1} s_i^T s_i \tag{7.26}$$

This eigenvalue reflects the amount of variance explained by $PC_i$, as the average sum of squares of the distances between the project points of observations on the PC and the origin.

**Step 3: Choosing and ranking principal components**

The chosen PCs for ranking should be orthogonal, meaning that there is no repeating information among these components and that they are not correlated with one another, namely different PCs are related to different factors. These PCs are calculated as linear combinations of the RIFs, and they are ranked as $PC_1$ with the highest eigenvalues, $PC_2$ with the second highest eigenvalues, and so on. The highest eigenvalue denotes that $PC_1$ includes the most variance for the exploratory information and the residual is conducted to contain less covariance.

By selecting the top PCs that explain the majority of the variance (e.g., 90%), the factors with significant influences are identified, and the dimensionality of the field data can then be reduced. The factors that are not critical for DU failures can be ignored in the following analysis. It should be noted that with a lower threshold for top PCs, there is a more obvious dimensionality reduction, and less information is kept.

### 7.5.3 Principal Component Regression and Partial Least Squares Regression

After identifying the critical RIFs through PCA, PCR or PLSR can be used to evaluate the relationship between observed performance metrics, such as time to DU failures, and the RIFs of interest.

PCR, based on the similar considerations of PCA, is particularly valuable if strong correlations exist among the RIFs. These RIFs can be linearly combined as PCs following the PCA approach. Since PCs have been given orthogonal, we can rely on regression analysis methods, like GLMs, to estimate how the time to DU failures of barriers is impacted by the PCs.

PCR is typically regarded as an unsupervised learning method, with a primary focus on explaining the variance in explanatory variables or factors. However, the effectiveness of PCR may be weakened if there are some factors that are not related to the observed performance metrics.

In contrast, PLSR can be considered a supervised extension of PCR, also using a regression-based statistical approach that considers the weighting of different variables during estimation. In PLSR, we need to build two matrices: $\boldsymbol{X} = [X_1, X_2, \ldots X_k]^T$ for describing the explanatory RIFs, as well as $\boldsymbol{Y} = [Y_1, Y_2, \ldots Y_k]^T$ for describing responses, here the time to DU failures for safety barriers. Their relationship can be expressed as follows:

$$\boldsymbol{Y} = \boldsymbol{X}\boldsymbol{A} + \boldsymbol{B} \tag{7.27}$$

where $\boldsymbol{A}$ and $\boldsymbol{B}$ are vectors of parameters that need estimation in PLSR. Both matrices of $\boldsymbol{X}$ and $\boldsymbol{Y}$ can be decomposed into scoring matrix and loading matrix:

$$\boldsymbol{X} = \boldsymbol{S}_X \boldsymbol{L}_X^T + \boldsymbol{E}_X \tag{7.28}$$

$$\boldsymbol{Y} = \boldsymbol{S}_Y \boldsymbol{L}_Y^T + \boldsymbol{E}_Y \tag{7.29}$$

where $\boldsymbol{S}_X$ and $\boldsymbol{S}_Y$ are score matrices, $\boldsymbol{L}_X$ and $\boldsymbol{L}_Y$ are the loading matrices for $\boldsymbol{X}$ and $\boldsymbol{Y}$, respectively, and $\boldsymbol{E}_X$ and $\boldsymbol{E}_Y$ are partial least squares (PLS) residuals corresponding. The weights provided in $\boldsymbol{L}_X$ and $\boldsymbol{L}_Y$ reflect the correlations between $\boldsymbol{X}$ and $\boldsymbol{Y}$. In PLSR, $\boldsymbol{X}$ and $\boldsymbol{Y}$ are decomposed to maximize the covariance between $\boldsymbol{S}_X$ and $\boldsymbol{S}_Y$, which is defined pair by pair, meaning that the covariance of column $i$ of $\boldsymbol{S}_X$ with the column $i$ of $\boldsymbol{S}_Y$ is maximized. We can effectively combine the variables in a manner that can explain both the factors (RIFs) and the responses (reliability metrics). Then, ordinary least square estimation is employed to determine the values of $\boldsymbol{A}$ and $\boldsymbol{B}$. Since least square estimation is only used for some coefficients in this method (not for $\boldsymbol{S}_X$ and $\boldsymbol{S}_Y$), it is called the PLSR.

Comprehensive introduction about PCA and PCR can be found in the textbook by Jolliffe (2002). Readers who have interests in these methods can also discover numerous introductory videos on YouTube.

### 7.5.4 Machine Learning

ML means to offer a machine or a computer system with the ability to learn from data automatically without being explicitly programmed, and it has emerged as a powerful tool in various fields, including barrier engineering. According to Murphy (2012), ML involves a set of methods that can detect or learn patterns in data and use them for prediction and decision-making under uncertainty. ML is a broad field of study that includes various techniques and algorithms, and most of them are from applied statistics but the solutions rely on the computing capacities. For example, both linear regression methods and PCA can be used in ML.

ML is well-suited for the situation where amount of data is collected from the operation of barriers. Like in other applications, in barrier engineering, the basic elements for ML include the datasets about barrier design property, operation, performance, and failures, for training and testing, a model (non-parametric or parametric) for explaining the above data and an objective function to optimize, e.g., the likelihood function. ML models require relevant features or input variables to make accurate predictions. In barrier engineering, these features include design parameters of barriers, and environmental and operational conditions.

Depending on the specific problem, different algorithms can be employed in ML. In general, these algorithms can be classified as *unsupervised learning* and *supervised learning*:

- *Unsupervised learning*: The algorithm is presented with unlabeled data, and its purpose is to find patterns, structures, or relationships within the data without explicit guidance or supervision. The applications of unsupervised learning in barrier engineering can include general data clustering (grouping the observations that share some similarities, without predefined categories) and anomaly detection in barrier operation. PCA is an unsupervised learning technique for dimensionality reduction of field data, which can be understood as a task in data clustering.

- *Supervised learning*: The algorithm learns a mapping from input data to target output by training on a labeled dataset. In training, parameters of the algorithm are adjusted to minimize the difference between its predictions and the true labels in the training dataset. The so-called supervision is from a set of input-output pairs. The applications of supervised learning in barrier engineering can include safe/dangerous failure classification of barriers, decision-making analysis in barrier operation, and regression analysis and Bayesian analysis for parameter estimation of the measures related to safety barriers, e.g., time to DU failure and remaining useful life.

In Chapter 11, we will discuss artificial intelligence in barrier engineering, where ML plays an important role. However, due to the page limitation, we do not discuss more ML methods, e.g., semi-supervised learning and enforcement learning, and more applications of ML in barrier engineering in this book. For those readers interested in exploring this subject, we recommend them to read the comprehensive review by Xu and Saleh (2021) on ML in reliability and safety applications.

In this chapter, we have explored the roles of different types of data analysis methods in barrier engineering for different types of data. In practice, the integration of these approaches is important, to significantly enhance our ability in predicting failure likelihood, ultimately leading to safer and more reliable barrier systems in various industries.

## REFERENCES

Brissaud, F., Charpentier, D., Fouladirad, M., Barros, A., & Bérenguer, C. (2010). Failure rate evaluation with influencing factors. *Journal of Loss Prevention in the Process Industries, 23*(2), 187–193.

Cox, D. R. (1972). Regression models and life tables. *Journal of the Royal Statistical Society, B34*, 187–220.

Hauge, S., Kråknes, T., Håbrekke, S., & Jin, H. (2013). *Reliability Prediction Methods for Safety Instrumented Systems, PDS Method Handbook*. SINTEF.

ISO14224. (2016). Petroleum, petrochemical and natural gas industries—Collection and exchange of reliability and maintenance data for equipment. Geneva, Switzerland: International Organization for Standardization.

Jolliffe, I. T. (2002). *Principal Component Analysis* (2nd ed.). Springer-Verlag.

Kumar, A. (2016). *Prediction of Failure Rates for Subsea Equipment*. Norwegian University of Science and Technology.

MIL-HDBK-217F. (1991). Reliability prediction of electronic equipment. Washington DC, USA: U.S. Department of Defense.

Murphy, K. P. (2012). *Machine Learning: A Probabilistic Perspective*. The MIT Press.

OREDA. (2009). *Offshore Reliability Data* (5th ed.). OREDA Participants.

Rahimi, M., & Rausand, M. (2013). Prediction of failure rates for new subsea systems: A practical approach and an illustrative example. *Proceedings of the Institution of Mechanical Engineers, Part O: Journal of Risk and Reliability, 227*(6), 629–640.

Rausand, M., Barros, A., & Hoyland, A. (2020). *System Reliability Theory: Models, Statistical Methods, and Applications* (3rd ed.). Wiley-Blackwell.

Vassiliou, P., & Mettas, A. (2001). *Understanding accelerated life-testing analysis. 2001* Annual Reliability and Maintainability Symposium. Philadelphia, Pennsylvania USA.

Walpole, R., Ye, K., Myers, R., & Myers, S. (2016). *Probability & Statistics for Engineers & Scientists, Global Edition*. Pearson Education.

Xie, L., Håbrekke, S., Liu, Y., & Lundteigen, M. A. (2019). Operational data-driven prediction for failure rates of equipment in safety-instrumented systems: A case study from the oil and gas industry. *Journal of Loss Prevention in the Process Industries, 60*, 96–105.

Xu, Z., & Saleh, J. H. (2021). Machine learning for reliability engineering and safety applications: Review of current status and future opportunities. *Reliability Engineering & System Safety, 211*, 107530.

Yang, G. (2007). *Life Cycle Reliability Engineering*. John Wiley & Sons.

Zhang, J., Liu, Y., Lundteigen, M. A., & Bouillaut, L. U. (2016). Using Bayesian networks to quantify the reliability of a subsea system in the early design. 2016 European Safety and Reliability Conference, Glasgow, Scotland.

# 8 Structural and Operational Analysis

## 8.1 INTRODUCTION

Based on the modeling approaches introduced in Chapter 5 and the performance measures, especially safety integrity measures, discussed in Chapter 6, this chapter analyzes the impacts of different voting structures and operational modes on the performance of safety barriers, particularly on the calculation of $PFD_{avg}$.

In this chapter, components are assumed to fail following an exponential distribution. Given the focus on voting structures and operational modes, we employ a simple test and maintenance strategy for safety barrier systems (BSs): Detected dangerous (DD) failures of any component in a BS are immediately identified by the integrated diagnostic module upon occurrence, and proof tests are simultaneously conducted on all barrier components at constant intervals of $\tau$ to detect undetected dangerous (DU) failures. Proof tests are considered perfect, allowing all DU-failures to be detected and subsequently repaired. Degradation issues of barrier components, and more comprehensive test and maintenance strategies will be discussed in Chapter 9.

Additionally, it is assumed that all components in the BSs are independent of each other. Issues of dependency and interdependency will be explored in Chapter 10.

## 8.2 INTEGRITY ANALYSIS FOR DIFFERENT VOTING STRUCTURES

In Chapter 6, we discussed how to calculate $PFD_{avg}$ and PFH, and how to determine the SIL for a single barrier component (or 1-out-of-1 system) when introducing these performance measures. In this section, we will extend our analysis to more complex BSs, particularly those employing 1-out-of-2 (1oo2) and 2-out-of-3 (2oo3) voting structures. As defined in Chapter, an 1oo2 voting structure includes two channels, and each of channel can be either a single component or a combination of several components that work together to perform a specific function.

We make several assumptions regarding the voting structures in this section:

- All channels and/or components are independent in terms of failures and repairs;
- Times-to-failures and time-to-repairs of all components follow the exponential distribution;
- The components can be restored to an as-good-as-new state after their failures are revealed;
- Proof tests are carried out on all channels simultaneously (or sequentially with ignorable interval) with a constant time interval, and test duration is very short.


DOI: 10.1201/9781003245636-8

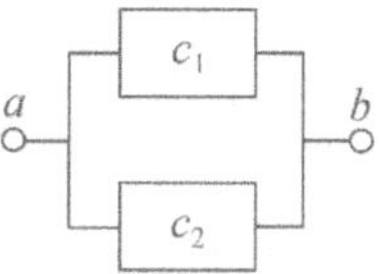


**FIGURE 8.1** Structure of a 1-out-of-2 barrier subsystem.

## 8.2.1 Approximation Formulas of PFD$_{avg}$

### 8.2.1.1 1oo2 Voting Structure

Consider a 1oo2 barrier subsystem operating in the low-demand mode, as illustrated in Figure 8.1. This diagram can also be interpreted as a reliability block diagram. To obtain the approximation formular of PFD$_{avg}$ of this system, we focus solely on DU-failures in this analysis. The DU-failure rates of these components are $\lambda_{DU,1}$ and $\lambda_{DU,2}$, respectively. First, we calculate the system reliability of this barrier group based on Equation (5.62) as follows:

$$R_{1oo2}(t) = R_1(t) + R_2(t) - R_1(t)R_2(t) = e^{-\lambda_{DU,1}t} + e^{-\lambda_{DU,2}t} - e^{-(\lambda_{DU,1}+\lambda_{DU,2})t}$$

Then, the average PFD of this barrier subsystem within the proof test interval of $(0, \tau]$ is calculated as:

$$\mathrm{PFD}_{avg}^{1oo2} = 1 - \frac{1}{\tau}\int_0^{\tau} R_{1oo2}(t)\,dt = 1 - \frac{1}{\lambda_{DU,1}\tau}\left(1 - e^{-\lambda_{DU,1}\tau}\right) - \frac{1}{\lambda_{DU,2}\tau}\left(1 - e^{-\lambda_{DU,2}\tau}\right) + \frac{1}{\left(\lambda_{DU,1} + \lambda_{DU,2}\right)\tau}\left[1 - e^{-(\lambda_{DU,1}+\lambda_{DU,2})\tau}\right] \tag{8.1}$$

Using the Taylor series $e^x = \Sigma_{n=0}^{\infty}\frac{x^n}{n!}$, given that $\lambda\tau > 0$ is small (e.g., $< 0.1$, see the explanation in Note 8.1), there is an approximation that $e^{-\alpha\tau} \approx 1 - \lambda\tau + \frac{(\lambda\tau)^2}{2} - \frac{(\lambda\tau)^3}{6}$, namely $1 - e^{-\alpha\tau} \approx \lambda\tau - \frac{(\lambda\tau)^2}{2} + \frac{(\lambda\tau)^3}{6}$. $\mathrm{PFD}_{avg}^{1oo2}$ can thus be approximated as:

$$\mathrm{PFD}_{avg}^{1oo2} \approx \frac{\left(\lambda_{DU,1}\lambda_{DU,2}\right)\tau^2}{3} \tag{8.2}$$

If the two barrier components are identical, $\lambda_{DU,1} = \lambda_{DU,2} = \lambda_{DU}$, the average PFD of this barrier subsystem is:

$$\mathrm{PFD}_{avg}^{1oo2} = \frac{\left(\lambda_{DU}\tau\right)^2}{3} \tag{8.3}$$

### 8.2.1.2 2oo3 Voting Structure

Similarly, consider a 2-out-of-3 (2oo3) voting structure for a barrier subsystem operating in low-demand mode, as illustrated in Figure 8.2.

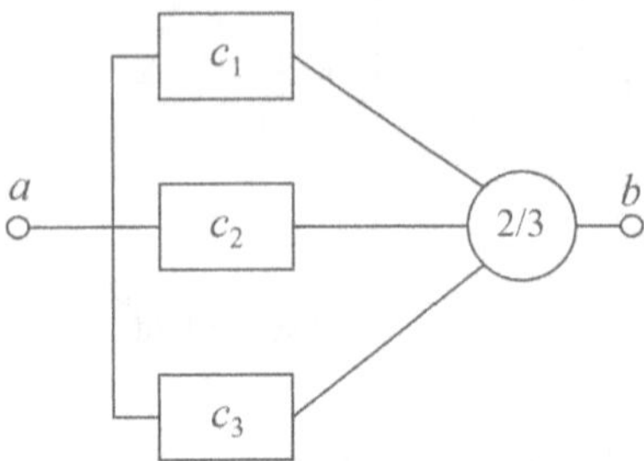


**FIGURE 8.2** Structure of a 2-out-of-3 barrier subsystem.

This structure is equivalent to a parallel configuration of three channels: $c_1c_2$, $c_2c_3$, and $c_1c_3$, as shown in Figure 5.9. Consequently, the reliability of this voting structure, based on Equation (5.67), is:

$$R_{2oo3}(t) = R_1(t)R_2(t) + R_2(t)R_3(t) + R_3(t)R_1(t) - 2R_1(t)R_2(t)R_3(t)$$

The $\text{PFD}_{\text{avg}}$ of this barrier subsystem within the interval (0, $\tau$] is

$$\text{PFD}_{\text{avg}}^{2oo3} = 1 - \frac{1}{\tau}\int_0^{\tau}\left(e^{-\lambda_{\text{DU},1}\lambda_{\text{DU},2}t} + e^{-\lambda_{\text{DU},2}\lambda_{\text{DU},3}t} + e^{-\lambda_{\text{DU},1}\lambda_{\text{DU},3}t} - 2e^{-\lambda_{\text{DU},1}\lambda_{\text{DU},2}\lambda_{\text{DU},3}t}\right)dt$$

Using the Taylor series, replace $e^{-\lambda_{\text{DU},i}\tau}$ with $1 - \lambda_{\text{DU},i}\tau + \frac{(\lambda_{\text{DU},i}\tau)^2}{2}$ when $\lambda_{\text{DU},i}\tau > 0$ is small. Therefore, we can derive the following:

$$\text{PFD}_{\text{avg}}^{2oo3} \approx \frac{\left(\lambda_{\text{DU},1}\lambda_{\text{DU},2} + \lambda_{\text{DU},2}\lambda_{\text{DU},3} + \lambda_{\text{DU},1}\lambda_{\text{DU},3}\right)\tau^2}{3} \tag{8.4}$$

To simplify the calculation, we assume that all three barrier components are identical, with a DU-failure rate of $\lambda_{\text{DU}}$. The reliability of this barrier subsystem, based on Equation (5.67), can be calculated as:

$$R_{2oo3}(t) = 3e^{-2\lambda_{\text{DU}}t} - 2e^{-3\lambda_{\text{DU}}t}$$

The $\text{PFD}_{\text{avg}}$ of this barrier subsystem within the proof test interval of (0, $\tau$] is given by:

$$\text{PFD}_{\text{avg}}^{2oo3} = 1 - \frac{1}{\tau}\int_0^{\tau}\left(3e^{-2\lambda_{\text{DU}}t} - 2e^{-3\lambda_{\text{DU}}t}\right)dt$$

$$= 1 - \frac{3}{2\lambda_{\text{DU}}\tau}\left(1 - e^{-2\lambda_{\text{DU}}\tau}\right) + \frac{2}{3\lambda_{\text{DU}}\tau}\left(1 - e^{-3\lambda_{\text{DU}}\tau}\right)$$

Using the Taylor series, replace $e^{-\lambda_{\text{DU}}\tau}$ with $1 - \lambda_{\text{DU}}\tau + \frac{(\lambda_{\text{DU}}\tau)^2}{2}$ when $\lambda_{\text{DU}}\tau > 0$ is small, and we have:

$$\text{PFD}_{\text{avg}}^{(2oo3)} \approx \left(\lambda_{\text{DU}}\tau\right)^2 \tag{8.5}$$

#### 8.2.1.3 k*oo*n Voting Structure

Next, we consider more general configurations, specifically *k*-out-of-*n* (*koon*) structures with independent and identical components. In such a structure, the system remains operational as long as at least *k* of its *n* components are functioning. Conversely, the system fails when at least $n-k+1$ components have failed. Therefore, for a *koon* voting structure, we can derive a general formula for the average PFD:

$$\mathrm{PFD}_{\mathrm{avg}}^{1\mathrm{oo}(n-k+1)} = \frac{\left(\lambda_{\mathrm{DU}}\tau\right)^{n-k+1}}{n-k+2} \tag{8.6}$$

Then, for *koon* voting structures, a generic formula also can be obtained as:

$$\mathrm{PFD}_{\mathrm{avg}}^{koon} \approx \begin{pmatrix} n \\ n-k+1 \end{pmatrix} \frac{\left(\lambda_{\mathrm{DU}}\tau\right)^{n-k+1}}{n-k+2} \tag{8.7}$$

Based on these approximation formulas, $\mathrm{PFD}_{\mathrm{avg}}$ calculation results of some commonly used safety barrier configurations are summarized in Table 8.1.

These general formulas and specific examples are essential for understanding the reliability performance of various safety barrier configurations under different operational conditions.

### 8.2.2 Mean Downtime of a DU-failure

As discussed in Section 6.3, the mean downtime for a barrier component due to a DU-failure is $\tau/2$. This is based on the assumption that the time to failure follows an exponential distribution, implying that the probability of failure occurring at any given moment within the proof test interval $(0, \tau]$ remains constant. On average, the failure is expected to happen at the midpoint of the interval, namely the time to failure is $\tau/2$, leading to a mean downtime of $1-\tau/2=\tau/2$. The $\mathrm{PFD}_{\mathrm{avg}}$ of this barrier component can be regarded as the product of its failure rate and the mean downtime after a failure occurs.

Applying similar considerations to a 1oo2 voting structure with two independent and identical barrier channels, we assume that the two exponentially distributed failures have an equal probability of occurring at any time within a proof test interval. Therefore, if the barrier group is found to be faulty during a proof test, it is

**TABLE 8.1**
**Approximation Formulas of *koon* Voting Structures of Barriers**

| *k*/*n* | 1 | 2 | 3 | 4 |
|---|---|---|---|---|
| 1 | $\lambda_{\mathrm{DU}}\tau/2$ | $(\lambda_{\mathrm{DU}}\tau)^2/3$ | $(\lambda_{\mathrm{DU}}\tau)^3/4$ | $(\lambda_{\mathrm{DU}}\tau)^4/5$ |
| 2 | – | $\lambda_{\mathrm{DU}}\tau$ | $(\lambda_{\mathrm{DU}}\tau)^2$ | $(\lambda_{\mathrm{DU}}\tau)^3$ |
| 3 | – | | $\lambda_{\mathrm{DU}}\tau$ | $2(\lambda_{\mathrm{DU}}\tau)^2$ |
| 4 | – | | | $2\lambda_{\mathrm{DU}}\tau$ |

reasonable to assume that the two failures on its channels are evenly distributed over the preceding period. On average, one failure occurs after $\tau/3$ from the last test or restoration, and the other failure occurs by $2\tau/3$. Consequently, the mean downtime for the group is calculated as $1-2\tau/3=\tau/3$.

Then, $PFD_{avg}$ of this barrier group can be understood as follows: The probability that one channel fails within a proof test interval is the product of the failure rate and the duration of the period. $PFD_{avg}$ is then calculated as the product of this probability, the failure rate of the second channel, and the mean downtime of the group after the second failure occurs.

For a *koon* voting structure with independent and identical channels, given that $n-k+1$ faulty channels are found during the proof test, a similar analysis can be applied. The $n-k+1$ failures evenly divide the proof test interval into $n-k+2$ sections, resulting in a mean downtime of the structure as $\tau/(n-k+2)$. Using the same consideration for calculating $PFD_{avg}$ for a 1oo2 barrier subsystem, $PFD_{avg}$ for such a *koon* voting structure is the product of the probability that $n-k$ channels have failed during the period as $(\lambda_{DU}\tau)^{n-k}$, the failure rate of another channel as $\lambda_{DU}$, and the mean downtime of the group after the $n-k+1$ failure occurs as $\tau/(n-k+2)$. This calculation result aligns with the generic formula in Equation (8.6).

### 8.2.3 IEC61508 Formulas

IEC61508-6 provides several formulas for the integrity analysis of safety barriers with different voting structures. In this subsection, we present these formulas, allowing readers to compare them with the approximation formulas introduced earlier. The main ideas of these formulas are to calculate $PFD_{avg}$ of a voted structure/group (G) of channels as if the group were a single item (here the term of group is used since it aligns with IEC 61508):

$$PFD_{avg}^{(G)} = \lambda_{D,G} t_{GE} \tag{8.8}$$

where

- $\lambda_{D,G}$ is the average dangerous group failure rate.
- $t_{GE}$ is the group-equivalent mean downtime.

In calculation, it is necessary to consider the mean downtime of a channel that has experienced a D failure ($t_{CE}$), which can be calculated as,

$$t_{CE} = \frac{\lambda_{DU}}{\lambda_D}\left(\frac{\tau}{2} + \text{MRT}\right) + \frac{\lambda_{DD}}{\lambda_D}\text{MTTR} \tag{8.9}$$

It can be found that $t_{CE}$ represents the weighted sum of the downtime due to a DU-failure and the downtime due to a DD-failure. In this formula:

- MRT denotes the mean repair time of the channel that has failed fault.
- $\frac{\tau}{2}+\text{MRT}$ is the mean downtime before the channel is restored during a proof test.

- $\lambda_{DU}/\lambda_D$ represents the percentage of downtime due to a DU-failure within the total downtime. $\lambda_{DD}/\lambda_D$ is the percentage of downtime due to a DD-failure within the total downtime.
- MTTR is the mean time to repair for a DD-failure, which includes the time from the moment the failure is automatically diagnosed until the subsequent repair is completed.

In the case of a 1oo2 voting structure, where the two channels are independent and identical with the same failure rate $\lambda_D$, a group failure must start from a failure on one of two channels. Thus, the rate of this event is $2\lambda_D$, considering that either of the channels may fail first. $t_{CE}$ represents the duration from the initial channel failure to the group failure. The probability of the remaining channel failing during this period can be calculated as $1-e^{-\lambda_D t_{CE}}$, and thus we have the following equation:

$$\lambda_{D,G} = 2\lambda_D\left(1-e^{-\lambda_D t_{CE}}\right) \approx 2\lambda_D\lambda_D t_{CE} = 2\left(\lambda_D\right)^2 t_{CE} \tag{8.10}$$

For this 1oo2 voting structure, to experience a D group failure, one of the channels must first incur a D failure. When this channel is not functioning with a D fault, the other channel must also fail with a D failure. If the second failure is a DU-failure, the down of the 1oo2 group will be approximately $\tau/3+\text{MRT}$. If the second failure is a DD-failure, the downtime will be MTTR. The group mean downtime is thus:

$$t_{GE} = \frac{\lambda_{DU}}{\lambda_D}\left(\frac{\tau}{3}+\text{MRT}\right)+\frac{\lambda_{DD}}{\lambda_D}\text{MTTR} \tag{8.11}$$

Then, $\text{PFD}_{\text{avg}}^{(G)}$ can be calculated using Equation (8.8).

For a 2oo3 voting structure of three independent and identical channels with the same failure rate, $\lambda_D$, a dangerous group failure also starts with a D failure of one of the three channels. The rate of this event is $3\lambda_D$, because either of the three channels can fail. When one channel has failed, it will be down for a period of $t_{CE}$ as calculated by Equation (8.9).

To incur a dangerous group failure, at least one of the remaining two channels fail within the downtime of the already-failed channel. Thus,

$$\lambda_{D,G} = 3\lambda_D\left(1-e^{-2\lambda_D t_{CE}}\right) \approx 3\lambda_D 2\lambda_D t_{CE} = 6\left(\lambda_D\right)^2 t_{CE} \tag{8.12}$$

For a *koon* voting structure, the group mean downtime is:

$$t_{GE} = \frac{\lambda_{DU}}{\lambda_D}\left(\frac{\tau}{n-k+2}+\text{MRT}\right)+\frac{\lambda_{DD}}{\lambda_D}\text{MTTR} \tag{8.13}$$

For a *koon* voting structure, the frequency of first D failure is $n\lambda_D$, and a dangerous group failure occurs if at least $n-k$ of the remaining $n-1$ channels fail while the first failed channel is down. We can calculate the probability of at least $n-k$ failures, and then have the following equation:

$$\lambda_{\mathrm{D,G}} \approx n\begin{pmatrix} n-1 \\ k-1 \end{pmatrix}\lambda_{\mathrm{D}}^{n-k+1} t_{\mathrm{CE}}^{n-k} \tag{8.14}$$

The average unavailability is then calculated as,

$$\mathrm{PFD}_{\mathrm{avg}}^{koon} = P_{n-k+1}^{n} \lambda_{\mathrm{D}}^{n-k+1} \prod_{i=1}^{n-k+1} \mathrm{MDT}_{1ooi} \tag{8.15}$$

where $P_{n-k+1}^{n} = n!/(k-1)!$ and $\mathrm{MDT}_{1ooi} = \frac{\lambda_{\mathrm{DD}}}{\lambda_{\mathrm{D}}}\mathrm{MTTR} + \frac{\lambda_{\mathrm{DU}}}{\lambda_{\mathrm{D}}}\left(\frac{\tau}{i+1} + \mathrm{MRT}\right)$.

### 8.2.4 Calculation with the Markov Model

Markov models are effective for modeling and analyzing the performance of a BS with a specific voting structure. By identifying the potential states in the Markov model, we can account for various conditions where all barrier channels are functioning (working), some channels are functioning while others have failed, and scenarios where all channels have failed.

Figure 8.3 illustrates a simple Markov model for a 1oo2 barrier subsystem, considering only DU-failures. In this model, 2W means that two channels are working, 1W1F means that one channel is working while the other is failed, and 2F means that two channels are failed.

Similarly, Figure 8.4 presents a Markov model for a 2oo3 barrier subsystem, again considering only DU-failures. The naming principle of the states follows the same convention as in Figure 8.3: 3W denote that all three channels are working, 2W1F denotes that two channels are working while one has failed, 1W2F denotes that one channel is working while two have failed, while 3F denotes that all three channels have failed.

#### Example 8.1:

#### *Comparison of the Results by Different Methods*

Suppose a 1oo2 actuator subsystem in a BS. The subsystem consists of two identical channels, each with a constant failure rate of $1.0 \times 10^{-5}$ per hour, and the diagnostic

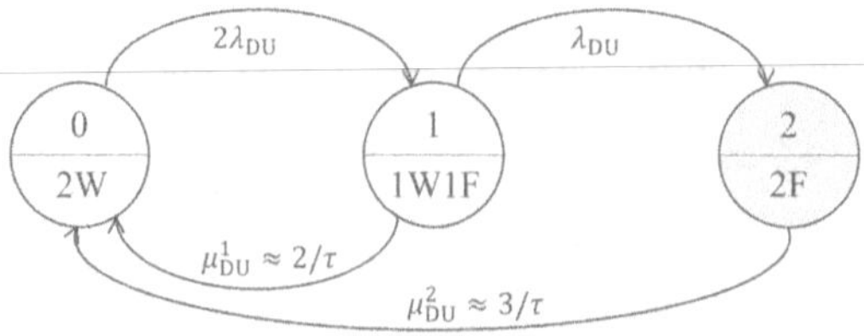


**FIGURE 8.3** The Markov model of a 1oo2 barrier subsystem only with DU-failures.

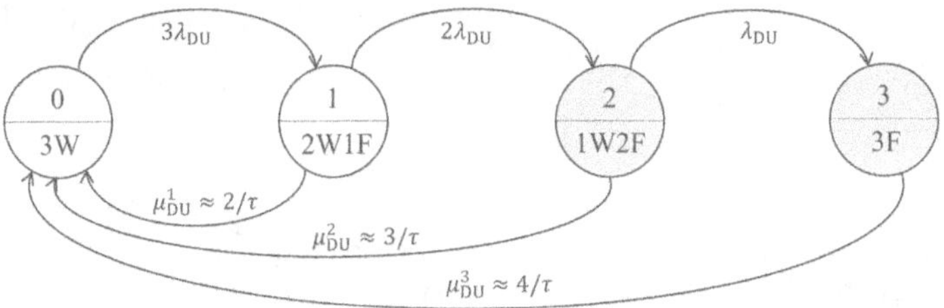


**FIGURE 8.4** The Markov model of a 2oo3 barrier subsystem only with DU-failures.

coverage is 90%. To fix a failure, whether it is DD or DU, it is necessary to stop normal production for 8 hours, meaning MTTR = MRT = 8. The proof test is carried out once per year ($\tau = 8760$ hours), and the duration in the test can be ignored.

Using the approximation formula of Equation (8.3), the average unavailability of this barrier subsystem is calculated as:

$$\text{PFD}_{\text{avg}}^{(\text{G:1oo2})} = \frac{(\lambda_{\text{DU}}\tau)^2}{3} = \frac{(0.1\times 10^{-5}\times 8760)^2}{3} \approx 2.5579\times 10^{-5}$$

With the IEC61508 formula, the mean downtime of a channel due to a D failure is obtained by Equation (8.10):

$$t_{\text{CE}} = \frac{\lambda_{\text{DU}}}{\lambda_{\text{D}}}\left(\frac{\tau}{2}+\text{MRT}\right)+\frac{\lambda_{\text{DD}}}{\lambda_{\text{D}}}\text{MTTR} = 0.1\times\left(\frac{8760}{2}+8\right)+0.9\times 8 = 446$$

Then,

$$\lambda_{\text{D,G}} = 2(\lambda_{\text{D}})^2 t_{\text{CE}} = 8.92\times 10^{-8}$$

$t_{\text{GE}}$ is obtained with Equation (8.10) as 300 hours, and $\text{PFD}_{\text{avg}}$ is calculated as:

$$\text{PFD}_{\text{avg}}^{(\text{G})} = \lambda_{\text{D,G}}t_{\text{GE}} = 8.92\times 10^{-8}\times 300 = 2.676\times 10^{-5}$$

Using the Markov method, a transition matrix can be constructed based on the model in Figure 8.3 as follows:

$$\mathcal{A} = \begin{bmatrix} -2\lambda_{\text{DU}} & 2\lambda_{\text{DU}} & 0 \\ 2/\tau & -(\lambda_{\text{DU}}+2/\tau) & \lambda_{\text{DU}} \\ 3/\tau & 0 & -3/\tau \end{bmatrix}$$

We solve the following set of equations:

$$-2\lambda_{\text{DU}}p_0 + \frac{2}{\tau}p_1 + \frac{3}{\tau}p_2 = 0$$

$$\lambda_{\text{DU}}p_1 - \frac{3}{\tau}p_2 = 0$$

$$p_0 + p_1 + p_2 = 1$$

$PFD_{avg}$ can be then calculated by capturing the sojourn probability of the system in state 2:

$$\text{PFD}_{\text{avg}} = p_2 = \frac{2\lambda_{\text{DU}}^2\tau^2}{6+9\lambda_{\text{DU}}\tau+2\lambda_{\text{DU}}^2\tau^2} \approx 2.2557\times10^{-5}$$

It can be observed that although all three methods lead to the same conclusion on the SIL 4 of the barrier subsystem, the unavailability calculation result using the IEC 61508 formula is the highest. This is because the IEC 61508 method includes the contribution of DD-failures to PFD. The calculation result of the Markov method is the least conservative because it considers the possibility that one failed channel can be detected during a proof test and subsequently restored. ■

### 8.2.5 PFH Calculation Formulas

For a series structure with $n$ identical and independent components operating in the high-demand mode, if one of the components fails, the entire BS fails. Considering DU-failures within the interval $(0, \tau]$, according to Equation (6.37), we have:

$$\text{PFH} \approx \frac{F_{\text{DU}}(\tau)}{\tau} = \frac{1-e^{-n\lambda_{\text{DU}}\tau}}{\tau} \approx \frac{n\lambda_{\text{DU}}\tau}{\tau} = n\lambda_{\text{DU}} \tag{8.16}$$

Innal et al. (2010), Jin et al. (2011), and Wang and Rausand (2014) have discussed the PFH calculation and approximation formulas for different configurations. For a *koon* voting structure, the number of channels found with DU-faults in a proof test, $N(\tau)$, is binomially distributed as:

$$\Pr\left[N(\tau)=i\right]=\binom{n}{i}\left(1-e^{-\lambda_{\text{DU}}\tau}\right)^{i}\left(e^{-\lambda_{\text{DU}}\tau}\right)^{n-i} \tag{8.17}$$

Considering that $1-e^{-\lambda_{\text{DU}}\tau} \approx \lambda_{\text{DU}}\tau$, and $e^{-\lambda_{\text{DU}}\tau} \approx 1$, the above equation can be approximated by:

$$\Pr\left[N(\tau)=i\right]\approx\binom{n}{i}\left(\lambda_{\text{DU}}\tau\right)^{i} = \frac{n!\left(\lambda_{\text{DU}}\tau\right)^{i}}{(n-i)!i!} \tag{8.18}$$

The probability that the BS is found to have failed due to independent DU-failures in this proof test interval is:

$$\Pr\left[N(\tau)\geq n-k+1\right]=\sum_{i=n-k+1}^{n}\Pr\left[N(\tau)=i\right] \tag{8.19}$$

For $i = n-k+1$,

$$\Pr\left[N(\tau)=n-k+1\right]=\frac{n!\left(\lambda_{\text{DU}}\tau\right)^{n-k+1}}{(k-1)!(n-k+1)!}$$

It can be found that:

$$\Pr\left[N(\tau)=n-k+2\right]=\frac{(k-1)\lambda_{\mathrm{DU}}\tau}{n-k+2}\Pr\left[N(\tau)=n-k+1\right]$$

In practice, since $\lambda_{\mathrm{DU}}\tau$ is very small, $\Pr[N(\tau)=n-k+2]$ is much smaller than $\Pr[N(\tau)=n-k+1]$, and thus we can approximate:

$$\Pr\left[N(\tau)\geq n-k+1\right]\approx\Pr\left[N(\tau)=n-k+1\right] \tag{8.20}$$

The PFH of a *koon* voting structure thus can be approximated as

$$\mathrm{PFH}_{koon}\approx\frac{F_{\mathrm{DU}}(\tau)}{\tau}\approx\frac{\Pr\left[N(\tau)=n-k+1\right]}{\tau}=\frac{n!\left(\lambda_{\mathrm{DU}}\tau\right)^{n-k}}{(k-1)!(n-k+1)!} \tag{8.21}$$

## 8.3 IMPORTANCE ANALYSIS OF BARRIER COMPONENTS

In Section 5.3.3, we have introduced several importance measures for reliability analysis. Here, we will consider the low-demand operational mode of BSs and discuss how the measures are used in barrier analysis when PFD($t$) and $\mathrm{PFD}_{\mathrm{avg}}$ are the performance metrics. It is noticed that only DU-failures are considered in the following discussions.

### 8.3.1 Birnbaum's Measure

According to Equation (5.67), the Birnbaum's measure of barrier component $i$ in a barrier subsystem (SS) within a specific proof test interval can be written as:

$$I^{\mathrm{B}}(i|t)=\frac{\partial R_{\mathrm{SS}}(t)}{\partial R_i(t)}=\frac{-\partial\left[1-F_{\mathrm{SS}}(t)\right]}{-\partial\left[1-F_i(t)\right]}=\frac{\partial \mathrm{PFD}_{\mathrm{SS}}(t)}{\partial \mathrm{PFD}_i(t)} \tag{8.22}$$

Birnbaum's measure also can be calculated with Equation (5.71) as a method for sensitivity analysis, thus we have:

$$\begin{aligned} I^{\mathrm{B}}(i|t)&=R_{\mathrm{SS}}\left[t|X_i(t)=1\right]-R_{\mathrm{SS}}\left[t|X_i(t)=0\right]\\ &=\mathrm{PFD}_{\mathrm{SS}}\left[t|X_i(t)=0\right]-\mathrm{PFD}_{\mathrm{SS}}\left[t|X_i(t)=1\right] \end{aligned} \tag{8.23}$$

$X_i=0$ means that the barrier component $i$ is unavailable, while $X_i=1$ means that the component is functioning. Then, it is possible to calculate the average value of the Birnbaum's measure within a proof test interval, as

$$\begin{aligned} I^{\mathrm{B}}_{\mathrm{avg}}(i)&=\frac{1}{\tau}\int_0^{\tau} I^{\mathrm{B}}(i|t)dt=\frac{1}{\tau}\int_0^{\tau}\left\{\mathrm{PFD}_{\mathrm{SS}}\left[t|X_i(t)=0\right]-\mathrm{PFD}_{\mathrm{SS}}\left[t|X_i(t)=1\right]\right\}dt\\ &=\frac{1}{\tau}\int_0^{\tau}\mathrm{PFD}_{\mathrm{SS}}\left[t|X_i(t)=0\right]dt-\frac{1}{\tau}\int_0^{\tau}\mathrm{PFD}_{\mathrm{SS}}\left[t|X_i(t)=0\right]dt\\ &=\mathrm{PFD}_{\mathrm{avg,SS}}\left[X_i=0\right]-\mathrm{PFD}_{\mathrm{avg,SS}}\left[X_i=1\right] \end{aligned} \tag{8.24}$$

### Example 8.2:

### *Case Studies of Birnbaum's Measure*

Birnbaum's measure for the importance of a barrier channel/component in a 1oo3 subsystem with identical channels can be calculated following the approximation formulas in Table 8.1, if only DU-failures are considered. Given that the channel is unavailable ($X_i = 0$), the subsystem will fail on demand when both of the other two channels fail. In other words, the rest part of the subsystem is equivalent with a 1oo2 configuration, whose $\text{PFD}_{\text{avg}}$ is $(\lambda_{\text{DU}}\tau)^2/3$. When the channel is perfect ($X_i = 1$), the 1oo3 subsystem is always functioning, and its $\text{PFD}_{\text{avg}}$ is 0. Thus,

$$I_{\text{avg}}^{\text{B}}(i) = \frac{(\lambda_{\text{DU}}\tau)^2}{3} - 0 = \frac{(\lambda_{\text{DU}}\tau)^2}{3}$$

If the channel is within a 2oo4 barrier subsystem, its importance is calculated as:

$$I_{\text{avg}}^{\text{B}}(i) = (\lambda_{\text{DU}}\tau)^2 - \frac{(\lambda_{\text{DU}}\tau)^3}{4}$$

The above calculation is based on the consideration when the channel is unavailable ($X_i = 0$), the subsystem is failed when more than one of the other three channels fail, and its $\text{PFD}_{\text{avg}}$ is equivalent to that of a 2oo3 configuration. When the channel is perfect ($X_i = 1$), the subsystem is failed when more than two of the other three channels fail, and its $\text{PFD}_{\text{avg}}$ is equivalent to that of a 1oo3 configuration. ■

If the reliability or failure probability of a barrier component is focused, the testing interval $\tau$ can be regarded as a constant for the whole subsystem. Using the chain rule for derivatives, Birnbaum's measure of the barrier component with respect to failure rate can be obtained based on Equation (8.15) as:

$$I^{\text{B}}(\lambda_i|t) == \frac{\partial \text{PFD}_{\text{SS}}(t)}{\partial \text{PFD}_i(t)} \cdot \frac{\partial \text{PFD}_i(t)}{\partial \lambda_i(t)} \tag{8.25}$$

For an item with the exponential failure distribution, $\text{PFD}_i(t) = F_i(t) = 1 - e^{-\lambda_i t}$, and thus, taking the derivative with respect to $\lambda_i$, Equation (8.25) can be written as:

$$I^{\text{B}}(\lambda_i|t) = \frac{\partial \text{PFD}_{\text{SS}}(t)}{\partial \text{PFD}_i(t)} e^{-\lambda_i t} t \tag{8.26}$$

When the whole BS is considered, we can use $I_{\text{avg}}^{\text{B}}(i,j)$ to denote the Birnbaum's measure of barrier component $i$ within the subsystem $j$,

$$I_{\text{avg}}^{\text{B}}(i,j) = \text{PFD}_{\text{avg,S}}\left[X_{i,j} = 0\right] - \text{PFD}_{\text{avg,S}}\left[X_{i,j} = 1\right] \tag{8.27}$$

All the importance measures for a barrier component within the entire BS can be calculated using the same criteria. Therefore, we will omit such formulas for the other measures in the following subsections.

### 8.3.2 Improvement Potential

Improvement potential (IP) of a barrier component can be explained as the difference of the unavailability between the actual BS and a system where the component is perfect or is replaced by a perfect one. The measure of IP ($I^{\mathrm{IP}}$) of a barrier component in a subsystem is:

$$I^{\mathrm{IP}}(i|t) = \mathrm{PFD_{SS}}(t) - \mathrm{PFD_{SS}}(X_i = 1, t) \tag{8.28}$$

The average value of IP within a proof test interval is:

$$\begin{aligned} I^{\mathrm{IP}}_{\mathrm{avg}}(i) &= \frac{1}{\tau}\int_0^\tau I^{\mathrm{IP}}(i|t)\,dt = \frac{1}{\tau}\int_0^\tau \left[\mathrm{PFD_{SS}}(t) - \mathrm{PFD_{SS}}(X_i = 1, t)\right]dt \\ &= \mathrm{PFD_{avg,SS}} - \mathrm{PFD_{avg,SS}}(X_i = 1) \end{aligned} \tag{8.29}$$

**Example 8.3:**

*Case Studies of Improvement Potential*

Re-consider a barrier component in a 1oo3 subsystem with identical channels. When the component is replaced by a perfect one ($X_i = 1$), the 1oo3 subsystem is always functioning, and its $\mathrm{PFD_{avg}}$ is 0. Thus,

$$I^{\mathrm{IP}}_{\mathrm{avg}}(i) = \frac{(\lambda_{\mathrm{DU}}\tau)^3}{4} - 0 = \frac{(\lambda_{\mathrm{DU}}\tau)^3}{4}$$

If the barrier component is in a 2oo4 configuration and given that it is perfect ($X_i = 1$), the subsystem is equivalent to a 1oo3 configuration. The IP of this component is:

$$I^{\mathrm{IP}}_{\mathrm{avg}}(i) = (\lambda_{\mathrm{DU}}\tau)^3 - \frac{(\lambda_{\mathrm{DU}}\tau)^3}{4} = \frac{3(\lambda_{\mathrm{DU}}\tau)^3}{4} \quad \blacksquare$$

### 8.3.3 Risk Achievement Worth

The risk achievement worth (RAW) of a barrier component is obtained by calculating the ratio of the subsystem unavailability if the component is always failed ($X_i = 0$) with the actual subsystem unavailability.

$$I^{\mathrm{RAW}}(i|t) = \frac{\mathrm{PFD_{SS}}(X_i = 0, t)}{\mathrm{PFD_{SS}}(t)} \tag{8.30}$$

It is not very meaningful to calculate the integral of the ratio of the two instantaneous PFD functions. Therefore, in this subsection, the average value of RAW within a proof test interval is approximated as the ratio of the average PFD of the subsystem when the component is always failed with the actual average PFD within the interval, namely,

$$I_{avg}^{RAW}(i) = \frac{PFD_{avg,SS}(X_i = 0)}{PFD_{avg,SS}} \tag{8.31}$$

**Example 8.4:**

*Case Studies of Risk Achievement Worth*

Re-consider a barrier component in a 1oo3 subsystem with identical channels. When the component is always faulty ($X_i = 0$), the subsystem is equivalent with a 1oo2 system. The RAW of this component is:

$$I_{avg}^{RAW}(i) = \frac{\frac{(\lambda_{DU}\tau)^2}{3}}{\frac{(\lambda_{DU}\tau)^3}{4}} = \frac{4}{3\lambda_{DU}\tau}$$

If the barrier component is in a 2oo4 configuration and given that is always faulty ($X_i = 0$), the subsystem is equivalent to a 2oo3 configuration. The RAW of this component is:

$$I_{avg}^{RAW}(i) = \frac{(\lambda_{DU}\tau)^2}{(\lambda_{DU}\tau)^3} = \frac{1}{\lambda_{DU}\tau} \quad \blacksquare$$

### 8.3.4 Risk Reduction Worth

The risk reduction worth (RRW) of a barrier component is obtained by calculating the ratio of the actual subsystem unavailability with the unavailability of a subsystem where the barrier component is replaced by a perfect one.

$$I^{RRW}(i|t) = \frac{PFD_{SS}(t)}{PFD_{SS}(X_i = 1, t)} \tag{8.32}$$

Based on the same consideration in Section 8.3.3, the average value of RRW can be approximated as the ratio of the actual average PFD with the average PFD of the subsystem when the component is perfect within a proof test interval:

$$I_{avg}^{RRW}(i) = \frac{PFD_{avg,SS}}{PFD_{avg,SS}(X_i = 1)} \tag{8.33}$$

**Example 8.5:**

*Case Studies of Risk Reduction Worth*

Re-consider a barrier component in a 1oo3 subsystem with identical channels. When the component is replaced by a perfect one ($X_i = 1$), the subsystem is always functioning, meaning that $PFD_{avg,SS}(X_i = 1) = 0$, and the RAW is infinite.

If the barrier component is in a 2oo4 configuration and the component is replaced by a perfect one ($X_i = 1$), the subsystem is equivalent to a 1oo3 configuration. The RAW of this component is:

$$I_{avg}^{RRW}(i) = \frac{(\lambda_{DU}\tau)^3}{\frac{(\lambda_{DU}\tau)^3}{4}} = 4 \blacksquare$$

**Example 8.6:**

### *Importance Measures of Heterogeneous Barrier Channels*

It is natural to consider that identical channels within a redundant barrier subsystem have the same importance, and the calculation results also can verify such a conclusion. It is also interesting to evaluate the importance of barrier components with different reliability levels.

Here we consider a 1oo2 barrier subsystem with two heterogeneous channels (1 and 2), which DU-failure rates are $1.0 \times 10^{-5}$ and $2.0 \times 10^{-5}$ per hour, respectively. Suppose that the two components are proof tested simultaneously and their test interval is 8760 hours.

For such a subsystem, its instantaneous unavailability or failure probability at time $t$ is:

$$PFD_{SS}(t) = F_{SS}(t) = \left(1 - e^{-\lambda_1 t}\right)\left(1 - e^{-\lambda_2 t}\right)$$

The average PFD of this subsystem can be calculated based on Equation (8.2).

If one of the barrier components fails, the subsystem becomes a single barrier component. While if one of the barrier components is perfect, the subsystem is always functioning, meaning that $PFD_{SS}(t) = 0$.

We can calculate their average Birnbaum's measures, IPs, RAWs, and RRWs, using Equations (8.16), (8.22), (8.24), and (8.26), respectively. It should be noted that for RRW, since $PFD_{avg,SS}(X_i = 1) = 0$, it cannot be calculated. The calculation results of the other three measures are listed in Table 8.2.

It can be found that Birnbaum's measure and RAW can reflect the difference between reliabilities of the two barrier components. With both of the measures, the stronger barrier component (with lower failure rate) is regarded more important. IP does not distinguish the two barrier components in the parallel structure.

**TABLE 8.2**
**Importance Measures of Two Barrier Components with Different Failure Rates**

| Comp. | Birnbaum | IP | RAW |
|---|---|---|---|
| 1 | $8.76 \times 10^{-2}$ | $5.12 \times 10^{-3}$ | 17.12 |
| 2 | $4.38 \times 10^{-2}$ | $5.12 \times 10^{-3}$ | 8.56 |

**TABLE 8.3**
**Importance Measures of Two Barrier Components with Different Test Intervals**

| Comp. | Birnbaum | IP | RAW |
|---|---|---|---|
| 1 | $8.76\times10^{-2}$ | $1.28\times10^{-3}$ | 55.44 |
| 2 | $4.38\times10^{-2}$ | $1.28\times10^{-3}$ | 27.72 |

In addition, we can consider two barrier components with the same failure rate but different proof test intervals. Suppose that both barrier components have a failure rate of $1.0\times10^{-5}$ per hour, but the proof test interval of component 1 is 4380 hours, while the proof test interval of component 2 is 8760 hours. The values of Birnbaum's measures, IP, and RAW are listed in Table 8.3.

The observed outcomes are consistent with what are presented in Table 8.2. When the proof test interval is shorter, it results in a reduced PFD or an increased system availability, indicating a stronger barrier component. This finding establishes a strong correlation between importance measurements utilizing Birnbaum's measure and RAW.

From a managerial perspective, these findings offer valuable insights. Enhancing the integrity of a barrier subsystem can be optimized by focusing efforts on components that already exhibit higher performance or lower PFD. This enhancement can be achieved through either increasing reliability or more frequent proof tests. The choice between these approaches can be driven by cost considerations and the practical challenges associated with implementation. ■

## 8.4 IMPACTS OF REDUNDANCY STRATEGIES

As we have mentioned in Section 3.3.3, redundant design is often used for technical safety barriers, and redundancy can be realized in different ways: hot redundancy, cold redundancy, and warm redundancy. All the models so far in this book for analyzing redundant BSs are for hot redundancy, meaning that all the parallel channels are active. In this section, we will discuss the modeling issues of cold and warm redundancies and compare the performances using these strategies with that of the hot redundant system.

### 8.4.1 Cold-backup Redundancy with a Primary Channel

In a two-channel barrier subsystem employing the cold redundancy strategy, one of the barrier components serves as the primary channel, which is in-use since the starting point and designated as component 1 (C1) for the sake of clarity, while the other component remains in a standby, non-active state during normal operations, referred to as component 2 (C2). C2 is only switched to the operational mode when the C1 fails. We assume that the failed barrier C1 can be repaired and restored to the functioning state. Consequently, the repaired C1 continues to serve as the primary channel, and C2 is only in-use during the downtime of C1.

**TABLE 8.4**
**States of a Cold-Backup 1oo2 Barrier Subsystem with a Primary Channel**

| State | Description |
|---|---|
| 0 | C1 functioning, C2 backup |
| 1 | C1 failed with a DD-failure, C2 in-use and functioning |
| 2 | C1 failed with a DD-failure, SW failed |
| 3 | C1 failed with a DD-failure, C2 failed during in-use |
| 4 | C1 failed with a DU-failure |

In the cold backup situation, we can assume that C2 has no failure probability while in its backup phase since it is not exposed to any workload. However, the failing probability of C2 (due to both DD and DU) arises once it becomes in-use.

When considering the mechanism for switching between these two components, the switch (SW) is equipped with a continuous detection function monitoring the status of the barrier components. The failures of barrier components that can be detected are thus DD-failures. In an ideal scenario where the switch operates perfectly, the occurrence of failure on the primary channel C1 immediately triggers the activation of C2, resulting in no downtime for the BS. However, in cases where the switch is not perfect, there exists a possibility that it may fail to respond when a DD-failure is observed in C1. Consequently, the barrier function of C2 cannot be realized, even though it is fully functioning.

Therefore, the following states can be identified in Table 8.4 for the 1oo2 barrier subsystem employing the cold redundancy strategy:

Here, we actually assume that the switch time is very short and can be ignored. Based on these states, a Markov diagram can be plotted in Figure 8.5.

The transitions depicted in Figure 8.5 can be explained as follows: The transition from state 0 to state 1 indicates the occurrence of a DD-failure on C1 given that the switch responses well, and the associate transition rate is the DD-failure rate of C1 multiplying with the availability of the switch ($A_{SW}$). This availability can be a function of time and can be restored to 1 if the switch is also renewed after each proof test. Here we do not have more discussions on the operational mode of the switch. If

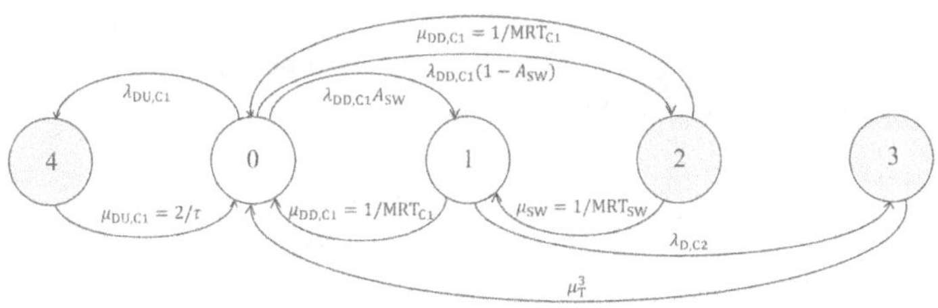


**FIGURE 8.5** The Markov model of a cold-backup 1oo2 barrier subsystem with a primary channel.

the switch fails to respond to the failure of C1, the Markov process goes to state 2, as a faulty state. The transition rate from state 0 to state 2 is the DD-failure rate of C1 multiplying with the unavailability of the switch $(1-A_{SW})$.

State 1 is a state with a potentially prolonged period during which C2 temporarily takes over the role of C1 to maintain the barrier function. Following the activation of C2, there is a possibility that C2 also fails, leading the system entering faulty state 3, with the failure rate of C2. In many practices, since C1 is the primary channel, and it is typically and engineered with higher reliability compared to C2. Then, the system can return from state 1 to state 0 with the repair rate of C1, which can be calculated by the reciprocal of the mean repair time ($\mathrm{MRT}_{C1}$).

States 2 and 3 represent the unavailability of the barrier subsystem. In the scenario where the subsystem can be restored before an actual demand for the barrier function arises, the transitions leaving from state 2 depend on which of C1 and the switch can be repaired firstly. If C1 is repaired at first, the system comes back to state 0 with the rate of $1/\mathrm{MRT}_{C1}$. On the other hand, if the switch is repaired at first, the system comes to state 1 with the repair rate of SW ($1/\mathrm{MRT}_{SW}$).

Given that we consider C1 as the primary channel, it can have the priority in repair than C2. However, the leaving rate from state 3 depends on how C2 fails. If C2 occurs a DD-failure, the transition rate from state 3 to state 0 is $1/\mathrm{MRT}_{C1}$. If C2 occurs a DU-failure, the rate back to state 0 can be the reciprocal of half of the rest time of the proof test interval from the DD-failure of C1.

State 4 is the faulty state due to a DU-failure on C1, and the leaving rate from this state is the reciprocal of the half of a proof test interval. The switch is not activated since such a failure is not detected, and no further action is initiated. The average PFD of the barrier subsystem can thus be calculated by capturing the sojourn probabilities of states 2, 3, and 4 (shadowed in Figure 8.5).

The transition matrix of the presented Markov model is:

$$\mathcal{A}=\begin{bmatrix} -\lambda_{D,C1} & \lambda_{DD,C1}A_{SW} & \lambda_{DD,C1}(1-A_{SW}) & 0 & \lambda_{DU,C1} \\ \mu_{DD,C1} & -(\mu_{DD,C1}+\lambda_{D,C2}) & 0 & \lambda_{D,C2} & 0 \\ \mu_{DD,C1} & \mu_{SW} & -(\mu_{SW}+\mu_{DD,C1}) & 0 & 0 \\ \mu_T^3 & 0 & 0 & -\mu_T^3 & 0 \\ \mu_{DU,C1} & 0 & 0 & 0 & -\mu_{DU,C1} \end{bmatrix}$$

We can follow the standard procedure to calculate the steady state probabilities of the Markov process. To simplify calculation, we assume that $\mu_T^3 \approx \mu_{DD,C1} \approx \mu_{SW}$, and the $\mathrm{PFD}_{avg}$ is obtained by approximation as:

$$\mathrm{PFD}_{avg} = p_2 + p_3 + p_4 \approx \frac{\lambda_{DD,C1}\overline{A_{SW}}}{\mu_{SW}+\mu_{DD,C1}} + \frac{\lambda_{DU,C1}}{\mu_{DU,C1}}$$

In this approximation, the contribution of state 3 to $\mathrm{PFD}_{avg}$ is omitted since it is much smaller than that by state 2 or state 4. The first item on the right-hand side of the above equation describes the contribution by the unavailability of the switch

**TABLE 8.5**
**States of a Cold-Backup 1oo2 Barrier Subsystem with Two Identical Channels**

| State | Description |
|---|---|
| 0 | In-use functioning, backup functioning |
| 1 | In-use functioning, backup in-repair |
| 2 | SW failed when switching is needed |
| 3 | Both channels failed with DD-failures |
| 4 | In-use failed with a DU-failure |

($\overline{A_{SW}}$), which results in PFD given that the primary channel has failed. The second item of the right-hand side of the equation is the contribution of the DU-failure occurring on the primary channel.

### 8.4.2 Cold-Backup Redundancy with Identical Channels

Given that the two barrier components in the subsystem are identical, after the originally active channel (C1) fails, it is repaired and then starts to serve as the backup channel when it comes back to the barrier application. The other component (C2) is activated and then operated until it has a failure when C1 can be activated again. Since cold backup strategy is discussed in this subsection. It is assumed that neither of the two components has failures in their backup states.

Because the two channels are identical, it is not necessary to distinguish them in the identified states, as shown in the following Table 8.5.

The associate Markov model is illustrated in Figure 8.6.

The transitions depicted in Figure 8.6 can be explained as follows: The transition from state 0 to state 1 indicates the occurrence of a DD-failure on the in-use channel given that the switch responses well, and the transition rate is determined by multiplying the DD-failure rate ($\lambda_{DD}$) with $A_{SW}$. Subsequently, the transition from state 1 back to state 0 occurs with the repair rate the failed barrier component, denoted as 1/MRT.

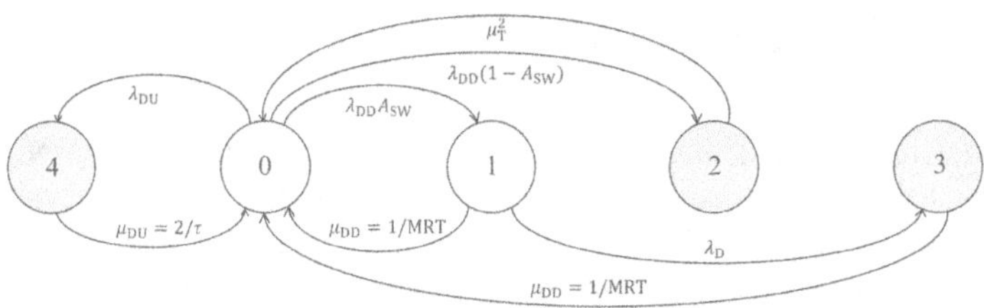


**FIGURE 8.6** The Markov model of a cold-backup 1oo2 barrier subsystem with two identical channels.

Similar with the model in Figure 8.5, if the switch fails to respond to the DD-failure, the process goes to state 2, with a rate of $\lambda_{DD}(1-A_{SW})$. It is essential to notice that this switch failure is also a DU-failure. While the switch may experience a DD-failure, it is not included in this model since it can typically be fixed without impacting the performance of the barrier subsystem, and the probability of simultaneous DD-failures of the switch and the in-use barrier component is very low. A transition exists from state 2 to state 0, but the rate ($\mu_T^2$) depends on the priority strategy in repair. If the failed barrier component is prioritized to be repaired, the rate is 1/MRT. Alternatively, if it is more convenient to fix the failed switch, the rate is the reciprocal of the mean repair time of the switch.

The process can transition from state 1 to state 3 when the second barrier component fails due to a DD-failure during the repair period of the first failed component. In state 3, once one barrier component is restored, the Markov process returns to state 0, and the transition rate remains at 1/MRT.

If the in-use channel experiences a DU-failure, the switch is unable to detect the failure and respond to it. The barrier subsystem falls in an unavailable state, denoted as state 4 in Figure 8.6, until the next proof test.

The average PFD of this barrier subsystem is then calculated by capturing the sojourn probabilities in states 2, 3, and 4 (shadowed in Figure 8.6).

It should be noted that the Markov model in Figure 8.6 omits a potential state in which a DU-failure has occurred on the in-use barrier channel when the other channel is still in the repair process. The sojourn probability of the system in this state is very low, but if this state, referred to as state 5, is added in the Markov model, a transition connecting state 1 with it has the rate of $\lambda_{DU}$, and an outgoing transition of state 5 directs to state 4 with the rate of 1/MRT.

The transition matrix of the presented Markov model in Figure 8.6 is:

$$\boldsymbol{A} = \begin{bmatrix} -\lambda_D & \lambda_{DD}A_{SW} & \lambda_{DD}(1-A_{SW}) & 0 & \lambda_{DU} \\ \mu_{DD} & -(\mu_{DD}+\lambda_D) & 0 & \lambda_D & 0 \\ \mu_T^2 & 0 & -\mu_T^2 & 0 & 0 \\ \mu_{DD} & 0 & 0 & -\mu_{DD} & 0 \\ \mu_{DU} & 0 & 0 & 0 & -\mu_{DU} \end{bmatrix}$$

Similar with the last subsection, we can obtain the approximation $\text{PFD}_{\text{avg}}$ as,

$$\text{PFD}_{\text{avg}} = p_2 + p_3 + p_4 \approx \frac{\lambda_{DD}\overline{A_{SW}}}{\mu_T^2} + \frac{\lambda_{DU}}{\mu_{DU}}$$

The contribution of state 3 to $\text{PFD}_{\text{avg}}$ is still ignored. The first item on the right-hand side of the above equation describes the contribution by the unavailability of the switch ($\overline{A_{SW}}$), which results in PFD given that the in-operation channel has failed. The second item of the right-hand side of the equation is the contribution of the DU-failure occurring on the in-use channel.

Comparing the models depicted in Figures 8.5 and 8.6 can yield some valuable insights. It is intuitive that the primary channel in a BS should be equipped with higher reliability than the backup channel. However, a question can arise: To what extent must the failure rate difference between the two channels be in order to achieve higher performance with the former redundancy strategy over the latter one? Furthermore, if the difference in reliability also means substantial difference in costs, the subsequent question is about how the cost function, when factored into the reliability equation, can impact the selection between these two redundant strategies.

### 8.4.3 Warm-backup Redundancy with a Primary Channel

In the warm-backup redundancy strategy, the backup channel remains running, actively receiving and sending data, but it does not withstand any shocks as the primary operational channel. As a result, this backup channel may fail, while the failure probability is lower than that in the operational state. In a BS, the failure can be a DU-failure or a DD-failure, because since the status of the channel status, being engaged in receiving and sending data, can be partially monitored.

Meanwhile, one of the key objectives of employing a warm-backup strategy is to enhance response speed and success rates. Hence, in the simplified modeling in this subsection, we have the assumption that the switch operates perfectly, and every attempt to switch to the backup channel is guaranteed to be successful.

In this subsection, we also consider a 1oo2 subsystem with a primary channel and that with two identical channels. For a warm-backup 1oo2 redundant barrier subsystem, C1 is set as the primary channel, with three states:

- In-use and functioning,
- Unavailable/in-repair after a DD-failure, and
- In-use with a DU-failure.

Meanwhile, C2 is the backup channel, with six possible states:

- Backup and functioning,
- Backup but unavailable/in-repair,
- Backup with a DU-failure,
- In-use and functioning,
- Unavailable/in-repair, and
- In-use with a DU-failure.

Considering all possible combinations of the states of the two channels, we can identify a total of nine distinct states of the subsystem. Each state of C1 corresponds to three states of C2. In a warm-backup context, the failure rate of C2 in the backup phase may differ but generally being lower than its failure rate after activation.

Involving nine states can lead to a complex Markov mode, and thus simplification is needed. For example, the status of C2 when C1 is hidden failed does not impact the performance of the barrier subsystem, so we only need to keep one state where C1 is hidden failed. The situations in which C2 is in a DD failed state and a DU failed state can be merged considering that the priority is given to repairing C1, and

**TABLE 8.6**
**States of a Warm-Backup 1oo2 Barrier Subsystem with a Primary Channel**

| State | Description |
|---|---|
| 0 | C1 functioning, C2 backup and functioning |
| 1 | C1 functioning, C2 backup but hidden failed |
| 2 | C1 failed with a DD-failure, C2 in-use and functioning |
| 3 | C1 failed with a DD-failure, C2 in-use but failed |
| 4 | C1 hidden failed |

unavailability of the subsystem in these situations mainly relies on when C1 can be fixed. Then, we keep five identified states in Table 8.6.

It should be noted that in this case, if C2 has already failed due to a DU-failure in the backup phase, it can still be switched to operational status, but it remains in a faulty state and cannot perform the barrier function in response to a demand. Another assumption is that both the primary and backup channels are examined in proof tests and then restored to the as-good-as-new state.

The associated Markov diagram is illustrated in Figure 8.7.

The transitions depicted in Figure 8.7 can be explained as follows: The transition from state 0 to state 1 indicates the occurrence of a DU-failure of C2 when the channel is in backup (with the rate of $\lambda^*_{DU,C2}$). State 1 is not an unavailable state for the barrier subsystem, the DU-failure can be revealed in the next proof test, and thus the transition back to state 0 from state 1 is $2/\tau$. Another outgoing transition from state 0 directs to state 2, illustrating the DD-failure of C1.

When the subsystem is in state 1, and a DD-failure occurs on C1, the Markov process enters state 3, because C2 has already been in the unavailable state even though it is switched to the in-use mode. The process also can progress to state 3 from state 2, when C2 fails after it is put into operation. The transition rate from state 2 to state 3 is $\lambda_{D,C2}$, which is expected to be higher than the failure rate of C2 in the backup phase.

The average PFD of this barrier subsystem is calculated by capturing the sojourn probabilities in states 3 and 4 (shadowed in Figure 8.7).

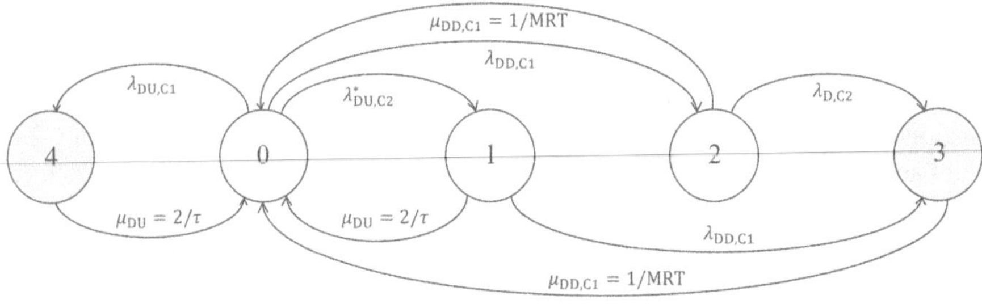


**FIGURE 8.7** The Markov model of a warm-backup 1oo2 barrier subsystem with a primary channel.

The transition matrix of the presented Markov model is:

$$\mathcal{A} = \begin{bmatrix} -\left(\lambda_{D,C1} + \lambda^*_{DU,C2}\right) & \lambda^*_{DU,C2} & \lambda_{DD,C1} & 0 & \lambda_{DU,C1} \\ \mu_{DU} & -\left(\mu_{DD,C1} + \lambda_{D,C2}\right) & 0 & \lambda_{D,C2} & 0 \\ \mu_{DD,C1} & 0 & -\left(\lambda_{D,C2} + \mu_{DD,C1}\right) & \lambda_{D,C2} & 0 \\ \mu_{DD,C1} & 0 & 0 & -\mu_{DD,C1} & 0 \\ \mu_{DU} & 0 & 0 & 0 & -\mu_{DU} \end{bmatrix}$$

Similar with the last subsection, we can obtain the approximation of $\text{PFD}_{\text{avg}}$ as:

$$\text{PFD}_{\text{avg}} = p_3 + p_4 \approx \frac{\lambda^*_{DU,C2}}{\mu_{DU}} \cdot \frac{\lambda_{DD,C1}}{\mu_{DD,C1}} + \frac{\lambda_{DU,C1}}{\mu_{DU}}$$

The first item on the right-hand side of the above equation describes the contribution of the DU-failure on the backup channel to $\text{PFD}_{\text{avg}}$, and the second item of the right-hand side of the equation is the contribution of the DU-failure occurring on the in-use channel. The contribution of the potential DD-failures is about $\frac{\lambda_{DD,C1}}{\mu_{DD,C1}} \cdot \frac{\lambda_{D,C2}}{\mu_{DD,C1}}$. Typically, this element is of lesser significance, but it should not be underestimated, particularly if the DD-failure of the primary channel requires a long repair time or if $\lambda_{D,C2}$ is high.

It is also meaningful to compare the models in Figure 8.5 and in Figure 8.7, taking into account the difference between the cold-backup and warm-backup strategies. In this context, a warm-backup strategy implies a lower failure probability of the switching mechanism (to 0 in the Markov model), but a higher failure probability of the backup channel itself ($\lambda_{D,C2}$). Therefore, this raises the following question: To what extent does the difference in failure rates between a cold-backup channel and a warm-backup channel need to be to offset the impact of the increased failure rate of the switching mechanism on the unavailability of the barrier subsystem?

### 8.4.4 Warm-Backup Redundancy with Identical Channels

For a warm-backup 1oo2 redundant barrier subsystem with two identical channels, each of the channels has two operational phases: in-use and backup. Each channel has five potential states:

- Backup and functioning (a),
- Backup with a DU-failure (b),
- In-use and functioning (c),
- Unavailable/in-repair (d), and
- In-use with a DU-failure (e).

It is not necessary to distinguish the two identical channels, but they cannot sojourn in the same operational phase at any time (except (d)). When a DD-failure is observed on a channel, it is automatically switched to the backup mode, meaning

**TABLE 8.7**
**States of a Warm-Backup 1oo2 Barrier Subsystem with Two Identical Channels**

| State | Description |
|---|---|
| 0 | In-use functioning, backup and functioning (ac) |
| 1 | In-use functioning, backup with a DD-failure (ad, cd) |
| 2 | Two unavailable/in-repair (bd, dd) |
| 3 | In-use functioning, backup with a DU-failure (bc) |
| 4 | In-use with a DU-failure (ae, be) |

that the combination of states (a) and (d) is a transitional state without any sojourn time. Based on these considerations, we can identify seven combined states in total (ac, ae, bc, bd, be, cd, dd), but two states can be omitted in the Markov diagram: When one barrier channel is in repair due to a DD-failure, another channel has a DU-failure (bd). Since the repair duration of DD-failure is relatively short, the DD failed channel can be fixed soon, and thus the state bd be merged into the state where two channels are unavailable (dd). The situation of two DU-failure (be) is also merged into state of 1 DU failed, 1 backup (ae), since when the in-use channel is failed due to a DU-failure, the status of the backup channel has no impact. The identified states are listed in Table 8.7.

The associated Markov diagram is illustrated in Figure 8.8.

The transitions depicted in Figure 8.8 can be explained as follows: The transition from state 0 to state 1 indicates the occurrence of a DD-failure, but the DD-failure can occur on the in-use channel or on the backup channel, and thus the transition rate is the sum of $\lambda_{DD}$ (DD-failure rate of the in-use channel) and $\lambda^*_{DD}$ (DD-failure rate of the backup channel). State 1 means that one channel is in-use, and the other channel is in repair, and if the in-use channel fails due to a DD or DU-failure, the subsystem enters state 2, as an unavailable state.

Given that at least one of the two failed channels is due to a DD-failure, the restoration rate can be 1/MRT, bringing the subsystem back to state 1. Here we can assume that maintenance team will check the DU-failure when they fix the DD-failure, to make the model more valid.

On the other side of the Markov model, when the backup channel has a DU-failure, the subsystem enters state 3, with the DU-failure rate in the backup phase. The barrier subsystem also can directly enter an unavailable state 4 from state 0, when a

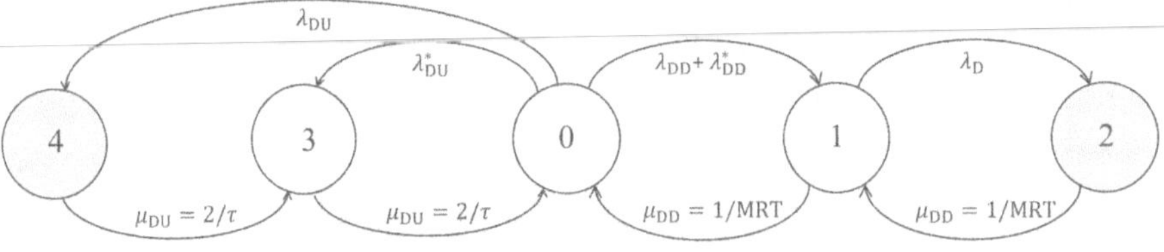


**FIGURE 8.8** The Markov model of a warm-backup 1oo2 barrier subsystem with two identical channels.

DU-failure occurs on the in-use channel. Since a DU-failure cannot be revealed until a proof test, the transitions rates from state 3 and state 4 to state 0 are $2/\tau$.

The average PFD of this barrier subsystem is calculated by capturing the sojourn probabilities in states 2 and 4 (shadowed in Figure 8.8).

The transition matrix of the presented Markov model is:

$$\mathcal{A} = \begin{bmatrix} -(\lambda_{\mathrm{D}} + \lambda_{\mathrm{D}}^{*}) & \lambda_{\mathrm{DD}} + \lambda_{\mathrm{DD}}^{*} & 0 & \lambda_{\mathrm{DU}}^{*} & \lambda_{\mathrm{DU}} \\ \mu_{\mathrm{DD}} & -(\mu_{\mathrm{DD}} + \lambda_{\mathrm{D}}) & \lambda_{\mathrm{D}} & 0 & 0 \\ 0 & \mu_{\mathrm{DD}} & -\mu_{\mathrm{DD}} & 0 & 0 \\ \mu_{\mathrm{DU}} & 0 & 0 & -\mu_{\mathrm{DU}} & 0 \\ \mu_{\mathrm{DU}} & 0 & 0 & 0 & -\mu_{\mathrm{DU}} \end{bmatrix}$$

Similar with the last subsection, we can obtain the approximation $PFD_{avg}$ as,

$$\mathrm{PFD}_{\mathrm{avg}} = p_2 + p_4 \approx \frac{\lambda_{\mathrm{DD}} + \lambda_{\mathrm{DD}}^{*}}{\mu_{\mathrm{DD}}} \cdot \frac{\lambda_{\mathrm{D}}}{\mu_{\mathrm{DD}}} + \frac{\lambda_{\mathrm{DU}}}{\mu_{\mathrm{DU}}}$$

The first item on the right-hand side of the above equation describes represents the impact of two sequent failures on the two channels, and the second item of the right-hand side of the equation is the contribution of the DU-failure occurring on the in-operation channel.

### 8.4.5 Rotating Strategy

The operational strategies discussed above can be extended by addressing a more general question regarding the rotating strategy of barrier subsystems: Given that a demand arises, one channel or multiple channels of a barrier subsystem have been activated to carry out the barrier function, it becomes necessary to determine whether the same channel(s) remain in-use for subsequent demands or whether a switch to alternative channel(s) is conducted for the next demand. In other words, the switch between different barrier channel does not need to wait for the failure of the in-use channel but can be performed after each actual demand.

Such a strategy can be useful when considering that after barrier components experience noticeable wear or other degradations due to the high workload of an actual demand, leading to an increase in the failure rate of these components. In practical applications, barrier components subjected to heavy workloads require following-up inspections and maintenance. For example, this kind of strategy for a warm-backup redundant barrier can be modeled in a simple manner by modifying the Markov model depicted in Figure 8.8. This modification involves adding the demand rate/frequency ($\lambda_{\mathrm{DE}}$) into the transition rate from state 0 to state 1.

In the context of a more complex barrier subsystem, such as the 2oo3 configuration, implementing a rotating strategy is not a straightforward process but rather requires optimization that considers the degradations of all channels. The aim is to minimize the average PFD over the long term.

## 8.5 IMPACTS OF DEMANDS

In this section, we will evaluate the impacts of actual demands, or process variations, on the performance of barriers.

### 8.5.1 Demand Rate and Demand Duration

In most practices, it is reasonable to assume that a demand arises randomly, and thus the time to the next demand follows an exponential distribution with the rate of $\lambda_{DE}$. Then, we can extend the Markov models for analyzing performance of barriers, for example, for the simplest model for a single barrier component only with DU-failures.

In Figure 8.9, a new state is added in a simple two-state Markov diagram for a barrier only considering DU-failure, as the hazard to the next barrier. The states of this model are listed in Table 8.8. This is reasonable when the safety barrier studied here is an intermediate barrier, rather than an ultimate one, and when a demand arrives when the barrier is unavailable, the result is a hazardous event to the next layer of protection. Such an assumption is made to avoid absorbing state in the Markov model. If the barrier is the ultimate one, its failure to response to the demand will lead to a severe accident, e.g., an explosion destroying the whole facility, where it is difficult to judge how long the asset can be restored. In such a case, it is not necessary to keep the transition from state 2 to state 0.

It can be found in Figure 8.9 that when the Markov process is in state 1, it can transit to state 2 when a demand arrives with the rate of $\lambda_{DE}$.

In this model, we firstly assume that the demand duration is rather short, and it is possible for the barrier to recovered from state 2 to state 0, so that there is no state for the barrier on demand and no absorbing state in the Markov model. $PFD_{avg}$ is calculated by summing up the sojourn probabilities of the barrier in states 1 and 2, while the Hazardous Event Frequency (HEF, see Section 6.4.5) can be obtained by capturing the sojourn probability of the process in state 3. This explains why $HEF = \lambda_{DE} PFD_{avg}$ in the low-demand mode. This model will be further investigated in the next subsection.

We are also motivated to evaluate the impacts of demand duration, since we have introduced durability as a performance measure in Section 6.3.3. Here, we consider unignorable duration of a demand and the potential failure of a barrier during the demand, for example, an earlier stop of a barrier function. A new state

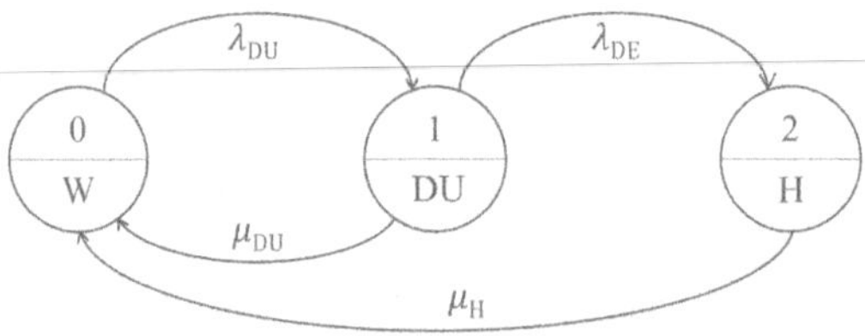


**FIGURE 8.9** The Markov model of a barrier component in consideration of demands.

**TABLE 8.8**
**States of the Markov Model in Figure 8.9**

| State | Description |
|---|---|
| State 0 | Normally functioning |
| State 1 | Failed due to a DU-failure |
| State 2 | Hazard to the next barrier |

is added in the model of Figure 8.9 (State 4 for the state in demand). In the model, $\mu_{DE}$ represents the completion rate of a demand, and $\lambda_H$ describes the rate at which the barrier encounters durability problems, making it unable to withstand the load after activation. In most cases, $\lambda_H$ is expected to be higher than the combined values $\lambda_{DD}$ and $\lambda_{DU}$, as the barrier component operates under significantly higher stress than during normal operation. Furthermore, given the uncertainty surrounding the restoration time following a hazardous event, we have set state 3 as an absorbing state in Figure 8.10. This means no outgoing transitions from the state, preventing the analysis from the influence of different restoration time. In the case that the restoration time from state 3 can be estimated, it is also possible to add a transition from state 3 to state 0.

Based on these settings, we can obtain the transition matrix of the Markov model in Figure 8.10 as:

$$\mathcal{A} = \begin{bmatrix} -(\lambda_{DE}+\lambda_{DU}) & \lambda_{DE} & \lambda_{DU} & 0 \\ \mu_{DE} & -(\mu_{DE}+\lambda_H) & 0 & \lambda_H \\ \mu_{DU} & 0 & -(\mu_{DU}+\lambda_{DE}) & \lambda_{DE} \\ 0 & 0 & 0 & 0 \end{bmatrix}$$

Let $S(t)$ be the probability by time $t$ that the barrier component sojourns in the states without a hazard, which is:

$$S(t) = p_0(t) + p_1(t) + p_2(t)$$

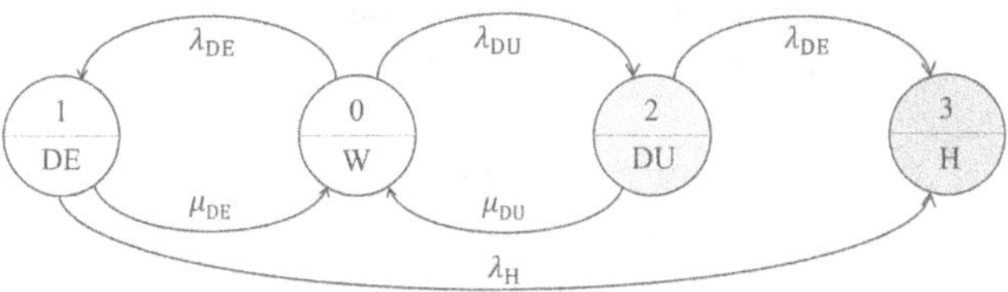


**FIGURE 8.10** The Markov model of a barrier considering demand duration.

According to Equation (5.110), the state equations of this model can be written as

$$[p_0(t), p_1(t), p_2(t), p_3(t)] \cdot \begin{bmatrix} -(\lambda_{DE}+\lambda_{DU}) & \lambda_{DE} & \lambda_{DU} & 0 \\ \mu_{DE} & -(\mu_{DE}+\lambda_H) & 0 & \lambda_H \\ \mu_{DU} & 0 & -(\mu_{DU}+\lambda_{DE}) & \lambda_{DE} \\ 0 & 0 & 0 & 0 \end{bmatrix}$$

$$= [\dot{p}_0(t), \dot{p}_1(t), \dot{p}_2(t), \dot{p}_3(t)]$$

All elements in the fourth row are equal to 0, and thus the above equation can be reduced, and by taking Laplace transforms, we can have:

$$[p_0^*(0), p_1^*(0), p_2^*(0)] \cdot \begin{bmatrix} -(\lambda_{DE}+\lambda_{DU}) & \lambda_{DE} & \lambda_{DU} \\ \mu_{DE} & -(\mu_{DE}+\lambda_H) & 0 \\ \mu_{DU} & 0 & -(\mu_{DU}+\lambda_{DE}) \end{bmatrix} = [-1,0,0]$$

The mean time to the hazard state 3 (MTTH) can be determined by

$$\text{MTTH} = S^*(0) = p_0^*(0) + p_1^*(0) + p_2^*(0)$$

HEF is then the reciprocal of MTTH, as:

$$\text{HEF} = \frac{\lambda_{DE}(\lambda_H\mu_{DU} + \lambda_{DE}\lambda_H + \lambda_{DU}\mu_{DE} + \lambda_{DU}\lambda_H)}{\mu_{DE}\mu_{DU} + \mu_{DE}\lambda_{DE} + \lambda_H\mu_{DU} + \lambda_{DE}\lambda_H + \lambda_{DE}\mu_{DU} + \lambda_{DE}\lambda_{DE} + \lambda_{DU}\mu_{DE} + \lambda_{DU}\lambda_H} \tag{8.34}$$

In general, we can assume that $\lambda_{DU} < \lambda_H \ll \mu_{DU}$. Then, we can make some approximations for different operational modes of the single barrier.

In the low-demand and short demand duration mode, $\lambda_{DE} \ll \mu_{DU} \ll \mu_{DE}$, and thus HEF can be further approximated as,

$$\text{HEF} \approx \frac{\lambda_{DE}\lambda_{DU}}{\mu_{DU}} = \lambda_{DE}\text{PFD}_{avg} \tag{8.35}$$

Such a result is aligned with Equation (6.38) in the definition of HEF.

In the low-demand mode where the demand duration can be ignored, $\mu_{DU}$ is not much less than $\mu_{DE}$, and HEF can be approximated as:

$$\text{HEF} \approx \frac{\lambda_{DE}(\lambda_H\mu_{DU} + \lambda_{DU}\mu_{DE})}{\mu_{DE}\mu_{DU}} = \frac{\lambda_{DE}\lambda_{DU}}{\mu_{DU}} + \frac{\lambda_{DE}\lambda_H}{\mu_{DE}} \tag{8.36}$$

This result can be explained by hazards occurring in two types of situations. Firstly, when the barrier becomes unavailable due to a DU-failure (with the probability of $\lambda_{DU}/\mu_{DU}$ in a test interval) and then a demand arrives (with the rate of $\lambda_{DE}$) before the next proof test, this can be expressed as $\lambda_{DE}\lambda_{DU}/\mu_{DU}$. Secondly, the barrier is successfully activated when a demand arises (with the rate of $\lambda_{DE}$) and then fails before the demand duration completes (with a probability of $\lambda_H/\mu_{DE}$), this can be explained as $\lambda_H\lambda_{DE}/\mu_{DE}$.

If the demand rate is rather high, meaning that $\lambda_{DE} \gg \mu_{DU}$, in such a case, $\mu_{DE}$ is also high because only a short demand duration is compatible with a high number of demands. Specifically, the inequality $\mu_{DE} \geq \lambda_{DE}\mu_{DU}$ should hold so that the barrier responds to at most only one demand at any given time. Equation (8.27) can then be approximated as follows:

$$\text{HEF} \approx \frac{\lambda_{DE}\lambda_H}{\lambda_{DE} + \mu_{DE}} \tag{8.37}$$

When $\mu_{DE} = \lambda_{DE}\mu_{DU}$ or the value of $\mu_{DE}$ is close to $\lambda_{DE}\mu_{DU}$, the barrier is almost always in a state of on-demand. In such a case, HEF $\approx \lambda_H$, implying that once the barrier fails, a hazard is generated immediately. This actually aligns with Equation (6.37) since when the barrier is always on-demand, $\lambda_H$ is the rate of the failure that may occur.

Liu and Rausand (2011) and Liu et al. (2012) provide more comprehensive discussions on the reliability analysis of BSs in different demand modes. However, it should be noted that these articles assume that the failure rate of a barrier remains the same before and during a demand.

### 8.5.2 Demands and Applicability of PFD

In Chapter 6, we have stated several times that PFD is used for the low-demand mode while PFH is for high-demand. As we have introduced in Chapter 3, IEC61508 (2010) classifies demand modes in a very practical approach:

- Low-demand mode: Frequency of demands is no greater than one per year.
- High-demand mode: Frequency of demands is greater than one per year.

The standard does not explain why the borderline between two kinds of demand modes is at once per year. In this section, we will evaluate this classification by exploring the applicability of PFD.

We can take demands into account of a very simplified Markov model of a single barrier component, coming back to the model in Figure 8.9 in the last subsection. We only consider DU-failures, and when the barrier is in a failed state due to a DU-failure (state 1). The safety barrier studied here is assumed to be an intermediate barrier, the demand without response of this barrier can be handled by the next layer of protection, and the failure of the current barrier can be identified and fixed in a certain time. In the model, we assume that both the time to demand and the time to restore from a hazardous event follow the exponential distribution. $\lambda_{DE}$ is used to denote the demand rate, and $\mu_H$ is the restoration rate from a hazardous event.

Since the Markov model in Figure 8.9 has no absorbing state, we can calculate the steady state probabilities by using the following equation:

$$[P_0, P_1, P_2] \cdot \begin{bmatrix} -\lambda_{DU} & \lambda_{DU} & 0 \\ \mu_{DU} & -(\lambda_{DE} + \mu_{DU}) & \lambda_{DE} \\ \mu_H & 0 & -\mu_H \end{bmatrix} = [0,0,0]$$

Then we have

$$\lambda_{DU} P_0 - (\lambda_{DE} + \mu_{DU}) P_1 = 0$$

$$\lambda_{DE} P_1 - \mu_H P_2 = 0$$

$$P_0 + P_1 + P_2 = 1$$

According to the definition, $PFD_{avg}$ of such a barrier can be obtained by calculating the sojourn probability in state 1, thus

$$PFD_{avg} = P_1 = \frac{\lambda_{DU}\mu_H}{\lambda_{DE}\mu_H + \mu_{DU}\mu_H + \lambda_{DU}\mu_H + \lambda_{DE}\lambda_{DU}} \quad (8.38)$$

**Example 8.7:**

*Numerical Example of Applicability of PFD*

We can use numerical values to replace the symbols in the Equation (8.38), to examine the changing trend of $PFD_{avg}$ along with the demand rate. We simply set $\lambda_{DU} = 1.0 \times 10^{-6}$, the mean recovery time from the hazardous event is 10 hours, namely $\mu_H = 0.1$, and assume that the proof test interval is 1 year, so that $\mu_{DU} \approx 2/\tau = 1/4380$. Then we plot $PFD_{avg}$ versus $\lambda_{DU}$ in Figure 8.11. It is noted that the values shown on the x-axis are $\log_{10} \lambda_{DE}$ for illustrating more information in the low-demand area.

It can be found in Figure 8.11 that when the demand rate is rather low, e.g., lower than $10^{-6}$ per hour, the value of $PFD_{avg}$ remains nearly constant. This observation aligns with the notion that $PFD_{avg,}$ serving as a performance metric for the safety barrier, should exhibit independence from demand frequency/rate. Therefore, when demand rate is low, $PFD_{avg}$ can be regarded as a credible measure. However, as demand rate increases, the value of $PFD_{avg}$ starts to decline, gradually reducing the effectiveness of $PFD_{avg}$ as the performance metric. ■

The profile of Figure 8.11 offers valuable insights into identifying the borderline between low- and high-demand modes. When demand rate is relatively low, the curve exhibits a sharp descent, followed by a noticeable change in slope after a turning point showing a decelerated decrease when the demand rate is relatively high. This implies the changing role of $PFD_{avg}$. This shift reflects a gradual reduction in the sojourn probability at the DU state, indicating that the asset is not damaged but lacks protection

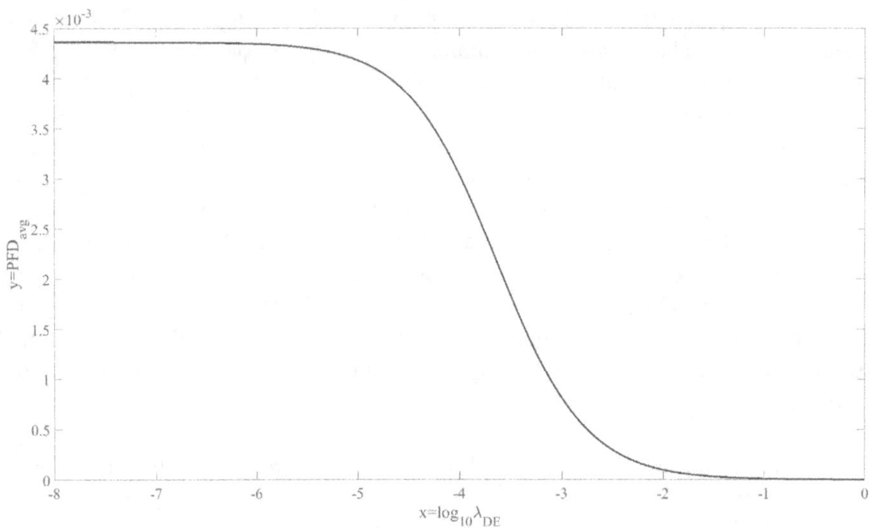


**FIGURE 8.11** $\mathrm{PFD_{avg}}$ versus demand rate for a barrier component, $\log_{10}$ scale on x-axis.

from the barrier. This reduction results from the rapid increase in the probability of a hazardous event to the asset or at least to the next barrier. When the demand rate increases to the value corresponding to the turning point, the rate of decline in $\mathrm{PFD_{avg}}$ reaches its maximum, and subsequently, the decrease becomes more gradual due to the diminished sojourn probability at the DU state. In other words, in this scenario, the occurrence probability of the hazardous state can no longer be disregarded.

This point can be observed as the turning point or *inflexion point* of the curve where the second-order derivative of $\mathrm{PFD_{avg}}$ with respect to $\log_{10}\lambda_{\mathrm{DE}}$ is equal to 0. Considering $\mu_{\mathrm{H}} \gg \mu_{\mathrm{DU}} \gg \lambda_{\mathrm{DU}}$, Equation (8.27) can be simplified as

$$\mathrm{PFD_{avg}} \approx \frac{\lambda_{\mathrm{DU}}}{\lambda_{\mathrm{DE}} + \mu_{\mathrm{DU}}} \tag{8.39}$$

Take the second-order derivative with respect to $\log_{10}\lambda_{\mathrm{DE}}$

$$\frac{d^2\left(\frac{\lambda_{\mathrm{DU}}}{\lambda_{\mathrm{DE}}+\mu_{\mathrm{DU}}}\right)}{d\left(\log_{10}\lambda_{\mathrm{DE}}\right)^2} = 0$$

Thus, we have

$$\frac{2\lambda_{\mathrm{DE}}^2\lambda_{\mathrm{DU}}(\ln 10)^2}{\left(\lambda_{\mathrm{DE}}+\mu_{\mathrm{DU}}\right)^3} - \frac{\lambda_{\mathrm{DE}}\lambda_{\mathrm{DU}}(\ln 10)^2}{\left(\lambda_{\mathrm{DE}}+\mu_{\mathrm{DU}}\right)^2} = 0$$

Finally, we can obtain

$$\frac{2\lambda_{\mathrm{DE}}}{\lambda_{\mathrm{DE}}+\mu_{\mathrm{DU}}} = 1 \Rightarrow \lambda_{\mathrm{DE}} = \mu_{\mathrm{DU}}$$

This finding, together the Markov model presented in Figure 8.11, offers valuable insights into the classification of demand modes. The barrier component presents two different pathways out of the DU state, where the sojourn probability in this state contributes to $PFD_{avg}$: It can transit to a hazard or undergo restoration. These two transitions are competing stochastic processes, and either of them may occur before the other one, but on average, the barrier is more likely to move to the state with a higher associated transition rate. If $\lambda_{DE}$ is less than $\mu_{DU}$, it becomes more probable that the failed barrier will be restored without leading to a hazardous event; conversely, when $\lambda_{DE}$ exceeds $\mu_{DU}$, a hazardous event is more likely to occur.

In terms of the adaptability of $PFD_{avg}$, when $\lambda_{DE}$ is less than $\mu_{DU}$, $PFD_{avg}$ is a valuable metric for illustrating a significant portion of the unavailability of the studied barrier. Therefore, the two demand modes can be classified in consideration of the frequency of proof tests:

- *Low-demand mode (2)*: Demand frequency is no greater than double of the frequency of proof tests.
- *High-demand mode (2)*: Demand frequency is greater than double of the frequency of proof tests.

For instance, if proof tests are conducted annually, the BS operates in a low-demand mode as long as the demand rate does not exceed twice per year. In Figure 8.11, we can find that the value of $PFD_{avg}$ starts to decline when the demand rate falls below the critical threshold, and thus to be more conservative, we can use $PFD_{avg}$ in scenarios with even lower demand rates. This finding validates the rationale behind the classification approach outlined in IEC61508 (2010), which sets an annual frequency as the threshold. This is particularly relevant as many safety barriers are indeed tested at one-year intervals.

Based on such consideration, we also can provide a definition for continuous mode as:

- *Continuous mode*: Demands occur continuously so that a hazardous event will occur if the barrier is failed.

The article by Liu (2014) has proved that the values of DU-failure rate, repair time and system configuration have no impact on this finding. If we consider DD-failures, we still can reach the same conclusion. Readers who have interest can read the above article for more detailed information.

## 8.6 ASSESSMENT OF SPURIOUS ACTIVATIONS

In Section 6.4.7, we have discussed the concept of validity and the associated performance metric spurious trip rate (STR). Here, we will introduce the calculation formulas of STR for different voting structures of barriers.

If we use $\lambda_{ST}$ to denote the spurious activation rate of one barrier component, and for a 1oo1 BS, its STR is equal to $\lambda_{ST}$. For a 1oo*n* voting structure, its STR is calculated as,

$$STR_{1oon} = n\lambda_{ST} \tag{8.40}$$

This is because when one of the channels wrongly performs its barrier function when it is not needed, the subsystem has wrongly performed the function. For example, when the barrier function is to close the flow in a pipeline, if any one of shutdown valves is closed, the flow will be stopped.

For a *koon*voting structure, the calculation of STR is more complicated. Firstly, we can calculate the frequency that one of the $n$ channels have a spurious activation, which is equal to $n\lambda_{ST}$. In most applications, such a failure can be noticed immediately, and then it can be fixed in a period time, which is in average $MTTR_{ST}$. For the rest $(n-1)$ channels, if there are at least $(k-1)$ are also activated incorrectly during this period, the whole barrier structure will have a spurious trip. The probability $p$ of a single barrier channel has a spurious activation during $MTTR_{ST}$ is:

$$p = 1 - e^{-\lambda_{ST} MTTR_{ST}} \approx \lambda_{ST} MTTR_{ST} \tag{8.41}$$

Then, STR of the structure is calculated as (Lundteigen & Rausand, 2008):

$$STR_{koon} = n\lambda_{ST} \sum_{m=k-1}^{n-1} \binom{n-1}{m} p^m (1-p)^{n-1-m} \tag{8.42}$$

### Example 8.8:

#### *PFD and STR of Simple Barrier Subsystems*

Based on the Equations (8.6) and (8.42), we can calculate $PFD_{avg}$, and STR of 1oo1, 1oo2, and 2oo3 voting structures, respectively. The results are shown in Table 8.9.

Since $MTTR_{ST}$ is rather short compared with a proof test interval, $(\lambda_{ST} MTTR_{ST})^2$ has a much smaller value that $\lambda_{ST}$. Consequently, it can be found that for the availability of the barrier subsystem, the 1oo2 voting structure is the most available one, followed by the 2oo3, and the 1oo1 is least available. Meanwhile, considering STR, the 2oo3 has the lowest spurious activation frequency, or the highest validity, followed by 1oo1, and the 1oo2 is the least valid voting structure. ■

Markov models for spurious activations can be extended from the models only for DU and DD-failures. For example, for a 1oo2 voting structure, the model can be extended from that in Figure 8.3 by adding a new state 3 for Safe state, as illustrated in Figure 8.12.

**TABLE 8.9**
**$PFD_{avg}$ and STR of 1oo1, 1oo2, and 2oo3 Voting Structures**

| Structure | $PFD_{avg}$ | STR |
|---|---|---|
| 1oo1 | $\lambda_{DU}\tau/2$ | $\lambda_{ST}$ |
| 1oo2 | $(\lambda_{DU}\tau)^2/3$ | $2\lambda_{ST}$ |
| 2oo3 | $(\lambda_{DU}\tau)^2$ | $9\lambda_{ST}(\lambda_{ST} MTTR_{ST})^2$ |

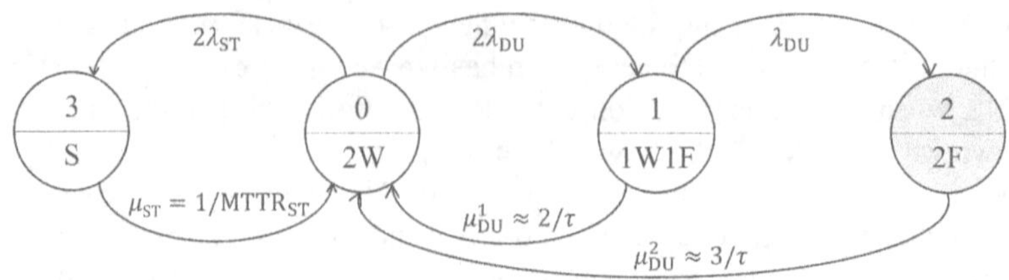


**FIGURE 8.12** A Markov model for a 1oo2 barrier subsystem considering spurious activations.

The transition rates from state 0 to state 3 is $2\lambda_{ST}$, aligning with Equation (8.40), and the transition rate from state 4 to state 0 can be calculated as the 1/$\text{MTTR}_{ST}$.

It is important to highlight that Equations (8.40)–(8.42), as well as the Markov model in Figure 8.12, assume that all spurious activations are independent, resulting from inherent failures of the barrier channels. This formula does not take into account the impact of common cause failures on multiple channels. More details about common cause failures can be found in Chapter 10. The article by Lundteigen and Rausand (2008) also propose that the calculation of STR can be based on the different the causes of spurious trips, which are from false demands, DD-failures or loss of utilities, and readers can refer to this article for a more comprehensive understanding on the calculation of STR.

## REFERENCES

IEC61508. (2010). Functional safety of electrical/electronic/programmable electronic safety-related systems. Geneva, Switzerland: International Electrotechnical Commission.

Innal, F., Dutuit, Y., Rauzy, A., & Signoret, J.-P. (2010). New insight into the average probability of failure on demand and the probability of dangerous failure per hour of safety instrumented systems. *Proceedings of the Institution of Mechanical Engineers Part O-Journal of Risk and Reliability*, *224*, 75–86.

Jin, H., Lundteigen, M. A., & Rausand, M. (2011). New PFH-formulas for k-out-of-n: F-systems. *Reliability Engineering & System Safety*, *111*, 112–118.

Liu, Y. (2014). Discrimination of low- and high-demand modes of safety-instrumented systems based on probability of failure on demand adaptability. *Proceedings of the Institution of Mechanical Engineers Part O-Journal of Risk and Reliability*, *228*(4), 409–418. https://doi.org/10.1177/1748006X14523263

Liu, Y., Lundteigen, M. A., & Rausand, M. (2012). *Reliability assessment of safety-instrumented systems: The influence of demand rates and demand durations.* The 11th Probabilistic Safety Assessment & Management Conference & European Safety and Reliability Conference 2012 Helsinki, Finland.

Liu, Y., & Rausand, M. (2011). Reliability assessment of safety instrumented systems subject to different demand modes. *Journal of Loss Prevention in the Process Industries*, *24*, 49–56. https://doi.org/10.1016/j.jlp.2010.08.014

Lundteigen, M., & Rausand, M. (2008). Spurious activation of safety instrumented systems in the oil and gas industry: Basic concepts and formulas. *Reliability Engineering & System Safety*, *93*(8), 1208–1217.

Wang, Y., & Rausand, M. (2014). Reliability analysis of safety-instrumented systems operated in high-demand mode. *Journal of Loss Prevention in the Process Industries*, *32*, 254–264.

# 9 Test and Maintenance

## 9.1 INTRODUCTION

Our quantitative analyses of technical safety barriers have so far been based on the assumption that a barrier system undergoes regular proof tests. In this context, faults arising from dangerous undetected (DU)-failures are exposed during the proof test and promptly rectified. Consequently, the barrier system is restored to a state comparable to its initial condition (as-good-as-new) after each proof test, leading to an incremental decrease in its instantaneous probability of failure on demand (PFD) over time, as depicted in Figure 6.2.

However, in this chapter, our focus will shift from constant proof test intervals to evaluating various testing approaches and also the combined influence of some approaches on the performance of safety barriers. In subsequent sections, we will explore diagnostic tests, tests induced by detected failures, staggering tests, and partial proof tests (PPTs).

Furthermore, this chapter will investigate diverse maintenance strategies employed for technical safety barriers. Contemporary detection technologies extend beyond mere fault identification, and they are also capable of discerning potential weaknesses, degradations, or vulnerabilities in safety barriers. The wealth of collected information enables the introduction of proactive maintenance or earlier replacement, aimed at preventing failures, mitigating the risk of barrier failure during critical situations, and minimizing costs while preserving the requisite barrier functions.

## 9.2 DIAGNOSTIC TEST

Diagnosis or diagnostic test means the detection procedure for recognizing faults, locating faults, identifying root causes of faults, and determining the current health state of a technical system (Jimenez et al., 2020). Diagnostics in the context of safety barriers involves monitoring, testing, and analyzing their condition and performance. These functionalities are often implemented by the equipped diagnosis modules, such as sensors and built-in tests (BIT) modules on the barrier components, or other self-checking mechanisms, followed by the timely maintenance or replacement of any components that are found to have failed.

In the traditional reliability framework of technical safety barriers, the role of diagnosis is to uncover/detect the faults capable of blocking barrier functions, meaning the dangerous faults. Therefore, the faults or failures we study in this section are classified as dangerous detected (DD), as explained in the previous chapters of this book.

In addition to fault diagnosis, detection technologies can be used together with an analysis of collected data to anticipate degradation-developing patterns and estimate time to failure of barrier components, collectively forming what is referred to as *prognosis* in this book. A comprehensive introduction of prognosis and its associated maintenance strategies will be in Section 9.7.

DOI: 10.1201/9781003245636-9

### 9.2.1 Diagnosis Coverage

*Diagnosis coverage* (DC) is used to describe the capability of a safety barrier system to automatically detect and identify faults or failures in its components with the equipped diagnosis modules or self-checking mechanisms. The failures within DC are thus DD failures, and the failures beyond DC are DU failures. As a result, we use $\alpha_{DD}$ to denote the DC, or the percentage of DD failures in the dangerous failures of a barrier component ($\alpha_{DD}$), as

$$\alpha_{DD} = \frac{\text{Number of DD failures}}{\text{Number of DD failures} + \text{Number of DU failures}} \tag{9.1}$$

When we assume that time to failures following the exponential distribution, in a certain time period, the numbers of DD- and DU-failures are linearly related to their failure rates, and thus,

$$\alpha_{DD} = \frac{\lambda_{DD}}{\lambda_{DD} + \lambda_{DU}} \tag{9.2}$$

This relationship also can be written as

$$\lambda_D = \lambda_{DD} + \lambda_{DU} = \alpha_{DD}\lambda_D + (1 - \alpha_{DD})\lambda_D \tag{9.3}$$

When the total dangerous failure rate $\lambda_D$ is a constant, an increased value of $\alpha_{DD}$ implies that a higher percentage of failures can be detected promptly. Given that the repair time associated with a DD-failure is relatively short, the expected downtime of the barrier system is thus reduced when $\alpha_{DD}$ increases. Consequently, a higher $\alpha_{DD}$ reflects an enhanced ability of the safety-related system to rapidly identify and manage faults, contributing to improving the overall safety integrity.

**Example 9.1:**

***Impacts of DD- and DU-failures on a single-barrier component***

Here we consider a single-barrier component with both DD- and DU-failures, as illustrated in Figure 9.1. In such a numerical example, we can set the total D-failure rate is $1 \times 10^{-5}$ per hour and aim to observe the impact of the change of value of $\alpha_{DD}$ on $PFD_{avg}$. The proof test interval is 1 year (8760 hours), and MRT is 8 hours.

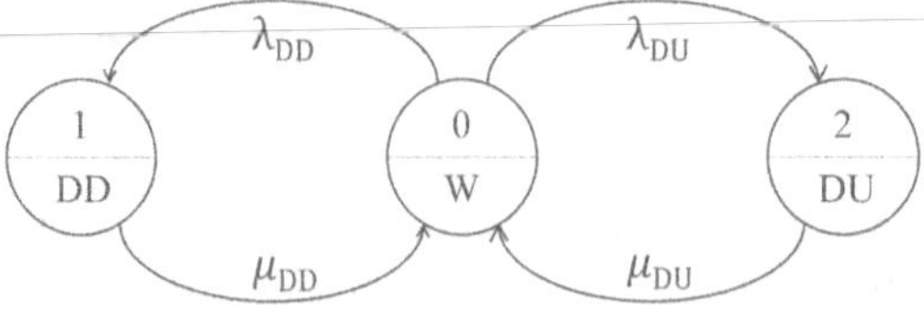


**FIGURE 9.1** The Markov model of a barrier component with DD- and DU-failures.

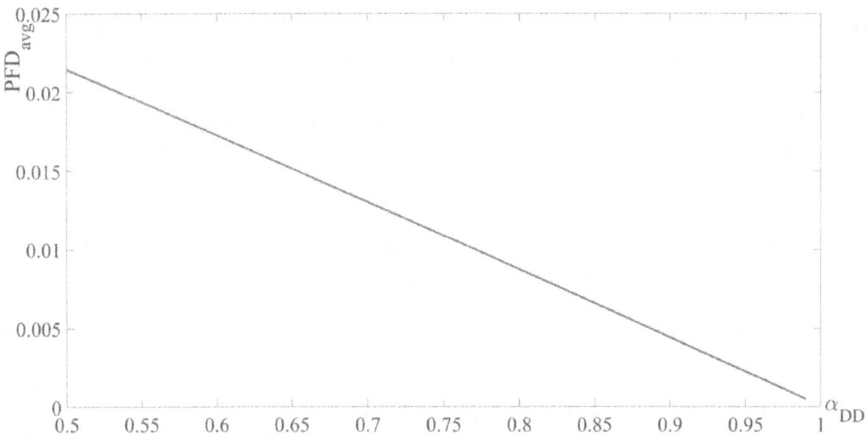


**FIGURE 9.2** Impact of $\alpha_{DD}$ on $PFD_{avg}$ of a single-barrier component.

$PFD_{avg}$ is calculated as the sum of the sojourn probabilities in states 1 and 2. The Markov transition matrix and calculation are skipped in this example. We change the value of $\alpha_{DD}$ from 0.5 to 0.99 and plot the relationship between $\alpha_{DD}$ and $PFD_{avg}$ in MATLAB®, shown in Figure 9.2.

It can be found that with increasing $\alpha_{DD}$, $PFD_{avg}$ keeps linearly decreasing, since the contribution of DU-failure to $PFD_{avg}$ is reducing. ■

For a barrier system with more than one component, the unavailability can also be calculated based on the combinations of different types of failures, supposing that all the failures are independent. For example, the failure probability of a 1oo2 barrier system at time $t$ can be contributed by two DDs, two DUs, and one DU with one DD,

$$F_S(t) = \left[F_{DD}(t)\right]^2 + \left[F_{DU}(t)\right]^2 + 2F_{DD}(t)F_{DU}(t) \tag{9.4}$$

The third item in the equation has a coefficient of 2, which can be explained by the existence of two scenarios: First, a DD-failure occurs on one channel at first, and a DU-failure occurs before the repair on the DD-fault is completed; and a DU-failure occurs on one channel; second, a DD-failure occurs before the next proof test. Assuming that time to failures follows the exponential distribution, for a certain time period, the equation can be written as

$$F_S(t) = \left[1 - e^{-\alpha_{DD}\lambda_D t}\right]^2 + \left[1 - e^{-(1-\alpha_{DD})\lambda_D t}\right]^2 + 2\left[1 - e^{-\alpha_{DD}\lambda_D t}\right]\left[1 - e^{-(1-\alpha_{DD})\lambda_D t}\right] \tag{9.5}$$

The average PFD within a proof test interval $[0, \tau)$ can be approximated by considering each contribution separately. For the unavailability due to two DD-failures, it is calculated as the rate of a DD-failure (as $\alpha_{DD}\lambda_D$) multiplied by the probability that the second DD-failure occurs during the repair time of the failed channel

($\alpha_{DD}\lambda_D$MTTR) and the downtime caused by the second DU-failure (MTTR). For the combined effect of DD- and DU-failures, PFD is approximated as follows:

- For the first scenario, it is the probability that a DD-failure occurs ($\alpha_{DD}\lambda_D\tau$) multiplied by the probability that a DU-failure occurs during the repair time of the failed channel ($(1-\alpha_{DD})\lambda_D$MTTR) for the first scenario.
- For the second scenario, it is the probability that a DU-failure occurs ($(1-\alpha_{DD})\lambda_D\tau$) multiplied by the fraction of time when another channel is down due to a DD-fault ($\alpha_{DD}\lambda_D\tau$MTTR/$\tau$).

Therefore, we have

$$\text{PFD}_{\text{avg,S}} = (\alpha_{DD}\lambda_D\text{MTTR})^2 + \frac{[(1-\alpha_{DD})\lambda_D\tau]^2}{3} + 2\alpha_{DD}(1-\alpha_{DD})\lambda_D^2\tau\text{MTTR} \quad (9.6)$$

The impact of $\alpha_{DD}$ on $\text{PFD}_{\text{avg,S}}$ can be analyzed using Equation (9.6), and the changing trend of $\text{PFD}_{\text{avg}}$ is plotted in Figure 9.3 based on the data provided in Example 9.1 (MTTR = MRT = 8 hour). It can be found that for a barrier system with more complex configurations, the overall trend of $\text{PFD}_{\text{avg}}$ consistently decreases with increasing $\alpha_{DD}$, similar to the behavior observed in the individual barrier component. However, the decrease is not necessarily linear.

Interested readers can further explore this by plotting curves for 1oo2 and 2oo3 voting structures, both with identical and distinct channels. Calculations based on Equations (8.8)–(8.12) of the IEC 61508 formulas can lead to the same conclusions.

It should be noted that the above approximation is based on the assumption that DD- and DD-failures and their associated testing activities are independent, meaning that repair on a DD-fault does not trigger any additional detection on a DU-fault. In the next subsection, we will release this assumption and discuss the DD-induced tests.

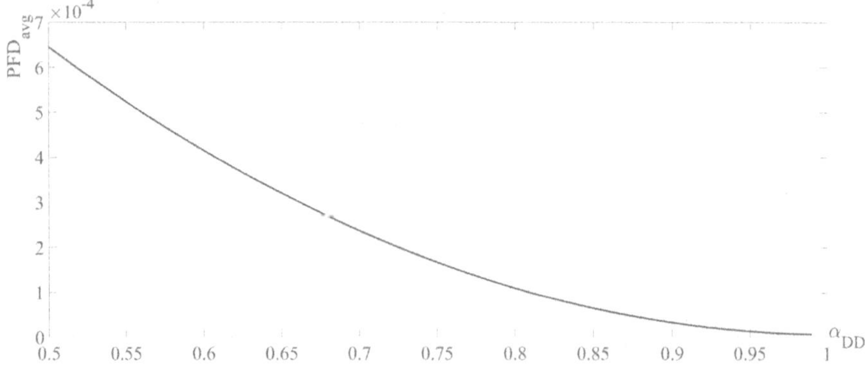


**FIGURE 9.3** Impact of $\alpha_{DD}$ on $\text{PFD}_{\text{avg}}$ of a 1oo2 barrier subsystem.

### 9.2.2 Diagnosis-induced Proof Test

DD-faults can be repaired online while the process continues to run normally. However, in many practical applications, the normal operation of the asset may be interrupted during the repair of safety barriers. Therefore, it is natural to minimize the number of repairs.

Moreover, even under the assumption of independence of DD- and DU-failures, it is not unreasonable to suppose that a barrier component with a DD fault may be more vulnerable to DU-failures. Given that DD- and DU-faults can coexist within a barrier component, when a DD-fault is identified, the testing and maintenance resources are readily available and justified in investigating the potential presence of a hidden DU-fault. This additional test does not significantly increase costs, particularly in scenarios characterized by remote operations, such as offshore and subsea operations, where inspection costs are much lower than logistical costs. We can call such an extra test the "insert proof test" or "diagnosis-induced proof test" (Liu & Rausand, 2016). Such tests are expected to mitigate the risks associated with the barrier and decrease the frequency of disruptions to normal operations.

Based on this consideration, we have two options after identifying and fixing a DD-fault:

Option 1: Do not perform any additional test for DU-fault;
Option 2: Perform an insert proof test for DU-fault in a short time.

As proof tests are routinely conducted with the scheduled intervals, an insert test may influence the original testing plan. As a result, following an insert test, decision-making is also needed regarding the timing of the next proof test, with the following options:

Option 2.1: Perform an extra proof test for DU-fault and keep the proof test schedule unchanged.
Option 2.2: Perform an extra proof test for DU-fault and postpone the next proof test.

For example, given that the planned proof test time is in August with the interval of one year and an insert test is conducted in March, the question arises whether the next proof test is still conducted in August or can be postponed to the next March with following the same test interval of year. These two options can result in different unavailability profiles of the barrier, as illustrated in Figure 9.4.

Since an insert proof test is not a part of the original testing plan, it can be regarded as a parallel process alongside the normal proof tests, and this scenario motivates us to use the RBD-driven Petri net here for modeling different options after an insert test and comparing their impacts on PFD. Readers can go back to Section 5.7.3 for recalling the terminologies of this method.

The three RBD-driven Petri net models in Figures 9.5–9.7 are used for modeling Options 1, 2.1, and 2.2, respectively. All of them have similar profiles with the example model in Figure 5.26, including three parts/blocks: The upper-left block is for modeling DD-failures and repairs, the lower-left block is for modeling DU-failures and restoration, and the right block is for proof tests.

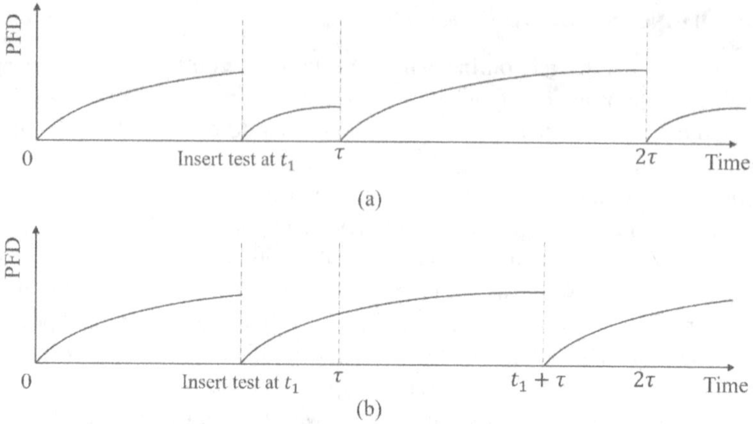


**FIGURE 9.4** PFD($t$) under the strategies of (a) keeping proof test schedule unchanged, and (b) postponing the next proof test.

Figure 9.5 illustrates the model of Option 1, where the occurrence of a DD-failure is reflected by the firing of the transition $tr_{DD}$ and removing the token in $pa_{DW}$, and then depositing a token in $pa_{DD}$. At the same time, we set a binary variable $X_1$ (true or false) to denote the state of the barrier component, whether it is failed due to a DD-failure. The firing of $tr_1$updates the values of $X_1$ as true, asserted as "!$X_1$". When the barrier

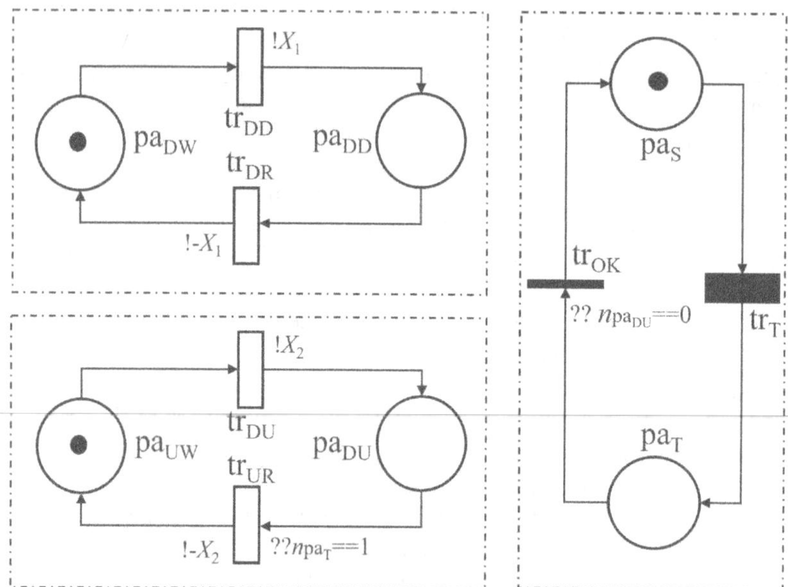


**FIGURE 9.5** RBD-PN model for Option 1.

component is repaired, the transition $tr_{DR}$ is fired to absorb the token in $pa_{DD}$ and release a token into $pa_{DW}$. The value of $X_1$ is then updated as false, asserted as "!-$X_1$".

A proof test is conducted when firing the transition $tr_T$, removing the token in $pa_S$, and depositing a token into $pa_T$. Since proof tests are performed with constant intervals, we use a filled rectangle to denote such a transition with the deterministic delay. When a token is in $pa_T$, it means that the proof test is ongoing. A proof test is completed when the staff can confirm that there is no DU fault on the barrier component, and it can happen when no DU-failure occurs on the component or when a revealed DU-fault is removed. As a result, firing of the transition $tr_{OK}$ needs a predicate "$??n_{paDU} == 1$" and then removes the token in $pa_T$ while depositing a token to $pa_S$. In this case, we ignore the time consumed in the proof test and thus use a bar to denote the transition with zero firing time.

On the other hand, the occurrence of a DU-failure is reflected by the firing of the transition $tr_{DU}$ and removing the token in $pa_{UW}$, and then depositing a token in $pa_{DU}$. We also set a binary variable $X_2$ (true or false) to denote the state of the barrier component, whether it is failed due to a DU failure. The firing of $tr_l$ updates the values of $X_1$ as true, asserted as "!$X_1$". The barrier component cannot be restored until a proof test is carried out, and thus firing of the transition $tr_{UR}$, as the restoration action, needs a predicate "$??n_{paT} == 1$". When $tr_{UR}$ is fired, the token in $pa_{DD}$ is absorbed, a token is released into $pa_{DW}$, and the value of $X_2$ is then updated as false, asserted as "!-$X_2$".

A global assertion in the model is "$!!X_S = X_1 + X_2$", meaning that when $X_1$ or $X_2$ is true, the barrier component is failed due to a DD-failure or a DU-failure, the variable $X_S$ is equal to 1. PFD can be then calculated based on the probability that $X_S = 1$.

In fact, Figure 9.5 is almost equivalent with the Markov model in Figure 9.1, which is even easier in modeling. However, the benefit of Petri net can be found in modeling the insert test. Figure 9.6 describes the behaviors of the barrier component with Option 2.1.

The model in Figure 9.6 is similar to that in Figure 9.5, but the predicate associated with the transition $tr_{UR}$ is extended as "$??n_{paT} == 1$ or $??n_{paDD} == 1$". This means that this transition can be fired in two scenarios: (1) There is one token in the place $pa_T$, meaning that a proof test is conducted, and (2) there is one token in the place $pa_{DD}$, meaning that a DD-failure occurs, and it triggered an insert test. This test does not have any influence on the right part of the Petri net model.

Figure 9.7 illustrates the behaviors of the barrier component with Option 2.2. When a DD-failure occurs, the transition $tr_D$ is fired by updating the value of $X_1$. It should be noted that there is a predicate of "$??T == 0$" for the transition and an assertion of "$!T = 1$". Such controlling statements introduce a new variable of T, and they can be used to link the DD-failure and proof test by assigning the resource for proof tests to a DD-induced test while eliminating the enabling condition of the transition $tr_T$ by adding a predicate "$??T == 0$".

The DD-fault is fixed by firing the transition $tr_{DR}$, which updates the value of $X_1$ to false and returns the value of T as 0. This assertation releases the resources for proof tests. In other words, the insert test on the DU-fault re-initiates the time-counting process for the next proof test.

It should be noted that the model in Figure 9.6 ignores the situation where a DD-failure takes place when a proof test is conducted. Since the probability of such a situation is very low, this simplification has little influence on the analysis.

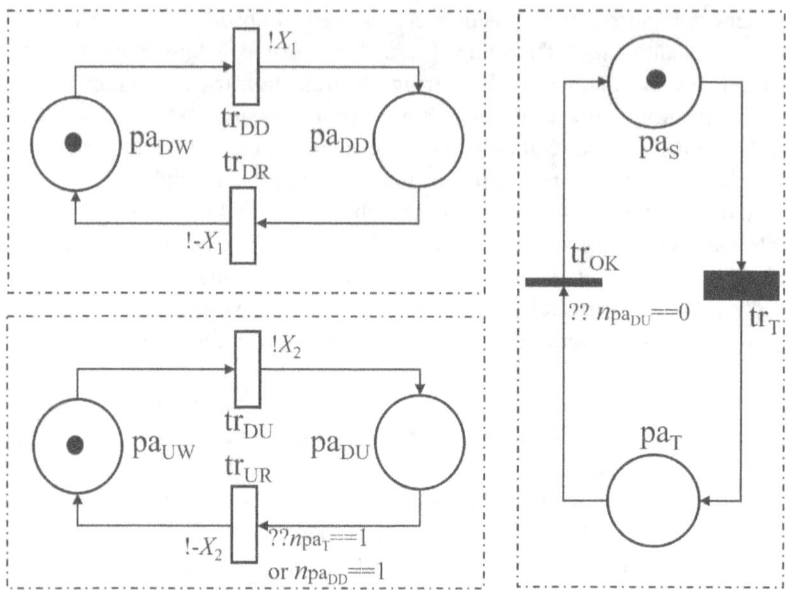


**FIGURE 9.6** RBD-PN model for Option 2.1.

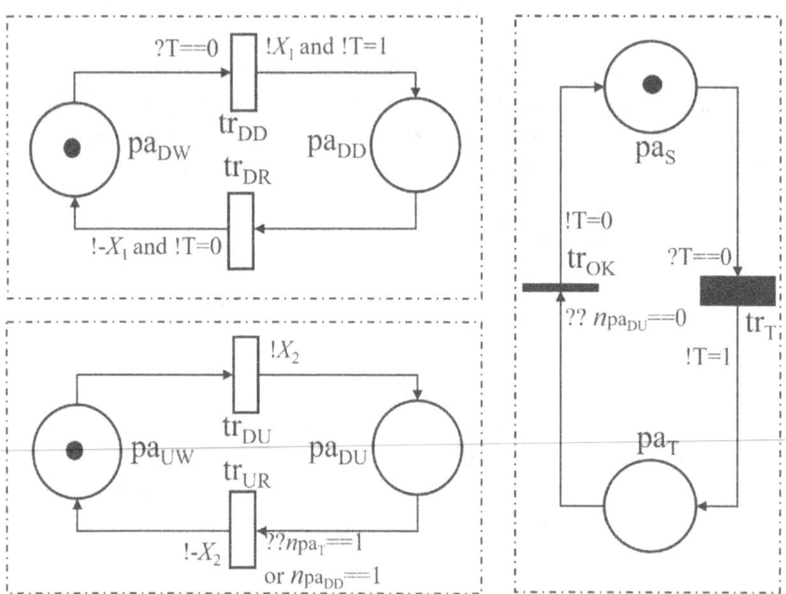


**FIGURE 9.7** RBD-PN model for Option 2.2.

**TABLE 9.1**
**$PFD_{avg}$ Due to DU-Failures for the Three DD-induced Testing Options Based on the Petri Net Models**

| | $PFD_{avg}$ due to DU Failures ($10^{-5}$/h) | | |
|---|---|---|---|
| $\lambda_{DD}$ ($10^{-5}$/h) | Option 1 | Option 2.1 | Option 2.2 |
| 2.0 | 8.70 | 8.21 | 8.30 |
| 6.0 | 8.70 | 7.39 | 7.83 |
| 10.0 | 8.70 | 6.61 | 7.30 |
| 20.0 | 8.70 | 5.27 | 6.19 |

*Source:* Adapted from Liu and Rausand (2016).

Liu and Rausand (2016) have studied the three options with numerical examples, where DU-failure rate is set as $2\times10^{-6}$ per hour, MRT for DD-fault is 8 hours, and proof test interval is 8760 hours. Based on the Monte Carlo simulation using the GRIF® Workshop program, the authors have compared the average PFD values due to DU failures with the three options when DD-failure rate is set differently, as illustrated in Table 9.1.

It is evident that for Option 1, DD- and DU-failures in the same components are mutually independent. Consequently, $PFD_{avg}$ due to DU-failures remains constant regardless of the value of $\lambda_{DD}$. In the cases of Options 2.1 and 2.2, an increase in the DD-failure rate has the potential to decrease the $PFD_{avg}$ due to DU-failures in consideration of the impacts of the insert tests.

Comparing the latter two options, it is apparent that Option 2.1 consistently yields a lower PFD value. This is because this option, featuring insert tests without a reduction in planned tests, provides more opportunities to detect DU failures.

The authors have also developed approximation formulas for Options 2.1 and 2.2. For Option 2.1, assuming that a DU-failure has occurred at $t_{DU}$, the hidden fault is not revealed until the new proof test or the repair on a DD-failure before the proof test. Therefore, the downtime due to the DU-failure is the minimum value of the time to the proof test and the time to the DD-failure. The mean downtime due to a DU-failure can thus be obtained as

$$\mathrm{MDT}_{DU}^{2.1} = \int_0^{\tau - t_{DU}} e^{-\lambda_{DD} t} dt = \frac{1}{\lambda_{DD}}\left[1 - e^{-\lambda_{DD}(\tau - t_{DU})}\right] \tag{9.7}$$

The occurrence time of the DU-failure ($T_{DU}$) is a random value within $[0, \tau)$, as $1-\mathrm{MDT}_{DU}^{2.1}$, and thus we can calculate the expected mean downtime due to a DU-failure as

$$E\left(\mathrm{MDT}_{DU}^{2.1}\right) = E\left[\int_0^{\tau} e^{-\lambda_{DD} t} f_{T_{DU}}(t) dt\right] = \frac{1}{\lambda_{DD}}\left[1 - \frac{1}{\lambda_{DD}}\left(1 - e^{-\lambda_{DD}\tau}\right)\right] \tag{9.8}$$

In alignment with Equation (6.29), the average PFD can be calculated as the expected mean downtime times with the rate of the failure that results in the downtime, and thus we have

$$\mathrm{PFD}_{\mathrm{DU}}^{2.1} \approx \lambda_{\mathrm{DU}} \cdot \mathrm{MDT}_{\mathrm{DU}}^{2.1} = \frac{\lambda_{\mathrm{DU}}}{\lambda_{\mathrm{DD}}}\left[1 - \frac{1}{\lambda_{\mathrm{DD}}}\left(1 - e^{-\lambda_{\mathrm{DD}}\tau}\right)\right] \tag{9.9}$$

The calculation results based on Equation (9.9) are slightly higher than those in Table 9.1 (no higher than 1%) considering that the approximation results are always more conservative.

For Option 2.2, after a DD-failure, the interval to the next proof test is re-initiated. The length $X$ of a test for DU-failures can be regarded as a stochastic variable with the probability density:

$$f_X(x) = \begin{cases} \lambda_{\mathrm{DD}} e^{-\lambda_{\mathrm{DD}} x} & \text{for } 0 < x < \tau \\ e^{-\lambda_{\mathrm{DD}}\tau} & \text{for } x = \tau \end{cases} \tag{9.10}$$

The mean length of a testing interval, or mean time between tests (MTBT) for DU-failures, is thus calculated as

$$E(X) = \int_0^{\tau} e^{-\lambda_{\mathrm{DD}} t} d = \frac{1}{\lambda_{\mathrm{DD}}}\left(1 - e^{-\lambda_{\mathrm{DD}}\tau}\right) \tag{9.11}$$

Based on the exponential distribution assumption of time to failures, since all testing intervals have the same stochastic properties, we can just study the first interval starting from $t = 0$ without losing generality. Supposing that a DU-failure occurs at $t_{\mathrm{DU}}$ even though its exact value is unknown, the time to the next test on a DU-failure is the minimum value of $\tau - t_{\mathrm{DU}}$ and the time to a DD-failure. The mean downtime due to a DU-failure for Option 2.2 is therefore conditional on $x$ and $t_{\mathrm{DU}}$:

$$\mathrm{MDT}_{\mathrm{DU}}^{2.2}(x, t_{\mathrm{DU}}) = \frac{1}{\lambda_{\mathrm{DD}}}\left[1 - e^{-\lambda_{\mathrm{DD}}(\tau - t_{\mathrm{DU}})}\right] \tag{9.12}$$

Given that the value of $t_{DU}$ is unknown, the mean downtime conditional on $x$ is

$$\begin{aligned} \mathrm{MDT}_{\mathrm{DU}}^{2.2}(x) &= \int_0^x \mathrm{MDT}_{\mathrm{DU}}^{2.2}(x, t_{\mathrm{DU}}) f_{T_{\mathrm{DU}}}(t)\,dt = \int_0^x \mathrm{MDT}_{\mathrm{DU}}^{2.2}(x, t_{\mathrm{DU}}) \frac{1}{x}\,dt \\ &= \frac{1}{\lambda_{\mathrm{DD}}} - \frac{e^{-\lambda_{\mathrm{DD}}\tau}}{\lambda_{\mathrm{DD}}^2 x}\left(e^{-\lambda_{\mathrm{DD}} x} - 1\right) \end{aligned} \tag{9.13}$$

Since the occurrence time of the DU-failure is uniformly distributed within the last test interval, given that a DU-fault is found in the proof test, the conditional mean occurrence time is in the middle of the period. Thus, the mean time to a DU-failure is equal to the mean downtime. As a result, with a certain value of $x$, we can determine the mean downtime due to a DU-failure by solving the following equation:

$$\frac{x}{2} = \frac{1}{\lambda_{\mathrm{DD}}} - \frac{e^{-\lambda_{\mathrm{DD}}\tau}}{\lambda_{\mathrm{DD}}^2 x}\left(e^{-\lambda_{\mathrm{DD}}x} - 1\right) \tag{9.14}$$

With MATLAB® or other programs, we can obtain the result based on the approximation formulas, which is also slightly higher than those obtained by Petri net simulation.

In practice, it is necessary to assess the costs, or the numbers of tests associated with the three options, to determine the more suitable strategy for proof tests. Considering a long-term of $n\tau$, the number of tests in Option 1 is the number of proof tests $n$. For Option 2.1, we need to consider the potential insert tests, which is approximately equal to $\lambda_{\mathrm{DD}}n\tau$ (ignoring the time used to fix DD-failures), and thus, the total number of tests is

$$N_{2.1} \approx n + \lambda_{\mathrm{DD}}n\tau \tag{9.15}$$

For Option 2.2, according to Equation (9.11), we have known the MTBT; the total number in the long term is thus approximated as (ignoring the time used to fix DD-failures):

$$N_{2.2} \approx \frac{n\tau}{\mathrm{MTBT}} \tag{9.16}$$

This subsection is based on the literature from the previous study of the author (Liu & Rausand, 2016), and it is restricted to a single-barrier component. Interested readers can try to study the barrier subsystems with multiple channels.

## 9.3 STAGGERED TESTING STRATEGY

According to the routine proof testing principle, when a barrier subsystem has a 1oo2 redundant structure, the two channels are assumed to undergo testing simultaneously up to this point in this book. In practice, they can also be tested at different times in consideration that the channels are geographically dispersed or the testing resources are with constraints. The latter approach is called *staggered proof test* (Torres-Echeverri et al., 2009).

### 9.3.1 Staggered Tests on 1oo2 Barrier of Identical Channels

We have known that the unavailability of a barrier is highly dependent on its proof test interval, as depicted in Figure 6.2. Here, we still focus on the impact of test interval. If the two channels of a barrier system are tested at different times but with the same test interval $\tau$, we can set without losing generality that Channel 1 is proof-tested at times $0, \tau, 2\tau, \ldots$, and Channel 2 is tested at $t_0, t_0 + \tau, t_0 + 2\tau, \ldots$ Time 0 here means the starting point to observe the barrier subsystem, and the time difference $t_0$ between the proof test on Channel 1 and that on Channel 2 is called the *staggered time*.

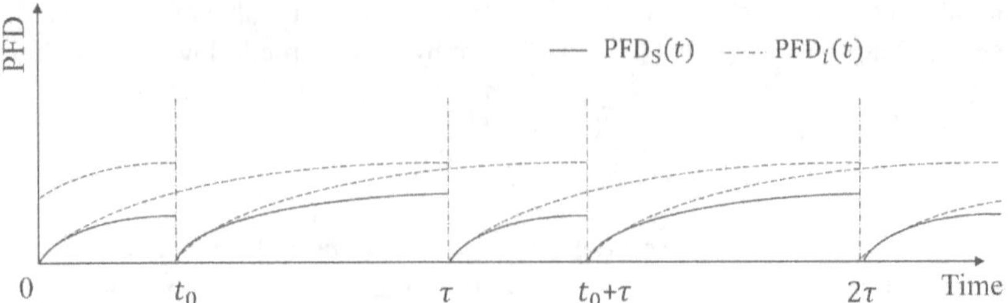


**FIGURE 9.8** PFD($t$) of a barrier system of two homogeneous channels under a staggered test.

After each proof test, faults in the tested barrier channels will be removed. As illustrated in Figure 9.8, within a proof test interval of Channel 1, e.g., $[\tau, 2\tau)$, $\mathrm{PFD}_2(t)$ of Channel 2 at time $\tau$ is not equal to 0, and actually it is equal to $1-e^{-\lambda_{\mathrm{DU}}(\tau-t_0)}$ because the barrier has been in operation before time 0, and the time from its last proof test is $\tau - t_0$. Then, $\mathrm{PFD}_2(t)$ is restored at time $t_0+\tau$, meaning that $\mathrm{PFD}_2(t_0+\tau)=0$. Within the proof test interval $[0, \tau)$, the changing of $\mathrm{PFD}_2(t)$ is same.

To simplify calculation and illustration, we only consider the impacts of DU-failures in $[0, \tau)$ that follow the exponential failure distribution. Both channels have a DU-failure rate $\lambda_{\mathrm{DU}}$, the instantaneous unavailability of the barrier subsystem can be calculated as the multiplicity of PFD($t$) of the two channels:

$$\mathrm{PFD_S}(t)=\begin{cases}\left[1-e^{-\lambda_{\mathrm{DU}}t}\right]\left[1-e^{-\lambda_{\mathrm{DU}}(t+\tau-t_0)}\right] & \text{for } 0<t<t_0\\ \left[1-e^{-\lambda_{\mathrm{DU}}t}\right]\left[1-e^{-\lambda_{\mathrm{DU}}(t-t_0)}\right] & \text{for } t_0<t<\tau\end{cases} \tag{9.17}$$

The average unavailability of the barrier subsystem is

$$\mathrm{PFD_{avg,S}}=\frac{1}{\tau}\int_0^{\tau}\mathrm{PFD_S}(t)\,dt=\frac{1}{\tau}\left[\int_0^{t_0}\mathrm{PFD_S}(t)\,dt+\int_{t_0}^{\tau}\mathrm{PFD_S}(t)\,dt\right] \tag{9.18}$$

Equation (9.18) can be deployed as

$$\begin{aligned}\mathrm{PFD_{avg,S}}&=1-\frac{2}{\lambda_{\mathrm{DU}}\tau}\left(1-e^{-\lambda_{\mathrm{DU}}\tau}\right)\\ &-\frac{1}{2\lambda_{\mathrm{DU}}\tau}\left[e^{-\lambda_{\mathrm{DU}}(t_0+\tau)}-e^{-\lambda_{\mathrm{DU}}(\tau-t_0)}+e^{-\lambda_{\mathrm{DU}}(2\tau-t_0)}-e^{-\lambda_{\mathrm{DU}}t_0}\right]\end{aligned} \tag{9.19}$$

It is a function of the staggered time $t_0$. The minimum value of $\mathrm{PFD_{avg,S}}$ can be determined by setting the derivative of the function with respect to $t_0$ equal to 0, and thus we have

$$e^{-\lambda_{\mathrm{DU}}(t_0+\tau)}+e^{-\lambda_{\mathrm{DU}}(\tau-t_0)}-e^{-\lambda_{\mathrm{DU}}(2\tau-t_0)}-e^{-\lambda_{\mathrm{DU}}t_0}=0 \tag{9.20}$$

Then, $t_0$ can be expressed as

$$t_0 = \frac{1}{2\lambda_{DU}} \ln\left(e^{\lambda_{DU}\tau}\right) = \frac{\tau}{2} \tag{9.21}$$

It means that when the staggered time is the half of the proof test interval, the average unavailability of the barrier system is lowest. In such a situation, according to Equation (9.19), we can make an approximation as

$$\text{PFD}_{\text{avg,S}} \approx \frac{5(\lambda_{DU}\tau)^2}{24} \tag{9.22}$$

Upon comparison, it is evident that the unavailability of a 1oo2 barrier subsystem under a simultaneous test strategy is higher than the $\text{PFD}_{\text{avg,S}}$ observed under staggered tests. The reduction is calculated by ((1/3-5/24))/(1/3), resulting in a 37.5% decrease in $\text{PFD}_{\text{avg,S}}$. It is important to note that in this section, we do not consider the costs associated with proof tests. However, in practical applications, it is reasonable to assess the cost-effectiveness of a staggered test strategy, considering the reduced risk and potential additional costs associated with its implementation.

The two barrier channels in a parallel structure can have different failure rates, and their proof test intervals are not necessarily the same; therefore, Liu (2014) has studied more situations of staggered tests, including,

- Failure rates of the two channels are the same, but their proof test intervals are different;
- Proof test intervals of the two channels are the same, but their failure rates are different;
- Neither the failure rate nor the proof test interval of the two channels are the same.

In the rest of this subsection, we will discuss the above three scenarios.

### 9.3.2 Staggered Tests on Channels with Different Testing Intervals

In this scenario, we can assume without losing generality that the proof test interval of Channel 1 ($\tau_1$) is more time-consuming than that of Channel 2 ($\tau_2$), and we can use $k$ to denote the ratio of $\tau_1/\tau_2$. Within the interval $[0, \tau_1)$, Channel 2 must be proof-tested at least once, and we can set the first proof test time for this channel at $t_0$. If Channel can be tested more than once within the interval, the next proof test time is $t_0 + \tau_2$. Generally, we can use $n_2$ to denote the testing times of Channel 2 within $[0, \tau_1)$, and the last proof test time is $t_0 + (n_2 - 1)\tau_2$, and it is known that:

$$t_0 + (n_2 - 1)\tau_2 < \tau_1 = k\tau_2$$

Thus, we can have

$$n_2 < k + 1$$

On the other hand, after the last proof test at $t_0+(n_2-1)\tau_2$, the time to the proof test on Channel 1 is no longer than that to the proof test on Channel 2, namely:

$$\tau_1-t_0-(n_2-1)\tau_2\le\tau_2$$

Replace $\tau_1$ with $k\tau_2$, and we can obtain:

$$n_2>k-1$$

If $k$ is also an integer, $n_2=k$.

Considering that Channel 2 is restored to the as-good-as-new state after each proof test, its failure probability at time $t$ is

$$F_2(t)=1-e^{-\lambda_{\mathrm{DU}}(t-n\tau_2)} \tag{9.23}$$

For the barrier subsystem, its average unavailability within the interval $(0,\tau_1]$ can be calculated and accumulated in different proof test intervals of Channel 2:

$$\begin{aligned}\mathrm{PFD}_{\mathrm{avg,S}}&=\frac{1}{\tau_1}\int_0^{\tau_1}F_1(t)F_2(t)dt=\frac{1}{\tau_1}\left[\int_0^{t_0}\left(1-e^{-\lambda_{\mathrm{DU}}t}\right)\left(1-e^{-\lambda_{\mathrm{DU}}(t+\tau_2-t_0)}\right)dt\right.\\&+\int_{t_0}^{t_0+\tau_2}\left(1-e^{-\lambda_{\mathrm{DU}}t}\right)\left(1-e^{-\lambda_{\mathrm{DU}}(t-t_0)}\right)dt+\cdots\\&+\int_{t_0+j\tau_2}^{t_0+(j+1)\tau_2}\left(1-e^{-\lambda_{\mathrm{DU}}t}\right)\left(1-e^{-\lambda_{\mathrm{DU}}(t-t_0-j\tau_2)}\right)dt+\cdots\\&\left.+\int_{t_0+(n_2-1)\tau_2}^{\tau_1}\left(1-e^{-\lambda_{\mathrm{DU}}t}\right)\left(1-e^{-\lambda_{\mathrm{DU}}(t-t_0-(n_2-1)\tau_2)}\right)dt\right]\end{aligned} \tag{9.24}$$

The general item within the period $(j\tau_2,(j+1)\tau_2]$ can be organized as

$$\begin{aligned}&\int_{t_0+j\tau_2}^{t_0+(j+1)\tau_2}\left(1-e^{-\lambda_{\mathrm{DU}}t}\right)\left(1-e^{-\lambda_{\mathrm{DU}}(t-t_0-j\tau_2)}\right)dt=\tau_2+\frac{1}{\lambda_{\mathrm{DU}}}\left[e^{-\lambda_{\mathrm{DU}}(t_0+(j+1)\tau_2)}-e^{-\lambda_{\mathrm{DU}}(t_0+j\tau_2)}\right]\\&+\frac{1}{\lambda_{\mathrm{DU}}}\left(e^{-\lambda_{\mathrm{DU}}\tau_2}-1\right)+\frac{1}{2\lambda_{\mathrm{DU}}}\left[e^{-\lambda_{\mathrm{DU}}(t_0+(j+2)\tau_2)}-e^{-\lambda_{\mathrm{DU}}(t_0+j\tau_2)}\right]\end{aligned} \tag{9.25}$$

The second part and fourth part in this item can be compensated with the counterparts in the previous- and subsequent items in sum-up calculation, and finally, Equation (9.24) can be written as

$$\begin{aligned}\mathrm{PFD}_{\mathrm{avg,S}}&=1+\frac{1}{\lambda_{\mathrm{DU}}\tau_1}\left(e^{-\lambda_{\mathrm{DU}}\tau_1}-1\right)+\frac{n}{\lambda_{\mathrm{DU}}\tau_1}\left(e^{-\lambda_{\mathrm{DU}}\tau_2}-1\right)\\&-\frac{1}{2\lambda_{\mathrm{DU}}\tau_1}\left[e^{-\lambda_{\mathrm{DU}}(\tau_1+\tau_2-t_0)}+e^{-\lambda_{\mathrm{DU}}(\tau_1+t_0)}-e^{-\lambda_{\mathrm{DU}}t_0}-e^{-\lambda_{\mathrm{DU}}(\tau_2-t_0)}\right]\end{aligned} \tag{9.26}$$

Similarly to the last subsection, the minimum value of $\text{PFD}_{\text{avg,S}}$ can be obtained by setting the derivative of the function with respect to $t_0$ equal to 0, and then we have

$$e^{-\lambda_{\text{DU}}(\tau_1+\tau_2-t_0)} - e^{-\lambda_{\text{DU}}(\tau_1+t_0)} + e^{-\lambda_{\text{DU}}t_0} - e^{-\lambda_{\text{DU}}(\tau_2-t_0)} = 0 \tag{9.27}$$

Finally, we can obtain

$$t_0 = \frac{\tau_2}{2} \tag{9.28}$$

The results mean that the optimal staggered time is half of the shorter interval when the proof test intervals of the two channels in a barrier system are different.

### 9.3.3 Staggered Tests on Channels with Different Reliability

When the two channels have distinct DU-failure rates ($\lambda_{\text{DU},1}$ and $\lambda_{\text{DU},2}$), they demonstrate different instantaneous PFD profiles, as depicted in Figure 9.9. Although the failure rate of Channel 1 is higher in the example, the relative reliability of each channel does not influence the analysis in this subsection.

The proof test interval ($\tau$) for the two channels is the same. Similar to the last subsection, the average unavailability of the barrier system is calculated as

$$\begin{aligned}\text{PFD}_{\text{avg,S}} &= \frac{1}{\tau}\int_0^{\tau_1} F_1(t)F_2(t)dt = \frac{1}{\tau}\left[\int_0^{t_0}\left(1-e^{-\lambda_{\text{DU},1}t}\right)\left(1-e^{-\lambda_{\text{DU},2}(t+\tau_2-t_0)}\right)dt\right. \\ &\quad \left. + \int_{t_0}^{\tau}\left(1-e^{-\lambda_{\text{DU},1}t}\right)\left(1-e^{-\lambda_{\text{DU},2}(t-t_0)}\right)dt\right]\end{aligned} \tag{9.29}$$

Equation (9.29) can be then organized as

$$\begin{aligned}\text{PFD}_{\text{avg,S}} &= 1 + \frac{1}{\lambda_{\text{DU},1}\tau}\left(e^{-\lambda_{\text{DU},1}\tau}-1\right) + \frac{1}{\lambda_{\text{DU},2}\tau}\left(e^{-\lambda_{\text{DU},2}\tau}-1\right) \\ &\quad - \frac{1}{\lambda_{\text{DU},1}+\lambda_{\text{DU},2}}\left[e^{-\lambda_{\text{DU},1}t_0-\lambda_{\text{DU},2}\tau} - e^{-\lambda_{\text{DU},2}(\tau-t_0)} + e^{-\lambda_{\text{DU},1}\tau-\lambda_{\text{DU},2}(\tau-t_0)} - e^{-\lambda_{\text{DU},1}t_0}\right]\end{aligned} \tag{9.30}$$

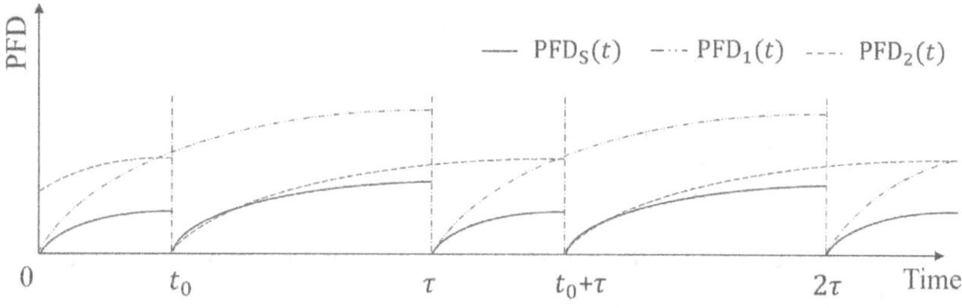


**FIGURE 9.9** PFD($t$) of a barrier system of two channels with different failure rates under a staggered test.

The minimum value of $PFD_{avg,S}$ is obtained by setting the derivative of the function with respect to $t_0$ equal to 0, and thus we have

$$e^{-\lambda_{DU,1}t_0-\lambda_{DU,2}\tau}+e^{-\lambda_{DU,2}(\tau-t_0)}-e^{-\lambda_{DU,1}\tau-\lambda_{DU,2}(\tau-t_0)}-e^{-\lambda_{DU,1}t_0}=0 \tag{9.31}$$

And $t_0$ is calculated as

$$t_0=\frac{1}{\lambda_{DU,1}+\lambda_{DU,2}}\ln\left(e^{\lambda_{DU,1}\tau}\cdot\frac{\lambda_{DU,1}}{\lambda_{DU,2}}\cdot\frac{e^{\lambda_{DU,2}\tau}-1}{e^{\lambda_{DU,1}\tau}-1}\right) \tag{9.32}$$

Given that $\lambda_{DU,i}\tau$ is rather small, where $i=1, 2$, $t_0$ can be approximated using the Taylor series as

$$t_0\approx\frac{1}{\lambda_{DU,1}+\lambda_{DU,2}}\ln\left(e^{\lambda_{DU,1}\tau}\cdot\frac{e^{\frac{\lambda_{DU,2}\tau}{2}}}{e^{\frac{\lambda_{DU,1}\tau}{2}}}\right)=\frac{\tau}{2} \tag{9.33}$$

The calculated staggered time for the minimum value of $PFD_{avg,S}$ remains consistent with the result obtained for the homogeneous barrier system. This suggests that the optimal staggered time is unaffected by the failure rates of barrier channels.

### 9.3.4 Staggered Tests on Channels with Different Reliability and Testing Intervals

In this subsection, we assume that Channel 1 has a longer proof test interval, namely $\tau_1=k\tau_2$, where $k>1$. Within a longer proof test interval $[0, \tau_1)$, the average unavailability is calculated as

$$\begin{aligned}PFD_{avg,S}&=\frac{1}{\tau_1}\int_0^{\tau_1}F_1(t)F_2(t)dt=\frac{1}{\tau_1}\Bigg[\int_0^{t_0}\left(1-e^{-\lambda_{DU,1}t}\right)\left(1-e^{-\lambda_{DU,2}(t+\tau_2-t_0)}\right)dt\\&+\int_{t_0}^{t_0+\tau_2}\left(1-e^{-\lambda_{DU,1}t}\right)\left(1-e^{-\lambda_{DU,2}(t-t_0)}\right)dt+\cdots\\&+\int_{t_0+j\tau_2}^{t_0+(j+1)\tau_2}\left(1-e^{-\lambda_{DU,1}t}\right)\left(1-e^{-\lambda_{DU,2}(t-t_0-j\tau_2)}\right)dt+\cdots\\&+\int_{t_0+(n_2-1)\tau_2}^{\tau_1}\left(1-e^{-\lambda_{DU,1}t}\right)\left(1-e^{-\lambda_{DU,2}(t-t_0-(n_2-1)\tau_2)}\right)dt\Bigg]\end{aligned} \tag{9.34}$$

We can use the similar simplification and approximation approaches in the last two subsections and obtain

$$t_0\approx\frac{\tau_2}{2} \tag{9.35}$$

This result is consistent with that for the two channels with different failure rates but the same proof test interval, implying that the failure rate has no impact on the optimal staggered time. More calculation details and case studies about the optimal staggered testing time can be found in the paper by Liu (2014).

## 9.4 PARTIAL PROOF TESTS AND IMPERFECT TESTS

We understand that the safety integrity of a barrier system, given that its reliability is constant, is influenced by the frequency of proof tests. Because such tests are conducted at regular intervals with the expectation of detecting all DU-faults on a barrier component, we call them the full proof tests (FPTs) or regularly full proof tests (see the introduction in Section 3.5).

PPTs of barrier components, particularly for the actuating subsystems, serve as supplements to FPTs. PPTs are employed to assess specific barrier functions, ensuring their functionality, but they are not designed to uncover all potential faults. In the process industries (especially when valves are the actuators), they are commonly referred to as *partial stroke tests* (PSTs). For instance, in a PPT/PST for a shutdown valve, the valve undergoes a partial stroke to verify its ability to operate and move without reaching the closed position, thereby avoiding production disruption. The minimized movement of the valve in such a PPT only has a small or even negligible impact on the process, preventing wear in the valve seat area. Moreover, since PPTs need fewer full valve closures, the likelihood of production disturbances in the events where the valves fail to open from the closed position is reduced (Lundteigen & Rausand, 2008).

For the shutdown valve example, PPTs can reveal the DU-failure mode of "fail to close on demand", but they cannot discover another DU-failure mode of "leakage in the closed position" since they do not finish the total travel. Therefore, PPTs cannot replace the FPTs on the valve or other safety barrier components.

PPT is also conducted with a time interval, different from diagnosis or diagnostic test, which can be performed continuously.

### 9.4.1 Analytical Formulas for Partial Proof Tests

Partial test coverage ($\alpha_{\text{PPT}}$) denotes the fraction of DU-failures the fault by which can be revealed detected by a PPT among all DU-failures. With the exponential distribution assumption, we have

$$\alpha_{\text{PPT}} = \frac{\lambda_{\text{DU,PPT}}}{\lambda_{\text{DU}}} \tag{9.36}$$

Equation (9.36) implies that DU-failure is actually split into two parts if PPT is considered: DU-failures that can be detected in partial tests (with the rate of $\lambda_{\text{DU,PPT}}$) and the remaining failures that are hidden in partial tests and only can be revealed in a full proof test (with the rate of $\lambda_{\text{DU,FPT}}$).

We also can define the PPT coverage ($\alpha_{PPT}$) as the conditional probability:

$$\alpha_{\text{PPT}} = Pr(\text{DU fault is revealed in a PPT} \mid \text{DU fault is presented}) \tag{9.37}$$

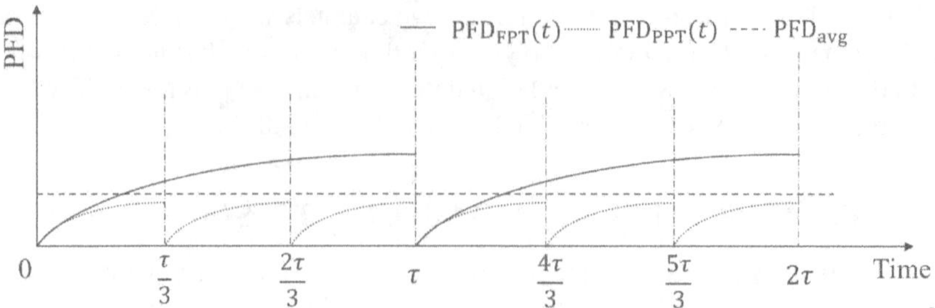


**FIGURE 9.10** PFD($t$) and $\text{PFD}_{\text{avg}}$ with partial tests of interval of $\tau/3$.

The value of $\alpha_{\text{PPT}}$ for a safety barrier should be determined based on specific conditions, such as barrier type, functionality, and operational and environmental conditions. For shutdown valves, Summers and Zachary (2000) have highlighted that their PPT coverage can be in the range of 60–70%. Lundteigen and Rausand (2008) have provided an approach to determine the value $\alpha_{\text{PPT}}$ in the cases of valves.

In this section, we assume that PPTs are periodically performed with a fixed interval of $\tau_{\text{PPT}}$ and $\tau_{\text{PPT}} = \tau/n$, where $n$ is an integer. With PPTs, PFD($t$) of a barrier component can change with time. Based on Figure 6.1, we can redraw PFD($t$) and $\text{PFD}_{\text{avg}}$ profiles with an example where the PPT interval is $\tau/3$, as depicted in Figure 9.10, where $\text{PFD}_{\text{FPT}}(t)$ is the unavailability of this barrier component due to the DU-fault that is only revealed in a full proof test, while $\text{PFD}_{\text{PPT}}(t)$ is the unavailability due to the DU-fault that can be detected in a PPT.

The analytical formulas of PFD with partial tests can be derived straightforwardly based on those for full proof tests. In consideration of DD-failures and two types of DU-failures, the instantaneous PFD of a barrier component (with the total failure rate of $\lambda$) can be described as

$$\begin{aligned}\text{PFD}(t) &= \text{PFD}_{\text{DD}}(t) + \text{PFD}_{\text{DU,PPT}}(t) + \text{PFD}_{\text{DU,FPT}}(t) \\ &= \left(1 - e^{-\alpha_{\text{DD}}\lambda_{\text{D}}t}\right)\left[1 - e^{-(1-\alpha_{\text{DD}})\alpha_{\text{PPT}}\lambda_{\text{D}}t}\right]\left[1 - e^{-(1-\alpha_{\text{DD}})(1-\alpha_{\text{PPT}})\lambda_{\text{D}}t}\right]\end{aligned} \tag{9.38}$$

Assuming that the probability that two or more failures occur at the same time is very low, the average PFD of a barrier component within the testing interval $[0, \tau)$ is

$$\begin{aligned}\text{PFD}_{\text{avg}} &= \text{PFD}_{\text{DD,avg}} + \text{PFD}_{\text{DU,PPT, avg}} + \text{PFD}_{\text{DU,FPT,avg}} = \text{PFD}_{\text{DD}} \\ &+ \frac{1}{\tau_{\text{PPT}}}\int_0^{\tau_{\text{PPT}}} \text{PFD}_{\text{DU,PPT}}(t)\,dt + \frac{1}{\tau}\int_0^{\tau} \text{PFD}_{\text{DU,FPT}}(t)\,dt = \alpha_{\text{DD}}\lambda_{\text{D}}\text{MTTR} \\ &+ \frac{1}{\tau_{\text{PPT}}}\int_0^{\tau_{\text{PPT}}} \left[1 - e^{-\alpha_{\text{PPT}}\cdot(1-\alpha_{\text{DD}})\cdot\lambda_{\text{D}}t}\right]dt + \frac{1}{\tau}\int_0^{\tau}\left[1 - e^{-(1-\alpha_{\text{PPT}})\cdot(1-\alpha_{\text{DD}})\cdot\lambda_{\text{D}}t}\right]dt\end{aligned} \tag{9.39}$$

Equation (9.39) can be approximated as

$$\mathrm{PFD_{avg}} \approx \alpha_{\mathrm{DD}}\lambda_{\mathrm{D}}\mathrm{MTTR} + \frac{(1-\alpha_{\mathrm{DD}})\alpha_{\mathrm{PPT}}\lambda_{\mathrm{D}}\tau_{\mathrm{PPT}}}{2} + \frac{(1-\alpha_{\mathrm{DD}})(1-\alpha_{\mathrm{PPT}})\lambda_{\mathrm{D}}\tau}{2} \tag{9.40}$$

The unavailability of a barrier system with more than one channel can also be contributed by different types of failures and their combinations, as in the example we have discussed in Section 9.2.1. Given that DU can be revealed in partial tests or in a full proof test, there are more combinations that need to be taken into account: two DDs, two DUs revealed in partial tests, two DUs revealed in a full proof test, one DD with one DU revealed in a partial test, one DD with one DU revealed in a full proof test, and two DUs of different types. Therefore, the average unavailability of a 1oo2 barrier system with identical and independent channels can be approximated as

$$\begin{aligned}\mathrm{PFD_{avg}} &\approx (\alpha_{\mathrm{DD}}\lambda_{\mathrm{D}}\mathrm{MTTR})^2 + \frac{[\alpha_{\mathrm{PPT}}(1-\alpha_{\mathrm{DD}})\lambda_{\mathrm{D}}\tau_{\mathrm{PPT}}]^2}{3} \\ &+ \frac{[(1-\alpha_{\mathrm{DD}})(1-\alpha_{\mathrm{PPT}})\lambda_{\mathrm{D}}\tau]^2}{3} + 2\alpha_{\mathrm{PPT}}(1-\alpha_{\mathrm{PPT}})(1-\alpha_{\mathrm{DD}})^2\lambda_{\mathrm{D}}^2\tau_{\mathrm{PPT}}\tau \\ &+ 2\alpha_{\mathrm{PPT}}\alpha_{\mathrm{DD}}(1-\alpha_{\mathrm{DD}})\lambda_{\mathrm{D}}^2\mathrm{MTTR}\,\tau_{\mathrm{PPT}} + 2\alpha_{\mathrm{DD}}(1-\alpha_{\mathrm{DD}})(1-\alpha_{\mathrm{PPT}})\lambda_{\mathrm{D}}^2\mathrm{MTTR}\,\tau\end{aligned} \tag{9.41}$$

For a more complex configuration, the length of formulas will increase exponentially since much more combinations need to be considered. Innal et al. (2016) propose a generic formula based on the method in IEC61508 (2010) for $\mathrm{PFD_{avg}}$ calculation of a *k*oo*n* voting structure with identical and independent barrier channels. According to Equation (8.14), we have known that for *k*oo*n* voting structure, the average unavailability can be calculated as

$$\mathrm{PFD}_{\mathrm{avg}}^{koon} = P_{n-k+1}^{n}\lambda_{\mathrm{D}}^{n-k+1}\prod_{i=1}^{n-k+1}\mathrm{MDT}_{1ooi} \tag{9.42}$$

where $P_{n-k+1}^{n} = n!/(k-1)!$ and

$$\mathrm{MDT}_{1ooi} = \alpha_{\mathrm{DD}}\mathrm{MTTR} + (1-\alpha_{\mathrm{DD}})\left(\frac{\tau}{i+1} + \mathrm{MRT}\right) \tag{9.43}$$

Considering two types of DU-failures, the above formula for $\mathrm{MDT}_{1ooi}$ can be generalized as follows:

$$\begin{aligned}\mathrm{MDT}_{1ooi} &= \alpha_{\mathrm{DD}}\mathrm{MTTR} + \alpha_{\mathrm{PPT}}(1-\alpha_{\mathrm{DD}})\left(\frac{\tau_{\mathrm{PPT}}}{i+1} + \mathrm{MRT_{PPT}}\right) \\ &+ (1-\alpha_{\mathrm{PPT}})(1-\alpha_{\mathrm{DD}})\left(\frac{\tau}{i+1} + \mathrm{MRT}\right)\end{aligned} \tag{9.44}$$

where $\text{MRT}_{\text{PPT}}$ is the mean repair time for a DU-fault revealed in a partial test. Then, the mean downtime for a *koon* voting structure can be calculated as

$$\text{MDT}_{koon} = \text{MDT}_{1oo(n-k+1)} = \alpha_{\text{DD}}\text{MTTR} + \alpha_{\text{PPT}}(1-\alpha_{\text{DD}})\left(\frac{\tau_{\text{PPT}}}{n-k+2} + \text{MRT}_{\text{PPT}}\right)$$
$$+(1-\alpha_{\text{PPT}})(1-\alpha_{\text{DD}})\left(\frac{\tau}{n-k+2} + \text{MRT}\right) \tag{9.45}$$

In terms of PPT, two factors are related to the availability of a barrier system: PPT coverage ($\alpha_{\text{PPT}}$) and PPT interval ($\tau_{\text{PPT}}$). PPT coverage is determined by system design, detection technology, and failure mode, and it can be determined based on historical knowledge and expert judgment. Lundteigen and Rausand (2008) have presented a procedure on determining PPT coverage. Users of barrier systems also need to specify PPT interval so as to comply with the integrity requirement, taking into consideration operational strategy and cost.

### 9.4.2 Petri Net Modeling for Partial Proof Tests

Innal et al. (2016) have built Markov models for analyzing the performance of barrier systems with different configurations under partial tests. We can also use RBD-driven Petri net to model barrier systems with PPTs. In the following models in Figure 9.11, we use DU1 to denote the DU-failure revealed in a partial test and DU2 to denote the DU-failure revealed in a full proof test to make the model more compact.

The model includes six parts, where the three parts on the left side are for modeling the behaviors of the barrier component of failure and restoration. When the token in place $\text{pa}_{\text{DW}}$ is absorbed by transition $\text{tr}_{\text{DD}}$, a DD-failure occurs. Similarly, absorbing the tokens in $\text{pa}_{\text{UW1}}$ and $\text{pa}_{\text{UW2}}$ by transitions $\text{tr}_{\text{DU1}}$ and $\text{tr}_{\text{DU2}}$ reflects the occurrences of DU-failures that can be revealed by a PPT and only by a FPT, respectively. It should be noted that we consider the repair time of DD failure to be exponentially distributed, so that $\text{tr}_{\text{DD}}$ is an empty rectangle. In this model, we introduce three binary variables, DD, DU1, and DU2, to denote the occurrences of different types of failures. When a failure occurs, the value of its associated variable is set as 1. When a fault is fixed, the value of its associated variable is set as 0.

On the right side of Figure 9.11, the upper two parts are for partial tests and full proof tests, respectively. Since the test intervals are constant, transitions $\text{tr}_{\text{PPT}}$ and $\text{tr}_{\text{FPT}}$ are represented with a filled rectangle. We here assume that after each of the tests, its fault is fixed.

The bottom block on the right side of the model is used to capture the barrier component unavailability. Given the predicate of transition $\text{tr}_{\text{SF}}$, if any of the DD-, DU1-, or DU2-failures occur, the barrier component enters the unavailable state with depositing a token in $\text{pa}_{\text{SF}}$. When all the faults are removed, the barrier returns to its functioning state.

One benefit of the RBD-PN model is the easiness of adopting the built model to more complex configurations without greatly increasing the model size. For example,

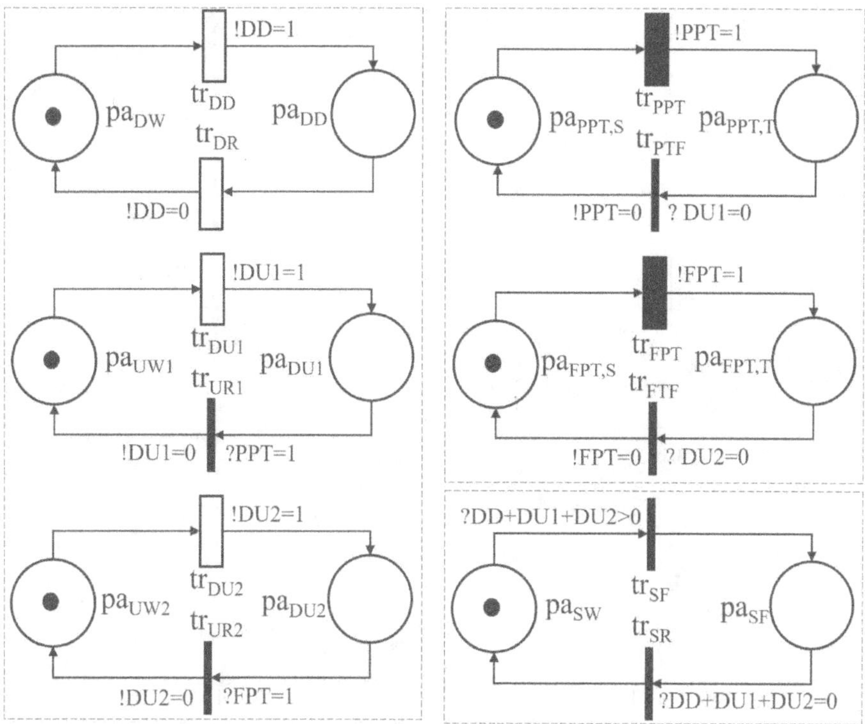


**FIGURE 9.11** RBD-PN for a barrier component with partial tests.

based on the model in Figure 9.11, we can build a model for a 2oo3 voting structure, as illustrated in Figure 9.12.

In the new model, upon the firing of $tr_{DD}$, $tr_{DU1}$, or $tr_{DU2}$, the corresponding variable DD, DU1, or DU2 undergoes a value change. It is important to note that the firing rates of these transitions depend on the number of tokens sojourning in their respective input places. For instance, in a scenario where all three channels within the barrier subsystem are functioning, the firing rate of $tr_{DD}$ is $3\alpha_{DD}\lambda_D$, and if one channel has been failed, the firing rate of $tr_{DD}$ decreases to $2\alpha_{DD}\lambda_D$.

It is noteworthy that this model is considered an approximation since it involves the possibility of more than three faults occurring simultaneously across the three channels. This approximation does not align with the assumption that only one type of fault can occur in one channel at a given time. Despite this simplification, its impact on the analysis is ignorable, given the low probability of such occurrences.

Unavailability of the 2oo3 barrier subsystem is also determined by the sojourn probability of a token in $pa_{SF}$. In Figure 9.12, transition $tr_{SF}$ becomes enabled when there are at least two failures, irrespective of whether they are DD, DU1, DU2, or a combination of them. On the other hand, when the number of faults is less than 2, transition $tr_{SR}$ becomes enabled, leading to the restoration of the barrier subsystem to a functioning state.

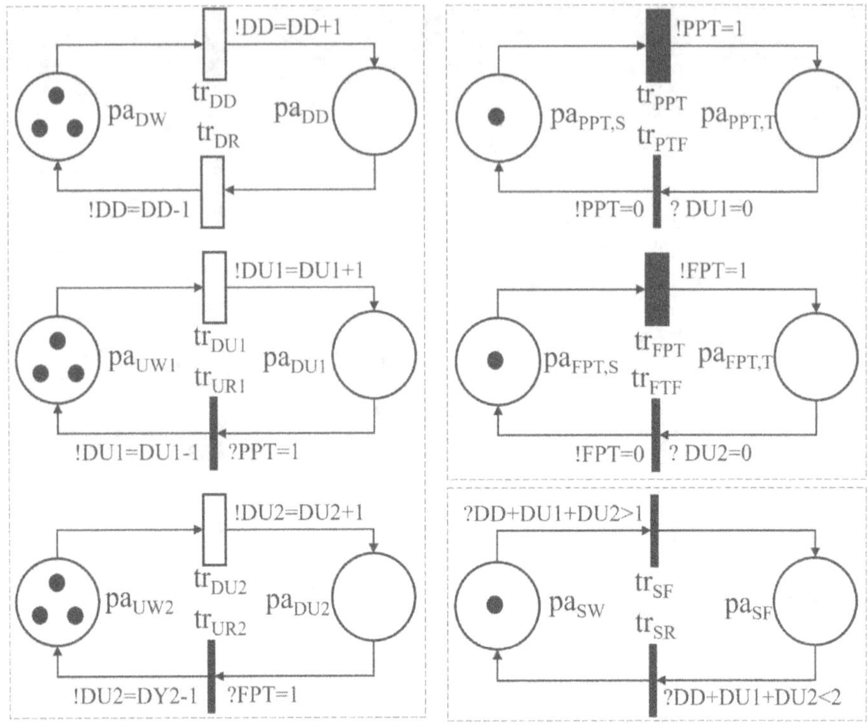


**FIGURE 9.12** RBD-PN for a 2oo3 barrier system with partial tests.

### Example 9.2:

### *Integrity of a 1oo2 Barrier Subsystem with Partial Tests*

Suppose there is a subsystem with a barrier subsystem with a 1oo2 voting configuration. A partial test is conducted every 30 days, and a full proof test is performed once every 5 years. The assumption is that these tests are efficiently executed on all channels within a very short timeframe. The failure rates for the two channels are set at $1.0 \times 10^{-5}$ per hour. The diagnostic coverage for the channels is 0.90, and the PPT coverage is 0.80.

To tailor the RBD-PN model shown in Figure 9.12 for this analysis, we remove one token each from $pa_{DD}$, $pa_{UW1}$, and $pa_{UW2}$, respectively. The predicates of $tr_{SF}$ and $tr_{SR}$ for the subsystem are modified as? DD + DU1 + DU2 > 1 and? DD + DU1 + DU2 > 2, respectively.

Using Equation (9.41), $PFD_{avg}$ can be calculated as $3.7131 \times 10^{-5}$, placing it in the SIL4. Both Markov analysis and RBD-PN simulation confirm the same integrity level. The primary contributor to the $PFD_{avg}$ is the coexistence of two DU-faults uncovered during the full proof test (68.89%), while the second most important contributor is the coexistence of one DU-fault revealed in the partial test and one DU-fault revealed in the full proof test (27.18%).

In the absence of partial proof testing ($\alpha_{PPT} = 0$), according to Equation (9.6), such a barrier subsystem has a $PFD_{avg}$ of $6.4579 \times 10^{-4}$. It can be found that introducing PPTs every 30 days reduces $PFD_{avg}$ by 94.25%. In fact, the resulting $PFD_{avg}$ is comparable to that of a barrier subsystem where only a proof test is conducted once per year ($2.6846 \times 10^{-5}$).

Therefore, in practical applications, it is important to compare the cost and risk associated with different testing methods. Evaluating which approach or combination of approaches is more cost-effective in achieving the required integrity level is very useful. ■

Wu et al. (2018) and Wu et al. (2019) have also used RBD-PN models for analyzing the integrity of barrier systems and evaluated the impacts of PPT coverage with case studies. Interested readers can reference these papers for more details. These models can also be used for situations of non-exponential distributions, and we will discuss these issues in Section 9.7.

### 9.4.3 Imperfect Tests

In this book, we define tests incapable of identifying all faults in a barrier system as *imperfect tests*. This classification encompasses both partial tests and diagnostic tests. However, the term "imperfect test" shows more significance when it is applied to the tests intentionally designed to overlook certain faults, either due to technical or cost constraints or without the awareness that some faults are being disregarded. Specifically, imperfect tests include three types:

- *Intended imperfect test*: This type of test is conducted with the specific purpose of examining only selected functions of a barrier component or system, deliberately omitting other functions. Diagnostic tests and PPT belong to this type of test. Imperfections of this type of test arise from limitations in cost, technology, or the need to consider risks during testing.
- *Unintended imperfect test*: This type of test fails to reveal all expected faults. The imperfections come from some constraints, such as limited testing time, operational limitations or errors, and an incomplete understanding of failure modes and mechanisms. Rausand et al. (2020) have discussed that part of the unavailability of a safety barrier is due to *systematic failure* that is not revealed by regular proof tests. Systematic failures are often related to defects in the design and errors in the specification and implementation of barrier hardware or software.
- *Harmful test*: This type of test poses a threat to the asset of interest or other property, leading to issues such as facility damage, harm to operators or the environment, or prolonged disruption to normal operations. Imperfections in this case result from poorly designed tests, incorrect testing procedures, and a lack of training.

We have introduced the quantitative analysis of the intended imperfect tests. For unintended imperfect tests, in the PDS handbook, Hauge et al. (2013) have presented the contributors to barrier unavailability beyond PFD, which include *probability of test-independent-failure* ($P_{TIF}$) and downtime unavailability (DTU), while DTU can

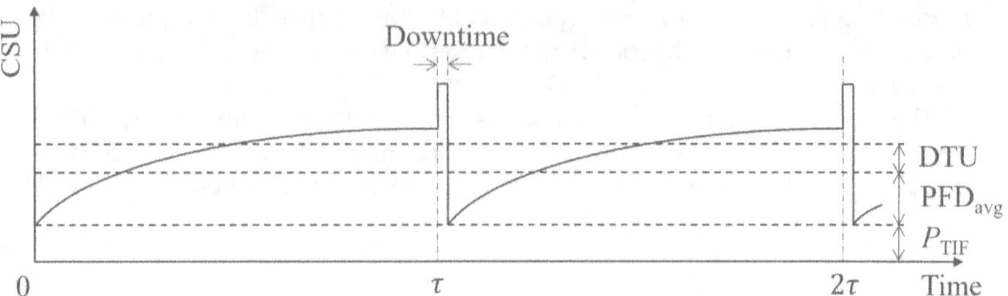


**FIGURE 9.13** Consideration of CSU and its elements.

be regarded as the fraction of the downtime in a proof test interval, and the downtime can be due to repair ($DTU_R$) and other planned downtime ($DTU_T$), such as planned testing time and maintenance time.

Then, the authors propose a new measure, *critical safety unavailability* (CSU), to quantify the overall loss of safety:

$$\text{CSU} = \text{PFD}_{\text{avg}} + P_{\text{TIF}} + \text{DTU} \tag{9.46}$$

The contributions of different parts of CSU can be explained by Figure 9.13. $P_{\text{TIF}}$ is a constant value independent from the operational time, and DTU is calculated as the fraction of downtime to the proof test interval $\tau$. Here we do not consider the partial test, but it is not difficult to divide PFD into two parts in such a calculation. Readers can derive the formulas by themselves.

## 9.5 MAINTENANCE STRATEGIES FOR BARRIERS

Until now, we simplified the maintenance activities in the integrity analysis of barriers, primarily emphasizing repair as the maintenance approach. The process of a repair involves addressing faults and restoring a failed barrier element to the as-good-as-new state. Nonetheless, real-world maintenance scenarios are always more complex considering the observed status of barriers and the requirements for cost reduction. This section will explore more frequently employed maintenance strategies for safety barriers.

In this book, the focus is on the maintenance of barriers with the purpose to ensure availability and satisfy performance requirements of barriers. Such maintenances are called *reliability-centered maintenances* (RCMs) in literature (Rausand & Høyland, 2004). Based on the time of conducting maintenance and decision-making approaches, maintenance can be classified into several types, as illustrated in Figure 9.14.

The *corrective maintenance* (CM), or *repair*, is the correction of the fault in an item after a failure has occurred. All the maintenance actions mentioned up to this point in the book are CMs, which are for both faults due to DD failures and DU failures on a barrier component.

*Preventive maintenance* (PM) is always needed to prevent or at least reduce the occurrences of failures and unexpected downtime. PM is thus conducted before

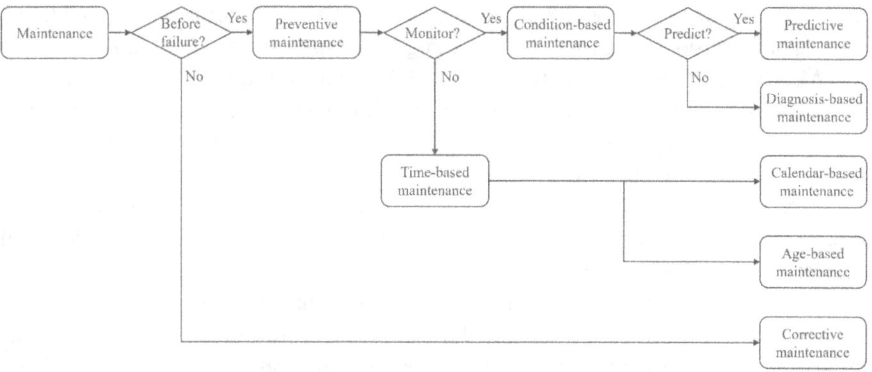


**FIGURE 9.14** Classification of maintenances.

failures and is according to a preset plan. A PM can be time-based (TPM) with the following two types:

- *Calendar-based PM (CPM)*: A PM is performed with a constant frequency or fixed calendar time interval, for example, once per year, regardless of the usage intensity and situation of the item. This strategy is similar to a regular proof test in terms of action timing.
- *Age-based PM (ABM)*: A PM is performed according to the usage age of an item rather than the calendar time lapsed. For example, a car requires maintenance after every 10,000 km driven, regardless of how much time has passed since its purchase.

In some cases, the coming PM can be determined in consideration of both dimensions: calendar time and usage age. However, in this book, we mainly consider CPM because this is the method used for safety barriers.

On the other hand, we have so far assumed that a barrier component only has two states: functioning or failing, and if the component is functioning, the performance is consistent no matter how much time it has been put into operation, namely, independent with time. However, such an assumption is not valid in consideration that many items, especially those mechanical components, are weaker and more vulnerable with time. For example, the long-term contact with an erosive medium can result in the erosion of the gate sealing area of a valve and then more leakages when the valve is closed, given the longer usage time. In this book, we regard the performance decrease with time as *degradation*.

It is possible to monitor the degradation process of an item given the development of sensor technologies. In this context, *condition monitoring* means the continuous or periodic observation on the system condition, and it is intended to measure the characteristics and parameters of the actual state of an item (Kobbacy & Murthy, 2008). Condition monitoring can be performed using human senses (such as view and smell), but more often relying on advanced sensing equipment. Assessment of the current situation is called *diagnosis*, but it should be noted that diagnosis in the previous chapters of this book only focused on the effort of finding a fault, and in this chapter, we consider diagnoses for detecting both faults and degradations.

PM activities can be proactively executed based on the insights obtained from condition monitoring and the assessment of the current state, commonly known as *diagnosis*. These types of maintenance actions are termed *condition-based maintenances* (CBMs). CBMs have obvious advantages since they can avoid the unnecessary interventions of TPMs, thereby reducing the risks of failures occurring prior to scheduled PMs. When maintenance is undertaken in response to a diagnostic result indicating necessity, it is referred to as *diagnosis-based maintenance*.

In practical scenarios, diagnostic results may not suggest immediate maintenance but imply a performance degradation that could lead to future failures. Maintenance crews are required to predict the potential degradation and determine when and how to conduct maintenance to avert impending issues. This approach is recognized as *predictive maintenance* (PdM), which can be defined as:

- *PdM*: A maintenance approach based on predictions or forecasts derived from the assessment of degradation of an item.

In reliability engineering, the term *prognosis* is frequently employed to denote the estimation of the remaining useful life (RUL), time to failure, failure probability, or the rate of a specific failure. This estimation is based on historical data and the prevailing conditions (Zhang, 2021). A closely related concept widely used in various applications is *prognosis and health management* (PHM), which essentially aligns with the concept of PdM.

PHM serves multiple objectives, one of which is the anticipation of the time remaining before a failure event or when a (barrier) system might lose its functionality. Another key goal is to predict the likelihood that an item can operate without encountering a failure up to a specified moment, such as the next proof test or planned PM. By providing these predictive insights, PHM facilitates timely and necessary interventions to address potential issues before they escalate into critical failures.

For further insights into the diverse types of maintenance, readers can refer to the works of Endrenyi et al. (2001) and Merizalde et al. (2019). The subsequent sections of this chapter will have a comprehensive discussion of how these maintenance strategies are considered, modeled, and analyzed within the context of barrier engineering. Furthermore, we will explore how these strategies influence the overall performance of barriers.

## 9.6 CORRECTIVE MAINTENANCE OF BARRIERS WITHOUT DEGRADATION

Our focus in this section is on the CMs, where repair time is no longer simplified as negligible. First, like the previous sections, we consider maintenance actions are performed after a fault is revealed. In the scenarios of barriers, these actions are needed in two situations considering the different failure modes:

- CM after a DD-fault is detected;
- CM after a DU-fault is revealed in a regular proof test.

We continue to adhere to the assumption of an exponential distribution concerning the time to failure of barriers. This assumption implies that the failure probability of these technical systems remains constant. Furthermore, during each proof test, the barrier is assumed to be in one of two states: either functioning or failed, with no intermediate states considered.

When certain barrier systems are deployed in remote or isolated locations, such as subsea environments or on islands, the duration between identifying a fault and completing the repair is often non-ignorable. In the case of subsea equipment, the calculation of MTTR for DD-fault or MRT for DU-fault necessitates a thorough consideration of various factors, including the lead time for spare parts, the travel time required for maintenance vessels, access time to subsea equipment, and seasonal time constraints. It is important to realize that the barrier function is lost throughout the entire CM duration considering the aforementioned factors. The DTU in Figure 9.13 in fact can come from such delays in CM, but the impacts of delays are not only the part reflected in the figure.

Based on the above considerations, we can extend the basic Markov model in Figure 9.1 for a single-barrier component by involving a new state (3) for modeling the noticeable duration in repair/CM, as illustrated in Figure 9.15.

When the barrier component is in state 2 (DU), the leaving rate from this state is also equal to $\mu_T = 2/\tau$ in average, and then the component enters state 3 (R). Given that the time in repair is not neglectable, the leaving rate from state 3 is regarded as the reciprocal of the mean time to repair for DU-fault, namely $\mu_{UR} = 1/\text{MRT}$. Since the mean time to restoration for a DU-fault is $\tau/2 + \text{MRT}$, $\mu_{DU}$ in Figure 9.1 is equal to $\mu_T\mu_{UR}/(\mu_T + \mu_{UR})$. The time to repair for a DD-fault also can be noticed, and it still is reflected by $\mu_{DD} = 1/\text{MTTR}$.

$\text{PFD}_{avg}$ is calculated as the sum of the sojourn probabilities in states 1–3, as

$$\text{PFD}_{avg} = \frac{\lambda_{DD}\mu_T\mu_{UR} + \lambda_{DU}\mu_{DD}\mu_{UR} + \lambda_{DU}\mu_{DD}\mu_T}{\mu_{DD}\mu_T\mu_{UR} + \lambda_{DD}\mu_T\mu_{UR} + \lambda_{DU}\mu_{DD}\mu_{UR} + \lambda_{DU}\mu_{DD}\mu_T} \tag{9.47}$$

When the duration of repair is very short, namely the value $\mu_{UR} \gg \mu_T$, this calculation can be approximated as the same formula for $\text{PFD}_{avg}$ in Section 6.4.3.2. However, if potential delay in repair is considered, a lower $\mu_{UR}$ or a longer repair delay can increase the value of $\text{PFD}_{avg}$. It is necessary to evaluate the impact of $\mu_{UR}$. For example, if it takes one month for preparing the repair, in other words, $\mu_{UR} = \frac{1}{\text{MRT}} = \frac{1}{30\times 24}$, we can re-calculate $\text{PFD}_{avg}$ using the same data in Example 6.3 for other parameters, as $8.92\times 10^{-3}$, 34.3% higher than the $\text{PFD}_{avg}$ when the repair time of DU-fault is 6 hours.

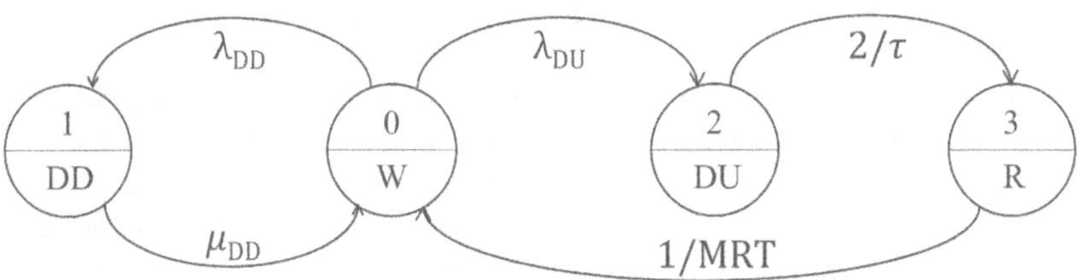


**FIGURE 9.15** The Markov model of a single barrier component with delayed repair.

In fact, the model in Figure 9.15 can be simplified and drawn as that in Figure 9.1, with considering $\mu_{DU} = \mu_T \mu_{UR} / (\mu_T + \mu_{UR})$. Since DU-fault is the main contributor, $PFD_{avg}$ of a single-barrier component can be approximated as

$$PFD_{avg} \approx \frac{MDT}{MTTF_{DU}} = \lambda_{DU}\left(\frac{\tau}{2} + MRT\right) \tag{9.48}$$

For a $koon$ voting structure, the mean downtime due the $(n-k+1)$th DU-failure can be approximated as $(\frac{\tau}{n-k+2} + MRT)$, and thus the general formula is

$$PFD_{avg}^{1oo(n-k+1)} = (\lambda_{DU}\tau)^{n-k} \cdot \left(\frac{\tau}{n-k+2} + MRT\right) \tag{9.49}$$

It can be found that the ratio of MRT and $\tau$ determines the contribution of duration of CM to the unavailability of barriers. This relationship suggests that, to lower $PFD_{avg}$, enhancing the reliability of barrier components may not be the only solution. An alternative strategy can be a thorough examination of the aforementioned factors that can influence MRT.

## 9.7 DEGRADATION OF SAFETY BARRIERS

In this section, we release the assumption of an exponential distribution concerning the time to failures. Barrier systems frequently incorporate mechanical components, the performance of which is influenced by factors such as material fatigue, corrosion, or other environmental stresses, and thus these components are susceptible to wear and aging, leading to a gradual decline in the performance of barrier functions. Consequently, a more reasonable modeling of failure time dynamics becomes necessary, moving beyond simplistic assumptions to consideration of the degradation processes within barrier systems.

### 9.7.1 Degradation Modes and Mechanisms

We define degradation of barrier components and systems as:

- *Degradation*: The process by which an (barrier) item becomes vulnerable with time and the performance deteriorates.

The possible degradation modes of a barrier component include:

- *Gradual degradation or normal degradation*: The continuous deterioration process resulting from mechanisms inherent in the operational context of a barrier. It is a persistent and ongoing process intrinsic to a barrier.
- *Abrupt degradation or sudden degradation*: A rapid decrease in the functionality of a barrier occurring within a short timeframe. This degradation is always triggered by unexpected external events.

For gradual degradation, the commonly potential degradation mechanisms include:

- *Corrosion*: Exposure to corrosive substances, atmospheric conditions, or chemicals can lead to the gradual deterioration of barrier components. Corrosion weakens materials, reducing their structural integrity and effectiveness.
- *Wear and tear*: Regular usage and operational stresses can cause mechanical components to experience wear and tear over time. Friction, stress, and repetitive movements contribute to the gradual degradation of surfaces and functional features.
- *Fatigue*: Repeated loading and unloading cycles, especially in dynamic actuating subsystems, can cause fatigue in materials. For example, this can result in microcracks, weakening the structural integrity of barrier components and potentially leading to failure.
- *Temperature variation*: Fluctuations in temperature, especially extreme variations, can impact the physical properties of materials. Thermal expansion and contraction may lead to stress, cracking, and other forms of degradation.
- *Aging*: Over time, even in the absence of external stressors, materials may undergo the aging processes that result in changes to their properties.

On the other hand, abrupt degradation occurs when a barrier component is subjected to demands for activating its barrier function and encounters severe environmental conditions, such as radiation, pollutants, higher temperature, pressure, or humidity. For instance, consider a shutdown valve with the task of stopping gas flow in a pipeline. During its operation under high pressure upon a demand, the potential misalignment between the valve gate and seat becomes a critical concern. This misalignment can further accelerate the existing wear process of the valve material. Over time, the wear can result in leaks or failure of the valve.

### 9.7.2 Degradation Modeling Approaches

In reliability engineering, degradation of an item is always studied in either of the three ways:

- *Physics-based modeling approach*: This modeling approach relies on physical parameters as the direct indicators of degradation, such as the length of a crack, trip time of a valve, or leakage rate. A predetermined threshold value for a physical parameter can be established to denote a failure. When degradation progresses to the defined threshold, a failure occurs. This approach involves monitoring and analyzing tangible aspects of the system that directly correlate with the physical degradation of barrier components.
- *Actuarial modeling approach or statistical approach*: In this approach, the foundation of analysis is based on probability distributions. Indicators used for quantification include reliability measures such as failure rate. The

fundamental assumption here is that as degradation intensifies, failure probability of the item increases. This approach relies on statistical principles to estimate the likelihood of failure based on historical data and probability distributions of relevant reliability measures.

- *Data-driven approach*: Measures used in reliability quantification are taken as indicators, such as MTTF. Unlike the actuarial modeling approach, this method does not make any presumptions about the probability distribution of these reliability measures. Instead, it relies on empirical data and statistical techniques to assess and predict reliability, making it particularly useful in situations where the underlying distribution is not easily discernible or subject to change over time.

In this section, we mainly discuss the physics-based modeling approach and the actuarial modeling approach. More specifically, we propose the following models for analyzing the impacts of degradation of barriers:

- *Discrete state model*: This model considers degradation as the intermediate state between functioning state and failed state. When an item is in the intermediate states, it can have different failure rates. In most cases, this model is based on the actuarial modeling approach.
- *Increasing hazard rate model*: It models degradation as increasing failure probability or failure rate along with time. This model is based on the actuarial modeling approach.
- *Incremental vulnerability model*: This model is a physics-based approach, regarding a closer performance to the failure threshold as degradation.

### 9.7.3 Discrete State Models

A simple method for modeling degradation is to add one "degraded" state (G) between the fully working state (W) and the failed state (DU). For example, based on the Markov method, Figure 9.16 can illustrate the development process of a barrier component from functioning to degraded and failed.

In this model, our focus is solely on degradation linked to a DU-fault. We use $\lambda_{UG1}$ and $\lambda_{UG2}$ to denote the transition rates from a fully functioning state to the degraded state and from the degraded state to the failed state, respectively. Introducing the possibility of a random event leading to the barrier failure, we also account for a

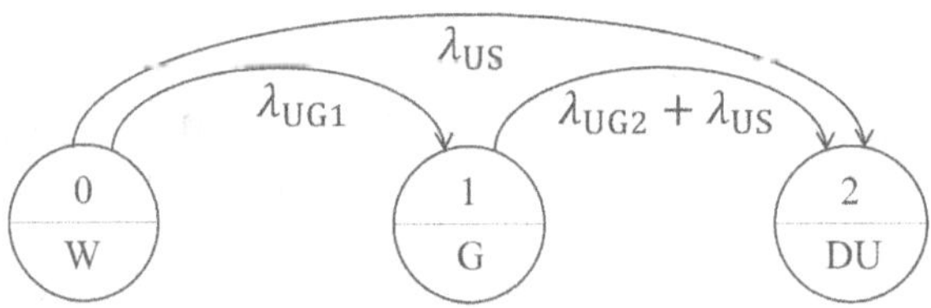


**FIGURE 9.16** The Markov model for degradation of a barrier component.

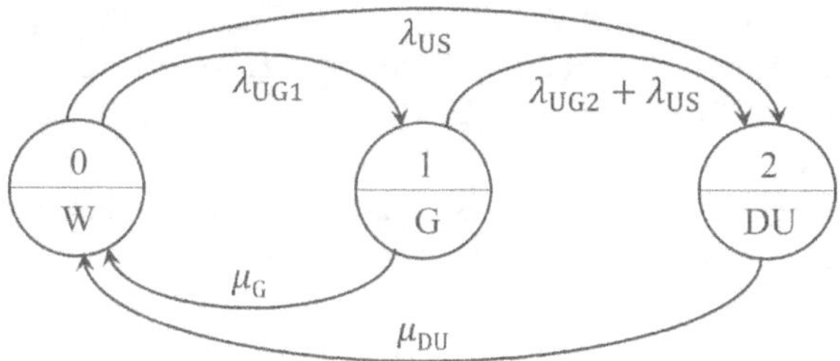


**FIGURE 9.17** The Markov model for degradation and maintenance of a barrier component.

direct transition from states 0 to 2 with a rate of $\lambda_{US}$. The same event can also occur when the barrier component is in the degraded state, so that the transition rate from states 1 to 2 is $\lambda_{UG2} + \lambda_{US}$.

For a barrier component, the DU-fault that can be revealed in a proof test. If we only rely on CM, meaning that the DU-fault is fixed after the proof test, as the arc from states 2 to 0 in Figure 9.17.

However, given that the degradation state has been included in the analysis, if we can perform PMs (more discussions in the next subsection), the degraded barrier component can be recovered to a fully working state before failure; there can be a transition from states 1 to 0, with a rate of $\mu_G$. In the context of barrier analysis, considering that the degradation is observed in a proof test, the barrier component can then be fixed with the transition rate $\mu_G = 2/\tau$.

Such a discrete state transition diagram is easy to understand. However, given that degradation represents a continuous process rather than a stable state, the designation of state 1 in Figures 9.16 and 9.17 is actually an artificial state, involving a performance range of a barrier, as depicted in Figure 9.18. This poses a challenge in specifying the exact performance level it denotes. To address this, it is essential to explicitly define the threshold values for entering state 1 and leaving state 1 (failure threshold). One simplified approach involves determining the center point between these two threshold values as the corresponding

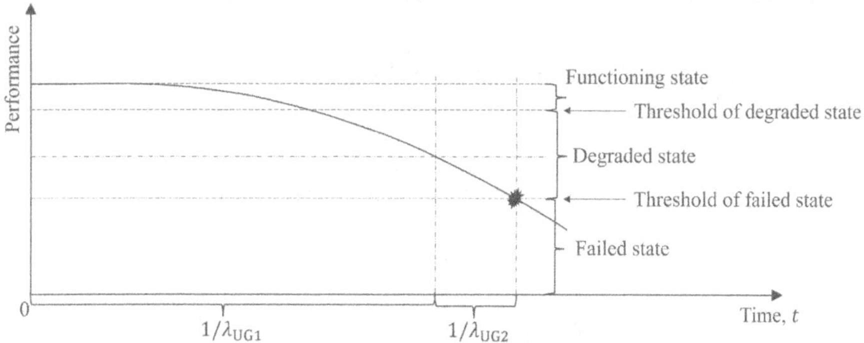


**FIGURE 9.18** Explanations of state 1 and the associated transition rates.

performance level for state 1. As illustrated in Figure 9.18, $\lambda_{UG1}$ is determined by the meantime from full functionality to the corresponding value of the center line of degraded performance, and $\lambda_{UG2}$ is the reciprocal of the meantime from the center line to failure.

The Markov model in Figure 9.17 also can be extended by introducing more intermediate states for modeling different phases of degradation.

### 9.7.4 Increasing Hazard Rate Model

Such kinds of models regard degradation as a process where failure probability keeps increasing. Under the assumption of exponential distribution, the failure rate of an item keeps as a constant independent with time. When we consider degradation, it is reasonable to allow the failure rate of an item to be a function of time. For example, the Weibull distribution (see Chapter 5) can be used for modeling such an increasing failure rate.

Wu et al. (2018) have provided approximation formulas for $\text{PFD}_{\text{avg}}$ of barrier components, the failure time of which follows the Weibull distribution $T$ ~ *Weibull*$(\alpha, \lambda)$. Here, the degradation of a barrier component implies that its failure rate increases with time. Considering DU-failure, according to Equation (5.35), $\lambda_{\text{DU}}(t) = \alpha\lambda_{\text{DU}}^{\alpha} t^{\alpha-1}$, and when $\alpha > 1$, $\lambda_{\text{DU}}(t)$ is a monotonic increasing function with time following the situation of degradation. The time-dependent unavailability in the first interval $(0, \tau)$ can be thus calculated as

$$\text{PFD}(t) = F(t) = 1 - e^{-(\lambda_{\text{DU}} t)^{\alpha}} = 1 - e^{-\frac{\lambda_{\text{DU}}(t)t}{\alpha}} \tag{9.50}$$

And the average failure rate $\lambda_{\text{avg}}(0, \tau)$ can be found as

$$\lambda_{\text{avg,DU}}(0, \tau) = \frac{1}{\tau}\int_0^{\tau} \lambda_{\text{DU}}(t)\,dt = \frac{1}{\tau}\int_0^{\tau} \alpha\lambda_{\text{DU}}^{\alpha} t^{\alpha-1}\,dt = \lambda_{\text{DU}}^{\alpha}\tau^{\alpha-1} \tag{9.51}$$

The average unavailability of a barrier component within $(0, \tau)$ is

$$\text{PFD}_{\text{avg}}(0, \tau) = \frac{1}{\tau}\int_0^{\tau}\left(1 - e^{-\frac{\lambda_{\text{DU}}(t)t}{\alpha}}\right)dt \approx \frac{1}{\tau}\int_0^{\tau} \frac{\lambda_{\text{DU}}(t)t}{\alpha}\,dt = \frac{\lambda_{\text{avg,DU}}(0, \tau)\tau}{\alpha + 1} \tag{9.52}$$

It should be noticed that the approximation above is valid when $\lambda_{\text{DU}}(t)t/\alpha$ is small, e.g., $< 0.1$.

Similarly, within the second proof test interval, the average PFD is

$$\text{PFD}_{\text{avg}}(\tau, 2\tau) = \frac{1}{\tau}\int_{\tau}^{2\tau} \text{PFD}(t)\,\text{d}t = \frac{1}{\tau}\int_{\tau}^{2\tau} \Pr(T \le t | T > \tau)\,dt \tag{9.53}$$

Then, we have

$$\mathrm{PFD}_{\mathrm{avg}}(\tau,2\tau)=\frac{1}{\tau}\int_{\tau}^{2\tau}\frac{\Pr(T\le t)-\Pr(T\le\tau)}{\Pr(T>\tau)}dt$$
$$=\frac{1}{\tau}\int_{\tau}^{2\tau}\frac{\left(1-e^{-\frac{z(t)}{\alpha}t}\right)-\left(1-e^{-\frac{z(t)}{\alpha}\tau}\right)}{e^{-\frac{z(t)}{\alpha}\tau}}dt \tag{9.54}$$

When $\lambda_{\mathrm{DU}}(t)t/\alpha$ is small, $\mathrm{PFD}_{\mathrm{avg}}(\tau,\,2\tau)$ can be approximated as

$$\mathrm{PFD}_{\mathrm{avg}}(\tau,2\tau)\approx\frac{1}{\tau}\int_{\tau}^{2\tau}\frac{\frac{\lambda_{\mathrm{DU}}(t)}{\alpha}t-\frac{\lambda_{\mathrm{DU}}(t)}{\alpha}\tau}{1-\frac{\lambda_{\mathrm{DU}}(t)}{\alpha}\tau}dt$$
$$=\frac{\frac{\lambda_{\mathrm{avg,DU}}(\tau,2\tau)}{\alpha+1}\cdot\frac{(2\tau)^{\alpha+1}-\tau^{\alpha+1}}{(2\tau)^{\alpha}-\tau^{\alpha}}-\frac{\lambda_{\mathrm{DU}}(\tau)\tau}{\alpha}}{1-\frac{\lambda_{\mathrm{DU}}(\tau)\tau}{\alpha}} \tag{9.55}$$

While for the $k$th test interval, the average failure rate will be

$$\lambda_{\mathrm{avg,DU}}(\tau_{k-1},\,\tau_k)=\frac{1}{\tau_k-\tau_{k-1}}\int_{\tau_{k-1}}^{\tau_k}\lambda_{\mathrm{DU}}(t)dt=\lambda_{\mathrm{DU}}^{\alpha}\frac{\tau_k^{\alpha}-\tau_{k-1}^{\alpha}}{\tau_k-\tau_{k-1}} \tag{9.56}$$

With the similar derivation processes above, the approximation formula of $\mathrm{PFD}_{\mathrm{avg}}$ within $((k-1)\tau,\,k\tau)$ can be obtained as (Wu et al., 2018)

$$\mathrm{PFD}_{\mathrm{avg}}\left((k-1)\tau,k\tau\right)\approx$$
$$\frac{\frac{\lambda_{\mathrm{avg,DU}}\left((k-1)\tau.k\tau\right)}{\alpha+1}\cdot\frac{(k\tau)^{\alpha+1}-\left((k-1)\tau\right)^{\alpha+1}}{(k\tau)^{\alpha}-\left((k-1)\tau\right)^{\alpha}}-\frac{\lambda_{\mathrm{DU}}\left((k-1)\tau\right)}{\alpha}\cdot(k-1)\tau}{1-\frac{\lambda_{\mathrm{DU}}\left((k-1)\tau\right)}{\alpha}\cdot(k-1)\tau} \tag{9.57}$$

It can be found that when $\alpha=1$, the Weibull distribution becomes an exponential distribution, and $\mathrm{PFD}_{\mathrm{avg}}$ in the in the $k$th interval is always equal to $\lambda_{\mathrm{DU}}\tau/2$.

An entire PM process includes proof testing and inspection and the following actions with the purpose to reduce the occurrences of failures and unexpected downtime. The above formulas can be used for analyzing the impacts of PM, especially the interval of PM ($\tau$), on the integrity of barrier systems with degradation. Here we actually assume that a PM can restore the barrier system to the as-good-as-new state.

Such approximation formulas are still very complex, and they are even more difficult to understand when applying for 1oo2 and 2oo3 voting structures. In the analysis of such systems, we can rely more on discrete state models. In more general way, the model of the degradation process can be generalized from the HPP to the non-homogeneous Poisson process (NHPP), where the failure rate of an item can be a function of time.

It should be noted that the parameters in the models in this subsection are design parameters. It is difficult to connect these parameters with the observed situations in inspections and tests during the operational phase since the collected data only shows the degradation progress, such as crack length and corrosion degree, but does not indicate failure rate. Such data cannot be used to estimate the parameters in the Weibull distribution, or NHPP. In addition, the impacts of external shocks are not taken into account in the above approximation formulas.

Based on the increasing hazard rate model, we can integrate aging degradation and random failures into one formula. In such a model, failure mode A is characterized by a random failure that adheres to the exponential distribution, and failure mode B is degradation-based failure that follows the Weibull distribution. A barrier component fails upon the occurrence of either failure mode, and thus the time until failure of the component is the minimum of the two times to failure for failure modes A and B.

The probability of failure of the barrier component by $t$ thus is

$$\begin{aligned} F(t) &= 1-\left[\Pr(T_A > t)\cdot \Pr(T_B > t)\right] = 1-\left(e^{-\lambda_{DU,A}t}\cdot e^{-\lambda_{DU,B}^{\alpha}t^{\alpha}}\right) \\ &= 1-e^{-\left(\lambda_{DU,A}t+\lambda_{DU,B}^{\alpha}t^{\alpha}\right)} \approx \lambda_{DU,A}t+\lambda_{DU,B}^{\alpha}t^{\alpha} \end{aligned} \tag{9.58}$$

Then, the average PFD is the interval $[0, \tau)$ is

$$\text{PFD}_{avg} = \frac{1}{\tau}\int_0^{\tau}\left(\lambda_{DU,A}t+\lambda_{DU,B}^{\alpha}t^{\alpha}\right)dt = \frac{\lambda_{DU,A}\tau}{2}+\frac{z_{avg}(0,\tau)\tau}{\alpha+1} \tag{9.59}$$

where $z_{avg}(0, \tau)$ is the average rate of the DU-failure due to degradation in the interval.

While for a 1oo2 configuration, each channel has two failure modes. Let $T_1$ and $T_2$ denote the time of channels 1 and 2, respectively. This is necessary because even though these two channels are identical, they may fail with different failure modes. The failure probability of this configuration is calculated as

$$F_S(t) = F_1(t)F_2(t) = \left(\lambda_{DU,A}t+\lambda_{DU,B}^{\alpha}t^{\alpha}\right)^2 \tag{9.60}$$

Similarly, the average PFD is the interval $[0, \tau)$ is

$$\begin{aligned} \text{PFD}_{avg} &= \frac{1}{\tau}\int_0^{\tau}\left(\lambda_{DU,A}t+\lambda_{DU,B}^{\alpha}t^{\alpha}\right)^2 dt = \frac{\left(\lambda_{DU,A}\tau\right)^2}{3} \\ &+ \frac{\left(z_{avg}(0,\tau)\tau\right)^2}{2\alpha+1}+\frac{2\lambda_{DU,A}z_{avg}(0,\tau)\tau^2}{\alpha+2} \end{aligned} \tag{9.61}$$

This formula can be explained as the average PFD of the 1oo2 barrier subsystem is the sum of $PFD_{avg}$ contributed by two random failures, and $PFD_{avg}$ contributed by degradation-based DU-failures, and $PFD_{avg}$ contributed by two failure modes.

### 9.7.5 Incremental Vulnerability Model

When a specific physical parameter can be observed as an indicator of performance, we can use the changes in parameter value to model a degradation process. The failure of a barrier is regarded as the event that the value of the selected parameter arrives at the threshold. In reliability studies, independent increment processes, such as the Wiener process,[1] Gamma processes, and inverse Gaussian processes, have been widely applied in degradation modeling (Nicolai et al., 2007). By nature, degradation processes are with fluctuations over time, and all the stochastic processes above are able to capture and quantify these uncertainties.

#### 9.7.5.1 Modeling Normal Degradation

For a degradation model based on the conventional Wiener process, we can let the degradation process $\{\text{RUL}(t), t \geq 0\}$ be formulated as

$$G(t) = g_0 + wt + \sigma_B B(t) \tag{9.62}$$

where $g_0$ represents the initial performance level of the parameter of interest, $w$ is the *drift coefficient* for capturing the degradation rate, $\sigma_B$ is called the *diffusion coefficient*, and $\{B(t), t \geq 0\}$ is a standard one-dimensional Brownian motion representing stochastic dynamics in the degradation process (Zhang et al., 2018). Thus, the Wiener process is also regarded as *Brownian motion with linear drift.*

The Wiener process can be more fitting for the non-monotonous degradation processes. In these instances, factors like maintenance interventions or self-healing mechanisms can potentially reverse the degradation process, allowing for intermittent improvements or recovery (Zhang et al., 2018). However, the basic Wiener process is not applicable for multi-phase degradation that can occur in a system involving multiple components, or in a barrier system considering multiple test intervals.

When degradation can be regarded as cumulative damage, the Gamma process is an appropriate model (Alaswad & Xiang, 2017). As we have discussed in Chapter 5, the Gamma process is a continuous-time stochastic process where the increments in disjoint time intervals are independent variables. According to Equation (5.29), the increment $G(t_2) - G(t_1)$ has the density function as

$$f_{\alpha(t_2-t_1),\lambda}(x) = \frac{\lambda^{\alpha(t_2-t_1)}}{\Gamma[\alpha(t_2-t_1)]}(\lambda x)^{\alpha(t_2-t_1)-1} e^{-\lambda x} \text{ for } g > 0 \tag{9.63}$$

where $\Gamma(\cdot)$ is called the *Gamma function*, and $\lambda$ here reflects the degradation rate. $\Gamma(\alpha)$ is defined as in Equation (5.30).

When a physical parameter is specified, we can use $G(t)$ to denote the time dependence of this parameter and $G_L$ to denote the threshold value of failure. $G(t)$ can be

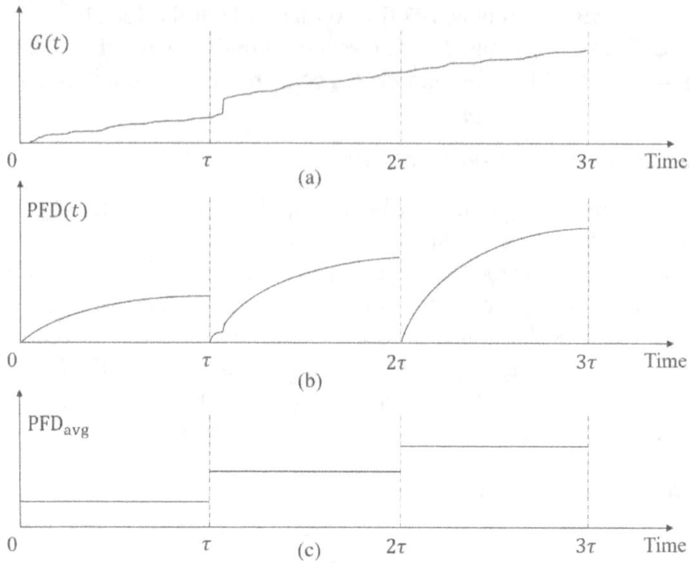


**FIGURE 9.19** Illustration of time-dependent degradation, PFD($t$) and $\text{PFD}_{avg}$.

a cumulative damage on a barrier component, as shown in Figure 9.19(a), and the cumulative density function (CDF) of $G(t)$ is

$$F_{G(t)}(G_L) = \Pr\left[G(t) < G_L\right] = \int_0^{G_L} f_{G(t)}(u)du = \frac{\gamma(\alpha t, \lambda G_L)}{\Gamma(\alpha t)} \tag{9.64}$$

where $\gamma$ is defined according to Equation (5.32) as

$$\gamma(\alpha, G_L) = \int_0^{G_L} x^{\alpha-1} e^{-x}\, dx,\; G_L \geq 0,\; \alpha > 0 \tag{9.65}$$

For the barrier, its instantaneous PFD($t$) within the first test interval $[0, \tau)$ can be described as

$$\text{PFD}_1(t) = 1 - \Pr\left[G(t) < G_L\right] = \Pr\left[G(t) \geq G_L\right] \tag{9.66}$$

The curve of PFD($t$) in such a scenario is illustrated in Figure 9.19(b). In the second interval, the degradation status observed in the first proof test ($G_1$) acts as the input information for estimating PFD. Thus, for $\tau \leq t < 2\tau$, the instantaneous unavailability can be calculated as:

$$\text{PFD}_2(t) = \Pr\left[G(t) \geq G_L \middle| G(\tau) = G_1, G_1 < G_L\right] \tag{9.67}$$

Similarly, for the $i$th interval, PFD($t$) is calculated as

$$\mathrm{PFD}_i(t) = \Pr\left[G(t) \geq G_L \middle| G((i-1)\tau) = G_{i-1}, G_{i-1} < G_L\right] \tag{9.68}$$

Then we can calculate the average PFD in each interval, as shown in Figure 9.19(c). It should be noted that Figure 9.19 is not derived from precise calculations, and its purpose is solely to illustrate the changing trends of the three measures.

The inverse Gaussian process is another stochastic model and is particularly useful when the degradation process exhibits a rapid increase initially, followed by a slower, long-term degradation. This model is suitable for capturing diverse degradation patterns.

It should be noted that the Gamma process and the inverse Gaussian process are well-suited for modeling a monotonous degradation process. In such scenarios, the progression of deterioration or damage is irreversible, implying that recovery or restoration is not possible. Thus, the Gamma process and the inverse Gaussian process can model the degradation process with some minimal repairs that do not change the damage level, but they cannot model the situation where an overhaul exists that resets an item to the as-good-as-new state.

Readers who have interests in the inverse Gaussian process and other models can reference the review paper by Shahraki et al. (2017).

#### 9.7.5.2 Modeling Sudden Degradation

The abrupt shift observed in the degradation curve depicted in Figure 9.19(a) represents the influence of a demand on the barrier component. This impact is similarly illustrated in the corresponding position on Figure 9.19(b). It is important to note that Figure 9.19(b) is a rough approximation as it does not effectively depict the variation in instantaneous unavailability with increasing degradation. However, it does capture the overall trend.

In such a case considering demands, $G(t)$ is the cumulation of the normal degradation by the time $G_N(t)$ and the sudden degradation by the time $G_S(t)$. Supposing that demands occur following a homogeneous Poisson process with the rate of $\lambda_{\mathrm{DE}}$, and the number of demands occur by time $t$ is $N_S(t)$, according to Equation (5.49), the probability that $n$ demands appear in the time interval of $[0, t)$ is

$$\Pr\left[N_S(t) = n\right] = \frac{(\lambda_{\mathrm{DE}} t)^n}{n!} e^{-\lambda_{\mathrm{DE}} t}$$

Then Equation (9.66) can be re-written by considering two types of degradations, as

$$\mathrm{PFD}_1(t) = \Pr\left[G(t) \geq G_L\right] = \sum_{i=0}^{\infty} \Pr\left[G_N(t) + G_S(t) \geq G_L \middle| N_S(t) = i\right] \Pr\left(N_S(t) = i\right) \tag{9.69}$$

If all demands are assumed to be non-negative (in terms of producing damage), independent, and Gamma-distributed, and the degradation in the $i$th demand is denoted as $g_i$.The cumulative damage or degradation by time $t$ is

$$G_S(t) = \begin{cases} 0, \ if \ N_S(t) = 0 \\ \sum_{i=1}^{N_S(t)} g_i, \ if \ N_S(t) > 0 \end{cases} \tag{9.70}$$

$G_S(t)$ also follows the Gamma distribution as $G_S \sim Gamma(\delta_i, \rho)$, where $\delta_i$ is the shape parameter for denoting the damage level.

Take a convolution integral in Equation (9.66), and let $f_{gi}^n$ as the probability density function of the sum of $n$ independent and identically distributed demands, then PFD($t$) in the interval of [0, $t$) can be calculated as the probability that total degradation by time $t$ is higher than the threshold value $G_L$:

$$\text{PFD}_1(t) = 1 - R_1(t) = \sum_{i=1}^{\infty} \left[ \int_0^{G_L} G_S(G_L - u, t) f_{gi}^n \, du \right] \frac{(\lambda_{\text{DE}} t)^i}{i!} e^{-\lambda_{\text{DE}} t} \tag{9.71}$$

For a 1oo2 configuration, the system fails when the degradations of both channels arrive at the threshold. We use $G^j(t)$ to denote the degradation of channel $j$ by time $t$, and so that in the first test interval of [0, $\tau$), we have

$$\text{PFD}_1(t) = \Pr\left[G^1(t) \geq G_L\right] \cap \Pr\left[G^2(t) \geq G_L\right] \tag{9.72}$$

While the survival probability is

$$R_1(t) = \Pr\left[G^1(t) < G_L\right] \cap \Pr\left[G^2(t) < G_L\right] \tag{9.73}$$

Assuming that each demand has the same size of damage on both channels, $R(t)$ can be calculated using Equation (9.71) as

$$\begin{aligned} R_1(t) &= \int_0^{G_L} \left[ 1 - \prod_{j=1}^{2} \left(1 - \Pr(G^j(t) < G_L) | N_S(t)\right) \right] f_{G_S}(x) dx \\ &= \left[ 1 - \left( 1 - \frac{\gamma(\alpha t, \lambda G_L)}{\Gamma(\alpha t)} \right)^2 \right] e^{-\lambda_{\text{DE}} t} \\ &+ \sum_{n=1}^{\infty} \int_0^{G_L} \left[ 1 - \left( 1 - \frac{\gamma(\alpha t, \lambda G_L)}{\Gamma(\alpha t)} \right)^2 \right] \cdot \frac{\rho^{n\delta} u^{n\delta - 1} e^{-\rho x}}{\Gamma(n\delta)} dx \cdot \frac{(\lambda_{\text{DE}} t)^n e^{-\lambda_{\text{DE}} t}}{n!} \end{aligned} \tag{9.74}$$

Some derivation steps in Equation (9.74) are skipped, and readers who have interest can refer to the paper by Zhang et al. (2019).

Then, the average PFD in the first proof test interval of $[0, \tau)$ can be calculated as

$$\mathrm{PFD}_{\mathrm{avg},1} = 1 - \frac{1}{\tau}\int_0^{\tau} R(t)\,dt = 1 - \frac{1}{\tau}\int_0^{\tau}\left\{\left[1-\left(1-\frac{\gamma(\alpha t, \lambda G_L)}{\Gamma(\alpha t)}\right)^2\right]e^{-\lambda_{\mathrm{DE}}t}\right.$$
$$\left. + \sum_{n=1}^{\infty}\int_0^{G_L}\left[1-\left(1-\frac{\gamma(\alpha t, (G_L - g)\lambda)}{\Gamma(\alpha t)}\right)^2\right]\cdot\frac{\rho^{n\delta}u^{n\delta-1}e^{-\rho x}}{\Gamma(n\delta)}dx\cdot\frac{(\lambda_{\mathrm{DE}}t)^n e^{-\lambda_{\mathrm{DE}}t}}{n!}\right\}dt \tag{9.75}$$

In the proof test conducted at time $\tau$, if a channel is found functioning but with unknown degradation level, its instantaneous PFD in the subsequent proof test interval is the conditional probability that its lifetime is longer than $\tau$, as

$$\mathrm{PFD}_2(t) = \Pr(T < t | T > \tau) = 1 - \frac{R(t)}{R(\tau)} \tag{9.76}$$

Then,

$$\mathrm{PFD}_{\mathrm{avg},2} = \frac{1}{\tau}\int_{\tau}^{2\tau}\mathrm{PFD}_2(t)\,dt = 1 - \frac{1}{\tau}\int_{\tau}^{2\tau}\frac{R(t)}{R(\tau)}dt \tag{9.77}$$

Similarly, for the $i$th interval of $[(i-1)\tau, i\tau)$, the instantaneous PFD is calculated as

$$\mathrm{PFD}_i(t) = \Pr\left[T < t | T > (i-1)\tau\right] = 1 - \frac{R(t)}{R\left[(i-1)\tau\right]} \tag{9.78}$$

And,

$$\mathrm{PFD}_{\mathrm{avg},i} = \frac{1}{\tau}\int_{(i-1)\tau}^{i\tau}\mathrm{PFD}_i(t)\,dt = 1 - \frac{1}{\tau}\int_{(i-1)\tau}^{i\tau}\frac{R(t)}{R\left[(i-1)\tau\right]}dt \tag{9.79}$$

Readers can find numerical examples of integrity analysis for 1oo2 barrier systems with normal and sudden degradations in the paper by Zhang et al. (2019). The authors also have explored the impacts of threshold value, demand frequency, and demand severity (damage level or acceleration in degradation) on barrier integrity and propose that a flexible proof test strategy can be introduced by reducing testing frequency at the early phase of the operational lifetime of the barrier to reduce testing cost while keeping a sufficient integrity level.

## 9.8 CONDITION-BASED MAINTENANCE OF BARRIERS

For barrier systems, monitoring equipment or routine proof tests can be employed to detect the degradation of components. Based on the observed condition, decisions can be made regarding the necessity and timing of PM. This methodology is known as CBM for barriers.

According to Tsang (1995), three types of decisions are needed to make in CBM, including:

- *Selecting the condition parameters to be monitored*: This involves a careful analysis of the critical aspects that directly impact the performance and integrity of a barrier system. The choice of condition parameters depends on the nature of the barrier system and the potential failure modes, and some examples of condition parameters include vibration level, leakage rate, and trip time.
- *Determining the monitoring methods and frequency*: The options include continuous and online monitoring, as well as regular and offline proof tests. This decision needs to ensure an effective approach to gathering data on the health status of a barrier component, but find a balance at the same time between monitoring effectiveness and operational continuity.
- *Establishing the threshold value of the parameters for triggering maintenance*: This is to define the limits beyond which the observed conditions indicate a need for intervention. It also needs to find an optimal point in setting the threshold to prevent unnecessary interventions while effectively addressing emerging issues.

For DD-failure-related degradation, the modeling and analysis methods are the same as those for normal systems. There are many literatures studying CBM; for example, readers can refer to the review papers and survey by Ahmad and Kamaruddin (2012), Shin and Jun (2015), and Alaswad and Xiang (2017) for the procedure, methods used, and recent developments in CBM. Here, we still focus on the degradation that can lead to DU-failures. In the last subsection, we assumed that degradation remains undetected in a proof test. However, if the status of a barrier component can be thoroughly examined during the test, certain actions should be taken to stop or alleviate the degradation process. Figure 9.20 illustrates the degradation process with mitigation through CBM following a proof test.

### 9.8.1 Maintenance Threshold

In Figure 9.20, the horizontal line close to $G_L$ represents the threshold for triggering PM. If during a proof test the degradation level is found to surpass this specified

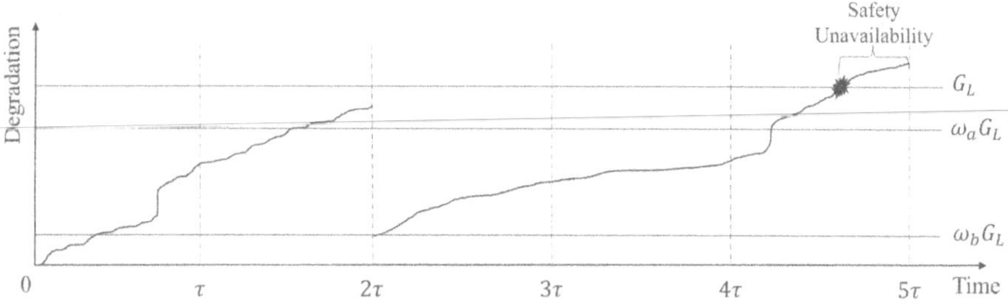


**FIGURE 9.20** Degradation with a barrier with regular proof test and maintenance.

value, the maintenance crew needs to promptly implement actions to rectify the barrier component and restore it to a healthier state with less degradation. We denote this *maintenance threshold* value as $\omega_a G_L$, where $0 < \omega_a < 1$. For a barrier component, given that in a proof test, it is observed that the degradation level $G(t)$ is less than the $\omega_a G_L$, no maintenance activity is needed. In such a situation, the operational cost is only related to the proof test, denoted as $C_{\text{PT}}$. While if the degradation is higher than $\omega_a G_L$, a PM is conducted. Let $C_{\text{PM}}$ represent the PM cost, and the total cost becomes to $C_{\text{PT}} + C_{\text{PM}}$. Normally, $C_{\text{PM}} > C_{\text{PT}}$.

It can be found that a lower value of $\omega_a$ indicates a more stringent control principle for degradation, leading to earlier or more frequent PM. While such a maintenance strategy can effectively lower the average PFD, it comes with the trade-off of increased interruptions and higher maintenance costs.

Another threshold value is $\omega_b G_L$, denoting the degradation level that the item goes back after a PM. It is essential to note that $0 \le \omega_b < \omega_a < 1$. When $\omega_b = 0$, the maintenance is perfect, indicating the restoration of the barrier to the as-good-as-new state. Otherwise, the maintenance is considered imperfect, resulting in a faster increase in the PFD post-maintenance compared to the initial interval. A lower value of $\omega_b$ always means a lower PFD but a higher PM cost $C_{\text{PM}}$, which can be expressed as a function of $\omega_b$, $C_{\text{PM}}(\omega_b)$.

The potential loss in an accident resulting from the failure of a safety barrier is $C_{\text{LA}}$, and thus the cost associated with the risk of barrier failure is calculated as the product of $C_{\text{LA}}$ and probability of accident. This probability, in turn, is determined by the product of downtime of the barrier and the demand rate.

In a barrier lifetime of $n$ proof test intervals, the total lifecycle cost ($C_{\text{TL}}$) can be described as a function of $\omega_a$ and $\omega_b$:

$$C_{\text{TL}}^{(\omega_a,\omega_b)} = nC_{\text{PT}} + C_{\text{PM}}(\omega_b)\sum_{i=1}^{n}\Pr\left[G(i\tau) \ge \omega_a G_L\right] + C_{\text{LA}}\sum_{i=1}^{n}\lambda_{\text{DE}}\tau\text{PFD}_{\text{avg},i} \quad (9.80)$$

$\text{PFD}_{\text{avg},i}$ in the above equation is also a function of $\omega_a$ and $\omega_b$. We calculate the instantaneous PFD in each proof test interval as the conditional probability that the barrier component is failed given that its degradation level in the last proof test is observed as $G_0$, with $G_0 \in [0, \omega_a G_L)$, namely for the $i$th proof test interval, there is no maintenance action, and thus we have

$$\text{PFD}(t) = \Pr\left[G(t) \ge G_L \middle| G(i-1)\tau = G_0\right], \text{for } (i-1)\tau < t < i\tau, G_0 \in \left[0, \omega_a G_L\right) \quad (9.81)$$

In other words, if the development of degradation ($G$) in this proof test interval, as a variable, is higher than $G_L - G_0$, the barrier will become unavailable. Integrating with Equation (9.79), the average unavailability in the proof interval considering both gradual degradation and can $k$ times of abrupt degradations be calculated as

$$\text{PFD}_{\text{avg},i} = \frac{1}{\tau}\int_{(i-1)\tau}^{i\tau}\text{PFD}(t)dt = \frac{1}{\tau}\sum_{k=1}^{\infty}\left[\int_{G_0}^{G_L} G_S(G_L - u, t) f_{gk}^{n}\, du\right]\frac{(\lambda_{\text{DE}}t)^k}{k!}e^{-\lambda_{\text{DE}}t} \quad (9.82)$$

In the case $G_0 \in [\omega_a G_L, G_L)$, the degradation will be reduced to the level of $\omega_b G_L$, and thus,

$$\text{PFD}(t) = \Pr\left[G(t) \geq G_L | G(i-1)\tau = \omega_b G_L\right], \text{ for } (i-1)\tau < t < i\tau,\ G_0 \in \left[\omega_a G_L, G_L\right) \tag{9.83}$$

The calculation is complex; Monte Carlo simulation can be used here by generating random events to obtain degradation scenarios. Zhang et al. (2023) have provided the details of steps for such a Mante Carlo simulation. The Markov model is also an alternative approach, which will be used in the next subsection.

The article by Zhang et al. (2020) discusses how to optimize CBM with setting the values of $\omega_a$ and $\omega_b$, and using a cost function like Equation (9.80). There are some useful findings:

- When $\omega_a$ is fixed, the lifecycle cost decreases with the higher failure threshold $G_L$, but reducing $G_L$ can result in increased downtime.
- When the failure threshold $G_L$ is given, $\text{PFD}_{\text{avg}}$ is reduced with $\omega_a$.
- Parameter $\omega_b$ has slight impact on $\text{PFD}_{\text{avg}}$.
- Non-periodically adaptive proof testing approach can be introduced for controlling maintenance cost while keeping the required integrity level. We will discuss this approach in the next subsection.
- The ratios of proof test cost $C_{\text{PT}}$, PM cost $C_{\text{PM}}$, and accident costs $C_{\text{LA}}$ have a significant impact on the lifecycle cost profile and thus CBM strategy.

More details and numerical case studies for the impacts of PM triggering threshold and PM completion degree can be found in Zhang et al. (2020).

### 9.8.2 Adaptive Proof Test

Given that degradation is in consideration, when the barrier is observed in a proof test that it is approaching failure due to degradation, the next proof test will be conducted earlier. Otherwise, if the degradation is found not serious yet, a longer interval before the next test can be accepted to reduce testing costs. Such an approach can be called *adaptive proof test*, or *condition-based test planning*.

In this subsection, we consider a barrier system characterized by $m+1$ states, within a state space $\Omega = \{0, 1, \ldots, m\}$, with these states being observed during proof tests. The sojourn probabilities in these states can be denoted as a vector $\mathbf{P}(t) = \{P_0(t), P_1(t), \ldots, P_m(t)\}$. The states of the barrier can be generally categorized into three distinct sets: Set $\Omega_W$ comprises the fully functioning states, set $\Omega_G$ encompasses the degraded states $G_L$), and set $\Omega_{DU}$ includes the DU-failed states (degradation level is higher than $G_L$). Each set may consist of one or more states, and the time-dependent probabilities of the barrier component residing in these state sets are denoted as $P_W(t)$, $P_G(t)$, and $P_{DU}(t)$.

Based on the consideration in the last subsection, if we consider the two thresholds of maintenance, it is reasonable to further divide the set $\Omega_G$ into two subsets: slight degradation $\Omega_{G1}$ (from $\omega_b G_L$ to $\omega_a G_L$) that does not trigger PM, and serious degradation $\Omega_{G2}$ (from $\omega_a G_L$ to $G_L$). For a 1oo2 configuration of barrier system, the state that both two channels are failed belongs to $\Omega_{DU}$, while we can regard the following three states: One channel is degraded + one channel is failed, one channel is fully functioning + one channel is failed, and two channels are degraded, belonging to $\Omega_{G2}$, and the state that one channel is fully functioning + one channel is degraded belongs to $\Omega_{G1}$, and the state that two channels are functioning belongs to $\Omega_W$.

We use $\tau_W$ and $\tau_G$ to represent the proof test intervals when the barrier is within the $\Omega_W$ and $\Omega_G$, respectively. It is inherent that even if $\tau_W \geq \tau_G$, the barrier can still be at the same integrity level considering the initial degradation level is low. If we further distinguish $\Omega_{G1}$ and $\Omega_{G2}$, the proof test interval within $\Omega_{G1}$ is $\tau_G$ but the proof test does not trigger a PM, meaning that the slight degradation can be tolerated. When the degradation state is observed within $\Omega_{G2}$ in a proof test, PM is initiated to restore the barrier to $\Omega_{G1}$ given the serious degradation. If the barrier falls in $\Omega_{DU}$, CM becomes necessary to restore it. If replacement is performed in case of failure, and the new state after the CM is within $\Omega_W$. If a minimal repair is conducted, the barrier item may be returned to the degraded state. In this subsection, we only consider the previous scenario and assume that the barrier is installed with a fully new state, where degradation $G(t=0)=0$. Duration of CM is assumed to be negligible. Figure 9.21 illustrates a degradation scenario with proof tests and maintenance given different observation results in the tests. Compared with the fixed proof test intervals in Figure 9.20, the intervals are adapted to the degradation level.

The above statement can be described mathematically. We use $\tau_i$ to denote the cumulative time by the $i$th proof test. It should be noted that $\tau_i$ is not necessarily same with $i \cdot \tau_1$, since each proof test interval can be different. $\tau_i^+$ is used to denote the moment that the $i$th proof test is just finished, $\tau_i^-$ is the moment just before the start of the proof test, and thus $G(\tau_i^-)$ is the degradation level that can be observed in the proof test. Using the adaptive testing policy, the subsequent testing interval after a proof test is dependent on the status of the barrier item observed in the test, and thus the length of the $(i+1)$th proof test interval is

$$\tau_{i+1} = \begin{cases} \tau_i + \tau_W, \text{ if } G(\tau_i^-) < \omega_b G_L \\ \tau_i + \tau_G, \text{ if } \omega_b G_L < G(\tau_i^-) < \omega_a G_L \\ \tau_i + \tau_G, \text{ if } \omega_a G_L < G(\tau_i^-) < G_L \\ \tau_i + \tau_W, \text{ if } G(\tau_i^-) > G_L \end{cases} \tag{9.84}$$

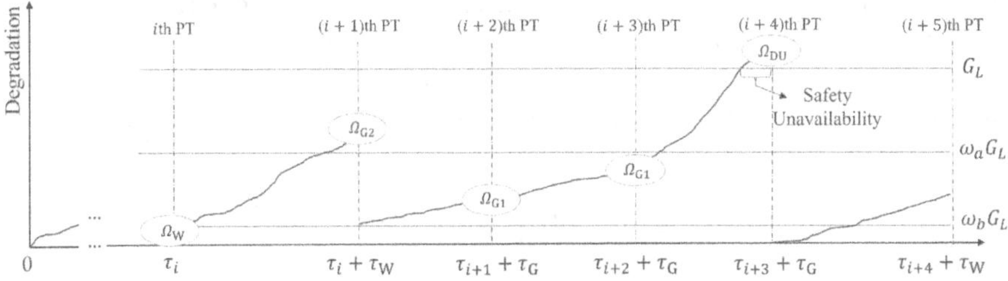


**FIGURE 9.21** A degradation scenario with different observations in proof tests.

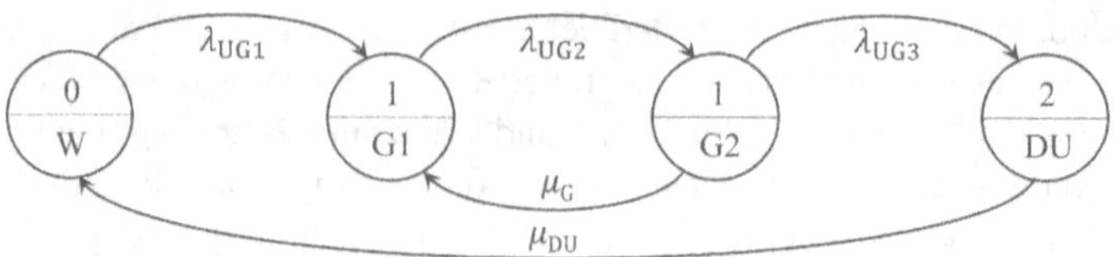


**FIGURE 9.22** The Markov model of a barrier component with two degradation states.

We can build a Markov state transition diagram to describe the dynamics of the barrier component, as shown in Figure 9.22. This Markov model only considers degradation but ignores the external shock that can lead to a failure immediately.

When the barrier is in the state G2, conducting a proof test has the capability to restore it to the state G1. The only possibility under which the barrier can transit to the DU state is if it was in either W or G1 during the preceding proof test and the degradation progression is notably rapid within the proof test interval. For a specific barrier component, its degradation rate, denoted as $\lambda_{UGi}$, can be regarded as constant. However, the values of the other two transition rates illustrated in Figure 9.22, namely $\mu_G$ and $\mu_{DU}$, depend on the initial state of the barrier following a proof test. The initiation state can either be W or G1. Thus, to conduct a thorough numerical analysis, the application of the multi-phase Markov method is very helpful in this scenario.

According to the approach proposed by Zhang et al. (2022), a transition matrix $\boldsymbol{\mathcal{A}}$ can be established for considering all states of the barrier system in a certain proof test interval according to Equation (5.102). In the first proof test interval, the system state probability at time $t$ is calculated as

$$\mathbf{P}(t) = P_0 \cdot \exp(\boldsymbol{\mathcal{A}}t) \tag{9.85}$$

In the first interval, $P_0(t=0)=1$ because the barrier is fully new when it is put in operation. Considering degradation, the probability of the barrier in the as-good-as-new state will be reduced in the following proof test intervals. For the state transition in the $i$th proof test, we have

$$\mathbf{P}(\tau_i^+) = \mathbf{P}(\tau_i^-) \cdot \Lambda \tag{9.86}$$

where $\Lambda$ is the probability matrix of different states after the $i$th proof test and the potential following-up maintenance. If the barrier is in state $j$ before the proof test, the probability that this item enters state $k$ is $\rho_{jk}$, and $\sum \rho_{jk} = 1$. It should be noted that $P_{DU}(\tau_i^+) = 0$, since the barrier is just examined and maintained. Based on the Markov model in Figure 9.22, the transition probability matrix $\Lambda$ can be written as

$$\Lambda = \begin{bmatrix} 1 & 0 & 0 & 0 \\ 0 & 1 & 0 & 0 \\ 0 & 1 & 0 & 0 \\ 1 & 0 & 0 & 0 \end{bmatrix}$$

Considering the four state sets, $\mathbf{P}(\tau_i^+)=\{P_W(\tau_i^+), P_{G1}(t)(\tau_i^+), P_{G2}(t)(\tau_i^+), 0\}$. Considering two potential initial states, the probability distribution that the barrier item in different states just before the $(i+1)$th proof test is

$$\mathbf{P}(\tau_{i+1}^-)=\mathbf{P}(\tau_i^+)\cdot\theta_W\cdot\exp(\tau_W\mathcal{A})+\mathbf{P}(\tau_i^+)\cdot\theta_G\cdot\exp(\tau_G\mathcal{A}) \quad (9.87)$$

where $\theta_W=[1, 0, 0, 0]$ and $\theta_{G1}=[0, 1, 0, 0]$ are vectors that mark functioning and degraded states, respectively, as 1, while other states as 0. They are used to obtain the corresponding state probabilities of the barrier item. The instantaneous unavailability is

$$\text{PFD}(t)=\mathbf{P}(t)\theta_{DU} \quad (9.88)$$

where $\theta_{DU}=[0, 0, 0, 1]$ for a single-barrier component. Since we have known $P_0(t=0)=1$, PFD calculation in the first proof test interval is the same as what we did previously. For the second proof test interval, we should start by calculating the state probabilities $P_W(\tau_1^+)$, $P_{G1}(t)(\tau_1^+)$, and $P_{G2}(t)(\tau_1^+)$ based on Equation (9.87).

The next step is to use the calculated results and Equation (9.84) to calculate the state probabilities of this barrier component in two scenarios: longer proof test interval and shorter proof test interval.

For the average unavailability within the $i$th proof test interval, we need to rely on the mathematical formula:

$$\text{PFD}_{\text{avg},i}=\frac{1}{\tau_i-\tau_{i-1}}\int_{\tau_{i-1}}^{\tau_i}\text{PFD}(t)\,dt \quad (9.89)$$

In the article of Zhang et al. (2022), the authors introduce a concept of virtual proof test that is equal to $\tau_W$. The purpose is to evaluate average PFD in different scenarios with the same time horizon. For example, consider the lifecycle of a barrier $L=n\tau_W$, which can include more $\tau_G$ dependent on degradation rate, maintenance strategies, and settings of $\omega_a$ and $\omega_b$. Due to the page limitation, we do not discuss more calculation details in this book. Readers can refer to this article for numerical case studies of using the multi-phase Markov model for barrier systems.

## NOTE

1. Named after the U.S. mathematician and philosopher Norbert Wiener (1894–1964).

## REFERENCES

Ahmad, R., & Kamaruddin, S. (2012). An overview of time-based and condition-based maintenance in industrial application. *Computers & Industrial Engineering, 63*(1), 135–149.

Alaswad, S., & Xiang, Y. (2017). A review on condition-based maintenance optimization models for stochastically deteriorating system. *Reliability Engineering & System Safety, 157*, 54–63.

Endrenyi, J. E. A., Aboresheid, S., Allan, R. N., Anders, G. J., Asgarpoor, S., Billinton, R., Chowdhury, N., Dialynas, E. N., Fipper, M., & Fletcher, R. H. (2001). The present status of maintenance strategies and the impact of maintenance on reliability. *IEEE Transactions on Power Systems*, *16*(4), 638–646.

Hauge, S., Kråknes, T., Håbrekke, S., & Jin, H. (2013). Reliability prediction methods for safety instrumented systems, PDS method handbook. SINTEF.

IEC61508. (2010). Functional safety of electrical/electronic/programmable electronic safety-related systems. Geneva, Switzerland: International Electrotechnical Commission.

Innal, F., Lundteigen, M. A., Liu, Y., & Barros, A. (2016). PFDavg generalized formulas for SIS subject to partial and full periodic tests based on multi-phase Markov models. *Reliability Engineering & System Safety*, *150*, 160–170. https://doi.org/10.1016/j.ress.2016.01.022

Jimenez, J. J. M., Schwartz, S., Vingerhoeds, R., Grabot, B., & Salaun, M. (2020). Towards multi-model approaches to predictive maintenance: A systematic literature survey on diagnostics and prognostics. *Journal of Manufacturing Systems*, *56*, 539–557. https://doi.org/10.1016/j.jmsy.2020.07.008

Kobbacy, K. A. H., & Murthy, D. N. P. (2008). *A Complex System Maintenance Handbook* (D. N. P. Murthy, Ed.). Springer-Verlag London. https://doi.org/10.1007/978-1-84800-011-7

Liu, Y. (2014). Optimal staggered testing strategies for heterogeneously redundant safety systems. *Reliability Engineering & System Safety*, *126*, 235–243.

Liu, Y., & Rausand, M. (2016). Proof-testing strategies induced by dangerous detected failures of safety-instrumented systems. *Reliability Engineering & System Safety*, *2016*(145), 366–372.

Lundteigen, M., & Rausand, M. (2008). Partial stroke testing of process shutdown valves: How to determine the test coverage. *Journal of Loss Prevention in the Process Industries*, *21*(6), 579–588. https://doi.org/10.1016/j.jlp.2008.04.007

Merizalde, Y., Hernández-Callejo, L., Duque-Perez, O., & Alonso-Gómez, V. (2019). Maintenance models applied to wind turbines. A comprehensive overview. *Energies*, *12*(2), 225.

Nicolai, R. P., Dekker, R., & van Noortwijk, J. M. (2007). A comparison of models for measurable deterioration: An application to coatings on steel structures. *Reliability Engineering & System Safety*, *92*(12), 1635–1650.

Rausand, M., Barros, A., & Hoyland, A. (2020). *System Reliability Theory: Models, Statistical Methods, and Applications* (3rd ed.). Wiley-Blackwell.

Rausand, M., & Høyland, A. (2004). *System Reliability Theory: Models, Statistical Methods, and Applications* (2nd ed.). Wiley.

Shahraki, A. F., Yadav, O. P., & Liao, H. (2017). A review on degradation modelling and its engineering applications. *International Journal of Performability Engineering*, *13*(3), 299–314.

Shin, J.-H., & Jun, H.-B. (2015). On condition based maintenance policy. *Journal of Computational Design and Engineering*, *2*(2), 119–127.

Summers, A., & Zachary, B. (2000). Partial-stroke testing of safety block valves. *Control Engineering*, *47*(12), 87–89.

Torres-Echeverri, A., Martorell, S., & Thompson, H. (2009). Modelling and optimization of proof testing policies for safety instrumented systems. *Reliability Engineering & System Safety*, *94*(4), 838–854.

Tsang, A. H. C. (1995). Condition-based maintenance: Tools and decision making. *Journal of Quality in Maintenance Engineering*, *1*(3), 3–17.

Wu, S., Zhang, L., Lundteigen, M. A., Liu, Y., & Zheng, W. (2018). Reliability assessment for final elements of SISs with time dependent failures. *Journal of Loss Prevention in the Process Industries*, *51*, 186–199. https://doi.org/10.1016/j.jlp.2017.12.007

Wu, S., Zhang, L., Zheng, W., Liu, Y., & Lundteigen, M. A. (2019). Reliability modeling of subsea SISs partial testing subject to delayed restoration. *Reliability Engineering & System Safety*, *191*, 106546.

Zhang, A. (2021). *Prognostics and health management of safety-instrumented systems: Approaches of degradation modeling and decision-making.* Norwegian University of Science and Technology. Trondheim. https://hdl.handle.net/11250/2726046

Zhang, A., Barros, A., & Liu, Y. (2019). Performance analysis of redundant safety-instrumented systems subject to degradation and external demands. *Journal of Loss Prevention in the Process Industries*, *62*, 103946.

Zhang, A., Hao, S., Xie, M., Liu, Y., & Yu, H. (2023). Inspection and maintenance optimization for heterogeneity units in redundant structure with non-dominated sorting genetic algorithm III. *ISA Transactions*, *135*, 299–308.

Zhang, Z., Si, X., Hu, C., & Lei, Y. (2018). Degradation data analysis and remaining useful life estimation: A review on wiener-process-based methods. *European Journal of Operational Research*, *271*(3), 775–796.

Zhang, A., Wu, S., Fan, D., Xie, M., Cai, B., & Liu, Y. (2022). Adaptive testing policy for multi-state systems with application to the degrading final elements in safety-instrumented systems. *Reliability Engineering & System Safety*, *221*, 108360.

Zhang, A., Zhang, T., Barros, A., & Liu, Y. (2020). Optimization of maintenances following proof tests for the final element of a safety-instrumented system. *Reliability Engineering & System Safety*, *196*, 106779.

# 10 Dependencies within Safety Barriers

## 10.1 INTRODUCTION

As we have discussed in Chapters 3 and 8, technical safety barriers are often designed with redundant structures to reduce the probability that the whole system fails. However, dependency may exist among the components performing the same function, meaning that an event occurring on one component may occur on the others or affect the performance of the others. For example, when two shutdown valves are installed one a pipeline to stop gas flow in emergency, closure of one valve can realize the barrier function. However, if one valve fails, the other must activate more frequently. It is reasonable to assume that frequent safety activations may lead to more degradations and more potential failures.

In this chapter, we consider the dependencies within one barrier system, mostly within one barrier subsystem/group. We will also discuss the failure propagation with the asset being protected by safety barriers, and their impacts on barrier performance.

## 10.2 DEPENDENCE

When we study complex systems, we are always referring to those systems with numerous components and interconnections or dependencies (Magee & De Weck, 2002). For a complex system, it cannot be understood as a simple combination of its individual components, and its performance is based on a composite functionality of components with interactions, cooperation, and coordination. Rausand and Haugen (2020) have discriminate a *complex system* with a *complicated system*, and they indicate that the existing knowledge can explain all interactions between components in a complicated system but cannot explain those in a complex system. However, we do not distinguish these two definitions in this book and only use the term of complex system.

In this chapter, we will focus on the dependency or dependence that is one of the main reasons making a system complex. For safety barriers, although independence is regarded extremely important to ensure that the systems are effective, dependence always exists between barrier components, between different barriers, and between barriers and the asset they are protecting (Johansen & Rausand, 2015).

In the Oxford dictionary, dependence is defined as the state of relying on or being controlled by someone or something else. Here, we regard dependence as the antonym of independence, which implies that the state in one component has no impact on the other component. In mathematics, two events $E_1$ and $E_2$ are statistically independent if

$$\Pr(E_1|E_2) = \Pr(E_1) \text{ and } \Pr(E_2|E_1) = \Pr(E_2)$$



DOI: 10.1201/9781003245636-10

Therefore, *independence* means that the occurrence probability of $E_1$ is not changed by knowing that $E_2$ has occurred and vice versa (Rausand et al., 2020). For dependence, we consider the above conditional probability can be different, namely:

$$\Pr(E_1|E_2) \neq \Pr(E_1) \text{ and } \Pr(E_2|E_1) \neq \Pr(E_2)$$

Taking this into account, we can define dependence in a multi-component technical system as:

- *Dependence*: A relationship between two or more components, where a state change of one component always coincides or leads to the change of states of one or more other components.

We do not use dependability in this book, but here we can define dependency as:

- *Dependency*: Strength of the dependence between two or more components.

Dependence does not necessarily mean causality, implying that dependence also lies in the situation where two components fail together due to the same reason but without a sequence. From this perspective, *correlation* is an appropriate word to describe the relationship in the definition above, but uncorrelation is not same as independence. When two variables (e.g., $X_1$ and $X_2$) are said to be uncorrelated, their *covariance* as the degree to which two variables vary together, calculated as $\text{cov}(X_1, X_2) = E([X_1 - E(X_1)][X_2 - E(X_2)])$, is equal to 0. We can find that two independent variables must be uncorrelated, but two uncorrelated variables can be dependent. For example, given that $X_1$ is a random variable taking values 1 or –1 with equal probability, and $X_2 = X_1^2$, then $X_1$ and $X_2$ are uncorrelated (because their covariance is zero), but they are not independent (as $X_2$ is determined by $X_1$).

Dependences existing between components in a technical system can be classified as several types according to their mechanisms and causes, including:

- *Structural dependence*: One component is structurally or physically close to one or more other components. If abnormal events occur on one component, the neighborhood components can be affected. For example, when fire is generated in a workshop, the fire can quickly spread to adjacent areas. In maintenance, the repair of a component may require another component to be dismantled at first.
- *Economic dependence*: Operational and maintenance cost of one component can be changed when state of one or more other components changes or some works are conducted on other components. For example, clustering the maintenances of several components leads to an economic dependence between these within the group, since such a combined work can lead to a lower cost than maintaining each component separately (Keizer et al., 2017).
- *Functional dependence*: Functionality of one component to the whole system is dependent on other components. In a functional sequence, the output of one function can be the necessary input of the next function. For example,

component A must successfully complete its function before component B can begin its operation. For a coffee machine, the function of heating water is dependent on first successfully importing and containing water. Without water, the heating function cannot initiate.

- *Resource dependence*: Multiple components rely on a shared resource in operation and maintenance. For example, the same maintenance team is responsible for repair and preventive maintenance of all components in a system. When the team is occupied with working on one component, the repairs of others will be delayed. Time is also a kind of resource. For some offshore facilities, the time window for maintenance in a year is limited. If the time is used for maintaining one facility, there is no time for working on the others.
- *Loading dependence*: Multiple components share a common working load. When one component is removed due to a failure or other reasons, the overall working load will be redistributed among the remaining components. This adjustment can impact the performance and lifetime of these components.

Any type of the above dependences can lead the dependence of two components in terms of their performance. If dependence exists between at least two components in a system, the system can be called as a *dependent system.*

In the book of Rausand and Haugen (2020), the authors highlight that dependence means event *A* can influence event *B*, while *B* does no influence *A*, and interdependence means the influences are bi-directional. However, in this chapter, we do not strictly distinguish the two terms of dependence and interdependence.

## 10.3 DEPENDENT FAILURES AND DEGRADATIONS

In some literature (e.g., Nicolai and Dekker (2008), Keizer et al. (2017)), researchers use *stochastic dependence* to describe the deterioration or failure processes of components that are dependent. It is reasonable that such dependent failures or deteriorations are assumed able to significantly weaken the performance of a system since they include more than one component.

There is no commonly accepted definition on dependent failures. According to IEC61508 (2010) and IEC61511 (2016), dependent failures are the failures whose probabilities cannot be expressed by unconditional probabilities of the individual event, while ISO26262 (2018) regards dependent failures as those that compromise the required independence between components. In this chapter, we define dependent failures occurring on a component with the purpose to distinguish them from independent failures (IFs):

- *Dependent failure*: Failure of a component that results from the state change of other components.

According to Chen et al. (2015), when one component fails in a system, it can trigger, accelerate, or inhibit the failures of other components within the same system, and it also can accumulate or reduce the damage on other components.

As the counterpart, an *IF* is defined here as:

- *IFs*: Failure of a component, the occurrence probability of which is not impacted by state change of other components.

We also call an IF as the *individual failure* or *self-failure* of a component, which can be a random failure or a failure due to the aging of the component itself. It is noted that an IF of one component can result in dependent failures or degradations of other components.

Similarly, *dependent degradation* can be defined as the degradation of one component influenced by the change of state of other components. *Independent degradation*, or *self-aging* of a component, is the degradation process not influenced by the states of other components. However, in this chapter, we focus on the impacts of dependent failures rather than dependent degradations.

To investigate a dependent failure occurring in more than one component, it is necessary to identify the *root cause* of this failure, namely the most fundamental reason of the negative event, as well as the *coupling factors* that can explain why several components are affected. A root cause may be inadequate design, deficiency in manufacture, human errors, lack of maintenance, and inappropriate procedure, etc. If the root cause can be addressed, similar failures can be avoided in the future. On the other hand, a coupling factor can be a design property, a kind of environmental condition or an operational strategy that makes multiple components susceptible to the root cause mentioned above. For example, structural and functional dependence can be coupling factors. Coupling factors are not hazards, nor are they the starting point of an accidental scenario. They can be regarded as conditions that change the direction of the accidental scenario. If the coupling factors are identified and eliminated, we prevent the development of an initiative event to escalate to dependent failures.

According to the manners that failures occur, dependent failures can be roughly divided into two types: *common-cause failures* (CCFs) and *cascading failures* (CAFs), and then the two types can be further divided into several sub-types. The remainder of this chapter will provide modeling and analyzing approaches for these two types of failures within barrier systems, respectively.

## 10.4 COMMON-CAUSE FAILURES

### 10.4.1 Definitions

In general, when two or more components in a system fail due to a shared cause, such failures can be regarded as CCFs. CCFs significantly weaken the redundant structures that are designed for higher fault tolerance, and CCFs make a classical reliability model with the independence assumption to underestimate the probability of a system failure. A survey conducted by OCED-NEA (1994) has shown that CCFs contribute between 20% and 80% of the unavailability of safety barrier systems within the nuclear reactors. For example, the Fukushima accident can be attributed to CCFs, where the tsunami simultaneously disabled the cooling pump power sources of all the reactors in the Fukushima Daiichi nuclear plant.

CCFs can be found in many applications. In IEC61508 (2010), CCFs are the failures resulting from one or more events, causing concurrent failures of two or more separate channels, while ISO13849-1 (2015) defines that CCFs are failures of different items resulting from a single event and highlights that these failures are not consequences of one another.

In this book, we propose the following definition of CCF in a barrier system:

- *CCFs*: Failures, as the direct result of a shared cause, occurring on two or more components, simultaneously or within a short time interval.

CCFs should be distinguished with the *common-mode failures*, given that failure mode is the way how a failure is observed. Multiple failures can be found with the same mode in two or more items, for example, when cars cannot start as required (common-mode), they can be resulted from different causes, such as low battery level, problems in start motor, problems in timing belt, broken or cracked distributor cap, or bad ignition coil.

Root cause and coupling factors need to be considered in CCF analysis. The relationship between CCF with its root cause and coupling factors is illustrated in Figure 10.1, where both failures of components 1 and 2 are due to a shared cause, and such failures occur under the condition where the coupling factors exist.

The root causes of CCFs on the components in a technical system mainly include:

- *Deficiencies in engineering and management*: During the design phase, an inadequate understanding of failure mechanisms, improper selection of hardware components and software, and latent software faults can lead to CCFs in the future (IEC62340, 2007; Lundteigen & Rausand, 2007). Additionally, inappropriate installation and wrong calibration, procedural deficiencies, organizational issues, poor management, lack of maintenance, and human errors are also unignorable root causes (Galán et al., 2007; Rahimi et al., 2011). These problems can be intentional actions or unintentional oversights, such as forgetting specific instructions, or inadequate implementation of prescribed instructions, such as improper adjustments, equipment miscalibration, or incorrect bypassing.

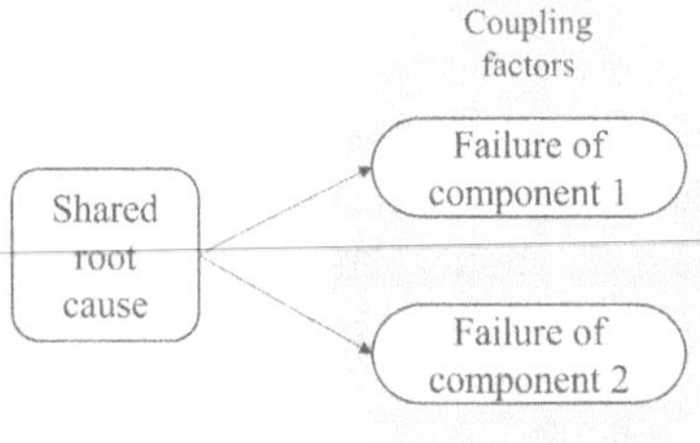


**FIGURE 10.1** Illustration of CCF mechanism.

- *External shocks*: Some natural disasters, such as earthquakes, tsunamis, thunderstorms, wildfires, and extreme temperatures, can destroy or make noticeable damages to all components in a system. Recent well-known cases include the eruption of volcanos in Iceland causing enormous disruption of air travels across Europe in 2010, the earthquake and following tsunami resulting in the Fukushima accident, and the pandemic of COVID-19, which stop the normal operations of many things. Deliberate attacks and vandalisms also can be external shocks for a technical system.
- *Operational conditions*: Commonly operational conditions can lead to common deterioration and failure mechanisms. For example, all the equipment in subsea oil and gas production is operated under high pressure in deep water, and they are exposed to sea salt and lack regular maintenances. Such conditions make corrosion far more likely to occur.

These causes are the hazards on the left side of a bow-tie diagram, and they are the start points of accidental scenarios, but they need the existence of coupling factors among multiple components to initiate CCFs. Coupling factors related to the CCFs in a barrier system can come from many aspects, including but are not limited to:

- Same/similar design (same hardware and software), appearance, and layout;
- Same manufacturer or similar manufacturing techniques;
- Same/similar installation approach and installation staffs;
- Same/similar maintenance procedure, maintenance schedule, and maintenance crews;
- Same location and same/similar operational environment.

CCFs can occur in different assets, and also can occur in different barrier components. We will discuss these measures for reducing CCFs in a technical system we will discuss these measures in Section 10.4.2, and then introduce modeling approaches for quantitative CCF analysis with the focus on barrier systems. In the rest part of this book, we mainly investigate the impacts of dependent failures on parallel structures, since the CCFs in a series structure do not bring much more difficulties in modeling and analysis.

More discussions on root causes and coupling factors of CCFs can be found in the studies of Parry (1991), and Lundteigen and Rausand (2007), and more examples of CCFs can be found in the report of OCED-NEA (2011).

### 10.4.2 Risk Reduction Measures of CCFs

The risk reduction measures of CCFs in this subsection are not only applied for barrier systems but also for normal production systems. IEC62340 (2007) has summarized some approaches against CCFs of equipment in the nuclear industry, and most of them are design precaution to reduce systematic faults or reduce triggering mechanisms of CCFs. In general, the risk of CCF can be reduced by avoiding the impacts of root cause and weakening the strength of coupling. For the first type of approaches, they are from the generic methods for hazard reduction in reliability design and safety management, as we have discussed in Chapters 2 and 3. Some examples include:

- *Simplicity in design*: A simpler design with less components and less interactions is often easier to be understood and analyzed, and also meaning less mistakes in installation, less random failures, and less interaction errors.
- *Robustness in design*: A robust design means a larger margin between the capacity and working load, as well as less weaker items, so that it is helpful to withstand the fluctuating environmental stresses and generate less hazardous events.
- *Reduction of manual interference*: Integrate more automatic procedures to reduce reliance on manual processes and implement technological solutions for monitoring and controlling safety barriers, reducing the chance of human error.

To deal with coupling factors in technical system design, operation and environment, the main approaches in literature (IEC61508, 2010; Rausand, 2014) can be summarized as:

- *Independence in design and operation*: A component should be designed and operated independently from the other components that perform similar or same functions. This means that each component is expected to have unique parts, rely on distinct input signals and controlling command, and operate with independent supporting systems and services. Interaction or interference among the components should be minimized.
- *Diversity in design and operation*: To reduce the impacts of coupling factors, different types of components in a system, especially those performing same functions, can be designed as differently as possible. For example, for two projectors in a classroom, one of them runs on normal AC power, and another one can be different by using a battery. When power outages in the building, the battery-powered projector is still functioning. Another type of design diversity lies in the different approaches to be functioning. Even though several barriers are installed to protect the same asset, but their working principles can be diverse, e.g., firewall and water sprinkler. In such a method, the barriers can be regarded as different layers of protection. Diversity also can be in operational strategies, and one example is that the valves in a barrier system are inspected and maintained at different times and by different persons, so as to avoid repeated human mistakes.
- *Separation or segregation*: To avoid the common shocks from a single event and to reduce couplings due to same location and same operational environment, components in a system can be deployed with separation from the others. Separation can be physical or spatial separation, meaning long distance between components, and it also can be electrical separation where there is no complete circuit through which shock current could flow.

In additional to the technical approaches, the following administrative measures are helpful in reducing CCFs:

- *Quality control and similarity check*: Rigorous quality control measures and similarity checks are needed to ensure the reliability of safety barriers and the independence of safety barrier with the asset. Regular audits and inspections can be conducted to identify the coupling factors.
- *Reduction of human and procedural errors*: Commonly used methods include regular training programs for personnel involved in safety procedures to enhance their understanding of potential risks, clear and concise guidelines ensuring that instructions are easy to understand and follow.
- *Reporting and documentation*: Each recorded failure event needs to be screened to identify potential CCFs. Such examinations should cover the detailed descriptions of failure causes, effects, and detection methods, to provide necessary information to evaluate whether or not a CCF has occurred. The examiners can use a set of prepared questions for field technicians during failure recording (Lundteigen & Rausand, 2007).
- *Verification and test*: According to Summers and Raney (1999), the most critical cause of CCFs in barrier design and implementation is the error or incompleteness of the safety requirement specification. Verification of design is thus needed, and it should be conducted by independent teams without conflict of interest.
- *Monitoring and review*: Such works need to rely on field experiences and data. The insights gained from practical situations can enhance the effectiveness of safety barriers and identify areas for continuous improvement.

More details about the risk reduction measures on CCFs can be found in the study by Lundteigen and Rausand (2007).

### 10.4.3 Explicit and Implicit Modeling Approaches

CCFs have received extensive attention when studying safety barriers in many applications, such as in nuclear plants and offshore production. Many probability models have been introduced when CCFs are in consideration, for example in recent years, Ma et al. (2017) have used the binomial failure rate (BFR) model for reliability analysis of digital reactor protection systems with CCFs in nuclear plants, and Wang et al. (2021) have combined the beta-factor for CCFs with Petri net in availability evaluation of subsea high integrity pressure protection system. In general, modeling approaches for CCFs can be categorized as two types, and we can use fault tree models for a butterfly valve system of two valves in Figures 10.2 and 10.3 to illustrate these two approaches. Both of the two models show that the valve system can fail when two individual failures have occurred on the two valves, but they are different in the CCF part:

- *Explicit modeling approach*: This approach seeks to identify and understand the root causes of CCFs and considers the parameters relevant with the root causes in models. For example, in Figure 10.2, either of events

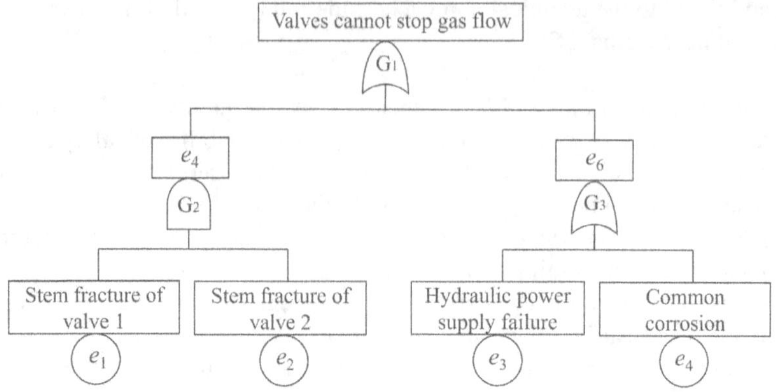


**FIGURE 10.2** An explicit model for the failure of a butterfly valve system with CCFs.

$e_3$ and $e_4$ can result in a CCF of the two valves. Based on such a model, we can then investigate the occurrence probabilities of these two events respectively for further quantitative fault tree analysis. In practices, coupling factors need to be studied, but it is sometimes not easy to study coupling factors as events of a fault tree.

- *Implicit modeling approach*: This approach regards all causes that can lead to CCFs as the residual causes and uses one event to represent all of them. As shown in Figure 10.3, no further analysis on the shared root causes and coupling factors is conducted. This approach can simplify the representation of models, but it should be remembered that which causes are included in the nodes of CCFs and to avoid double-counting them in the same model.

In many applications, the root causes of CCFs are difficult to be identified, and their relationships with coupling factors are difficult to be modeled explicitly. If it

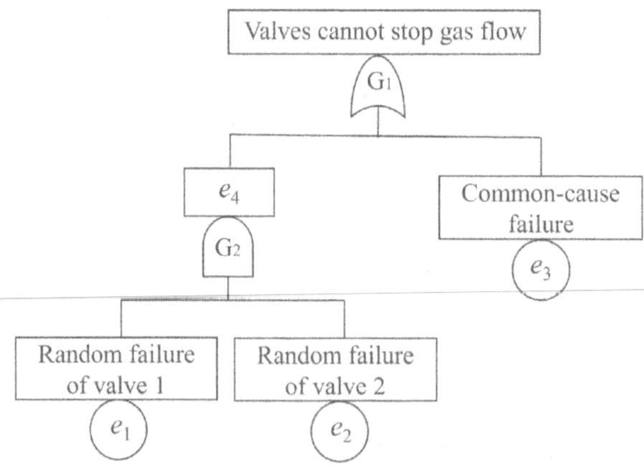


**FIGURE 10.3** An implicit model for the failure of a butterfly valve system with CCFs.

is possible to identify all the root causes, they can be simply connected with the TOP event via OR gates or regarded as series elements in reliability block diagrams (RBDs), and therefore, in rest of this book, we do not have more discussions on the explicit modeling, while focus on the implicit approaches.

For the introduction of implicit CCF models, we will focus on *KooN* barrier systems with identical barrier components. We follow the generic assumptions used in the previously relevant research (Fleming, 1975; Hokstad & Rausand, 2008; Lundteigen & Rausand, 2007):

- Each of the $n$ barrier component has a constant failure rate, and only DU-failure is considered for simplification;
- No CAFs, meaning that the failure of any one or more barrier components will not impact the other components, or change the failure rate/probability of other components;
- Only regular proof tests with the interval of $\tau$ are conducted to reveal the DU failures.

### 10.4.4 The Beta-Factor Model

Starting from the basic parameter model proposed by Fleming (1975), the *beta-factor model* is the most widely used method in the analysis of CCFs and is suggested by IEC61508 (2010).

#### 10.4.4.1 Beta-Factor Model for a Parallel Structure with Two Components

Figure 10.4 illustrates the multiplicity of faults and its distribution. We set $E_i$ as the event that component $i$ is failed by time $t$, and thus $E_i^*$ as the event that component is not failed by time $t$. As shown by the Venn diagram in Figure 10.4(a), two events $E_1$ and $E_2$ represent the IFs of two components. While in the Venn diagram of Figure 10.4(b), there are some overlapping areas between the sets of the two events, meaning that they can occur at the same time. Here, we suppose that all the $n$ components are identical, and let $p_{ioon}$ denote the probability that there are $i$ specific components that are failed out of $n$ components.

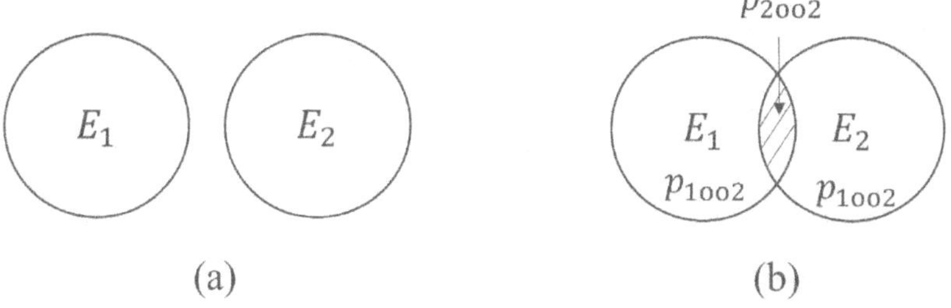


**FIGURE 10.4** (a) Independent failures of two components, (b) Probability of fault multiplicities.

Then, we can calculate the value of $p_{2oo2}$ as the shadow area in Figure 10.4 (b):

$$p_{2oo2} = \Pr(E_1 \cap E_2) = \Pr(E_1|E_2) \cdot \Pr(E_2) = \Pr(E_1|E_2) \cdot \Pr(E_2)$$
$$= \Pr(E_1|E_2) \cdot F(t) = \Pr(E_2|E_1) F(t)$$

Symbol $\beta$ is used here to denote $\Pr(E_2|E_1)$, and the meaning of $\beta$ is the conditional probability of a CCF on all the $n$ components when a failure has been observed on the specific component, namely

$$\beta = \Pr\left(\text{CCF}|\text{Failure of component}\right) \tag{10.1}$$

Thus, $p_{2oo2} = \beta$ in this case. Consider any of the two components, and $p_{1oo2}$ can be calculated as:

$$p_{1oo2} = \Pr(E_1^* \cap E_2) = \Pr(E_1^*|E_2) \cdot \Pr(E_2) = \left[1 - \Pr(E_1|E_2)\right] \cdot \Pr(E_2)$$
$$= \left[1 - \Pr(E_1|E_2)\right] \cdot F(t) = (1-\beta) F(t)$$

The total failure probability ($p$) of a component by time $t$ is:

$$p = F(t) = (1 - \beta + \beta) F(t) = p_{1oo2} + p_{2oo2}$$

Thus, $\beta$ can be explained as the fraction of CCF in the total failures:

$$\beta = \frac{p_{2oo2}}{p_{1oo2} + p_{2oo2}} \tag{10.2}$$

The failure in such studies can be DU-, DD-, or S-failure. For a specific failure, a component can be represented as a series structure with two blocks, as shown in Figure 10.5, where $\overline{\text{IF}}$ reflects the survivability of this component from an IF, and $\overline{\text{CCF}}$ reflects the survivability from the CCF.

Under the exponential distribution assumption, for such a series structure, failure rate of the system is the sum of those of the two elements: rate of IF (I), and rate of CCF, therefore, the overall failure rate is:

$$\lambda = \lambda^{\text{I}} + \lambda^{\text{CCF}} \tag{10.3}$$

According to Equation (10.2), the beta-factor, $\beta$, can be calculated as

$$\beta = \frac{\lambda^{\text{CCF}}}{\lambda} = \frac{\lambda^{\text{CCF}}}{\lambda^{\text{I}} + \lambda^{\text{CCF}}} \tag{10.4}$$

**FIGURE 10.5** RBD of a component as a series structure with two blocks.

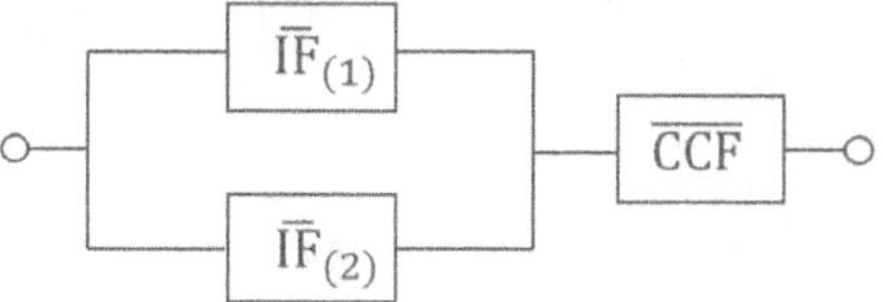


**FIGURE 10.6** RBD of a 1oo2 structure with CCF.

For the structure with two identical components, its RBD involving CCF is shown as in Figure 10.6, where $\overline{\text{IF}}_{(i)}$ denotes the survivability of component $i$ from the IF. The model in Figure 10.6 is actually equivalent with the fault tree in Figure 10.3 if we consider that the components are valves.

Equation (5.61) has given the survivor function of the parallel structure of IF. Consider the CCF as a series element, reliability of such a series-parallel system in Figure 10.6 is:

$$R_{S:1oo2}(t) = \left[2e^{-(1-\beta)\lambda t} - e^{-2(1-\beta)\lambda t}\right]e^{-\beta\lambda t} = 2e^{-\lambda t} - e^{-(2-\beta)\lambda t} \tag{10.5}$$

The mean-time-to-failure (MTTF) of this structure is:

$$\text{MTTF}_{1oo2} = \int_0^\infty tR'_{S:1oo2}(t)\,dt = \frac{2}{\lambda} - \frac{1}{(2-\beta)\lambda} \tag{10.6}$$

Consider this structure is for a barrier system with the proof test interval of $\tau$. Firstly, only DU-failures is considered, and the approximation of $\text{PFD}_{\text{avg}}$ is:

$$\begin{aligned}\text{PFD}_{\text{avg}}^{1oo2} &= 1 - \frac{1}{\tau}\int_0^\tau R_{S:1oo2}(t)\,dt = 1 - \frac{1}{\tau}\int_0^\tau \left[2e^{-\lambda_{\text{DU}}t} - e^{-(2-\beta)\lambda_{\text{DU}}t}\right]dt \\ &\approx \frac{\left[(1-\beta)\lambda_{\text{DU}}\tau\right]^2}{3} + \frac{\beta\lambda_{\text{DU}}\tau}{2}\end{aligned} \tag{10.7}$$

Such an approximation is based on the values in Table 8.1. We can see that the result of $\text{PFD}_{\text{avg}}$ includes two parts: the contribution of IFs and the contribution of CCF. In the IF part, the structure is a 1oo2 barrier group with the individual failure rate of $(1-\beta)\lambda_{\text{DU}}$, and in the CCF part, the element can be regarded as a single barrier component with the failure rate of $\beta\lambda_{\text{DU}}$.

### Example 10.1:

#### *CCF Contribution to PFD of a 1oo2 Barrier Subsystem*

Revisiting the emergency shutdown (ESD) ball valve in Example 7.2, the DU failure rate of one valve is $3.0\times10^{-6}$ per hour. Now we consider a 1oo2 voting structure, the proof test is six months (4,380 h), and thus $PFD_{avg}$ of this barrier system can be calculated as:

$$PFD_{avg}^{1oo2,no\ CCF} = \frac{(\lambda_{DU}\tau)^2}{3} = \frac{(3.0\times10^{-6}\times4380)^2}{3} = 5.76\times10^{-5}$$

It means that the barrier is very reliable satisfying the requirement SIL 4. However, if we consider the probability of CCF in this case, for example, $\beta = 0.01$, $PFD_{avg}$ will be re-calculated as:

$$PFD_{avg} = \frac{[(1-\beta)\lambda_{DU}\tau]^2}{3} + \frac{\beta\lambda_{DU}\tau}{2} = \frac{[0.99\times3.0\times10^{-6}\times4380]^2}{3} + \frac{0.01\times3.0\times10^{-6}\times4380}{2} = 1.22\times10^{-4}$$

In such a case, the contribution of IFs to $PFD_{avg}$ is $5.64\times10^{-5}$ (46.2%), while the contribution of CCF is $6.57\times10^{-5}$ (53.8%). It can be found that even though the fraction of CCF in the total failure is very low, its impact is noticeable in the system unavailability, which has been lifted to SIL 3 in this case.

When $\beta = 0.10$, the value of $PFD_{avg}$ becomes to $7.04\times10^{-4}$, which is about 10% $PFD_{avg}$ of a single barrier component ($6.57\times10^{-3}$). The contribution of IFs ($4.66\times10^{-5}$) only covers 6.6% of the total $PFD_{avg}$ and the CCF becomes to the dominator of the system unavailability with the contribution of 93.4%. With more numerical examples, it can be found that $PFD_{avg}$ of a redundant barrier system is very sensitive to the value of $\beta$. ■

In this book, when different failure modes of barrier systems are considered, we use $\beta$ to denote the CCF fraction of DU failures, $\beta_D$ for that of DD failures, and $\beta_S$ for that of safe failures. The overall CCF rate of a barrier component is thus

$$\lambda^{CCF} = \beta\lambda_{DU} + \beta_D\lambda_{DD} + \beta_S\lambda_S \tag{10.8}$$

#### 10.4.4.2 Beta-Factor Model for a Parallel Structure with Three Components

With the same consideration, for a structure with three identical components, following the same definition, $\beta$ is the fraction of CCF in the total failures of a component. The multiplicity of faults and distributions are shown by the Venn diagram in Figure 10.7, where we still set $E_i$ as the event that component $i$ is failed by time $t$, and $p_{ioo3}$ as the probability of $i$ specific components failing out of three components. The probability of a triple fault is:

$$p_{3oo3} = \Pr(E_1 \cap E_2 \cap E_3)$$

According to the definition of the beta-factor model, $p_{3oo3} = \beta F(t)$, and $p_{1oo3} = (1-\beta)F(t)$. Therefore, the probability of a double fault on a specific component $p_{2oo3} = F(t) - p_{1oo3} - p_{3oo3} = 0$. In other words, the beta-factor model skips the situations where more than one but not all components are failed.

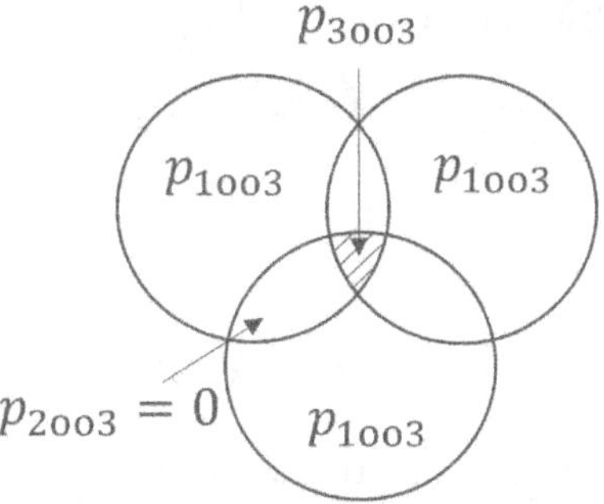


**FIGURE 10.7** Probabilities of fault multiplicities for three identical components.

Figures 10.8(a) and (b) illustrate the RBD models for 1oo3 and 2oo3 voting structures, respectively.

For the 1oo3 structure with CCF in Figure 10.8(a), suppose that all components are identical, and the system reliability is:

$$\begin{aligned} R_{S:1oo3}(t) &= \left[3e^{-(1-\beta)\lambda t} - 3e^{-2(1-\beta)\lambda t} + e^{-3(1-\beta)\lambda t}\right]e^{-\beta\lambda t} \\ &= 3e^{-\lambda t} - 3e^{-(2-\beta)\lambda t} + e^{-(3-2\beta)\lambda t} \end{aligned} \tag{10.9}$$

The MTTF of this 1oo3 structure is:

$$\text{MTTF}_{1oo3} = \frac{3}{\lambda} - \frac{3}{(2-\beta)\lambda} + \frac{1}{(3-2\beta)\lambda} \tag{10.10}$$

Consider this structure is for a barrier subsystem with DU-failures and the proof test interval of $\tau$, $\text{PFD}_{\text{avg}}$ is approximated by dividing RBD model into independent part and CCF part as:

$$\text{PFD}_{\text{avg}}^{1oo3} = \frac{\left[(1-\beta)\lambda_{\text{DU}}\tau\right]^3}{4} + \frac{\beta\lambda_{\text{DU}}\tau}{2} \tag{10.11}$$

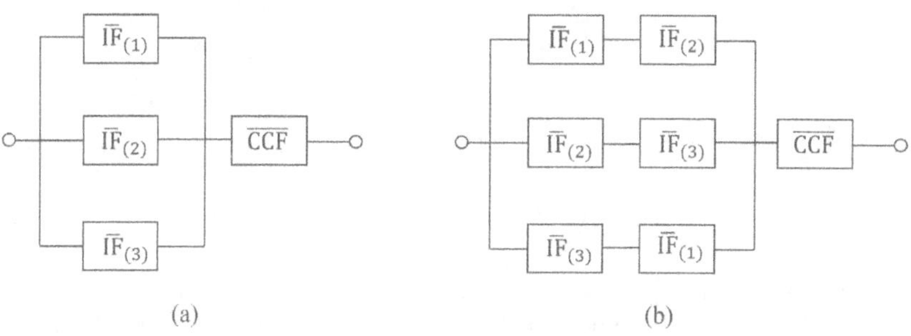


**FIGURE 10.8** RBDs of 1oo3 (a) and 2oo3 (b) voting structures with CCFs.

While for the 2oo3 structure with CCF in Figure 10.8(b), the system reliability is:

$$R_{S:2oo3}(t) = \left[3e^{-2(1-\beta)\lambda t} - 2e^{-3(1-\beta)\lambda t}\right]e^{-\beta\lambda t} = 3e^{-(2-\beta)\lambda t} - 2e^{-(3-2\beta)\lambda t} \quad (10.12)$$

The MTTF of this 2oo3 structure is:

$$\text{MTTF}_{2oo3} = \frac{3}{(2-\beta)\lambda} - \frac{2}{(3-2\beta)\lambda} \quad (10.13)$$

Consider this structure is for a barrier subsystem with DU-failures and the proof test interval of $\tau$, $\text{PFD}_{avg}$ is approximated by dividing RBD model into independent part and CCF part as:

$$\text{PFD}_{avg}^{2oo3} = \left[(1-\beta)\lambda_{DU}\tau\right]^2 + \frac{\beta\lambda_{DU}\tau}{2} \quad (10.14)$$

**Example 10.2:**

***CCF Contributions to PFD of a 1oo3 and 2oo3 Barrier Subsystems***

We consider using the same type of ESD valves of Example 10.1 in 1oo3 and 2oo3 structures (just for illustration, 2oo3 is not a common configuration for valve systems), with the same proof test of six months on all valves simultaneously. $\text{PFD}_{avg}$ of the 1oo3 barrier system without CCF can be calculated according to Table 8.1 as:

$$\text{PFD}_{avg}^{1oo3,\text{no CCF}} = \frac{(\lambda_{DU}\tau)^3}{4} = 5.67 \times 10^{-7}$$

$\text{PFD}_{avg}$ of the 2oo3 barrier system without CCF is:

$$\text{PFD}_{avg}^{2oo3,\text{no CCF}} = (\lambda_{DU}\tau)^2 = 1.73 \times 10^{-4}$$

It is obvious that the 1oo3 barrier system is much more reliable than the 2oo3 system.

When CCFs can occur in these systems with $\beta = 0.01$,

$$\text{PFD}_{avg}^{1oo3} = \frac{\left[(1-\beta)\lambda_{DU}\tau\right]^3}{4} + \frac{\beta\lambda_{DU}\tau}{2} = 6.63 \times 10^{-5}$$

and

$$\text{PFD}_{avg}^{2oo3} = \left[(1-\beta)\lambda_{DU}\tau\right]^2 + \frac{\beta\lambda_{DU}\tau}{2} = 2.35 \times 10^{-4}$$

The value of $\text{PFD}_{avg}$ for the 1oo3 barrier system increases significantly by more than 100 times, most of unavailability of this system is from CCF. While for the 2oo3 barrier system, the value of $\text{PFD}_{avg}$ increases by 36%, and the fraction of the contribution from CCF to the total system unavailability is about 26%. It can be found that the effectiveness of higher redundancy is greatly weakened by CCFs. In this example, the contribution of CCF to $PFD_{avg}$ of the 1oo3 voting structure is 99.10% ($6.57\times10^{-5}$), while the contribution of CCF to $PFD_{avg}$ of the 2oo3 voting structure is 27.76%.

When $\beta = 0.1$,

$$PFD_{avg}^{1oo3} = \frac{\left[(1-0.1)\lambda_{DU}\tau\right]^3}{4} + \frac{0.1\lambda_{DU}\tau}{2} = 6.57\times10^{-4}$$

and

$$PFD_{avg}^{2oo3} = \left[(1-0.1)\lambda_{DU}\tau\right]^2 + \frac{0.1\lambda_{DU}\tau}{2} = 7.97\times10^{-4}$$

The values of $PFD_{avg}$ for the two barrier systems have been much closer, and they are also close to that of 1oo2 system in Example 10.1. The reason of closeness lies in the dominating contribution of CCFs to system unavailability in both cases: The contribution of CCF to $PFD_{avg}$ of the 1oo3 voting structure is more than 99.99%, while the contribution of CCF to $PFD_{avg}$ of the 2oo3 voting structure is 82.43%. ■

For *koon* voting structures, when CCFs are taken into account, the generic approximation formula for $PFD_{avg}$ can be extended based on Equation (8.7) as:

$$PFD_{avg}^{koon} \approx \binom{n}{n-k+1} \frac{(\lambda_{DU}\tau)^{n-k+1}}{n-k+2} + \frac{\beta\lambda_{DU}\tau}{2} \tag{10.15}$$

When there are more than three identical components to perform the same barrier function, the probability that more than one component is independently failed within a proof test interval is very low, and the main contribution to system unavailability will be CCFs. Therefore, $PFD_{avg}$ of *koon* voting structures (when $n-k \geq 2$) can be simply approximated as $\beta\lambda_{DU}\tau/2$.

When the repair duration on any barrier component is not ignorable, as we have discussed in Section 9.6, in some remote operations, the interval from the moment when faults are found to the restoration time is long, the downtime of such a barrier system can be approximated as $\left(\frac{\tau}{2}+\text{MRT}\right)$, and thus the general formula in Equation (9.43) can be approximated as:

$$PFD_{avg} \approx \frac{\text{MDT}}{\text{MTTF}_{DU}} \approx \beta\lambda_{DU}\left(\frac{\tau}{2}+\text{MRT}\right) \tag{10.16}$$

For a series structure with $n$ identical components with a constant failure rate $\lambda_{DU}$, it can be regarded as a *noon* configuration. Based on the analyses with Figures 10.5, 10.6 and 10.8, we can see that CCF modeling for a series structure means to add a new block in the sequence. Its survivor function can be written as

$$R_{S:noon}(t) = e^{-n(1-\beta)\lambda_{DU}t} \cdot e^{-\beta\lambda_{DU}t} = e^{-[n-(n-1)\beta]\lambda_{DU}t} \tag{10.17}$$

$\text{PFD}_{\text{avg}}$ of such a *noon* barrier system with a proof test interval of $\tau$ can be obtained based on Equation (10.15) as:

$$\text{PFD}_{\text{avg}}^{noon} \approx \frac{n\lambda_{\text{DU}}\tau}{2} + \frac{\beta\lambda_{\text{DU}}\tau}{2} = \frac{(n+\beta)\lambda_{\text{DU}}\tau}{2} \quad (10.18)$$

In fact, it can be found from the cases above that the beta-factor model defines a CCF as the failure occurring on all components in a system or the event that can fail the entire system. Similar to what consider in Equation (10.16), if MRT cannot be skipped, Equation (10.18) can be extended as:

$$\text{PFD}_{\text{avg}}^{noon} \approx (n+\beta)\lambda_{\text{DU}}\left(\frac{\tau}{2} + \text{MRT}\right) \quad (10.19)$$

However, based on the definition in this book, a CCF also can influence several but not all the components, and we will discuss such situations in the following subsections.

#### 10.4.4.3 Markov Analysis Based on Beta-Factor Model

We also can build Markov transition diagram for the barrier systems mentioned in the Sections 10.4.3.1 and 10.4.3.2.

Figure 10.9 is a Markov model of a 1oo2 barrier subsystem with identical components and independent DU-failures and common-cause DU-failures.

Here we use the circle filled with grey to highlight the failed state of the system. Like the model for a 1oo2 barrier without CCF in Figure 8.3, this model also has three states as listed in the following table:

| State | Description |
|---|---|
| 0 | 0 component is failed, 2 components are functioning |
| 1 | 1 component is failed, 1 component is functioning |
| 2 | 2 components are failed, 0 component is functioning |

The transitions in this model are different. Since CCFs can occur, there is a transition directly from state 0 to state 2 with the rate of $\beta\lambda_{\text{DU}}$. The transition rate

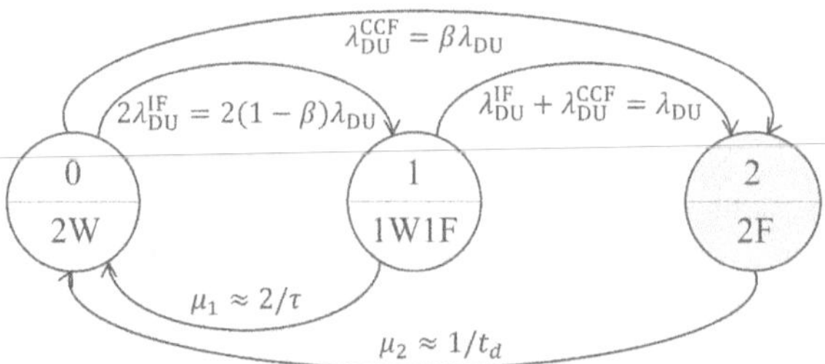


**FIGURE 10.9** The Markov model of a 1oo2 barrier system with CCFs.

from state 0 to state 1 is 2 times of the IF rate as $2(1-\beta)\lambda_{DU}$, since either of the two components can fail. From state 1 to state 0, only component can be failed, but the failure can be an IF or can be due to the common cause, even though the other component has been failed. Thus, the transition rate from state 1 to state 2 is the total DU failure rate.

When one failed component is revealed in a proof test, it can be restored to the functioning state, and the repair time is ignored in the model of Figure 10.9 since it is assumed much shorter than the proof test interval. If the assumption is not reasonable in some practices, for example, in some remote operations, delivery of spare part needs several weeks, and the total period from the revealing of a fault to restoration is not neglectable. Based on the exponential distribution assumption for DU-failures, the failure can occur at any moment with the same possibility in the whole period, the expected downtime of the failed component is half of the proof test interval ($\tau/2$). The transition rate from state 1 to state 0, or the restoration rate after one component is failed, is thus approximated as $\mu_1 = 2/\tau$.

When two components are found faulty in the proof test, the expected sojourn at state 2 (downtime: $t_d$) is calculated based on the expected downtime due to a CCF from state 0 and the expected downtime due to a failure from state 1. For the latter situation, we need approximations in quantitative analysis. Since the value of $\beta$ is often small ($\leq 0.1$), we suppose that the occurrence times of individual failures of the two components are evenly distributed within the proof test interval, even though the second failure actually can be due to a common cause. With this assumption, the expected downtime in the latter situation can approximated as be $\tau/3$. The expected sojourn time at state 2 is:

$$t_d \approx P_{0\to2}\frac{\tau}{2} + P_{0\to1\to2}\frac{\tau}{3}$$

where $P_{0\to2}$ is the possibility that arrival of the system at state 2 is directly from state 0 when the system is found failed in a proof test, and $P_{0\to1\to2}$ is the possibility that the system arrives at state 2 via state 1 when the system is found failed in a proof test. $P_{0\to2}$ is the ratio of CCF probability to the total failure probability, while $P_{0\to1\to2}$ is the ratio that the fault is resulted from two sequential failures. We have known the probability of CCF in a proof test interval is approximated as $\beta\lambda_{DU}\tau$, and the probability of two sequential failures is approximated as $(1-\beta)(\lambda_{DU}\tau)^2$. Then the expected downtime is:

$$t_d \approx \frac{\beta\lambda_{DU}\tau}{\beta\lambda_{DU}\tau + (1-\beta)(\lambda_{DU}\tau)^2}\frac{\tau}{2} + \frac{(1-\beta)(\lambda_{DU}\tau)^2}{\beta\lambda_{DU}\tau + (1-\beta)(\lambda_{DU}\tau)^2}\frac{\tau}{3}$$

Therefore, the transition rate from state 2 to state 0, namely the restoration rate when two components are failed, is approximated as $\mu_2 = 1/t_d$. Then, we can calculate the sojourn probabilities of the system at different states, where the sojourn probability at state 2 is equal to $\text{PFD}_{\text{avg}}$.

**Example 10.3:**

*Markov Analysis for CCF*

We can conduct a quantitative analysis based on the Markov model and the data we use in Example 10.1. Keep DU-failure rates of components and the proof test interval same, and we consider the situation where $\beta = 0.01$. In this case, $\lambda_{DU}\tau = 3.0\times10^{-6}\times4380 = 0.01314$, and thus $t_d$ can be calculated as:

$$t_d \approx \frac{0.01\times0.01314}{0.01\times0.01314+0.99(01314)^2}\frac{\tau}{2}+\frac{0.01\times0.01314+0.99(01314)^2}{0.01\times0.01314+0.99(01314)^2}\frac{\tau}{3}\approx\frac{2}{5}\tau$$

The transition rate matrix of the Markov model in Figure 10.9 is:

$$\mathcal{A} = \begin{pmatrix} -2\lambda_{DU}+\beta\lambda_{DU} & 2(1-\beta)\lambda_{DU} & \beta\lambda_{DU} \\ 2/\tau & -(\lambda_{DU}+2/\tau) & \lambda_{DU} \\ 5/2\tau & 0 & -5/2\tau \end{pmatrix}$$

The following equations can be obtained as:

$$2(1-\beta)\lambda_{DU}p_0 - (\lambda_{DU}+2/\tau)p_1 = 0$$

$$\beta\lambda_{DU}p_0 + \lambda_{DU}p_1 - 5p_2/2\tau = 0$$

$$p_0 + p_1 + p_2 = 1$$

Then, we can resolve the equations and get the solution as:

$$p_0 = 0.9872,\ p_1 = 0.01276,\ p_2 = 1.189\times10^{-4}$$

This calculation can be realized with the following codes in Matlab:

```
syms p0 p1 p2;
e1=2*(1-0.01)*3e-6*p0-(3e-6+2/4380)*p1;
e2=0.01*3e-6*p0+3e-6*p1-(5/(2*4380))*p2;
e3=1-p0-p1-p2;
[p00,p01,p02]=solve(e1,e2,e3,p0,p1,p2);
vpa([p00,p01,p02],4)
```

In this case, $p_2$ is equal to $PFD_{avg}$, with the value close to but a bit lower than that calculated by approximation formulas ($1.22\times10^{-4}$) in Example 10.1. The reason is the departure rate from the failed state is higher ($5/2\tau$ compared with $2/\tau$). It also can be found that $PFD_{avg}$ is not linearly related with the departure rate from the failed state. Such results are valid since approximation formulas always need more conservation.

When the value of $\beta$ is higher, the contribution ratio of CCF to the system downtime is higher, and $t_d$ is closer to $\tau/2$. For example, when $\beta = 0.1$, the ratio of CCF in the total failure will be increased to 88.49%, and $t_d$ in this case is equal to 0.4808 $\tau$. Then, $PFD_{avg}$ is obtained as $6.973\times10^{-4}$, compared with $7.04\times10^{-4}$ by the approximation formula in Example 10.1. ■

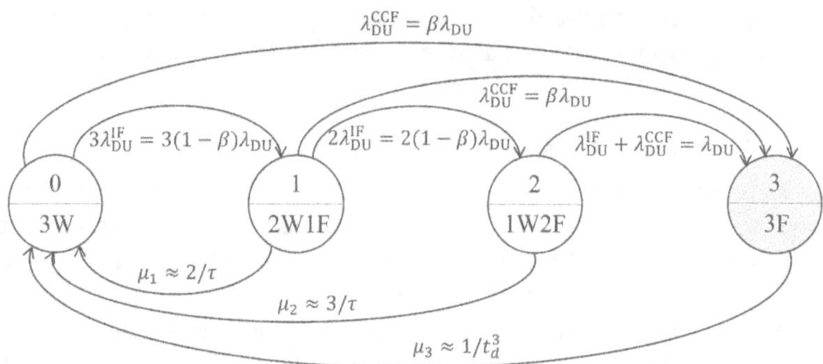


**FIGURE 10.10** The Markov model for a 1oo3 barrier system with CCFs.

For barrier systems with three identical components with independent DU-failures and common-cause DU-failures, we can build a Markov model based on the similar considerations. Figure 10.10 illustrates such a model for a 1oo3 barrier group. Based on the beta-factor model, $p_{2oo3} = 0$, thus there is no transition from state 0 to state 2.

States of the Markov model are described in the following table:

| State | Description |
|---|---|
| 0 | 0 channel is failed, 3 channels are functioning |
| 1 | 1 channel is failed, 2 channels are functioning |
| 2 | 2 channels are failed, 1 channel is functioning |
| 3 | 3 channels are failed, 0 channel is functioning |

The system can transit from state 0 directly to state 3 with the rate of $\beta\lambda_{DU}$. On the other hand, since three channels exist, the IF of any of them can result in a transition from state 0 to state 1, so that the transition rate of $3(1-\beta)\lambda_{DU}$. Then in state 1, the IF of any of the remaining two channels can lead to a transition to state 2, meaning a transition rate of $2(1-\beta)\lambda_{DU}$. Failure of the last remaining channel can be independent or due to the common cause, and thus the transition rate from state 1 to state 2 is $\lambda_{DU}$.

In terms of restoration, the departure rate from state 1 is equal to $2/\tau$, and that from state 2 is $3/\tau$ since the system falls in this state due to two IF. The transition rate from state 3 is calculated based on the expected downtime in this state, which should be an algorithmic average of the downtime by different failure sequences, as:

$$t_d^3 \approx P_{0\to3}\frac{\tau}{2} + P_{0\to1\to3}\frac{\tau}{3} + P_{0\to1\to2\to3}\frac{\tau}{4}$$

$\text{PFD}_{\text{avg}}$ of this barrier system can be obtained by capturing the sojourn probability of this Markov process in state 3.

For the 2oo3 configuration with the same three identical channels, its Markov transition diagram can be same with that in Figure 10.10, but the $\text{PFD}_{\text{avg}}$ is calculated by the sum of the sojourn probabilities of the Markov process in state 2 and state 3. Since the beta-model assumes that a CCF is the failure occurring in all channels, there is no transition from state 0 to state 2.

Interested readers can use the data in Example 10.2 to calculate $PFD_{avg}$ of 1oo3 and 2oo3 configurations and compare the results with those based on the approximation formulas.

#### 10.4.4.4 Beta-Factor Model for Nonidentical Channels

All the cases of parallel structures with CCFs above are consisted of identical channels. Now we consider barrier subsystems with nonidentical channels with different failure rates. Hauge et al. (2013) have proposed an approach for analyzing such systems by defining $\beta$ as a fraction of the geometric mean (GM) of the failure rates of the channels in study. For a parallel structure of $n$ nonidentical channels vulnerable to CCFs, we can set $\lambda_i$ as the failure rate of channel $i$, for $i = 1, 2, \ldots, n$. This failure rate can be for DD-, DU-, or total failure rate. Then, the GM failure rates of all the $n$ nonidentical channels is:

$$\lambda_{GM} = \left( \prod_{i=1}^{n} \lambda_i \right)^{\frac{1}{n}} \tag{10.20}$$

Since $\beta$ is defined as a fraction of the GM, the rate of IF of component $i$ becomes:

$$\lambda_i^{IF} = \lambda_i - \beta\lambda_{GM} \tag{10.21}$$

Rausand et al. (2020) indicate that when all the failure rates of components in a parallel structure are in the same order of magnitude, this approach can be acceptable. Otherwise, if the failure rates are very different, such an approach will result in unrealistic results. For example, consider a simple parallel structure of two components, and their failure rate is $\lambda_1 = 10^{-2}$, and $\lambda_2 = 10^{-6}$, respectively. The GM can be calculated with Equation (10.18) as $\lambda_{GM} = 10^{-4}$. If we suggest $\beta = 0.10$, the CCF rate is $0.10 \times 10^{-4} = 10^{-5} > \lambda_2$, but we know that the CCF rate cannot be higher than the total failure rate of a component. In such a case, we may set a very low $\beta$ to make $\beta\lambda_{GM}$ lower than the smaller failure rate, but this is not often coinciding with the observation on the component with the higher failure rate.

#### 10.4.4.5 Checklist for Determining the Beta Factor

The next question is how to obtain the value of $\beta$ in practice. Both IEC61508 (2010) and IEC62061 (2021) (standard for machinery safety) provide a checklist for the beta factor estimation, partly based on the measures taken to prevent CCFs. Seven types of factors are in consideration:

- *Separation/segregation*: Degree of physical separation and communications between components.
- *Diversity*: Difference of technologies and physical principles adopted by components.
- *Complexity/maturity*: It is related to cross-connection of components, the number of inputs and outputs of the system, and experience of the relevant technologies.

- *Assessment/analysis and feedback of data*: It is related to how much analysis has been conducted for avoiding CCFs.
- *Procedures and human interface*: This category is only covered by IEC61508 (2010). If a clear procedure of maintenance is in place and well documented, it is supposed to reduce CCFs.
- *Competence/training/safety culture*: It is related to whether designers understand the causes and consequences of CCFs, and whether the company has the culture to avoid CCFs.
- *Environmental control*: It is related to whether the system is always operated within harsh environment, including extreme temperature, humidity, corrosion, dust, vibration as well as electromagnetic interference.

Analysts can score on those items in the checklist, and then according to their scores to determine the value of $\beta$. Normally, a careful consideration for adopting risk reduction measures for CCFs, a higher score can be obtained. For in IEC62061 (2021), the full score for automatic diagnosis is 10, and if a machine is installed with such a diagnostic module that conducts tests with the frequency more than 1/min, a very high score of 9–10 can be issued. In IEC62061 (2021), the total score for 14 items can be up to 100, and the categorization in Table 10.1 can be used to determine the value of $\beta$.

The scoring procedure in IEC61508-6 is a bit more complicated than that in IEC62061 (2021) but is based on the same principle. Diagnosis test efficiency (including interval and coverage) is used in the scoring formula of IEC61508 (2010) for adjusting the score before the final determination of the value of $\beta$.

#### 10.4.4.6 Data-Driven Estimation of the Beta Factor

The U.S. Nuclear Regulatory Commission (NRC) has released several reports on guiding the estimation of the beta factor based on collected field data (NUREG/CR-5485, 1998; NUREG/CR-6268, 2007). According to these reports, the estimator of the beta factor is firstly defined as:

$$\widehat{\beta_1} = \frac{N_{\text{DU,CCF}}}{N_{\text{DU}}} \tag{10.22}$$

**TABLE 10.1**
**Criteria for Estimating the Value of $\beta$**

| Overall Score | $\beta$ |
|---|---|
| ≤ 35 | 0.10 |
| 36–65 | 0.05 |
| 66–85 | 0.02 |
| 86–100 | 0.01 |

*Source:* **Adopted from IEC62061 (2021).**

where $N_{DU}$ is the observed number of DU-failures occurring in a homogenous population of barrier components over a period of time, while $N_{DU,CCF}$ is the number of barrier components that have experienced a CCF in the period.

Since this estimation does not consider whether the components experiencing CCFs belong to the same CCF group or not, it inclines to a conservative result when the population is large. Therefore, an alternative formula has been proposed with less conservative results:

$$\widehat{\beta_2} = \frac{2 \cdot N_{CCF}}{N_{DU,I} + 2 \cdot N_{CCF}} \tag{10.23}$$

where $N_{DU,I}$ is number of independent DU-failures, and $N_{CCF}$ is the number of CCF events. This estimation assumes that each CCF event occurs with failures of two components even though it is not always the fact. When different components fail due to the same cause, this is regarded as a CCF event. It should be noted that the beta factor here refers to the fraction of a components failure that is a CCF. For instance, for two components, A and B, three types of failures are observed: one IF on A, one IF on B, and one CCF event (both A and B fail). However, in this case, both A and B are observed failed twice, and the beta factor is estimated according to Equation (10.22) as 0.5, even there is only one CCF event (Hauge et al., 2016).

**Example 10.4**

*Estimating the Beta Factor Based on Data*

This example is based on the case study in the paper of Hauge et al. (2016). Suppose a total population pressure transmitters with the accumulated operational time of 10 million h. A total of 100 DU failures are recorded in the observation period, and ten of them occur in CCF events. After the analysis, it can be found that four CCF events have occurred.

According to Equations (10.22) and (10.23), we have

$$\widehat{\beta_1} = \frac{N_{DU,CCF}}{N_{DU}} = \frac{10}{100} = 0.1$$

$$\widehat{\beta_2} = \frac{2 \cdot N_{DU,CCF}}{N_{DU,I} + 2 \cdot N_{CCF}} = \frac{2 \cdot 4}{(100 - 10) + 2 \cdot 4} = 0.082 \ \blacksquare$$

### 10.4.5 Other CCF Models

#### 10.4.5.1 C-Factor Model

Evans et al. (1984) have presented a *C-factor model* for analyzing CCF, where CCF is regarded as a fraction of the IF, namely, $\lambda^{CCF} = C\lambda^{IF}$. The total failure rate is then

$$\lambda = \lambda^{IF} + \lambda^{CCF} = (1 + C)\lambda^{IF} \tag{10.24}$$

This model looks like the beta-factor model, but it sometimes can interpret the relationship between CCF and the overall failure in a different way. In the beta-factor model, when determining the value of $\beta$, a harsh environment with possible more external shocks always means a higher $\beta$, but it is uncertain how the IF rate change as $(1-\beta)\lambda$.

**Example 10.5:**

*Limitation of Beta Factor Model*

For a component in a system with CCFs, it has an overall failure rate of $2.0\times10^{-6}$ per hour, and $\beta = 0.05$. We can calculate that its CCF rate is $1.0\times10^{-7}$ per hour. If the component is re-located in a harsher operational environment, its CCF rate becomes to $2.0\times10^{-7}$, if the IF rate of this component keeps unchanged, the overall failure rate becomes to $2.1\times10^{-6}$, and $\beta = 0.095$. In most applications, the IF will also increase under the harsher environment, but to ensure the value of $\beta$ is no less than 0.05, the IF rate cannot be more than $3.8\times10^{-6}$, but it is not valid to suppose that the increase of IF rate can be limited by the change of CCF. ■

With $C$-factor model, the change of value of $C$, namely the CCF rate, has no influence on the IF rate. For the case in Example 10.4, $\lambda^{IF} = 1.9\times10^{-6}$, and thus the original value of $C$ is 0.053. In the harsher environment, the new value of $C$ has no relation with the new value of $\lambda^{IF}$.

### 10.4.5.2 The Alpha-Factor Model

Mosleh (1987) proposes the *alpha-factor model* for the structures of $n$ identical components. In this model, given that a failure is observed at time $t$, it can be an IF or a multiple failure. The alpha-factor of a component is defined as:

$$\alpha_j = \Pr\left(Exactly\ j\ components\ fail \middle| \text{failure occurs}\right) \text{ for } j = 1,2,\ldots,n \quad (10.25)$$

If we still use $p$ to denote the overall failure probability of a component, the equation can be written as

$$\alpha_j = \frac{p_{joon}}{p} \quad (10.26)$$

We can find that $\sum_{j=1}^{n} \alpha_j = 1$.

For a structure of two identical components, $\alpha_2 = \beta$ in the beta-factor model, while $\alpha_1 = 1-\beta$. For a structure of more than two components, if $\alpha_n = \beta$, $\alpha_1 = 1-\beta$, and $\alpha_j = 0$ for $j = 2,\ldots,n-1$, the alpha-factor model can be transferred to a beta-factor model.

### 10.4.5.3 Binomial Failure Rate Model

The *BFR* model proposed by Vesely (1977) regards CCFs of components in a system are results of external shocks. The occurrences of shocks follow an HPP process with the rate of $\lambda_{ES}$, and when a shock occurs, each of the $n$ identical components in the system has a probability of $p$ to fail. The overall number of components that are failed due to the shock is therefore binomially distributed $X$~bin($n, p$).

The BFR model assumes that:

- Shocks and IFs occur independently of each other.
- The repair time on any failures is neglectable short.

Therefore, in a certain period, the numbers of both IFs and shocks are Poisson distributed. The overall failure rate of a component is obtained as the sum of IF rate and CCF rate as:

$$\lambda = \lambda^{IF} + \lambda^{CCF} = \lambda^{IF} + p\lambda_{ES} \quad (10.27)$$

As highlighted by Rausand et al. (2020), one limitation of the BFR model lies in the assumption that components fail independently when a shock occurs, and the problem can be partly mitigated by defining some of the shocks are "lethal" shocks, meaning that these shocks endanger the components in study and fail them, while the other shocks do not result in the failures of the components in study. The fraction of lethal shocks is equal to $p$.

**Example 10.6:**

***BFR Model for a 2oo3 Voting Structure***

Consider a 2oo3 voting structure of identical components with IF rate is $2.0 \times 10^{-6}$ per hour. Random shocks can occur according to an HPP process with the rate of $1.0 \times 10^{-5}$ per hour on the system (around once per year), and the probability of lethal shock is $p = 0.1$. Based on the BFR model, the failure rate of one component is $\lambda^{IF} + p\lambda_{ES} = 2.0 \times 10^{-6} + 0.1 \times 1.0 \times 10^{-5} = 3.0 \times 10^{-6}$per hour. We can use the binomial distribution formulas in Equation (5.46) to calculate the probabilities of different numbers of failures

$$\Pr(X = 0) = \binom{3}{0} 0.1^0 (1 - 0.1)^{3-0} = 0.729$$

$$\Pr(X = 1) = \binom{3}{1} 0.1^1 (1 - 0.1)^{3-1} = 0.243$$

$$\Pr(X = 2) = \binom{3}{2} 0.1^2 (1 - 0.1)^{3-2} = 0.027$$

$$\Pr(X = 3) = \binom{3}{3} 0.1^3 (1 - 0.1)^{3-3} = 0.0001$$

For this 2oo3 voting structure, when two or more components fail, the system fail, and thus the system failure probability is $\Pr(X=2)+\Pr(X=3)=0.028$. Then, the CCF rate that can result in the system failure is $0.028\times 1.0\times 10^{-5}=2.8\times 10^{-7}$ per hour, namely the frequency is once per 400 years. ■

#### 10.4.5.4 The Multiple Greek Letter Model

For a system consisting of more than two components, the beta-factor model does not consider the situations where several but not all components are failed at the same time. As a response, the *multiple Greek letter* (MGL) model has been developed (Fleming & Kalinowski, 1983) for generalizing the beta-factor model and including more possibilities of CCFs.

We can re-draw the Venn diagram of Figure 10.7 for three identical components in Figure 10.11 to consider the situation where $p_{2oo3}$ is not equal to 0. In this Figure, each circle represents the failure probability of one component, in other words the occurrence probability of event $E_i$. Let $M$ be the multiplicity of the fault, meaning the number of faulty components, and without losing generality consider the failure of component 1 ($E_1$), and then we can define as in the beta-factor model:

$$\beta=\Pr\left(M\geq 2|E_1\right)$$

It means that $\beta$ is the probability that there are at least one more fault given the condition we have found that component 1 is failed. Since the system is symmetric, $\beta$ can be written as:

$$\beta=\Pr\left(M\geq 2|M\geq 1\right) \tag{10.28}$$

Then we introduce a new Greek letter $\gamma$ and define $\gamma$ as the probability that there are at least one more fault given the condition we have found that two components are failed, as:

$$\gamma=\Pr\left[M\geq 3|M\geq 2\right] \tag{10.29}$$

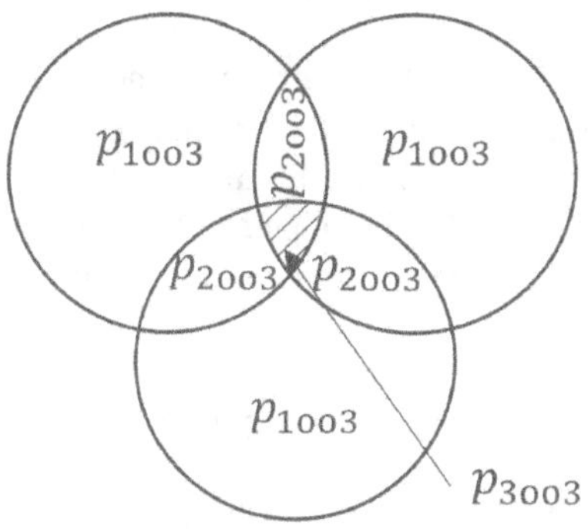


**FIGURE 10.11** Probabilities of fault multiplicities for three identical components (re-drawn).

For a structure with more than three components, we can even introduce more Greek letters for higher multiplicities of faults, for example:

$$\delta = \Pr\left[M \geq 4 \middle| M \geq 3\right] \tag{10.30}$$

Therefore, the beta-factor model can be regarded as a special case of the MGL model when $n = 2$. As shown in Figure 10.11, the probability of a fault with the multiplicity of $i$ on a specific component is $p_{io o3}$, and the overall failure probability of one component is:

$$p = \Pr(E_1) = p_{1oo3} + 2p_{2oo3} + p_{3oo3} \tag{10.31}$$

Thus, the parameter $\beta$ can be written as:

$$\beta = \Pr\left(M \geq 2 \middle| M \geq 1\right) = \frac{\Pr(M \geq 2)}{\Pr(M \geq 1)} = \frac{2p_{2oo3} + p_{3oo3}}{p} \tag{10.32}$$

Since only three components are in the system in Figure 10.11, $\gamma$ can be written as:

$$\gamma = \Pr\left[M = 3 \middle| M \geq 2\right] = \frac{\Pr(M = 3)}{\Pr(M \geq 2)} = \frac{p_{3oo3}}{2p_{2oo3} + p_{3oo3}} \tag{10.33}$$

Combining Equations (10.32) and (10.33), we have:

$$p_{3oo3} = \beta\gamma p \tag{10.34}$$

Replacing $p_{3oo3}$ in Equation (10.25), we can obtain:

$$p_{2oo3} = \frac{(1-\gamma)}{2}\beta p \tag{10.35}$$

Then,

$$p_{1oo3} = 1 - 2p_{2oo3} - p_{3oo3} = (1-\beta)p \tag{10.36}$$

The probability that one of the three components have IFs is thus equal to $3p_{1oo3} = 3(1-\beta)p$, and the probability that a double fault exists is $3(1-\gamma)\beta p/2$.

For a 1oo3 voting structure, the system fails when all the three components fail, so that the system failure can be developed in three scenarios: (1) three IFs, (2) a double failure and an independent failure, or (3) a triple failure, and the system failure probability is

$$F_{S:1oo3} = \left[(1-\beta)p\right]^3 + \frac{3(1-\gamma)\beta}{2}p^2 + \beta\gamma p \tag{10.37}$$

The RBD model for this 1oo3 structure is shown in Figure 10.12, where $\overline{\text{CCF}}_{(ij)}$ denotes the survivability from the double fault involving components $i$ and $j$.

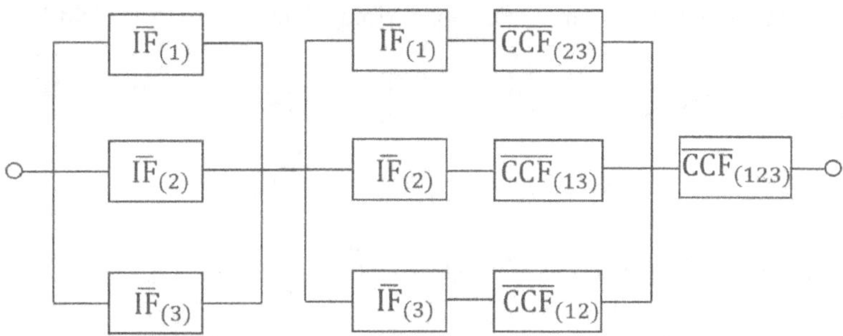


**FIGURE 10.12** RBD for a 1oo3 voting structure based on the MGL model.

Consider this structure is for a barrier subsystem with three identical components with constant DU failure rates ($\lambda_{\text{DU}}$) and the proof test interval of $\tau$, $\overline{\text{IF}}_{(k)}$ $\overline{\text{CCF}}_{(ij)}$ is same for any $i$, $j$, and $k$, and thus second items in Equation (10.30) also can be regarded as a 1oo3 voting structure with a channel reliability of $e^{-(1-\beta)\lambda_{\text{DU}}t} \cdot e^{-\frac{(1-\gamma)}{2}\beta\lambda_{\text{DU}}t} = e^{-\left(1-\frac{\beta}{2}-\frac{\beta\gamma}{2}\right)\lambda_{\text{DU}}t}$. Then, $\text{PFD}_{\text{avg}}$ of the whole system can be approximated according to Table 8.1 as:

$$\text{PFD}_{\text{avg}}^{\text{1oo3}} = \frac{\left[(1-\beta)\lambda_{\text{DU}}\tau\right]^3}{4} + \frac{\left[\left(1-\frac{\beta}{2}-\frac{\beta\gamma}{2}\right)\lambda_{\text{DU}}\tau\right]^3}{4} + \frac{\beta\gamma\lambda_{\text{DU}}\tau}{2} \tag{10.38}$$

While for a 2oo3 voting structure, it fails due to (1) two IFs, (2) a double failure, or (3) a triple failure, and thus the system failure probability is:

$$F_{\text{S:2oo3}} = \binom{3}{2}\left[(1-\beta)p\right]^2 + \frac{3(1-\gamma)\beta}{2}p + \beta\gamma p = 3\left[(1-\beta)p\right]^2 + \frac{(3-\gamma)\beta p}{2} \tag{10.39}$$

The RBD model for this system is shown in Figure 10.13.

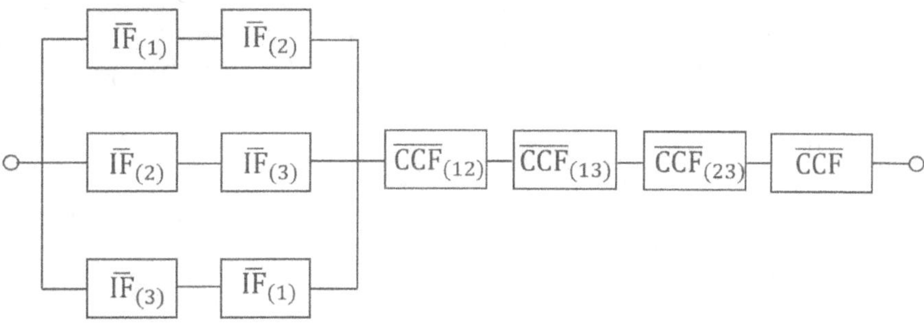


**FIGURE 10.13** RBD for a 2oo3 voting structure based on the MGL model.

$\text{PFD}_{\text{avg}}$ of the whole system can be approximated according to Table 8.1 as

$$\begin{aligned}\text{PFD}_{\text{avg}}^{2\text{oo}3} &= \left[(1-\beta)\lambda_{\text{DU}}\tau\right]^3 + \left[3(1-\gamma)\beta + \beta\gamma\right]\frac{\lambda_{\text{DU}}\tau}{2} \\ &= \left[(1-\beta)\lambda_{\text{DU}}\tau\right]^3 + \frac{(3-2\gamma)}{2}\beta\lambda_{\text{DU}}\tau\end{aligned} \tag{10.40}$$

We also can use the Markov method to analyze 1oo3 and 2oo3 voting structures based on the MGL model.

With simple comparisons between the formulas based on the beta-factor model and the MGL, we can find that the latter one can result in more conservation estimation on the integrity of safety barriers.

#### 10.4.5.5 The Multiple Beta-Factor Model

Another model developed based on the beta-factor model is the *multiple beta-factor (MBF)* model (Hauge et al., 2013; Hokstad & Rausand, 2008). In the MBF model, the probability of a *koon* voting structure of identical components being in the failed state due to a CCF is given by

$$F_{koon} = C_{koon}\beta p \tag{10.41}$$

where $p$ is still the overall failure probability of a component, $\beta$ is the fraction of double failures, and $C_{koon}$ is the correction factor relying on the structure, in other words, the fraction of multiple failures. The MBF model is developed on the basis of experiences from the oil and gas industry, $C_{koon}$ is determined with consensus on the impact of redundant structures, and its value can be updated with input data. When $C_{koon} = 1$, the MBF model becomes to a beta-factor model.

To provide an explicit formular for $C_{koon}$, the MBF model introduce $\beta_j$ (so that MBF) as the probability that the $(j+1)$th component fails given that components $1, 2, \ldots, j$ have just been failed due to a CCF. We use a structure of three components to present the meaning of $\beta_j$, where $C_{koon} = \beta_2$, as shown in Figure 10.14.

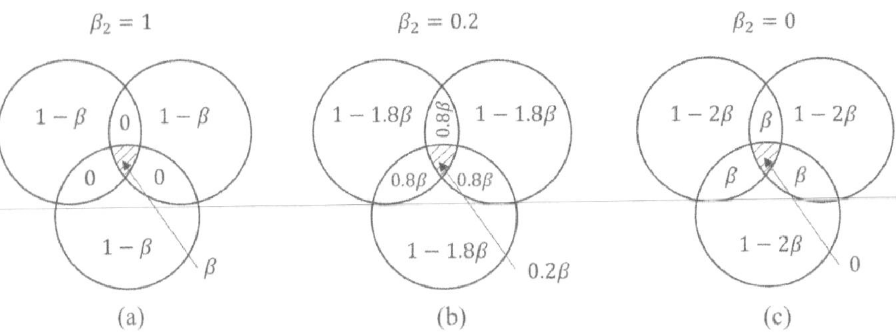


**FIGURE 10.14** (a–c) Illustration of the MBF model for a system of three components when $\beta_2$ is set as different values.

It can be found that when $\beta_2 = 1$, the MBF is equivalent with the beta-factor model for the system of three components as shown in Figure 10.14(a), and when $\beta_2 = 0$, there is no CCF of triple fault related to this system, as shown in Figure 10.14(c). Between these two extreme situations, $0 < \beta_2 < 1$, and in Figure 10.14(b), $\beta_2 = 0.2$, meaning that probability that a CCF damages all the three components given that two components are failed is 0.2.

Let $p_{ioon}$ denote the probability that exactly $i$ components are failed out of $n$ components, and it can be the function of $\beta_j$. As shown in Figure 10.14(b), for the structure of three components,

$$p_{1oo3} = 3\left[1 - (2 - \beta_2)\beta\right]p$$

$$p_{2oo3} = 3(1 - \beta_2)\beta p$$

$$p_{3oo3} = \beta_2 \beta p$$

It is necessary to repeat that $\beta$ here is the fraction of double failures, not totally same with the meaning of that in the beta-factor model.

For a 1oo3 voting structure, its failure probability due to a CCF is:

$$F_{1oo3} = p_{3oo3} = \beta_2 \beta p$$

For this perspective, we also can find that $C_{1oo3} = \beta_2$.

While for a 2oo3 voting structure, its failure probability due to a CCF is:

$$F_{2oo3} = p_{2oo3} + p_{3oo3} = (3 - 2\beta_2)\beta p$$

Thus, $C_{2oo3} = 3 - 2\beta_2$.

The knowledge for estimating $\beta_2$ is very limited, but in the PDS Method Handbook published by SINTEF[1] (Hauge et al., 2013), $\beta_2 = 0.3$ is taken as the base case for the offshore industry. It means that $C_{1oo3} = 0.3$, and $C_{2oo3} = 2.4$. The PDS Method Handbook also presents the generic formulas for calculating $C_{koon}$ as:

$$C_{koon} = \beta_2 \sum_{j=n-k+1}^{n} \binom{n}{j} \beta_3^{j-3} (1 - \beta_3)^{n-j} \quad \text{for } j = 1, 2, \ldots, n-2 \tag{10.42}$$

and

$$C_{(n-1)oon} = \binom{n}{2}\left(1 - \frac{\beta_2}{\beta_3}\right) + \beta_2 \sum_{j=2}^{n} \binom{n}{j} \beta_3^{j-3} (1 - \beta_3)^{n-j} \tag{10.43}$$

For the deduction process of these formulas, readers can reference the paper by Hokstad and Corneliussen (2004).

Hauge et al. (2016) have proposed a method that utilizes the NUREG framework for estimating parameters within the MBF model. This approach specifies $n$

as the number of components within CCF groups affected, where $n = 2$ denotes a scenario only involving the channels of a duplicated structure. Let $K$ represent the total observed number of failure events, comprising both IFs and CCFs. It may be determined that two CCF groups are affected, despite that the failures seem to occur "simultaneously", each group is considered a separate CCF event in such instance. Let $Y_j$ denote the number of components that have failed during the $j$th failure event $(j = 1, 2, \ldots, K)$. The MLE of $\beta$ is:

$$\hat{\beta}_K = \frac{\sum_{j=1}^{K} Y_j \left(Y_j - 1\right)}{(n-1)\sum_{j=1}^{K} Y_j} \tag{10.44}$$

where $\sum_{j=1}^{K} Y_j = N_{DU}$ is the total number of DU-failed components. Consider only one failure event (which is most probable if the proof test interval is not too long), and $Y_1$ (out of $n$) components failed in this event, $\hat{\beta}_1 = (Y_1 - 1)/(n - 1)$. This means the conditional probability that given that one channel has failed, exactly $(Y_j - 1)$ of the remaining $(n - 1)$ channels fail. For a 1oo2 structure, $\hat{\beta}_1 = 0$ when $Y_1 = 1$ and $\hat{\beta}_1 = 1$ when $Y_1 = 2$. While for a subsystem with three channels, $\hat{\beta}_1 = 0$ when $Y_1 = 1$ and $\hat{\beta}_1 = 0.5$ when $Y_1 = 2$.

#### 10.4.5.6 Markov Analysis Based on the MBF Model

Here we can consider a barrier subsystem with three channels, given that a DU-fault has been revealed in a proof test, based on the analysis in the last subsection, the probability that it is a triple fault is $p_{3oo3} = \beta_2 \beta p$, the probability of the fault existing in two channels is $p_{2oo3} = 3(1 - \beta_2)\beta p$, while the probability of an individual fault is $p_{1oo3} = 3[1 - (2 - \beta_2)\beta]p$. The Markov diagram in Figure 10.10 can thus be modified, as shown in Figure 10.15.

In Figure 10.15, a transition exists from state 0 to state 2, this is different from the Markov diagram based on the beta model, indicating that a CCF involving two channels is possible. Similarly, state 1 can transit to state 3, with a rate of $(1 - \beta_2)\beta$ because only two channels are still functioning in state 1.

It is also noted that since the subsystem enters state 2 due to two possible reasons: a CCF of two channels with the probability of $3(1 - \beta_2)\beta p$, or two individual

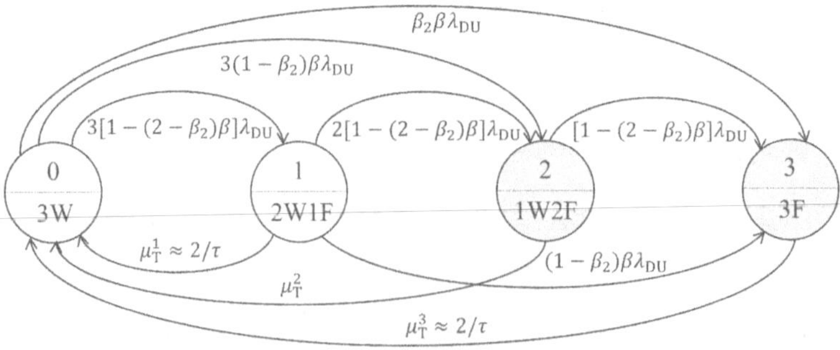


**FIGURE 10.15** The Markov model of a 2oo3 barrier subsystem with CCFs based on MBF model.

DU-failures with the probability of $3[1-(2-\beta_2)\beta]^2 p^2$, there are potentially two different sojourn times in state 2. The values of restoration rate $\mu_T^2$ from state 2 to state 0 can be calculated based on the total probability formula where the above two situations considered (ignoring the time in repair):

$$\mu_T^2 \approx \frac{3(1-\beta_2)\beta p \cdot \frac{2}{\tau} + 3\left[1-(2-\beta_2)\beta\right]^2 p^2 \cdot \frac{3}{\tau}}{3(1-\beta_2)\beta p + 3\left[1-(2-\beta_2)\beta\right]^2 p^2}$$

In most cases, since $p$ is small (e.g. <0.01), $\mu_T^2$ can be further approximated as $2/\tau$ if $(1-\beta_2)\beta$ is not very small, meaning that when a double DU-fault is revealed in a proof test, it is most probability resulted from a CCF, instead of two sequential individual failures.

Similarly, we need to consider four possible ways that the subsystem enters state 3: (1) a CCF occurring in three channels, (2) an individual DU-failure, and then a CCF occurring in the rest two channels, (3) a CCF occurring in two channels and then an individual DU-failure, and (4) three DU-failures. The calculation of $\mu_T^3$ is based on the four situations, but mostly, its value can be approximated as $2/\tau$ because there is high probability the triple fault is due to a CCF in three channels.

Readers interested in CCFs are also recommended to read the report by Hauge et al. (2015), where more application cases and estimation methods of the beta-factor can be found.

## 10.5 CASCADING FAILURES

### 10.5.1 Definitions and Classifications

*CAFs* are another type of dependent failures, they start from one component and then spread to the others. In literature, researchers use different terms to describe CAFs with different focuses, such as *induced failures* (Murthy & Nguyen, 1985), *domino effect* (Cozzani et al., 2005), *propagated failures* (Levitin & Xing, 2010), *and failure interactions* (Liu et al., 2015).

CAFs can be triggered by either unintentional causes/events, such as failures of equipment, human errors, or natural disasters, or intentional causes/events, such as cyberattacks or sabotage (Iaiani et al., 2022). In this book, the first event in the chain is called as the *initiating event*, and if the event is an equipment failure, it is called as the *initiating failure*. The propagation of failure to more components in a system is the *cascading effect*. Then, we propose the specific definition of CAF in a socio-technical system:

- *CAFs*: Chain failures in two or more components, resulted from an initiating failure in one of the components.

This definition can be understood as a subset of domino effect, which is a propagation sequence as the escalation of the primary/initiating event (Cozzani et al., 2005). In the studies for domino effects, events in the propagation sequence can be environmental change or wrong decision-making, not necessarily to be failures or deteriorations of hardware components. In our context, we only consider the state changes of barrier components as events, and such a definition is more similar with the propagated failure by Levitin and Xing (2010).

The root cause of CAFs should be tracked to the cause of the initiative failure in the firstly failed component. The cause can be a deficiency in design or engineering of this component or can be an external shock or environmental impact, such as a natural disaster, lack of maintenance, or an intentional attack. This initial cause is not necessary a shared one for all components in cascading. For example, aging in a power supply cable results in the short circuit in a condenser, igniting the polyurethane insulation materials on the equipment, and then the fire spreads to the other equipment in the same workshop. In this case, the shared failure cause for most of the failed equipment is the fire, but the root cause is the aging of the cable.

The coupling factors are the similar properties and dependencies between components. For instance, in the previous example, the similar property of all equipment in the workshop is their vulnerability to fire or high temperature, while the dependency lies in their geographic closeness. In addition, structural proximity, functional connection, and workload and resource sharing also can be the coupling factors of CAFs. For example, given that one maintenance team serves several barrier components, the team will be occupied if one component fails, perhaps delaying the planned maintenance works on other components and increasing their failure probabilities. The relationship between CAF with its root cause and coupling factors is illustrated in Figure 10.16.

Based on the different failure mechanisms, CAFs can be categorized as:

- *Failure-induced damage*: The failure of one component leads to a directly hazardous event, such as a fire, explosion, or corrosive gas, to its neighbor components, which then fail as the consequence of the hazardous event. Epidemic of infectious diseases among people has the similar principle with such kind of failure mechanism. In some cases, the initiative failure does not result in damages on others immediately, but the failure can change the working condition, for example, produce a lot of heat, water vapor or vibration, and then accelerate the degradation of other components.
- *Load re-distributed CAFs*: As discussed in Section 10.2 regarding dependence, multiple components in a system may share a workload. The failure of one component results in the redistribution of the workload within the group, leading to the overload on other components. The increased workload may fail these components or accelerates their degradations. In some

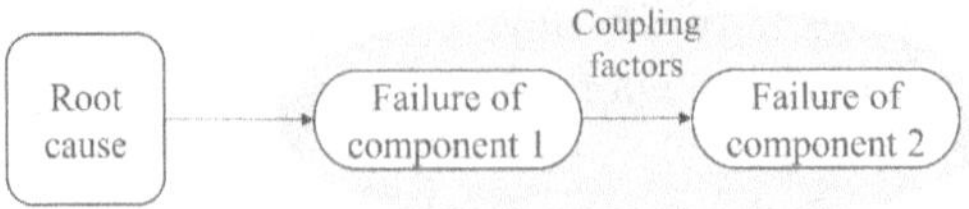


**FIGURE 10.16** Illustration of CAF mechanism.

parallel systems and networked systems, like power grids and urban transportation networks, such workload sharing phenomena are common. When one item fails, for example, one road section is blocked, drivers will select alternative paths, and make traffic jams on other roads. Similarly, within a redundant barrier system, when one channel is failed, the remaining channels must be used more frequently, which can result in faster degradation or higher failure rate. The failure of these channels after the first channel failure can be regarded as a CAF.

In addition, there is another scenario, called as functional impact, where the failure of one component blocks the functioning of other components. For example, in a series structure, if one component fails, other components cannot realize their functions even though they have no failures. Another case is from the situation where the start of task for one component relies on the finish of the task of another component, and if the latter one fails, the previous has no opportunity to be started. It can be regarded as a CAF or not, because the component affected does not function as intended, but it does not have a technical failure.

### 10.5.2 Comparisons of CCFs and CAFs

Both CCFs and CAFs are due to some common vulnerabilities of multiple components, but there are differences between the two types of failures that should be noted:

- *Initiation of failures*: The occurrence of a CCF is attributed to a shared root cause, which can be an inherent defect or an external shock. For instance, when several components are subjected to high temperatures or corrosive materials, they may fail either simultaneously or within a short interval. In this scenario, the high temperature or corrosive materials serve as the common cause, and the exposure is the triggering event. Conversely, a CAF originates from the failure of a specific component, which then acts as the catalyst for subsequent failures. The other components may not be directly exposed to the same conditions (such as high temperature or corrosive materials), but they may fail due to the increased workload resulting from the initial failure.
- *Propagation of failures*: Propagation refers to the evolution of failures following the initial event. A CCF is a first-in-line failure, resulting immediately from the root causes, with no further progression or spread (Xie et al., 2018). This is in contrast to CAF, which involves a complex series of interactions that not only escalates but also compounds the severity of the situation, affecting additional components over time. Therefore, a CCF typically affects several items almost instantaneously or within a brief period, while the progression of a CAF can be either rapid or prolonged, depending on the nature and extent of the interactions among the failed components.
- *Mechanisms of failures*: A CCF often results from a single, identifiable cause that impacts multiple components in a similar manner. On the other hand, the mechanisms behind CAFs are more dynamic and interactive,

with the initial failure leading to a domino effect where mechanisms of the subsequent failures can vary significantly triggered by the altered operational conditions rather than a single shared cause.

### 10.5.3 Risk Reduction Measures for CAFs

Because the differences listed above, the risk reduction and management measures for CAFs are also not same with those for CCFs. Here, we can provide some generic measures in barrier design and operation for reducing the likelihood of CAFs and mitigating the consequence when the cascading process has started. The main preventive methods that can be used in design and operation include:

- *Modularization*: Modularization means to design a system into discrete, self-contained modules that can perform specific functions independently of one another. Each module is designed to operate as a standalone unit, thus a failure in the module is isolated and less likely to affect the operation of others. In addition, fault diagnosis and maintenance is easier for a modular system, minimizing downtime of the failed item and reducing the potential load re-distributed CAF.
- *Decentralization*: Designing a system with a decentralized architecture can help limit the spread of failures. In barrier engineering, this means that barrier equipment should not be contained in a small area, and barrier functions should not be integrated into a central piece of equipment. Such design can reduce the potential for a single failure to damage the nearby items and reduce the speed of failure propagation.
- *Load balancing*: This means to evenly distribute workload across multiple components or channels. Such an approach can effectively prevent overloading and thereby reduce the risk of CAFs.

Additionally, some design and administrative strategies employed to mitigate CCFs can also be effectively adapted to minimize the risk of CAFs. For instance, implementing diversity in design is an effective approach, which is capable of eliminating the coupling factors among components that fulfill similar functions.

On the other hand, the mitigation methods after the initiating event of a cascading process has occurred include:

- *Separation and segmentation*: The failed item should be separated from the others to stop the cascading process. For the asset to be protected, safety barriers sometimes are installed with these functions. For example, firewalls are used to stop the spread of fire from one place to other facilities. Within a barrier system, isolation of failed component or channel sometimes is also needed, which can be removed to ensure the normal operation of other channels are not impacted.
- *Load management*: Following the same principle of load balancing, load management means the strategic redistribution of workloads in the initiating event of CAF, ensuring that even distribution of workload is maintained.

- *Dynamic risk assessment*: Given that CAFs are revealed over time, it is possible to implement dynamic risk assessment tools to adapt to varying conditions and pinpointing potential risks in real time. Consequently, adaptive safety measures can be promptly enacted to mitigate these emerging risks throughout the duration of the cascading process.

### 10.5.4 Modeling Approaches for CAFs

In general, CAFs also can be modeled with two types of approaches:

- *Explicit modeling approach*: Failure propagation paths among components are explicitly modeled and such an approach is mainly for the barrier systems suffering failure-induced damages. We will introduce one of such explicit models in the Section 10.5.6.
- *Implicit modeling approach*: The changes of system states are illustrated in these models, but they do not directly reflect how failures propagates. Such an approach can be used for analyzing CAFs due to failure-induced damages and load redistributions. Markov models in the subsection 10.5.5 and event trees belong to this type.

### 10.5.5 Implicit Modeling for CAFs

For a 1oo2 configuration with two identical channels, Chapter 9 introduces a discrete state transition model for degradation analysis, which idea can be used here for modeling the impact of CAF in a simple Markov model, as illustrated in Figure 10.17. This belongs to the implicit modeling approach because it does not explicitly simulate the CAF path.

The impact of CAFs can be captured by modifying the DU-failure rates of the channels. Initially, both channels are functioning with a normal failure rate, symbolized by a transition rate of $2\lambda_{DU\_N}$ from state 0 to state 1. In case the failure of one channel, the system enters state 1, while the DU-failure rate of the remaining functioning channel becomes to $\lambda_{DU\_V}$, as the combination of $\lambda_{DU\_N}$ and an additional rate attributed to CAF, which may result from failure-induced damage or overload.

The recovery rate from state 1 to state 0 is determined as $2/\tau$, since of a single DU-fault is revealed during a proof test. The transition rate from state 2 back to state 0 is more complex because the occurrences of the two failures cannot be simply assumed evenly distributed in the proof test interval. Considering $\lambda_{DU\_V}$ is higher

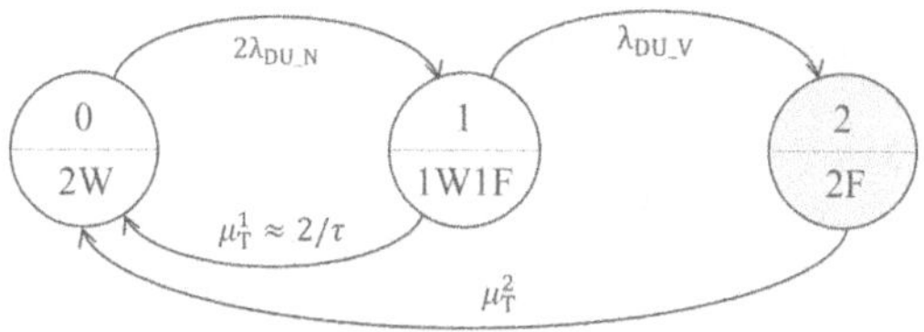


**FIGURE 10.17** The Markov model of a 1oo2 barrier subsystem with CAFs.

than $\lambda_{DU\_N}$, the expected system downtime extends beyond $\tau/3$. However, it is challenging to make a reasonable approximation for the restoration rate.

For more complex configurations, the modeling approaches are similar. We do not provide more examples.

### 10.5.6 Cascading Intensity Model for Analyzing Failure-Induced Damages

To investigate more details of the failure propagation, we can use the explicit modeling approach, where we can have deeper understandings on how cascading effects occur within a barrier system. This approach needs a topological model of the studying system, and in this chapter, we use reliability block diagram (RBD) as the foundation.

In the following analysis of CAFs, we need some assumptions:

- A barrier component can lose its functioning state due to an IF or a cascading effect;
- Propagation duration of CAFs is rather short and can be ignored;

In this section, we introduce a new parameter of *cascading intensity*, denoted as $\gamma_{ij} \in [0,1]$ $(i \neq j)$, to represent the easiness that the failure of component $i$ affects component $j$ in a certain period, namely

$$\gamma_{ij} = \Pr\left(j \text{ fails due to the failure of } i \middle| i \text{ has failed}\right) \tag{10.45}$$

The cascading effects due to both failure-induced damage and load-sharing failures can be reflected by cascading intensity. Such a model is thus called as the *gamma-model*, proposed by Xie et al. (2021), but it is noted that this is an explicit modeling approach.

A higher $\gamma_{ij}$ implies that failure propagation from $i$ to $j$ is more likely to occur. Like other parameters in barrier engineering, the value of $\gamma_{ij}$ can be estimated based on experimental data or historical data, with the approaches introduced in Chapter 6.

Next, we consider the CAFs within a barrier subsystem with the $k$-out-of-$n$ ($k$oo$n$) configuration. For example, Figure 10.18 illustrates in a 1oo3 voting structure with CAFs among barrier channels.

For a group with $n$ barrier components, we can establish a cascading matrix $\Gamma$ to describe all the potential failure propagations among components.

$$\Gamma = \begin{bmatrix} 1 & \gamma_{12} & \cdots & \gamma_{in} \\ \gamma_{21} & 1 & \cdots & \gamma_{2n} \\ \vdots & \vdots & & \vdots \\ \gamma_{i1} & \gamma_{i2} & \ddots & \gamma_{in} \\ \vdots & \vdots & & \vdots \\ \gamma_{n1} & \gamma_{n2} & \cdots & 1 \end{bmatrix} \tag{10.46}$$

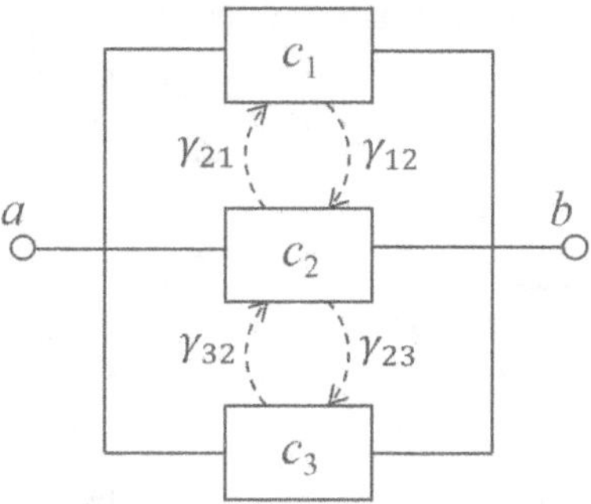


**FIGURE 10.18** RBD-based model for a 1oo3 structure with cascading failures.

All the elements on the diagonal of this matrix are set as 1. Then, $\delta_{ij} = 1 - \gamma_{ij}$ is used to denote the escaping probability of component $j$ from the impact of failure of component $i$. For the example in Figure 10.18, its cascading matrix is:

$$\Gamma = \begin{bmatrix} 1 & \gamma_{12} & 0 \\ \gamma_{21} & 1 & \gamma_{23} \\ 0 & \gamma_{32} & 1 \end{bmatrix}$$

It means that cascading effect can occur from component 1 to component 2, from 2 to 3, from 2 to 1, and from 3 to 2, with the associated cascading intensities.

Xie et al. (2021) have proposed a recursive approach for failure probability analysis of the barrier subsystem with a *koon* voting structure. Suppose that all barrier components are new when they are put into operation, and as good as new after a proof test. Let $n_C$ as the number of cascading effects from the initiating failure, and we can find that in the structure when $n_c \geq n - k$, the group fails, and when $n_c \leq n - k - 1$, failure of the group depends on how many of the rest components fail independently. In anyway, if more than $n - k$ components fail, the group fail. Here, we skip the cascading effects on the failed components.

$F_i(t_i)$ is set as the failure probability of component $i$ due to an IF by time $t_i$. Let $F_G(t_i, t)$ denote the probability that the barrier group G (G = [1,2,…,$n$]) fails in the interval of $(t_i, t)$ given that component $i$ fails at time $t_i$, and $F_G(t)$ or $F_G(0, t)$ denote the probability that the barrier group fails by time $t$ given that all the components are put into operation at time 0 and functioning. We introduce the following recursive aggregation-based formula for calculating the failure probability of the barrier group as:

$$F_G(t) = \sum_{i \in G} \int_0^t F_G(t_i, t) \prod_{i \neq j, j \in G} R_j(t) dF_i(t_i) \tag{10.47}$$

The approach is called recursive since we need to calculate $F_G(t_i,t)$ as:

$$\begin{aligned} F_G(t_i,t) &= \Pr(n_c=0)F_{G-\{i\}}(t_i,t) + \sum_{j_1\in G-\{i\}} \Pr(n_c=1)F_{G-\{i,j_1\}}(t_i,t) \\ &+\ldots+ \sum_{j_1,j_2\ldots j_{n-k-1}\in G-\{i\}} \Pr(n_c=n-k-1)F_{G-\{i,j_1,j_2\ldots j_{n-k-1}\}}(t_i,t) \\ &+ \Pr(n_c \geq n-k) \end{aligned} \tag{10.48}$$

In the above equation, $F_{G-\{i\}}(t_i,t)$ denotes the probability that the failures within the other components except $i$ result in the system failure in the interval $(t_i,t)$, and thus $\Pr(n_c=0)F_{G-\{i\}}(t_i,t)$ is the probability that no CAF starts from component $i$ but the system fails. In the second item of Equation, $F_{G-\{i,j_1\}}(t_i,t)$ is the probability that the failures within the components except $i$ and $j_1$ result in the system failure given that $j_1$ has been affected by the failure of component $i$, and then, $\sum_{j_1\in G-\{i\}}\Pr(n_c=1)F_{G-\{i,j_1\}}(t_i,t)$ means the sum of all probabilities that one component is failed by the cascading effect from component $i$ and the system fails. When $n_c=n-k-1$ cascading effects occur, only one more failure in the rest of components can lead to system to fail, but there are multiple combinations of components that are affected by the cascading process starting from component $i$. $\Pr(n_c \geq n-k)$ denotes the probability that the system fails due to CAFs.

For most barrier systems, we assume that all components in a group are identical, and therefore we can set $\gamma_{ij}=\gamma$, and $\delta_{ij}=\delta$. Then, the probability that there are $n_c$ CAFs is:

$$\Pr(n_c) = \begin{pmatrix} n_c \\ n-1 \end{pmatrix} \delta^{n-n_c-1}\gamma^{n_c} \tag{10.49}$$

With the exponential distribution assumption, when some components have failed, the time to failure of the rest component $G-\{i,j_1,j_2\ldots\}$ is still following the exponential distribution. The failure probability of the barrier group can be written as

$$F_G(t) = n\left[\begin{array}{c} \delta^{n-1}F_{n-1}(t) + \begin{pmatrix} 1 \\ n-1 \end{pmatrix}\delta^{n-2}\gamma F_{n-2}(t)+\cdots \\ + \begin{pmatrix} n-k-1 \\ n-1 \end{pmatrix}\delta^{n}\gamma^{n-k-1}F_{n-(n-k-1)}(t) \\ + \left(\begin{pmatrix} n-k \\ n-1 \end{pmatrix}\delta^{k-1}\gamma^{n-k} + \begin{pmatrix} n-k-1 \\ n-1 \end{pmatrix}\delta^{k-2}\gamma^{n-k+1}+\cdots+\begin{pmatrix} k-1 \\ n-1 \end{pmatrix}\gamma^{n-1}\right) \end{array}\right] \tag{10.50}$$

For any subset in G, the failure probability can be obtained in a similar way. Such recursive aggregations will continue until $(n-k-1)$ CAFs are found. The corresponding failure probability of the subset is $F_{G-(n-k-1)}(t) = 1 - e^{-k\lambda t}$.

Convolution and Laplace transformation can be used here to facilitate integration in Equation. (10.47). More details can be in studies by Xie et al. (2020) and Xie et al. (2021). The authors use Monte Carlo simulation to verify the above analytical formulas, and we do not present the procedure in this book due to page limitation.

Then, we consider two typical configurations in safety barriers: 1oo3 and 2oo3 voting structures. We assume that all components having a constant DU failure rate $\lambda_{DU}$ as the main contributor. According to Equation (10.48), reliability of a 2oo3 configuration can be obtained as:

$$R(t)^{(2oo3)} = 1 - F_G(t) = 3\delta^2 e^{-2\lambda_{DU}t} + \left(1 - 3\delta^2\right)e^{-3\lambda_{DU}t}$$

$PFD_{avg}$ of this 2oo3 barrier group can be calculated as:

$$PFD_{avg}^{(2oo3)} = 1 - \frac{1}{\tau}\int_0^\tau R(t)^{(2oo3)}\,dt = 1 - \int_0^\tau \left(3\delta^2 e^{-2\lambda_{DU}t} + \left(1 - 3\delta^2\right)e^{-3\lambda_{DU}t}\right)dt$$

$$= 1 - \frac{3\delta^2}{2\lambda_{DU}\tau}\left(1 - e^{-2\lambda_{DU}\tau}\right) - \frac{\left(1 - 3\delta^2\right)}{3\lambda_{DU}\tau}\left(1 - e^{-3\lambda_{DU}\tau}\right)$$

Since barrier components are highly reliable, $\lambda_{DU}\tau > 0$ is smaller than 0.1, we can replace $e^{-2\lambda_{DU}\tau}$ and $e^{-3\lambda_{DU}\tau}$ by using their Taylor series:

$$\begin{aligned} PFD_{avg}^{(2oo3)} &= 1 - 3\delta^2\left(1 - \frac{2\lambda_{DU}\tau}{2} + \frac{(2\lambda_{DU}\tau)^2}{3!}\ldots\right) \\ &\quad - \left(1 - 3\delta^2\right)\left(1 - \frac{3\lambda_{DU}\tau}{2} + \frac{(3\lambda_{DU}\tau)^2}{3!}\ldots\right) \\ &\approx 3\left(1 - \delta^2\right)\frac{\lambda_{DU}\tau}{2} = 3\gamma(2 - \gamma)\cdot\frac{\lambda_{DU}\tau}{2} \end{aligned} \quad (10.51)$$

Similarly, reliability of a 1oo3 barrier group can be calculated as:

$$\begin{aligned} R(t)^{(1oo3)} &= 3\delta(1 - \delta\gamma)e^{-\lambda_{DU}t} + 3\delta^2(2\gamma - 1)e^{-2\lambda_{DU}t} \\ &\quad + \left(1 - 3\delta(1 - \delta\gamma) - 3\delta^2(2\gamma - 1)\right)e^{-3\lambda_{DU}t} \end{aligned}$$

Using their Taylor series deployment, $PFD_{avg}$ of the 1oo3 configuration can be approximated as:

$$PFD_{avg}^{(1oo3)} \approx 3\gamma^2\frac{\lambda_{DU}\tau}{2} \quad (10.52)$$

**TABLE 10.2**
**Approximation Formulas for $PFD_{avg}$ of *KooN* Configurations with CAFs**

| *K/N* | $N=1$ | $N=2$ | $N=3$ | $N=4$ |
|---|---|---|---|---|
| $K=1$ | $\lambda_{DU}\tau/2$ | $2\gamma\lambda_{DU}\tau/2$ | $3\gamma^2\lambda_{DU}\tau/2$ | $4\gamma^3\lambda_{DU}\tau/2$ |
| $K=2$ | – | $\lambda_{DU}\tau$ | $3\gamma(2-\gamma)\lambda_{DU}\tau/2$ | $4\gamma^2(3-2\gamma)\lambda_{DU}\tau/2$ |
| $K=3$ | – | – | $3\lambda_{DU}\tau/2$ | $4\gamma(3-3\gamma+\gamma^2)\lambda_{DU}\tau/2$ |
| $K=4$ | – | – | – | $2\lambda_{DU}\tau$ |

We can list all the approximation formulas of $PFD_{avg}$ for barrier groups with *KooN* ($N \le 4$) configurations, as shown in Table 10.2.

Given that $\gamma$ is small (e.g., $\le 0.2$), $\gamma^2$, $\gamma^3$, and $\gamma^4$ in Table 10.2 can be negligible, and thus we can further simplify these formulas for $PFD_{avg}$ in Table 10.3.

Observing the $PFD_{avg}$ values in Table, we can summarize a general approximation formular for *KooN* configurations with CAFs as

$$PFD_{avg}{}^{(KooN)} = \begin{pmatrix} N-1 \\ K-1 \end{pmatrix} N\gamma^{N-K} \frac{\lambda_{DU}\tau}{2} \tag{10.53}$$

Here, we can use a 3oo5 configuration as an example to examine the formula. We have:

$$R(t)^{(3oo5)} = \left(10\delta^3\gamma + 10\delta^7\right)e^{-3\lambda t} + \left(5\delta^4 - 20\delta^7\right)e^{-4\lambda t}$$
$$+\left[1 - \left(10\delta^3\gamma + 10\delta^7\right) - \left(5\delta^4 - 20\delta^7\right)\right]e^{-5\lambda t}$$

Then, according to Equation (10.6), $PFD_{avg}$ of a 3oo5 configuration can be obtained as:

$$PFD_{avg}{}^{(3oo5)} = 1 - \frac{1}{\tau}\int_0^\tau R(t)^{(3oo5)}\,dt = 5\gamma^2\left(6 - 8\gamma + 3\gamma^2\right)\frac{\lambda\tau}{2} \approx 30\gamma^2 * \frac{\lambda\tau}{2}$$
$$= \begin{pmatrix} 5-1 \\ 3-1 \end{pmatrix} 5\gamma^{5-3}\frac{\lambda\tau}{2}$$

**TABLE 10.3**
**Simplified Approximation Formulas for $PFD_{avg}$ of *KooN* Configurations with CAFs**

| *K/N* | $N=1$ | $N=2$ | $N=3$ | $N=4$ |
|---|---|---|---|---|
| $K=1$ | $\lambda_{DU}\tau/2$ | $2\gamma\cdot\lambda_{DU}\tau/2$ | $3\gamma^2\cdot\lambda_{DU}\tau/2$ | $4\gamma^3\cdot\lambda_{DU}\tau/2$ |
| $K=2$ | – | $\lambda_{DU}\tau$ | $6\gamma\cdot\lambda_{DU}\tau/2$ | $12\gamma^2\lambda_{DU}\tau/2$ |
| $K=3$ | – | – | $3\lambda_{DU}\tau/2$ | $12\gamma\cdot\lambda_{DU}\tau/2$ |
| $K=4$ | – | – | – | $2\lambda_{DU}\tau$ |

Such a result coincides with the value based on the general formula in Equation (10.53).

### 10.5.7 Load-sharing CAFs

Considering the impact of load-sharing, when one channel is failed in the above configuration, the other channels have to be used more often, and they may degrade faster and have a higher failure rate. Given that a demand arises, one channel or multiple channels are activated to perform the barrier function, to reduce the accumulative damage on specific channels, it is natural to consider activating an alternative channel(s) for the next demand. In such a rotating activation strategy, barrier channels are sharing the overall workload in the long term. We have some discussions on this in Section 8.4, the cumulative degradations of two channels in a rotating activation strategy can be illustrated in Figure 10.19.

During the five proof test intervals, four demands arrive at $t_1$, $t_2$, $t_3$ and $t_4$, respectively. For the first demand, barrier channel $B_1$ is activated, so that its cumulatively degradation increases suddenly. Then, for the next demand at $t_2$, it is $B_2$ that is activated.

Some other activation strategies also can be applied, for example, channel 1 is activated twice and then channel 2, or the channel with less degradation is activated. Zhang (2021) has investigated the impacts of channel activation strategies on the cumulative degradation of barrier actuators, to seek the optimal activation sequence of channels.

Suppose that both channels working at $t = 0$, and their normal aging is assumed as a homogeneous Gamma process, demands occur following a homogeneous Poisson process with rate of $\lambda_{DE}$, and let $N(t)$ denote the number of all demands on the barrier system by time $t$. Let $g_{ij}$ $(i = 1, 2, \ldots, N(t))$, as the damage by the $j$ th demand on channel $i$ ($i = 1,2$ for a 1oo2 configuration), $y_{ij}$ is gamma distributed with parameters $(\alpha, \lambda)$, and $G_i(t)$ is cumulative damage on channel $i$ due to demands by time $t$. Reliability of this 1oo2 configuration by time $t$ is the probability that the cumulative degradation of at least one channel is less than the threshold $G_L$, calucalted as in Equation (9.66).

Consider a certain time interval, the probability that channel 1 is activated to withstand $k$ demands within all $n$ demands follows the binominal distribution as:

$$f(k,n,p) = \binom{n}{k} p^k (1-p)^{n-k} \tag{10.54}$$

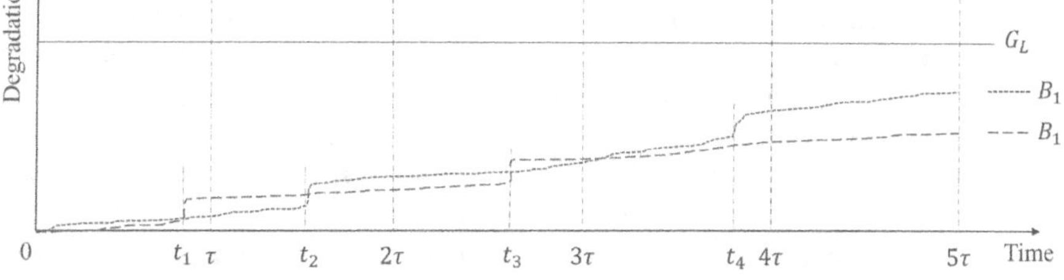


**FIGURE 10.19** Degradation of two barrier channels with sharing load.

Obviously, this probability is equal to that channel 2 withstands $(n-k)$ demands in the interval.

We can calculate the reliability of channel 1 by time $t$ with $N(t)=n$, as:

$$\begin{aligned} R_1(t) &= \sum_{k=0}^{n} f(k,n,p)\cdot \Pr\left(G_{N1}(t)+G_{S2}(t)<G_L | N(t)=k\right) \\ &= \sum_{k=0}^{n} f(k,n,p)\cdot \int_0^{G_L} G(G_L-u,t) f_{gi1}^{k}(u)du \end{aligned} \tag{10.55}$$

where $G_{N1}(t)$ is the degradation of channle 1 due to normal aging, and $G_{S2}(t)$ is the degradation of the channel due to sudden demands.

Similarly, reliability of channel 2 by time $t$ is calculated as:

$$\begin{aligned} R_2(t) &= \sum_{k=0}^{n} f(k,n,p)\cdot \Pr\left(G_{N2}(t)+G_{S2}(t)<G_L \middle| N(t)=n-k\right) \\ &= \sum_{k=0}^{n} f(k,n,p)\cdot \int_0^{G_L} G(G_L-u,t) f_{gi2}^{n-k}(u)du \end{aligned} \tag{10.56}$$

Then, for the barrier subsystem, we have:

$$\begin{aligned} R_S(t) &= 1-\left[1-R_1(t)\right]\left[1-R_2(t)\right] \\ &= \sum_{n=0}^{\infty}\left[1-\left[1-R_1(t)\right]\left[1-R_2(t)\right] | N(t)=n\right]\Pr\left[N(t)=n\right] \end{aligned} \tag{10.57}$$

Zhang (2021) also has derived the formulas for 2oo3 configuration and *KooN* configuration, which are skipped in the chapter due to the page limitation. When the value of $p$ in Equation (10.55) is changed, it means that the activation strategy is changed. If $p=0.5$, the two channels equally share the workload, while if $p=0$ or $p=1$, it means that one channel burdens all workloads. The author finds that at least for the 1oo2 configuration of barrier system, the equal sharing of workload can ensure the lowest PFD in average.

### 10.5.8 Closure

What we have discussed so far in this chapter are the CAFs within barriers. Safety barriers are always deployed in a large system serving as preventative measures to halt the failure propagation across different assets. A typical example of such applications is in controlling the spread of fires from one facility to another. Landucci et al. (2015) and (2016) have highlighted the substantial benefits that safety barriers offer in managing the risks associated with domino scenarios or CAFs. They provided a comprehensive framework for the main metrics and quantitative methods used to assess the performance of these safety barriers. The metrics discussed encompass aspects of barrier effectiveness as well as barrier integrity. However, due to page limitation, this book will not further investigate the modeling and quantitative analysis of safety barriers in such contexts. Readers interested in exploring this topic in greater detail are encouraged to refer to the aforementioned studies.

## NOTE

1. SINTEF one of Europe's largest independent research organizations, with the headquartered in, Trondheim Norway.

## REFERENCES

Chen, Y., Yang, L., Ye, C., & Kang, R. (2015). Failure mechanism dependence and reliability evaluation of non-repairable system. *Reliability Engineering & System Safety, 138*, 273–283.

Cozzani, V., Gianfilippo, G., Antonioni, G., Spadoni, G., & Zanelli, S. (2005). The assessment of risk caused by domino effect in quantitative area risk analysis. *Journal of Hazardous Materials, 127*(1–3), 14–30. https://doi.org/10.1016/j.jhazmat.2005.07.003

Evans, M. G. K., Parry, G. W., & Wareathall, J. (1984). On the treatment of common-cause failures in system analysis. *Reliability Engineering, 9*, 107–115.

Fleming, K. N. (1975). A reliability model for common mode failures in redundant safety systems. General Atomic Company, San Diego, CA.

Fleming, K. N., & Kalinowski, A. M. (1983). An extension of the beta factor method to systems with high levels of redundancy. Pickard, Lowe and Garrick Inc. Newport Beach, CA.

Galán, S. F., Mosleh, A., & Izquierdo, J. M. (2007). Incorporating organizational factors into probabilistic safety assessment of nuclear power plants through canonical probabilistic models. *Reliability Engineering & System Safety, 92*(8), 1131–1138.

Hauge, S., Hoem, Å. S., Hokstad, P., Håbrekke, S., & Lundteigen, M. A. (2015). Common cause failures in safety instrumented systems. SINTEF, Trondheim, Norway.

Hauge, S., Hokstad, P., Håbrekke, S., & Lundteigen, M. A. (2016). Common cause failures in safety-instrumented systems: Using field experience from the petroleum industry. *Reliability Engineering & System Safety, 151*, 34–45.

Hauge, S., Kråknes, T., Håbrekke, S., & Jin, H. (2013). Reliability prediction methods for safety instrumented systems, PDS method handbook. SINTEF.

Hokstad, P., & Corneliussen, K. (2004). Loss of safety assessment and the IEC 61508 standard. *Reliability Engineering & System Safety, 83*, 111–120.

Hokstad, P., & Rausand, M. (2008). Common cause failure modeling: Status and trends. In K. B. Misra (Ed.), *Handbook of Performability Engineering* (pp. 621–640). Springer.

Iaiani, M., Tugnoli, A., & Cozzani, V. (2022). Risk of cascading effects in digitalized process systems. In F. I. Khan, H. Pasman & M. Yang (Eds.), *Chemical Process Safety: Methods to Assess and Manage Process Safety in Digitalized Process System*. Elsevier.

IEC62340. (2007). Nuclear power plants – Instrumentation and control systems important to safety – Requirements for coping with common cause failure (CCF). Geneva, Switzerland: International Electrotechnical Commission.

IEC61508. (2010). Functional safety of electrical/electronic/programmable electronic safety-related systems. Geneva, Switzerland: International Electrotechnical Commission.

IEC61511. (2016). Functional safety – Safety instrumented systems for the process industry sector. Geneva, Switzerland: International Electrotechnical Commission.

IEC62061. (2021). Safety of machinery – Functional safety of safety-related control systems. Geneva, Switzerland: International Electrotechnical Commission.

ISO13849-1. (2015). Safety of machinery—Safety-related parts of control systems—Part 1: General principles for design. Brussels, Belgium: CEN-CENELEC Management Centre.

ISO26262. (2018). Road vehicles road vehicles—Functional safety. Switzerland: International Standard.

Johansen, I. L., & Rausand, M. (2015). Barrier management in the offshore oil and gas industry. *Journal of Loss Prevention in the Process Industries*, *34*, 49–55.

Keizer, M. C. A. O., Flapper, S. D. P., & Teunter, R. H. (2017). Condition-based maintenance policies for systems with multiple dependent components: A review. *European Journal of Operational Research*, *261*(2), 405–420.

Landucci, G., Argenti, F., Spadoni, G., & Cozzani, V. (2016). Domino effect frequency assessment: The role of safety barriers. *Journal of Loss Prevention in the Process Industries*, *44*, 706–717.

Landucci, G., Argenti, F., Tugnoli, A., & Cozzani, V. (2015). Quantitative assessment of safety barrier performance in the prevention of domino scenarios triggered by fire. *Reliability Engineering & System Safety*, *143*, 30–43.

Levitin, G., & Xing, L. (2010). Reliability and performance of multi-state systems with propagated failures having selective effect. *Reliability Engineering & System Safety*, *95*(6), 655–661.

Liu, B., Wu, J., & Xie, M. (2015). Cost analysis for multi-component system with failure interaction under renewing free-replacement warranty. *European Journal of Operational Research*, *243*(3), 874–882.

Lundteigen, M. A., & Rausand, M. (2007). Common cause failures in safety instrumented systems on oil and gas installations: Implementing defense measures through function testing. *Journal of Loss Prevention in the Process Industries*, *20*(3), 218–229.

Magee, C., & De Weck, O. (2002). An attempt at complex system classification, Massachusetts Institute of Technology.

Ma, Z., Yoshikawa, H., & Yang, M. (2017). Reliability model of the digital reactor protection system considering the repair time and common cause failure. *Journal of Nuclear Science and Technology*, *54*(5), 539–551.

Mosleh, A., & Siu, N. O. (1987). A multi-parameter common cause failure model., 9th International Conference on Structural Mechanics in Reactor Technology, Lausanne, Switzerland.

Murthy, D. N. P., & Nguyen, D. (1985). Study of two-component system with failure interaction. *Naval Research Logistics Quarterly*, *32*(2), 239–247.

Nicolai, R., & Dekker, R. (2008). Optimal maintenance of multi-component systems: A review. In K. A. H. Kobbacy & D. N. P. Murthy (Eds.), *Complex System Maintenance Handbook* (pp. 263–286). Springer-Verlag.

NUREG/CR-5485. (1998). Guidelines on modeling common-cause failures in probabilistic risk assessment. U.S. Nuclear Regulatory Commission, Washington DC.

NUREG/CR-6268. (2007). Common-cause failure databases and analysis system: event data collection, classification, and coding. U.S. Nuclear Regulatory Commission, Washington DC.

OCED-NEA. (1994). Results of recent risk studies in France, Germany, Japan, Sweden and the United States. OECD Nuclear Energy Agency, Paris, France.

OCED-NEA. (2011). International common cause failure data exchange (ICDE) – General coding guidelines. OECD Nuclear Energy Agency, Paris, France.

Parry, G. W. (1991). Common cause failure analysis: A critique and some suggestions. *Reliability Engineering & System Safety*, *34*(3), 309–326.

Rahimi, M., Rausand, M., & Lundteigen, M. A. (2011). Management of factors that influence common cause failures of safety instrumented system in the operational phase. European Safety and Reliability Conference (ESREL 2011), Troyes, France.

Rausand, M. (2014). *Reliability of Safety-Critical Systems: Theory and Applications.* John Wiley & Sons.

Rausand, M., Barros, A., & Hoyland, A. (2020). *System Reliability Theory: Models, Statistical Methods, and Applications* (3rd ed.). Wiley-Blackwell.

Rausand, M., & Haugen, S. (2020). *Risk Assessment: Theory, Methods and Applications.* Wiley.

Summers, A. E., & Raney, G. (1999). Common cause and common sense, designing failure out of your safety instrumented systems (SIS). *ISA Transactions, 38*(3), 291–299.

Vesely, W. E. (1977). Estimating common cause failure probabilities in reliability and risk analyses_ Marshall-Olkin specialization. In J. B. Fussell & G. R. Burdick (Eds.), *Nuclear Systems Reliability Engineering and Risk Assessment* (pp. 314–341). SIAM.

Wang, C., Gou, J., Tian, Y., Jin, H., Yu, C., Liu, Y., Ma, J., & Xia, Y. (2021). Reliability and availability evaluation of subsea high integrity pressure protection system using stochastic petri net. *Proceedings of the Institution of Mechanical Engineers, Part O: Journal of Risk and Reliability, 236*(3), 508–521.

Xie, L., Lundteigen, M. A., & Liu, Y. (2018). Common cause failures and cascading failures in technical systems: Similarities, differences and barriers. In S. Haugen, A. Barros, C. Gulijk, T. Kongsvik & J. E. Vinnem (Eds.), *Safety and Reliability – Safe Societies in a Changing World.* CRC Press.

Xie, L., Lundteigen, M. A., & Liu, Y. (2020). Reliability and barrier assessment of series-parallel systems subject to cascading failures. *Proceedings of the Institution of Mechanical Engineers, Part O: Journal of Risk and Reliability, 234*(3), 455–469. https://doi.org/10.1177/1748006X19899235

Xie, L., Lundteigen, M. A., & Liu, Y. (2021). Performance assessment of K-out-of-N safety instrumented systems subject to cascading failures. *ISA Transactions, 118*, 35–43. https://doi.org/10.1016/j.isatra.2021.02.015

Zhang, A. (2021). Prognostics and health management of safety-instrumented systems: Approaches of degradation modeling and decision-making Norwegian University of Science and Technology. Trondheim. https://hdl.handle.net/11250/2726046

# 11 Digitalization and Artificial Intelligence in Barrier Engineering

## 11.1 INTRODUCTION

In this chapter, we aim to illustrate the applications of digitalization and artificial intelligence (AI) techniques in the field of barrier engineering. Our intention is to provide a broad overview rather than an in-depth exploration of specific technologies. In fact, each of these technologies needs a dedicated introduction and a distinct knowledge base, which are beyond the scope of this book.

The typical applications of digitalization and AI technologies in the context of barrier engineering include the following:

- Digital models and digital technologies for supporting barrier design, assessment, and operation, such as the software programs for safety integrity level (SIL) analysis. This is a traditional application that has been reflected in the previous chapters.
- Digital twin approach for barrier performance evaluation and estimation to enhance the understanding of the vulnerability of barrier design, situation monitoring, real-time consequence estimation and decision-making, remaining useful life (RUL) prediction, and maintenance optimization.
- Machine learning methods for dealing with amount of sensing data to improve the recognition capability of a barrier system for the abnormal situations and to enhance condition-based maintenance.
- AI-based algorithms for action controlling to improve the automatic or autonomous decision-making capability of a barrier system.

## 11.2 DIGITALIZATION AND DIGITALIZED BARRIER SYSTEMS

Recent advancements in digital techniques are helpful for barrier engineering, particularly in the context of performance evaluation and estimation. Although Chapter 7 has presented data-driven methods for barrier engineering, it is important to recognize that these methods represent only a part of the applications of digital techniques.

In a broad sense, a so-called *digitalized system* includes digital elements and handles information as numerical data within a digital model, and *digitalization* refers to the process of integrating digital technologies and elements into a sociotechnical system or process. The typical benefits of digitalization include increased operational efficiency and flexibility, higher production quality, lower energy consumption, and higher safety.


DOI: 10.1201/9781003245636-11

Based on the considerations above, digitalization in barrier engineering can be in both assets and barriers, namely *digitalized assets* and *digitalized barriers*. The former implies that the systems under protection may generate a large amount of data, and this data should be captured and analyzed by the affiliated barriers. Chapter 7 has presented methodologies of the data-driven approaches for analyzing such data. For the latter type, a digitalized barrier system should be designed to process sensing information as the inputs of a digital model that reflects the safety control philosophy and controls the action of actuators. It can be found that most technical safety barriers are digitalized to a certain degree. Although a digital system can be viewed as the counterpart to an analog system in some situations (e.g., for control circuits), it is not always necessary to categorize barriers as digitalized or non-digitalized. However, it is worthwhile to explore the implications of using more complex digital models in the control subsystem, and other new challenges and opportunities arise from such implementations to the barrier systems that previously rely on simple signals.

Firstly, the collected information of barrier systems from the digitalized assets can be more complex and various. For instance, this information includes early degradation warning signals and corrosion monitoring data. Even though this is not directly related to the immediate activation of barriers, it is critical in fault detection and diagnosis. Such information can be analyzed within a digital framework to generate insights for further monitoring and adjustments in operational strategies, such as the optimization of proof test intervals. However, the intricate nature of the collected information proposes challenges in sensing and analyzing functions of barriers. In response to these challenges, many machine learning-based methods have been developed in recent years. For example, Arunthavanathan et al. (2021) demonstrate the applications of the convolutional neural networks (CNN) and long short-term memory (LSTM) methods in detecting faults within chemical processes, based on sensing data.

Secondly, the sophistication of digital models for control functions requires much higher computational capacity. This evolution can lead to a transition of the controlling subsystem of a safety barrier, from simple programmable logic controllers (PLCs) to industrial computers, which offer the advanced processing power. The complexity of the computer systems actually increases the susceptibility to failures and may reduce the overall integrity of barriers. In addition, the integration of these advanced computing resources requires significant investment in both hardware and the training of personnel and thus increases the lifecycle cost of barriers.

Digitalization can reduce human interruptions in barrier operations and so that reduce human errors, but at the same time, the digital transformation of barrier systems elevates the risk of human errors in the human-machine interactions (Pasman et al., 2022). This is because operators need to face abundance of data and more complex control interfaces. This can lead to cognitive overload, where the human ability to effectively monitor and interpret critical information is compromised. The challenge is further compounded by the increased reliance on digital technologies, which may inadvertently lead to a decline in situational and safety awareness among human operators.

On the other hand, cybersecurity is emerging as a paramount concern in such an era when barrier systems are increasingly interconnected with other digital infrastructures.

We will discuss the implications of cybersecurity threats and the strategies for mitigating such risks in the context of digitalized barrier systems in Chapter 12.

## 11.3 DIGITAL TWIN APPROACH IN BARRIER ENGINEERING

*Digital twin* (DT) is recently accepted as an effective approach for technical system modeling, status monitoring, performance evaluation, and estimation. DT has been defined as a virtual representation of a physical product containing its information (Grieves, 2014). However, to distinguish DT from the traditional digital model, DT is regarded in this book as follows:

- *Digital twin*: An up-to-date representation of a physical equipment in operation or a process.

In our understanding, a traditional digital model is always a static simulated version of a physical equipment, and it is not updated when the situation of the physical equipment has changed. A DT, on the other hand, is dynamic, receiving real-time updates from the physical world, and reflecting the changes in the equipment as immediate response. Thus, DT is expected to provide a more comprehensive picture of the current status of the physical world. Furthermore, the function of a DT is not only limited to being a *digital shadow* of the real-time change in the physical world. In the ideal situation, a DT is expected to real-time control its physical counterpart, meaning that any action in the digital twin can be realized immediately in the physical world. Figure 11.1 describes the relationship between a digital twin and its physical counterpart.

It should be noted that a DT can be a graphic and schematic simulation of its physical counterpart, such as those based on 3D CAD models, and DT also can be numerical functions and indicators that can reflect the updates of some aspects of the physical world.

In barrier engineering, DT can be used for monitoring the status of technical barriers, such as chronic degradation, and assessing and estimating barrier performance based on the sensed data (fully dynamic) or measured data in tests (partially dynamic).

In practices, constructing a digital twin should be based on a comprehensive understanding of the structural, physical, and operational attributes of barrier

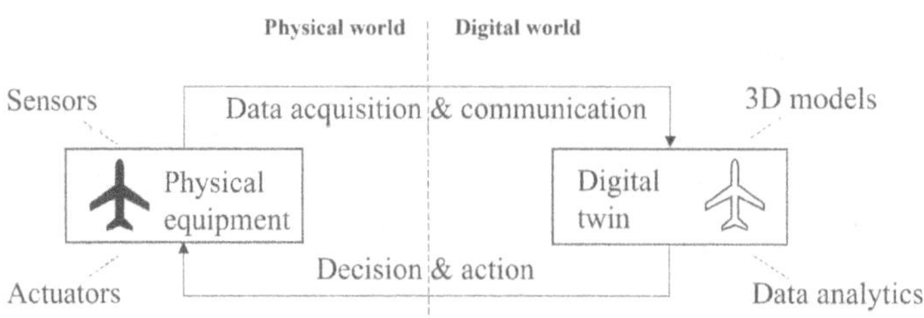


**FIGURE 11.1** Physical asset and digital twin.

equipment. This methodology is commonly known as the *physics-based approach*. The main procedure of physics-based approach of digital twin analysis in barrier engineering includes the following:

**Step 1: Development of a simulation model for technical barriers**

The first step is to build a simulation model for a technical barrier for identifying the potential scenarios and challenges that the barrier equipment may face. Such a model needs to be integrated with the simulation model of the asset. If necessary, the simulation model of the barrier system can be further refined for subsystems to ensure comprehensive analysis.

To construct a digital twin, it is essential to ensure that sensors collecting data from the physical world have effective interfaces with the digital model to facilitate smooth data exchange. This ensures that the simulation model can serve as a dynamic and responsive representation of the physical system.

**Step 2: Refinement and verification of the model**

The next step is to refine the model to improve its accuracy and reliability. This includes parameter estimation using sensed and measured data from the physical world. This step is pivotal in evolving a simulation model into a digital twin. The actual performance and operational conditions of the physical barrier equipment should be mirrored immediately in the digital twin, or at least with an acceptable delay. When the simulation outcomes cannot align with the observed behavior of the actual system, the model parameters should be adjusted to ensure a closer match.

A DT model is expected to own data analysis and decision-making functions to issue commands to physical equipment based on its condition and operational requirements. This analytical functionality needs to meet strict timing constraints for real-time control. Consequently, when twinning complex systems, a *surrogate model* is helpful, which can mimic the behavior of the comprehensive DT model with more efficient computational functions and approximate complex models and perform computations more cost-effectively and quickly. Linear regression, fault tree, and event tree methods introduced in the previous chapters can be used as the surrogate models.

For the verification of DTs, J. Liu et al. (2023) have proposed five criteria for evaluating the quality of a digital twin, including fidelity, standard compliance, integration, timeliness, and smartness. Here, *fidelity* refers to the extent to which a DT accurately mirrors the reality in the physical world, *standard compliance* evaluates whether a DT can follow the established standards of digitalization and safety, *integration* means the capability of a DT to connect with physical objects, the other systems, data sources, processes, and *timeliness*, as we discussed in the last paragraph, reflects how fast a DT can response to the physical changes, while *smartness* indicates the ability of a DT to adapt itself as the requirement for intelligence level increases. These five criteria may conflict in some instances, and thus it is necessary to find a balance during the development of a DT for specific applications.

**Step 3: Integration of failure modes into the model**

The next step is to integrate failure modes in the DT and perform a detailed scenario analysis. This needs the engagement of domain experts to explore various "what-if" scenarios related to the operation of barrier equipment. Utilizing simulation tools, these experts can assess the potential impacts of many hypothetical situations, including abnormal operation, degradation, and failures. This includes not only failures of the asset but also failures of the barriers in their protective functions. The DT should be capable of reflecting the main failure modes of each barrier component and degradation modes such as erosion, corrosion, and aging, with a particular focus on those related to dangerous undetected (DU) failures, showing how these changes in parameters affect barrier performance over time. Thus, the DT for barrier engineering can be referred to as a *probabilistic digital twin*, involving changes and unexpected events within its logic, accounting for the uncertainties of such events in terms of their likelihood and impact.

It is important that the DT model can simulate the frequency and severity of shocks and their damage to the barrier system. Scenario generation and simulation are an iterative process to provide a robust framework for qualitatively evaluating the potential issues and observing the usefulness of mitigation strategies. Interested readers can refer to the paper by Markwirth et al. (2021), where the authors discussed how to insert failures in the DT of technical safety barriers.

**Step 4: Performance assessment of barrier**

The final step is to analyze different scenarios based on system performance data to calculate probability of failure on demand (PFD) under specific operational conditions. Digital twin has been found useful in performance assessment and condition-based maintenance planning of technical systems with degradations (Errandonea et al., 2020). This step uses the collected data from the physical asset to evaluate and optimize the scheduling of tests and maintenance activities in the DT model, ensuring that performance requirements, such as SIL, are met effectively.

In comparison to the DTs utilized in production systems, there are several specialized considerations to account for in the modeling and analysis of DTs for barrier engineering:

- *Integration of DTs for both barriers and the asset.* DTs in barrier engineering should capture and evaluate the immediate effects of unexpected events on both the asset and barriers. The DT for a safety barrier should be able to integrate with the twin for the asset, ensuring interoperability and real-time data exchange between the two models. The challenge mainly comes from the responsiveness analysis for safety barriers, which relies on a risk analysis philosophy, but it cannot be straightforwardly represented in the DT of the asset, as a production system. Integrating these distinct analytical frameworks requires the modeling strategies that are able to accurately reflect the risk-based decision-making inherent in safety barrier analysis while maintaining alignment with the operational dynamics of the production system.

- *Time-dependent accidental scenarios*: The DT of a barrier should be able estimate time-dependent accidental scenarios for the asset, presenting different potential outcomes. This also implies that barriers, serving as risk mitigation measures, should be integrated into the digital twin of the asset they protect. The challenge lies in how to reflect the mostly dormant state of barrier elements within a dynamic model of an asset-keeping change. Given the diversity of barriers in terms of their operational mechanisms and functions, it is also challenging to represent these differences accurately.
- *Twinning of tests and activations of barriers*: Proof tests are significant for maintaining barrier integrity, but they are often omitted from DT modeling approaches. Currently, it is feasible to evaluate the impact of test intervals within models designed for barriers. However, when it comes to integrating barriers with the asset in one DT, mirroring the tests or actual activations within the integrated model is very challenging.

A volume of literature on the development and application of digital twins across various industries has been published. For those interested in a foundational understanding of the DT method, several review papers are highly recommended. For example, Qi et al. (2021) provide a comprehensive overview of the evolution and enabling technologies of digital twins, VanDerHorn and Mahadevan (2021) investigate the concepts of digital twins and explore their interrelations, Tao et al. (2022) present different digital twin modeling approaches, and Agnusdei et al. (2021) examine the applications of digital twins within the domain of industrial safety.

## 11.4 ARTIFICIAL INTELLIGENCE IN BARRIER ENGINEERING

In Section 7.5.4, we have briefly the role of machine learning techniques in data analysis, which are critical components of AI. AI encompasses a wider area of computer science and engineering, with the aim to create intelligent machines that can mimic human cognitive functions like perception, learning, reasoning, and problem-solving.

AI can be used in various aspects of barrier engineering and management. Referring to the barrier management processes outlined in Figure 2.7, AI has potential in nearly every step, including

- *In Step 1*: Identification of assets and hazards, AI technologies can be used in monitoring to analyze vast datasets to early detect potential issues in the process requiring protection, such as failures and degradation, and evaluate their impact on production and safety. Additionally, the natural language processing (NLP) function of AI can be utilized to analyze safety reports, incident logs, and maintenance records, and to extract valuable insights about failure causes and the performance deviation of safety barriers.
- *In Steps 3 and 7*: Risk assessment of the critical assets and updates, AI can enhance the ability to predict future incidents or failures by leveraging historical and current data, offering faster and deeper insights. It also can support the simulation of complex scenarios and a thorough evaluation of the potential impacts of different risk factors.

- *In Steps 4 and 8*: Identification, installation, and implementation of safety barriers, AI can aid in decision-making for the specification and placement of safety barrier solutions through sophisticated data analysis. For instance, AI algorithms can improve sensor placement strategies, recommend optimization methods for the operations of existing safety barriers, and suggest new implementations.
- *In Steps 5 and 6*: Assessment of non-technical and technical barriers, AI-driven simulation models can predict the performance of safety barriers under various conditions and scenarios, facilitating virtual tests of barrier effectiveness. This helps in determining the most suitable barriers for specific risks prior to their physical implementation. AI also can enhance the timeliness, fidelity, and intelligence of DT techniques, by enabling dynamic modeling and simulation of safety barrier systems.
- *In Step 9*: Operation and maintenance of safety barriers, AI supports decision-making by proposing optimal maintenance schedules and operational adjustments to improve safety. Like in Step 1, AI can detect anomalies in operational data and indicate potential barrier failures, thereby enabling condition-based maintenance.

In this section, we will present some examples of applying different AI techniques in barrier engineering.

### 11.4.1 Applications of Artificial Neural Network in Barrier Engineering

One of the most classic data-driven models in AI is the artificial neural network (ANN), the typical architecture of which mirrors the neural network in the human brain, as shown in Figure 11.2, an ANN involves an *input layer*, a single or a set of *hidden layers* (*intermediate layers*), and an *output layer* composed of neurons (denoted as nodes in Figure 11.2).

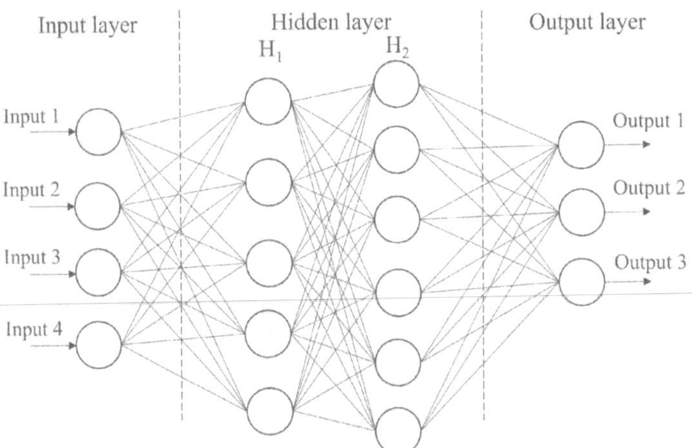


**FIGURE 11.2** An artificial neural network example with two hidden layers.

The process of information transfer within an ANN is designed to imitate the operation way of human nervous systems. *Neurons* in the hidden and output layers are used to calculate the weighted sum of inputs received from all neurons in the preceding layer and then apply a transfer function to this sum. The strength and influence of the connections between neurons are determined by *weights* and *biases* associated with each connection. The network fine-tunes these weights and biases during the training process to minimize prediction error and improve the accuracy of the output.

This adjustment of the internal structure in response to specific conditions makes ANNs versatile and powerful in barrier engineering. For example, a typical application of ANN is to estimate the RUL of barrier components (e.g., see Gebraeel et al., 2004; Mahamad et al., 2010; Tian, 2012). Given the measurable indicators of degradation can be observed, the ANN method is able to predict the RUL accurately by minimizing the discrepancy between predicted and actual values.

Each neuron within an ANN is equipped with an activation function, which enables the network to capture and model complex relationships between inputs and outputs. Activation functions can be different, from simple threshold logic units, where activation occurs when surpassing a predefined threshold, to more sophisticated linear or nonlinear functions. Linear activation functions ensure a direct proportionality between input and activation levels, while nonlinear functions introduce the flexibility necessary for modeling complex, nonlinear relationships inherent in real-world data.

Classical statistical models, such like linear regression analysis that we introduce in Chapter 7, can estimate the failure rates of barrier equipment under different operating conditions. However, these models are often designed with specific parameters for individual equipment, limiting their applicability to other types of equipment. In contrast, ANN is data-driven self-adaptive and can be generalized in different applications. ANNs do not need many prerequisite assumptions, given that the training datasets are sufficiently large. Such a characteristic enables ANNs to adapt to and learn from a broader array of situations, enhancing their usefulness in a variety of applications.

In barrier engineering, ANN has been proved very effective in prognosis of equipment (Yang et al., 2016), where DU-failure risk, RUL, or other indicators can be regarded as the only output of the network. Assuming that we can measure $m$ parameters or features related to equipment degradation, the number of the input nodes of ANN $n_{input} = m$. Otherwise, if we can have more observations on these parameters, for example, in the observation period $t$, we collect data with $n$ times, the number of total inputs is $mn$.

The next step is to determine on the number of hidden layers and the number of neurons in each hidden layer. A common practice is to start with one or two hidden layers. The number of neurons can be determined based on the complexity of the problem. For illustrative purposes, we will employ a single hidden layer with $n_h$ hidden nodes for RUL estimation.

Generally, we can let $w_{ij}^l$ denote the weight from neuron $i$ in the preceding layer to neuron $j$ in the layer $l$, and $b_j^l$ the bias of neuron $j$ in the layer $l$. In this illustrative example of RUL estimation, since there is only one hidden layer, its preceding layer is the input layer of the network. The initiation values of these parameters of the ANN should be set as small random numbers in typical practices.

For each layer, the sum of inputs should be calculated, and the activation function of each neuron will be considered to determine the output. For example, for the neuron $j$ in the layer $l$, its output $o_j^{\mathrm{H}}$ can be calculated as

$$o_j^l = g\left(\sum_{i=1}^{n_{l-1}} o_j^{l-1} w_{ij}^l + b_j^l\right) \tag{11.1}$$

where $g(\cdot)$ is the activation function, $n_{l-1}$ denotes the number of neurons in the $(l-1)$th layer. In this RUL estimation case, $n_{l-1} = m$.

These parameters are iteratively updated using the training dataset and an algorithm such as *backpropagation* (BP). Such an algorithm propagates the RUL estimation error backward through the network to adjust the weights and biases, guided by the gradient of the error with respect to each weight and bias, and scaled by a learning rate ($\eta$), which can control how much the weights are adjusted during each iteration. The update rule in the algorithm can be described as

$$w_{ij}^{l,\text{new}} = w_{ij}^l - \eta \cdot \frac{\partial E}{\partial w_{ij}^l} \tag{11.2}$$

where $E$ represents the error function or loss function, measuring the difference between the predicted output of the network ($\hat{y}$) and the actual value ($y$). In this case, the error is the difference between the predicted RUL and the actual RUL. If there are multiple ($N$) outputs of the network, we can the mean squared error (MSE) for the error function:

$$MSE = \frac{1}{N} \sum_{k=1}^{N} \left(y_k - \widehat{y_k}\right)^2 \tag{11.3}$$

The partial derivative of the error function with respect to the weight $w_{ij}^l$ in Equation (11.2) indicates how the changes in the weight can affect the overall error. This update rule is called *gradient descent method*, because based on the subtraction in Equation (11.2), the movement is against the gradient direction to reach the minimum error.

Such a BP process is repeated for each batch of training data until the prediction error of the ANN model converges to an acceptable level or a specified number of iterations is reached.

Once the model is trained, it can be used to predict the RUL of equipment by feeding the relevant condition monitoring data into the input layer. Since the output layer will have a single neuron for the RUL prediction, linear activation function can be for continuous RUL values. Then, in the case of only one hidden layer with $n_h$ neurons, we have

$$RUL = \sum_{j=1}^{n_h} w_{jo} o_j^l \tag{11.4}$$

where $w_{jo}$ is the weight connecting the $j$th node in the hidden layer to the node in the output layer.

ANN is a data-driven model, without need to know the failure mechanisms, but it requires a large amount of data for training to ensure an accurate prediction. In this subsection, we have introduced the fundamental principles of ANN models. For readers interested in exploring the use of ANNs for estimating RUL and other applications within barrier engineering, we encourage consulting the literature cited in this subsection.

It is also worthwhile mentioning that the deep learning methods, such as deep neural networks, recurrent neural networks, and CNN, are attracting significant attention in the domain of safety engineering. These advanced techniques can extract hidden features from big data and substantially improve computer vision and pattern recognition capabilities, which are invaluable for automated defect detection, fault diagnostics, and the identification of signs of degradation. For readers interested in exploring this topic further, the review by Hoang and Kang (2019) can be helpful.

### 11.4.2 Applications of Some Other Machine Learning Methods in Barrier Engineering

Bai et al. (2022) have presented an overview of the applications of machine learning methods in process safety, where commonly used methods include support vector machine (SVM), ridge regression, random forest, and extreme gradient boosting, in addition linear regression and ANN.

Due to the page limitation, in this subsection, we just have a very brief introduction on the selected methods and their potentials in barrier engineering.

#### 11.4.2.1 Support Vector Machine

SVM is a classical supervised machine learning model mainly for data classification and regression analysis. In SVM, an important concept is *hyperplane* that is a decision boundary that separates different classes of data in the feature space. The goal of SVM can be understood as finding the optimal hyperplane that maximizes the *margin* between classes. Margin in this context is the gap between the hyperplane and the nearest data points (*support vectors*) from both classes. Figure 11.3 explains the meanings of these concepts, where black points and white points represent different classes of data, $H_1$ does not separate the two classes, $H_2$ separates the two classes but with a small margin, and $H_3$ is the hyperplane with the maximized margin between the support vectors.

The hyperplane can be linear like $H_3$ in Figure 11.3. However, when data is not linearly separable, SVM uses kernel functions mapping the input features into a higher-dimensional space where a hyperplane can be used to separate the classes. Common kernels include the linear, polynomial, and radial basis function (RBF) or Gaussian kernel (Ben-Hur et al., 2001). Thus, SVM is well suited for the problems with high-dimensional spaces and cases where the number of dimensions exceeds the number of samples.

SVM has been used in a variety of applications, such as image classification and protein classification. In barrier engineering, SVM can be employed to classify process (asset) conditions as normal or abnormal, facilitating early

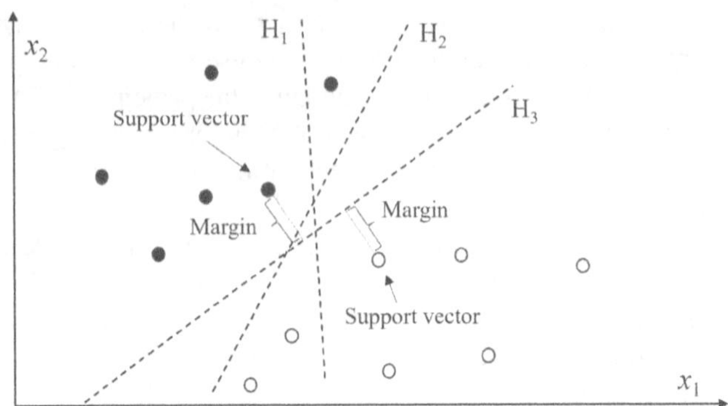


**FIGURE 11.3** Illustration of the concepts in the support vector machine model.

detection of degradation and potential failures. SVM also can perform well on smaller datasets, especially when the number of features is high compared to the number of samples. It can provide a relatively higher level of interpretability than other machine learning approaches, such as ANN, especially when using linear kernels.

### 11.4.2.2 Ridge Regression

*Ridge regression*, also known as *Tikhonov regularization*, is a technique used machine learning to analyze data that suffer from multicollinearity, where variables are highly correlated. In such situations, the variance of ordinary least square estimators becomes large, and the solution may become highly sensitive to random errors in the observed target, leading to overfitting. Ridge regression addresses this issue by introducing a penalty or regularization term to the loss function, so as to reduce and prevent overfitting. Generally, this method can improve efficiency in parameter estimation in exchange for a tolerable amount of bias (Gruber, 1998).

Such a property of Ridge regression can facilitate robust prediction in barrier engineering, where data can be complex and multicollinear due to the intricate relationships between variables that influence system reliability and safety. For example, Ridge regression can be used to predict equipment failures by analyzing operational data. The regularization helps deal with multicollinearity among the predictors, such as various sensor readings, to make more stable and reliable predictions, enabling timely maintenance actions before failures occur. Ridge regression also can handle the complex relationships among various factors, such as environmental conditions, equipment age, usage patterns, and maintenance history, and comprehensively evaluate their impacts on barrier performance.

### 11.4.2.3 Random Forest

*Random Forest* is a machine learning technique combining multiple decision trees, and it uses the mean prediction (regression) of the individual decision trees to enhance predictive accuracy and mitigate the risk of overfitting, which is a common problem with individual decision trees. This method belongs to the broader *ensemble learning* strategy, where multiple models are trained to solve the same problem and combined in some way to obtain better performance.

In principle, Random Forest generates multiple decision trees using a technique called *bootstrap aggregating* or *bagging*. This involves randomly selecting samples from the training dataset with replacement to train each tree. Given that each tree is exposed to different data subsets, it can uncover unique patterns, thereby reducing the likelihood of overfitting compared to a singular decision tree. Next, when constructing a tree and splitting a node, Random Forest selects a random subset of the features (variables) as candidates for the split. This increases the diversity among the trees, further reducing variance without significantly increasing bias. It can be used for classification and regression. For classification, the final prediction of Random Forest is made based on a majority vote among all trees, while for regression tasks, it averages the predictions of all trees.

As a result, Random Forest has higher accuracy in prediction. It works well with large and complex datasets, is able to deal with missing information smoothly, and is robust to noise in the data. In barrier engineering, Random Forest can be used for hazard identification, and prediction of equipment failure. The diverse and high-dimensional data can be collected by various sensors of barriers, on operational parameters, such as temperature, vibration, pressure, and rotational speed of machinery. Such data often exhibits nonlinear relationships and may contain noise and outliers due to sensor malfunction. This is an excellent application domain for Random Forest to provide generally reliable failure prediction and RUL estimation. For more information on Random Forest and its applications in the relevant fields, readers can reference to some relevant articles, e.g., by Yang et al. (2008) and Shen et al. (2018).

#### 11.4.2.4 Gradient Boosting and Extreme Gradient Boosting

*Gradient Boosting* also belongs to the ensemble learning methods. Unlike Random Forest that relies on averaging of models in the ensemble, Gradient Boosting adds new models to the ensemble sequentially. Decision trees are commonly used as the foundational learners in this approach. The process starts with a decision tree (or other simple model) and progressively incorporates new models that can correct the errors of the existing ensemble. Each additional tree is specifically trained on the gradient of the loss function, which quantifies the discrepancy between actual and predicted values (see the tutorial by Natekin and Knoll (2013)).

*Extreme Gradient Boosting* (XGBoost) further enhances the gradient boosting algorithm. XGBoost is advantageous for its speed and efficiency, achieved through various algorithmic optimizations, and incorporates regularization terms within its objective function. This inclusion not only helps prevent overfitting but also enhances model accuracy, as introduced in the report by Chen and He (2022).

XGBoost has shown its potentials in barrier engineering, especially due to its proficiency in predicting rare events and its versatility with different data types. Therefore, XGBoost can be used in the tasks like anomaly detection and predictive maintenance. The effectiveness of XGBoost in fault diagnosis is exemplified in a case study by Alfarizi et al. (2023).

### 11.4.3 Potentials of Natural Language Processing in Barrier Engineering

*Natural language processing* (NLP) is an interdisciplinary branch of AI that focuses on the interaction between computers and humans through natural language. NLP aims to enable computers to understand, interpret, and generate human language in a meaningful and useful way (Chowdhury, 2020; Nadkarni et al., 2011).

NLP focuses on two elements in language: syntax and semantics. Syntax is the arrangement of words in a sentence to make grammatical sense, which can be used in NLP to assess how the grammatical structure of a sentence conveys meaning, incorporating tasks like part-of-speech tagging, sentence breaking, and parsing. While semantics refers to the meaning that is conveyed by a text, which is used in NLP for understanding sentences, including the meanings of individual words and how those meanings combine to form the meanings of sentences.

The general tasks of NLP include (Nadkarni et al., 2011) sentence boundary detection, tokenization (breaking down text into smaller units, e.g., words or sentences), part-of-speech assignment to individual words, morphological decomposition of compound words, shallow parsing (identifying phrases from constituent part-of-speech tagged tokens), and segmenting text into meaningful groups. In addition, NLP is expected to identify spelling and grammatical errors and correct these errors and identify and classify key information in text into predefined categories.

The application of NLP within the realms of industrial safety and barrier engineering is still rather limited, but its potential can be clearly seen. For example, NLP can offer innovative ways to investigate vast amounts of textual data in those incident reports, standards, to regulations available online, to reveal potential hazards, and elucidate the root causes of past failures and incidents. The insights obtained from such studies are expected to be helpful not only in preventing future occurrences but also enhancing the overall barrier management. Similarly, NLP can identify compliance gaps based on comparisons between safety codes and actual barrier specifications and operational documents and recommend necessary adjustments, ensuring that all barriers are regular compliance and effective.

In accident investigation, NLP also can be a valuable tool, particularly in uncovering causes linked to human errors and organizational flaws. NLP technologies can be used to analyze various forms of communication, such as emails and meeting notes, to discover the relationships between some decisions and actions with accidents, reveal the weaknesses of safety culture, and find early warning signs of problems. Such investigation results can provide more insights that might be overlooked in conventional methods.

For training and safety education, as non-entitative barriers, NLP can help in scrutinizing educational materials and feedback systematically. By identifying knowledge gaps and understanding specific training needs, NLP enables the customization of educational programs. Tailoring training in this way ensures that individuals are equipped with the necessary knowledge and skills to maintain the effectiveness and integrity of barriers.

### 11.4.4 Potentials of Large-Language Models in Barrier Engineering[1]

A language model is a computational tool designed to predict the probability distribution of words within a sentence. The goal of using a large-language model (LLM) is to process, interpret, generate, and respond to human language through mathematical models. Recently, *large-language models* (LLMs), such as ChatGPT, which are built on an extensive scale of data and parameters, have captured global attention. LLMs represent a sophisticated and comprehensive iteration, trained on vast datasets to understand and produce text that mirrors human communication. These models can grasp context, generate coherent and diverse text, and even perform tasks they were not explicitly trained to do, thanks to their broad understanding of language and knowledge gleaned from their training data.

Similar to the applications of NLP technologies, LLMs also have promising potentials in barrier engineering, for example

- *Hazard-and-barrier identification*: LLMs can analyze extensive documents, including incident reports, research papers, standards and regulations, and safety guidelines written in different languages, to identify potential hazards and assess existing and potential barriers related to a certain type of social-technical system. By synthesizing information from diverse sources, LLMs can uncover less obvious risk factors and suggest mitigations. In practice, LLMs can be used as a brainstorming tool to guide failure mode, effect, and criticality analysis (FMECA) and Hazard and operability study (HAZOP) meetings.
- *Compliance review and documentation*: Currently, LLMs already have the capability to review images and reports and identify the hazards and non-safe actions. They can significantly facilitate the process of comparing project specifications against relevant safety and operational standards, identifying non-compliance areas, and recommending corrective measures. With the text generation functions, LLM is very powerful for drafting barrier management documents tailored to best practices and specific project needs and offers generic guidance on barrier deployment, installation, and maintenance.
- *Accident investigation and reporting support*: LLMs can greatly facilitate the accident investigation process by analyzing reports, communications, and data to identify contributing factors and root causes of incidents. They can assist in compiling comprehensive investigation reports, ensuring all relevant details are considered and adequately documented. The utility of LLMs (such as ChatGPT) in hazard analysis for safety-critical systems has been explored by Diemert and Weber (2023), where the authors find useful for supporting human analysts in simple systems, though the quality of responses diminished with system complexity.
- *Knowledge storage for training and education*: Acting as repositories of knowledge, LLMs can support knowledge retention and promote lifelong learning. For example, LLMs can be used to generate customized training materials, pre-assess training materials, and analyze training feedback and performance data to improve educational contents.

X. Liu et al. (2023) have examined the capabilities of ChatGPT 3.5 and GPT-4 in the context of the Certified Maintenance and Reliability Professional Exam, highlighting the potential of these technologies to support tasks in reliability engineering and maintenance, closely related to barrier engineering. While these models can provide self-consistent explanations, they exhibit limitations in arithmetic and handling domain-specific terminology. Their responses to troubleshooting queries are often wordy and lack specificity.

In the near future, LLMs are expected to play a more integral role in barrier engineering. However, the expertise and oversight of human professionals remain highly invaluable to guide the model and avoid the misleading, inaccurate, or even ridiculous outputs generated by LLMs.

### 11.4.5 Concerns of Using Digitalization and AI in Barrier Engineering

Although AI technologies are being fast applied in barrier systems and barrier management process, concerns always exist for such applications in the domain of industrial safety, where technical reliability and trustfulness are very emphasized.

Here are some of the main concerns associated with using AI in safety and barrier engineering:

- *Data security*: The integration of AI in barrier engineering typically needs the collection and processing of extensive datasets within the control subsystems of technical barriers. These datasets can contain sensitive information related to industrial operations and proprietary knowledge. A breach in data security could pose substantial risks to both individuals and the organization. We will further discuss this issue and potential measures to ensure data security in the next chapter.
- *Information accuracy and reliability*: The reliability and validity of predictions and decisions are extremely important in safety-critical applications, but as we have discussed in the last subsection, some AI models, such as LLMs, can generate inaccurate information. When the training dataset is not sufficient, the output of ANN is also with huge variability. When applying these technologies in fault diagnosis, the inaccuracies can result in either false alarms, leading to unnecessary maintenance, or, conversely, failing to warn about really dangerous situations.
- *Transparency and interpretability*: Many AI models, especially deep learning systems, are often seen as "black boxes" due to their complex and opaque nature. This lack of transparency and interpretability can be a significant concern in industrial safety, where understanding the rationale behind safety decisions and predictions is critical for trust and accountability.
- *Integration with existing systems and acceptance*: Integrating AI into existing barrier systems and industry safety management have some practical challenges, such as the compatibility issues of software programs. Furthermore, the adoption of AI technologies can be met with resistance from human operators, particularly if these technologies are perceived as complex or threatening to job security. In addition, skill gap also exists in

the workforce regarding the understanding of sophisticated AI systems, and thus there is an urgent need for training and adaptation efforts.

- *Human-machine interaction*: The balance of trust in AI by human operators is not simple. Over-reliance on AI for safety monitoring and decision-making can compromise human expertise and alertness, whereas insufficient trust can lead to excessive manual interventions, reducing operational efficiency and increasing costs.
- *Ethical and legal considerations*: The use of AI in the decision-making processes of barrier systems introduces complex ethical and legal dilemmas, such as liability in the event of failures or accidents and potential biases in AI decision-making. For example, when an accident happens because AI decides not to take a barrier action based on its collected information, it is difficult or too late to determine who should be responsible for the mistake. The tragic accidents involving Ethiopian and Indonesian flights in 2014 highlight the severe consequences and lead to long discussions of duties.

## NOTE

1. This subsection is partly based on the trial lecture of Nicola Tamascelli as a part of his PhD defense in January 2024. His PhD project is in the cotutelle agreement of NTNU and University of Bologna, Italy.

## REFERENCES


Agnusdei, G. P., Elia, V., & Gnoni, M. G. (2021). A classification proposal of digital twin applications in the safety domain. *Computers & Industrial Engineering*, *154*, 107137.

Alfarizi, M. G., Vatn, J., & Yin, S. (2023). An extreme gradient boosting aided fault diagnosis approach: A case study of fuse test bench. *IEEE Transactions on Artificial Intelligence*, *4*(4), 661–668.

Arunthavanathan, R., Khan, F., Ahmed, S., & Imtiaz, S. (2021). A deep learning model for process fault prognosis. *Process Safety and Environmental Protection*, *154*, 467–479.

Bai, Y., Xiang, S., Zhao, Z., Yang, B., & Zhao, J. (2022). Data-driven approaches: Use of digitized operational data in process safety. In F. Khan, H. Pasman, & M. Yang (Eds.), *Methods in Chemical Process Safety, Volume 6: Methods to Assess and Manage Process Safety in Digitalized Process System* (pp. 61–100). Elsevier.

Ben-Hur, A., Horn, D., Siegelmann, H., & Vapnik, V. N. (2001). Support vector clustering. *Journal of Machine Learning Research*, 2, 125–137.

Chen, T., & He, T. (2022). *Xgboost: Extreme gradient boosting. R package version 0.4-2*, *1*(4).

Chowdhury, K. R. (2020). Natural language processing. In *Fundamentals of Artificial Intelligence* (pp. 603–649). Springer.

Diemert, S., & Weber, J. H. (2023). *Can large language models assist in hazard analysis? In International Conference on Computer Safety, Reliability, and Security* (pp. 410–422). Cham: Springer Nature Switzerland.

Errandonea, I., Beltrán, S., & Arrizabalaga, S. (2020). Digital twin for maintenance: A literature review. *Computers in Industry*, *123*, 103316.

Gebraeel, N., Lawley, M., Liu, R., & Parmeshwaran, V. (2004). Residual life predictions from vibration-based degradation signals: A neural network approach. *IEEE Transactions on Industrial Electronics*, *51*(3), 694–700.

Grieves, M. (2014). *Digital twin: Manufacturing excellence through virtual factory replication*. White paper, 2014: 1–7.

Gruber, M. (1998). *Improving Efficiency by Shrinkage: The James–Stein and Ridge Regression Estimators*. CRC Press.

Hoang, D.-T., & Kang, H.-J. (2019). A survey on deep learning based bearing fault diagnosis. *Neurocomputing*, *335*, 327–335.

Liu, J., Liu, X., Vatn, J., & Yin, S. (2023). A generic framework for qualifications of digital twins in maintenance. *Journal of Automation and Intelligence*, *2*(4), 196–203.

Liu, X., Vatn, J., Yin, S., & Maithani, V. (2023). *Performance of ChatGPT on CMRP: Potential for assisting maintenance and reliability professionals using Large Language Models*. IECON 2023 – 49th Annual Conference of the IEEE Industrial Electronics Society, Singapore.

Mahamad, A. K., Saon, S., & Hiyama, T. (2010). Predicting remaining useful life of rotating machinery based artificial neural network. *Computers & Mathematics with Applications*, *60*(4), 1078–1087.

Markwirth, T., Jancke, R., & Sohrmann, C. (2021). *Dynamic fault injection into digital twins of safety-critical systems*. 2021 Design, Automation & Test in Europe Conference & Exhibition (DATE), Grenoble, France.

Nadkarni, P. M., Ohno-Machado, L., & Chapman, W. W. (2011). Natural language processing: An introduction. *Journal of the American Medical Informatics Association*, *18*(5), 544–551.

Natekin, A., & Knoll, A. (2013). Gradient boosting machines, a tutorial. *Frontiers in Neurorobotics*, *7*, 21.

Pasman, H., Sun, H., Yang, M., & Khan, F. (2022). Opportunities and threats to process safety in digitalized process systems—An overview. In F. Khan, H. Pasman, & M. Yang (Eds.), *Methods in Chemical Process Safety, Volume 6: Methods to Assess and Manage Process Safety in Digitalized Process System* (pp. 1–24). Elsevier.

Qi, Q., Tao, F., Hu, T., Anwer, N., Liu, A., Wei, Y., Wang, L., & Nee, A. Y. C. (2021). Enabling technologies and tools for digital twin. *Journal of Manufacturing Systems*, *58*, 3–21.

Shen, J., Wan, J., Lim, S.-J., & Yu, L. (2018). Random-forest-based failure prediction for hard disk drives. *International Journal of Distributed Sensor Networks*, *14*(11). https://doi.org/10.1177/1550147718806480

Tao, F., Xiao, B., Qi, Q., Cheng, J., & Ji, P. (2022). Digital twin modeling. *Journal of Manufacturing Systems*, *64*, 372–389.

Tian, Z. (2012). An artificial neural network method for remaining useful life prediction of equipment subject to condition monitoring. *Journal of Intelligent Manufacturing*, *23*, 227–237.

VanDerHorn, E., & Mahadevan, S. (2021). Digital twin: Generalization, characterization and implementation. *Decision Support Systems*, *145*, 113524.

Yang, Z., Baraldi, P., & Zio, E. (2016). *A comparison between extreme learning machine and artificial neural network for remaining useful life prediction*. 2016 Prognostics and System Health Management Conference (PHM-Chengdu), Chengdu, China.

Yang, B.-S., Di, X., & Han, T. (2008). Random forests classifier for machine fault diagnosis. *Journal of Mechanical Science and Technology*, *22*, 1716–1725.

# 12 Barrier Engineering and Security

## 12.1 INTRODUCTION

In this book, we mainly focus on the role of barrier functions in preventing and mitigating safety issues arising from technical failures, natural events, or unintentional errors. However, there is a growing recognition of threats to industrial processes and facilities from deliberate actions such as intended attacks, sabotage, theft, hacking, and more. These deliberate threats introduce additional harms to assets of interest, particularly as barrier systems increasingly incorporate information, digital, and AI technologies. We refer to the problems caused by these deliberate actions as *security issues*.

Numerous researchers have studied the differences and similarities between the safety and security of sociotechnical systems (e.g., Abdo et al., 2018). In earlier chapters, we have defined the safety of an asset as an acceptable level of risk associated with specific hazards. While for *security*, IEC62443-1-1 (2009a) has provided diverse definitions, including the following:

- Measures taken for protecting a system/asset, such as prevention of illegal intrusion or unauthorized access;
- Condition of a system with the establishment and maintenance of protective measures of the system;
- Condition of a system being free from unauthorized access and from unauthorized change, destruction, or loss.

Linking security directly with risk is sometimes challenging because the frequency or probability of deliberate actions is unpredictable. However, the connection remains valid in several respects. Security assessment can be based on the vulnerabilities of technical systems that are associated with risks, and a comprehensive risk management framework has potential to enhance security. To align with our definition of safety in the previous chapters, we define the security of an asset of interest as follows:

- *Security*: An acceptable level of risk that an asset is harmed by the deliberate actions of specific threats.

In such a context, the elements related to security include *threat* and *asset*. Using the same definition for asset in Chapter 1, the asset in this chapter can be *physical asset*, *logical asset*, and *human asset*. Among these three, logical asset is of an informational nature, such as intellectual property, algorithms, confidential data, and

DOI: 10.1201/9781003245636-12

proprietary and specific knowledge. Logical asset exists in the format of personal memory, paper-based documents, and electronic information contained in computers. Such kind of asset is not emphasized in traditional safety barrier engineering, but it is often the focus of security. To emphasize the objectiveness of attack in security study, we can also call asset as *target*.

In security study, we consider *threat* as the counterpart of hazard, with the definition:

- *Threat*: A source that has the potential and motivation to cause harm.

Since many security issues originate from human actions or organizations formed by humans, some literature refers to the sources of these threats as *threat actor*, *threat agent*, or *attacker* (see Rausand & Haugen, 2020). The motivations behind these threat actors are varied, including desires for power or political assertion, religious or economic motives, anger and revenge, and even personal enjoyment (Baybutt, 2002). The attack coming from the threat can be called a *threating event*, analogous to the pivotal event in safety risk assessment. In most cases, the threatening event is not the initiation point of an entire attack scenario targeting an asset but rather the development result of a series of preceding events. For example, terrorist attack on a metro station may start from purchasing materials to manufacture bombs, but such a transactional activity itself is not a threatening event that directly leads to harm. The impact of a threatening event can vary, resulting in different levels of harm depending on the countermeasures in place and the responses initiated. The attack path related to security issues can also be depicted using a Bow-Tie diagram, as illustrated in Figure 12.1.

When *vulnerabilities* of a sociotechnical system are exploited by a threatening event, harm occurs. These vulnerabilities are also known as *security holes*, as the weaknesses that can be utilized by the threat actors. Vulnerabilities can be designing flaws, programming errors, unqualified or negligent employees, or procedural lapses. Vulnerabilities of the asset and capabilities and motivations of threating actors can significantly influence the likelihood of a threating event.

In some situations, security is considered to have a broader spectrum that includes safety. For example, the U.S. Department of Homeland Security addresses both cyber-attacks and safety issues, such as industrial accidents. In the top 2% researchers report released by Stanford University, researchers of safety engineering and risk

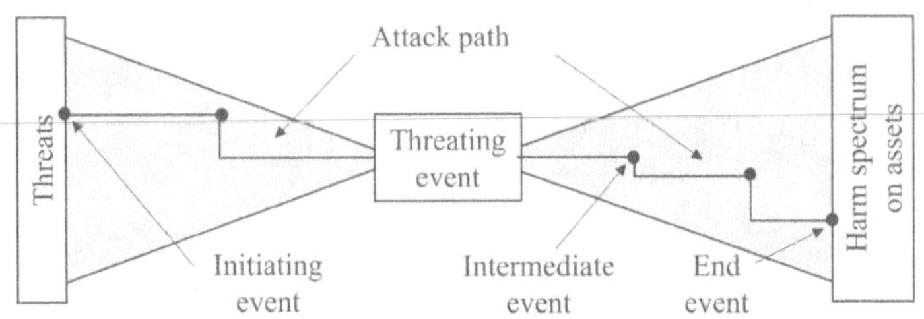


**FIGURE 12.1** Elements of an attack path.

analysis are categorized under the area of *Defense, Strategy, and Security*. However, in this book, we distinguish safety and security considering different types of threats.

Similar to our discussions on the performance assessment of safety barriers, we can investigate security issues related to barrier from two perspectives: the security issues inherent to the barrier itself (here barrier is regarded as an asset), and the role of a barrier in mitigating security risks within a larger sociotechnical system. Sections 12.2 and 12.3 will investigate the first category of issues, while Sections 12.4 and 12.5 will explore the latter topic.

## 12.2 SECURITY ISSUES OF BARRIER SYSTEMS

In this section, a barrier system is the study object. Security issues of a barrier are related to the threats that are intended to eliminate or weaken the barrier functions of protecting assets. The threating events on a barrier can also be physical attacks and logical attacks (mostly cyber-attacks). Considering that barriers can be entitative ones or non-entitative ones, physical attacks are mostly executed on the entitative barriers, while logical attacks can be harmful to both entitative- and non-entitative barriers.

Since barrier systems are often critical facilities, they are supposed to be stronger and more robust to external shocks than normal equipment, but for security issues, if threat actors can find the vulnerabilities of barrier systems, their attacks are often successful.

More specifically, the typical examples of threating events that can result in harms of barrier system security include the following:

- Physical attack events targeting barrier functions of hardware elements:
    - *Sabotage*: An intentional act aimed at damaging, destroying, or disrupting the operations of safety barrier systems. Sabotage can target critical barrier components, such as controller computer, or utility equipment, with the purpose to eliminate their barrier functions.
    - *Vandalism and theft*: The act of intentionally damaging or destroying property, and removal of critical components from safety barrier systems, such as sensors and cameras, and those equipment with expensive metal materials.
    - *Tampering*: The unauthorized modification of components in a barrier system, which can include altering settings, bypassing controls, or physically modifying parts of the system to impair barrier functions.
    - *Intrusion*: Effort to bypass physical barriers, such as fences, locked gates, or walls, without authorization. Intrusion aims to access those restricted areas, potentially leading to sabotage, theft, or other malicious activities.
    - *Arson*: The deliberate act of setting fire to facilities, equipment, or infrastructure, including components of safety barrier systems.
- Logical attack events targeting the barrier functions of digital, networked, and software components of barrier systems:
    - *Malware infection*: Attack involving malicious software designed to damage, disrupt, or gain unauthorized access to the data transmission

and controlling parts of a barrier system. Malware can include viruses, worms, Trojan horses, ransomware, and spyware.

- *Man-in-the-Middle (MitM) attack*: Such an attack also intercepts communication between two subsystems, allowing the threat actor to alter the information being exchanged. In the context of safety barrier systems, this attack can compromise the integrity of data or control commands.
- *Phishing attack*: Deceptive attempt to trick individuals to obtain sensitive information, such as passwords or network access credentials, which can then be used to gain unauthorized access to safety barrier systems.
- *Ransomware attack*: A specific type of malware that encrypts data, demanding a ransom to restore access. For safety barrier systems, this could immobilize critical safety functions when a real demand comes.
- *Denial-of-Service (DoS) attack*: Such an attack aims to overwhelm the computing resources of a barrier system, also making it unavailable or unresponsive upon demands.

To mitigate physical attacks, organizations can deploy surveillance systems, intrusion detection cameras, and security alarms. These tools assist in identifying unauthorized activities and ensure that all external areas are adequately illuminated to deter potential intruders. From an organizational management perspective, it is essential to employ specific security personnel tasked with patrolling the premises, responding to unexpected events, and overseeing surveillance operations. These security personnel can act swiftly to prevent a physical attack or minimize its impact. In terms of procedural management, conducting regular security audits and establishing emergency response plans are crucial.

To mitigate logical attacks, implementing robust cybersecurity measures is vital. These measures include conducting regular software updates, performing thorough security assessments, training employees to recognize phishing attempts, encrypting data both in transit and at rest, and applying rigorous access controls.

It should be noted that barrier systems are always not the ultimate targets for threat actors. Instead, the loss of a barrier function can be viewed as an intermediate event within an entire attack scenario, as illustrated in the Bow-Tie diagram in Figure 12.1.

## 12.3 SECURITY RISK ASSESSMENT OF BARRIER SYSTEMS

### 12.3.1 Bow-Tie Security Risk Assessment

The term *security risk* can be introduced to evaluate the criticality of a potential threat to a particular asset. In practice, due to the difficulties in precisely estimating the likelihood of attacks, some researchers think that the prioritization among risks can be solely based on the severity of the event consequences. We can adopt many approaches in safety risk assessment for security risk assessment.

The Bow-Tie diagram in Figure 12.1 can support a basic qualitative security assessment. According to Rausand and Haugen (2020), such security risk assessment includes the following steps:

1. *Scope analysis*: It is necessary to identify the study objects of interests. Here we use the term study object rather than asset to avoid confusion, because the study object in this section can be an asset or a barrier. The study objects should be important and can be harmed by some types of attacks. If a number of objects are identified, we need to rank them and spend more time on the most important ones.
2. **Threating event analysis**, with the following tasks:
   - Identifying the possible threating events on the selected study object;
   - Ranking the threating events based on the rough estimation on their impacts on the assets, event likelihoods, and the importance of the objects being impacted;
   - Describing the critical threating events and preparing Bow-Tie assessments for them. For each Bow-Tie diagram, only one threating event can be studied.
3. **Threat analysis**, with the following the tasks for a selected threating event:
   - Identifying the potential threats/threat actors that can lead to the threating event;
   - Analyzing the motivations of the threat actors;
   - Identifying the competence and resources that are needed by the threat actors in realizing the threating event;
   - Analyzing the modes of operation (i.e., attack priority and attack approach) of threat actors;
   - Ranking the relevant threat actors based on their capabilities and motivations.
4. **Vulnerability analysis**, with the following tasks:
   - Identifying vulnerabilities of the study object and analyzing which threat actors can exploit such vulnerabilities;
   - Estimating their likelihoods if necessary;
   - Identifying and plotting the attack paths and estimating their likelihoods.
5. **Consequence analysis**, with the following tasks:
   - Predicting possible consequences of the threating event;
   - Assessing the consequences and making a summary on the risk of the threating event;
   - Comparing the assessed risk with acceptance criteria.
6. **Countermeasure analysis**, which is like the barriers in Bow-Tie diagram of safety risk assessment, with the following tasks:
   - Identifying the measures that can reduce the potential or motivation for a threat actor to initiate an action;
   - Identifying the measures that can mitigate the consequences of the threating event;
   - Evaluating whether the existing countermeasures are sufficient.

There are some other frameworks for security risk assessment and management of sociotechnical systems, such as Process for Attack Simulation and Threat Analysis (PASTA) and Operationally Critical Threat, Asset, and Vulnerability Evaluation (OCTAVE). Interested readers can find detailed introductions to these methodologies that are available online.

### 12.3.2 Performance Measures

In the performance assessment of technical safety barriers, we consider measures that are used to evaluate barriers in the risk assessment of a larger system, and the measures that are used to evaluate barriers themselves in the design and development phases. When we talk about security of barriers, we are talking about barriers themselves. Availability and safety integrity are mostly used measures in industrial practices, and thus, here we consider the counterpart of safety integrity in security assessment, which is *security integrity.*

- *Security integrity*: The ability of a barrier to perform its required function under the specific threats.

For example, the security integrity can be used to describe whether a barrier can avoid unauthorized modification. Then, the security integrity level, or in short, *security level* (with the short name of SL or SEL) can be regarded as discrete levels of such an ability, as given in Table 12.1, and it is used in practice for evaluating the resistance against different attacks. IEC62443-1-1 (2009) provides a guidance of determining SLs for Industrial Automation and Control System (IACS) as follows, where technical barriers can often be regarded as a kind of IACSs.

The classification within the IEC 62443 framework is qualitative rather than quantitative, and therefore expert judgment and thorough evaluation of characteristics of the technical systems are important when determining SL. In general, we can see the following:

- At SL 1, threat actors have low skills and minimal resources. The required security integrity includes basic security policies and simple physical measures to mitigate risks.

**TABLE 12.1**
**Security Integrity Levels Specified in IEC62443-1-1 (2009)**

| SL | Description |
|---|---|
| SL 4 | Protection against intentional misuse using sophisticated means with extensive resources, specific knowledge, and high motivation. |
| SL 3 | Protection against intentional misuse by sophisticated means with moderate resources, specific knowledge, and moderate motivation. |
| SL 2 | Protection against intentional misuse by simple means with few resources, general skills, and low motivation. |
| SL 1 | Protection against unintentional or accidental misuse. |
| SL 0 | No special requirement or protection required. |

- At SL 2, threat actors own greater skills and motivation but limited resources. The security integrity demands more sophisticated countermeasures, such as user authentication for the controller of a barrier system, role-based access control, and event logging for security-relevant incidents within a barrier system.
- At SL 3, threat actors can gain physical access to the barrier system and are equipped with sophisticated tools. To realize the required security integrity, it needs a blend of physical and logical measures, such as encryption, intrusion detection systems, and comprehensive security management processes.
- At SL 4, threat actors possess substantial resources and skills, including state actors or highly organized crime groups. To satisfy the security integrity, highly advanced countermeasures are needed, including continuous security monitoring, in-depth inspection, and proactive emergency response strategies.

It is worth noting that such SLs are not only for barrier systems but also for industrial control systems and other technical systems, which can be assets being protected by barriers. We will see more discussions in the subsequent subsections.

## 12.4 USING PRINCIPLES OF BARRIER ENGINEERING FOR SECURITY ASSURANCE

As outlined in Section 12.2, assets can be non-entitative, such as test results, warranty data, algorithms, and proprietary knowledge. To protect such kind of assets from logical attacks, especially cyber-attacks, it is beneficial to implement protective methods based on those principles and systematic methodologies of barrier engineering and management. On the other hand, entitative assets, including machinery and personnel, are also susceptible to harm resulting from cyber-attacks. For instance, sophisticated malware can compromise operational technology systems, leading to the shutdown of manufacturing equipment or jeopardizing worker safety through the manipulation of control systems. Thus, it is reasonable to utilize the principles and philosophy of barrier management for assuring the security of the assets of interests in many applications.

Here, it is meaningful to define the term security barrier based on the concept given in IEC62443-1-1 (2009):

- *Security barrier*: Added-on measure or action that is expected to preclude attacks to the asset of interest or mitigate the consequence of these attacks.

For example, cyber firewall is a typical barrier philosophy in security management, which is installed as an inter-network connection device to restrict data communication traffic between two networks in case of need, to protect the network of interest from cyber-attacks. In such a case, data traffic is like energy flow in the previous scenarios.

Coming back to the bow-tie diagram, in Figure 12.2, security barriers can be modeled as the potential impedances on the attack path. When any one of these barriers performs its functions, the attack is stopped.

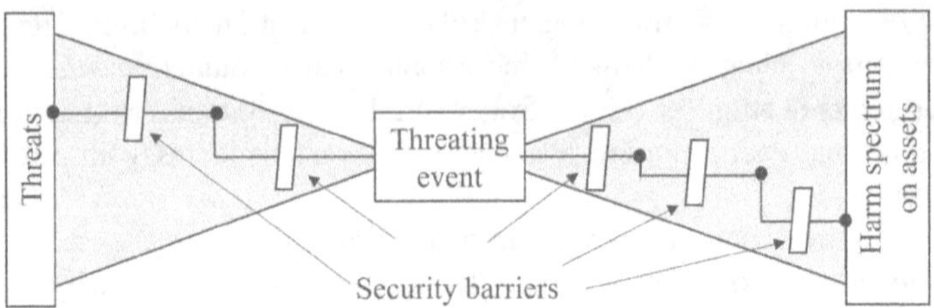


**FIGURE 12.2** Security barriers on an attack path.

Similar with safety barriers, security barriers can be entitative security barriers (e.g., locks controlling physical access to the assets for computation or communication) or logical/non-entitative barriers.

In cybersecurity assurance, logical or non-entitative barriers are more widely used, such as firewall. Some other examples of security barriers include the following:

- *Intrusion detection/prevention systems*: These tools can monitor network traffic to identify potentially dangerous activities and the affiliated log information. They can also be used to stop these threating events.
- *Antivirus and anti-malware software*: They are designed to scan the computer system to identify malicious software and malware, including viruses, worms, and Trojan horses. Once malware is detected, the software typically offers options to remove or quarantine the threats.
- *Encryption*: This method protects data confidentiality by using encoded information or encryption software as the barrier, making it accessible only to users who possess the decryption key. The unauthorized users are regarded as threats. Encryption is useful to preventing data breaches, cyber theft, and unauthorized data manipulation.
- *Access control and authentication*: This approach manages who can access which resources in a network. Access control systems are the barriers based on credentials, roles, or attributes and often require authentication mechanisms like passwords, biometrics, or security tokens.
- *User education and awareness training*: Like safety training, this is a generally preventive method, for informing users about the risks and appropriate security behaviors can significantly reduce security breaches caused by human errors or social tactics.

## 12.5 PERFORMANCE ASSESSMENT OF SECURITY BARRIERS

Similar to Chapter 6, we can use the measure of effectiveness to assess the performance of security barriers in a larger system. In this chapter, *effectiveness* of a security barrier implies the completeness of this barrier to provide protection against a specific threat.

In practice, *efficiency* is also used to assess security barriers, for comparing the resource usage of a barrier and its effectiveness in security assurance, to ensure the barrier is not adversely affecting system performance.

It is often challenging to conduct a quantitative assessment on effectiveness or efficiency of a security barrier. One method is *benchmarking*, namely, comparing the performance of security barriers with the industrial standards and best practices to evaluate relative effectiveness. Another type of method is to conduct drills and use model or simulation (like those in the next subsection) to test how effectively the system or organization can respond to threating events.

Performance assessment is not the ultimate goal of security engineering, but it is an important step for continuous security improvement. Such work uses the insights gained from assessments to make informed decisions about where to improve or adjust security strategies, to consider new and evolving cyber threats, and to integrate newer technologies or update existing ones to enhance security.

## 12.6 MODELING AND ASSESSMENT METHODS OF SECURITY BARRIERS

Assessment of security barriers is based on threat modeling, which has a purpose to simulate the relationship and interactions of threats and defenders. Many modeling methods have been developed, but in this chapter, we mainly mention two commonly used ones that are relevant with barriers.

### 12.6.1 Attack Tree

An attack tree is a proactive and structured methodology that helps in understanding the potential actions an attacker might undertake to utilize vulnerabilities within a sociotechnical system or asset (see Mauw & Oostdijk, 2006). Such a concept can be found firstly in the technical report by Salter et al. (1998).

This approach can be graphically represented in different ways, and one of them is like a fault tree, which is utilized for evaluating safety barriers (refer to Chapter 5). As depicted in Figure 12.3, an attack tree is organized hierarchically with nodes and gates. The *root node*, positioned at the top, signifies the primary goal or objective of an attacker. This goal is achieved through the occurrence of certain threatening or attack events ($e_i$), and the tree is constructed by breaking

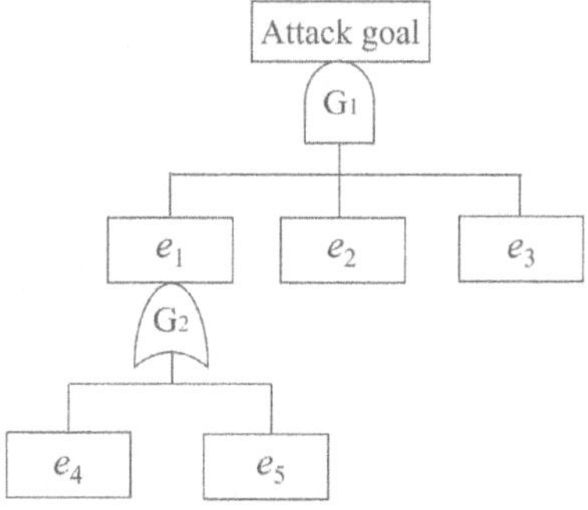


**FIGURE 12.3** An illustrative example of attack tree.

down these potential attack events into more detailed steps, represented as lower-level or input nodes.

Similar to a fault tree, an attack tree also describes the logical relationships between events using gates. In the example provided in Figure 12.3, $G_1$ is an AND-gate, indicating that the achievement of the output goal requires the occurrence of all input attack events. $G_2$ is an OR-gate, whereby the realization of the output goal is possible if any of the input events occur. If a node can be branched into more lower-level nodes, such a node is an *intermediate node*, such as $e_1$ in Figure 12.3. In contrast, if a node cannot be further branched, it is a *leaf node*, as a basic step or event of an attack, such as $e_2$, $e_3$, $e_4$, or $e_5$ in Figure 12.3.

Using an approach similar to that employed for identifying minimal cut sets in a fault tree, we can also determine the minimal cut sets of an attack tree. In the shown example, the minimal cut sets are $\{e_2, e_3, e_4\}$ and $\{e_2, e_3, e_5\}$.

Considering the vulnerabilities of the asset, a leaf node in the attack tree can be interpreted as a combination of a specific action ($a_i$) and its associated vulnerability ($v_i$). For example, the attack tree depicted in Figure 12.3 can be revised, as shown in Figure 12.4.

When introducing a security barrier, an additional event is incorporated as an input to an AND-gate, for instance, $e_2$ as the event where unauthorized access to a confidential database occurs. If a password control security barrier is implemented, it corresponds to $e_3$, which can be interpreted as the decryption of the password.

From the logical perspective, a leaf node in the attack tree illustrated in Figure 12.4 can be interpreted as the result of combining an associated action and the presence of a vulnerability, which serve as inputs to an AND-gate. However, given that actions and vulnerabilities hold distinct implications for system design and security assurance, it is reasonable to represent them with different symbols and analyze them separately.

The analysis based on an attack tree involves the following main steps:

1. *Defining the primary goal*: This initial step involves a clear understanding and description of the primary objective of a potential attacker, such as disrupting a safety function or stealing credit card information. Typically, such a goal is represented by some form of loss to the asset of interest,

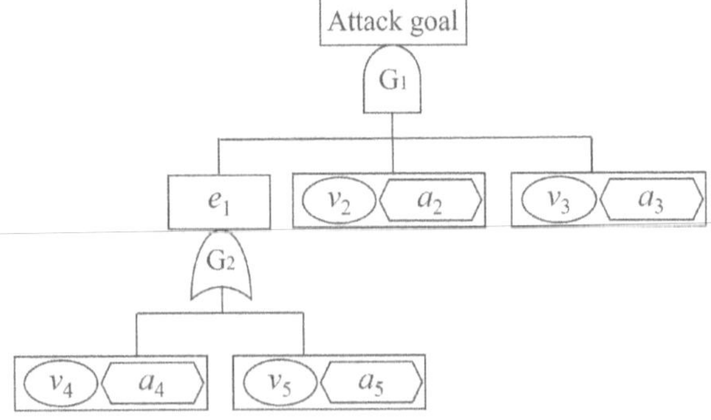


**FIGURE 12.4** An illustrative example of attack tree in consideration vulnerabilities.

highlighting the importance of precisely defining what the attacker aims to achieve.

2. *Breaking down the primary goal*: Once the primary goal is defined, it is divided into manageable sub-goals, or the necessary methods or tasks required to achieve the main objective. For instance, to steal credit card information, methods can be unauthorized access to a customer database or the use of malware. The identified sub-goal adds an additional level to the attack tree. This step requires evaluating whether a sub-goal can be achieved directly or if further decomposition is needed. This process continues iteratively until all sub-goals are decomposed into one-step tasks.
3. *Analyzing the attack tree*: The next step is to identify the minimal cut sets of the attack tree and uncover all potential paths an attack could take. It is important to ensure that each path is plausible and based on realistic attacking methods. The most likely or dangerous paths are identified here for the prioritization of these threats, focusing security efforts on the areas of greatest risk.
4. *Deploying and evaluating security barriers and measures*: Finally, an assessment is made regarding the adequacy of existing security barriers to counter identified threats. If current barriers are insufficient to withstand the most critical threats, additional security barriers should be implemented. This step involves a strategic deployment of security barriers, ensuring that the defenses are robust enough to mitigate significant vulnerabilities effectively. In addition, this step helps identify potential vulnerabilities and areas for improvement.

A simple approach for analyzing the effectiveness of security barriers is based on the principles of sensitivity analysis. This methodology involves assessing the likelihood of achieving a primary objective both in the absence and presence of a security barrier. By comparing these scenarios, we can know how significantly the specific security barrier enhances the overall security framework.

More information about the threat modeling and security analysis with attack tree can be found, such as in the articles by Mauw and Oostdijk (2006) and Saini et al. (2008).

### 12.6.2 Cyber Kill Chain Model

The cyber kill chain model, as adapted by Lockheed Martin Corporation from its original military concept, applies a structured approach to cybersecurity. As shown in Figure 12.5, this model is like the Domino model in risk management, reflecting

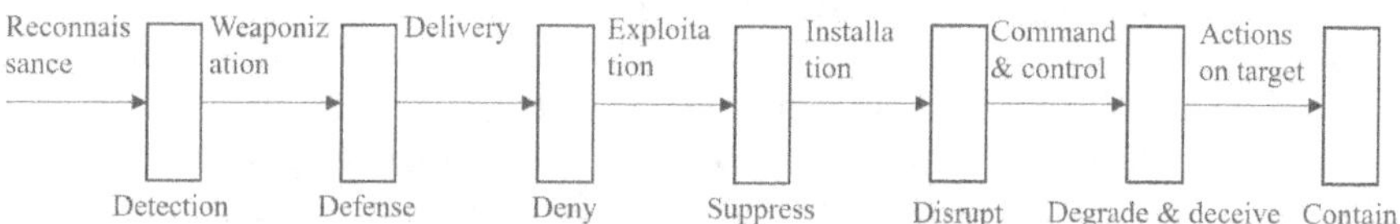


**FIGURE 12.5** Cyber kill chain model.

a cyber threat progressing through various stages, ultimately culminating in damage to an asset.

According to Yadav and Rao (2015), these stages include the following:

1. *Reconnaissance*: The attacker identifies a target asset, conducts thorough research, and locates vulnerabilities of the target.
2. *Weaponization*: The attacker creates a method, or "weapon", to exploit one or more identified vulnerabilities. In cyber-attack, this often involves crafting malware designed for remote access.
3. *Delivery*: The weapon is transmitted to the target asset, for example, via email attachments or USB drives in cyber-attack.
4. *Exploitation*: Once the weapon reaches the target, it can be activated. For instance, malware is triggered to infiltrate the system.
5. *Installation*: The weapon maintains its presence within the asset in a certain way to continue functioning. For example, the malware may establish a backdoor, facilitating access for the attacker.
6. *Command and control*: The attacker continues using the weapon on the target. In cyber-attack, it can maintain the control over the computer network through the access points established by the malware.
7. *Actions on target*: The attacker executes actions aimed at achieving their ultimate objectives, which may involve data theft, system disruption, or other harmful activities.

To counteract each phase of the attack path, specific security measures (as mitigation barriers) can be implemented:

1. *Detection*: Security systems, such as intrusion detection systems, are deployed to identify potential attacker presence.
2. *Defense*: It is important to analyze potential attack methods, assess system vulnerabilities regularly, guard against information leakage, and update protection approaches against newly developed exploits.
3. *Deny*: Efforts to block the delivery of threats in cybersecurity practices include enhancing access controls and employing advanced email filtering technologies to detect and block malicious attachments and links.
4. *Suppress*: This involves inhibiting the activation of the cyber weapon and thwarting exploitation attempts. For example, rigorous patch management policy is essential for applying security updates swiftly, thereby preventing malware from exploiting known vulnerabilities.
5. *Disrupt*: This involves removing or disabling the cyber weapon. Antivirus software can play a critical role in this phase.
6. *Degrade and deceive*: This step involves counteracting and disrupting the command and control operations of the attacker.
7. *Contain*: The final response aims to control and limit damage to an acceptable level. Techniques such as network segmentation can be effective in localizing and mitigating the impact of a cyber-attack.

All the security barriers above collectively form a comprehensive defense framework. Barrier assessment is needed at each stage to determine the effectiveness of each barrier or control measure. Additionally, an overall assessment is necessary, as these barriers often interact and depend on one another. This comprehensive evaluation determines whether the combined barriers adequately detect, prevent, and respond to specific threats. Another important task is to prioritize the implementation of these barriers based on the potential impact of an attack, ensuring that resources are allocated effectively to mitigate the most significant risks.

Similar to IEC 61508, which addresses functional safety, IEC 62443 is a series of international standards focused on the security of industrial networks and systems. It is also based on the concept lifecycle for security assessment and management. Instead of using the term "security barriers", the relevant standards refer to "countermeasures" or "security technologies". For further studies of the technologies employed in securing industrial networks and systems, readers are encouraged to refer to IEC62443-3-1 (2009).

## REFERENCES

Abdo, H., Kaouk, M., Flaus, J. M., & Masse, F. (2018). A safety/security risk analysis approach of Industrial Control Systems: A cyber bowtie – combining new version of attack tree with bowtie analysis. *Computers & Security*, *72*, 175–195. https://doi.org/10.1016/j.cose.2017.09.004

Baybutt, P. (2002). Assessing risks from threats to process plants: Threats and vulnerability analysis. *Process Safety Progress*, *21*(4), 269–275.

IEC62443-1-1. (2009a). Industrial communication networks – network and system security In *Part 1-1: Terminology, Concepts and Models*. Geneva, Switzerland: International Electrotechnical Commission.

IEC62443-3-1. (2009b). Industrial communication networks – network and system security. In *Part 3-1: Security Technologies for Industrial Automation and Control Systems*. Geneva, Switzerland: International Electrotechnical Commission.

Mauw, S., & Oostdijk, M. (2006). *Foundations of attack trees*. Information Security and Cryptology - ICISC 2005, Seoul, South Korea.

Rausand, M., & Haugen, S. (2020). *Risk Assessment: Theory, Methods and Applications*. Wiley.

Saini, V., Duan, Q., & Paruchuri, V. (2008). Threat modeling using attack trees. *Journal of Computing Sciences in Colleges*, *23*(4), 124–131.

Salter, C., Saydjari, O. S., Schneier, B., & Wallner, J. (1998). *Toward A Secure System Engineering Methodology*, National Security Agency.

Yadav, T., & Rao, A. M. (2015). Technical aspects of cyber kill chain. In J. Abawajy, S. Mukherjea, S. Thampi, & A. Ruiz-Martínez (Eds.), *Security in Computing and Communications: Third International Symposium, SSCC 2015, Kochi, India, August 10–13, 2015. Proceedings 3. Springer International Publishing*, Springer International Publishing.

# 13 Barrier Engineering and Sustainability

## 13.1 INTRODUCTION

The United Nations proposed 17 sustainable development goals (SDGs) in 2015, which have been adopted by all the member countries. These SDGs emphasize the interconnected environmental, social, and economic aspects of sustainable development by putting sustainability at their center.

The overall objective of barrier engineering is to protect critical assets, including humans, communities, and the environment, from various types of accidents. This aligns closely with multiple SDGs, such as SDG8 – decent work and economic growth, SDG9 – industry, innovation, and infrastructure, SDG12 – responsible consumption and production. In addition, the successful implementations of barrier engineering in diverse industries are also beneficial in the realization of SDG11 – sustainable cities and communities, SDG13 – climate action (SDG 13), SDG14 – life below water, and SDG15 – life on land.

For example, barrier engineering is expected to play an important role in ensuring the safe and smooth production, storage, transportation, and utilization of renewable energy, such as onshore and offshore wind power, lithium-ion battery, green and blue hydrogen, waste burning, biomass, and fuel cells. By preventing and mitigating accidents, barrier engineering contributes to reducing $CO_2$ emission and pollutions, safeguarding neighborhood residents of those process plants, and minimizing the potential environmental damages on land and in marine environments.

In literature (e.g., see Purvis et al. (2019)), there are three pillars of sustainability: economic, environmental, and social. The interactions between barrier engineering and these sustainability pillars can be viewed from two perspectives:

- *Sustainable barriers*: This means satisfying the requirements of economic-, environmental-, and social sustainability in the design, operation, and maintenance of safety barriers and in barrier management, while ensuring the safety requirement.
- *Barriers for sustainability*: This refers to the establishment of safety codes and the development of barriers within sociotechnical systems and processes aimed at achieving sustainability goals. The examples of these processes include reducing greenhouse gas emissions, minimizing waste and pollution, and enhancing the production, storage, and distribution of sustainable energy.



DOI: 10.1201/9781003245636-13

## 13.2 SUSTAINABILITY CONSIDERATIONS IN BARRIER DESIGN AND OPERATION

Being a part of sustainability efforts, it is natural to involve more sustainability considerations in barrier design, operation, and other phases of its lifecycle. The followings are some examples:

- *Eco-friendly materials*: Choosing materials that are less harmful to the environment is a very helpful in realizing sustainable barrier design. This includes selecting recyclable or biodegradable materials and those manufactured through environmentally friendly processes. Design should also aim to minimize waste and reduce the carbon footprint during the manufacturing stage.
- *Energy efficiency*: Barriers, especially active barrier systems, should be designed to operate with minimal energy consumption. For example, using energy-efficient components and light design can significantly reduce the overall energy consumption of barrier systems in their operations.
- *High reliability*: In fact, reliability is highly related to sustainability, since higher reliability can reduce the need for frequent replacements and thus lower the resource consumption in maintenance and waste generation over time. This involves using high-quality materials and robust design principles that can withstand environmental and operational stresses. High reliability also implies lower unexpected emission and pollution in accidents. However, high-reliability design, such as overdesign and redundancy use more raw materials, is not always friendly to sustainability, and thus optimization is needed.
- *High maintainability*: Barriers designed with high maintainability can allow for future modifications or upgrades without complete replacements, and thus reduce the material usage. Modular design also facilitates repairs and upgrades, which is more friendly to sustainability by extending the useful life of the barrier system.
- *Supplier management*: It is also necessary to consider the sustainability practices of suppliers, including their sourcing of materials, labor practices, and environmental management systems. Choosing suppliers with strong sustainability credentials can enhance the overall sustainability profile of the barrier system.
- *Eco-friendly operation and maintenance*: Effective operation and maintenance (O&M) strategy is importance to sustainability. This includes optimizing operational schedules and ensures that the barrier systems such like sensors and valves are only active when needed. Regular maintenance ensures that barrier systems operate efficiently, preventing excessive energy use and prolonging the life of the components.
- *End-of-life management*: Planning for the end-of-life of barrier systems needs to consider how they can be decommissioned and recycled. Developing and implementing a recycling or disposal plan for barrier systems can minimize their environmental impact once they are no longer in use.

## 13.3 SUSTAINABILITY ASSESSMENT OF BARRIERS

Considering the economic, environmental, and social requirements, a multi-criteria decision-making approach is always needed in barrier sustainability analysis. Specifically, the following inputs from different knowledge disciplines are typically utilized in the overall assessment of safety barrier systems and barrier management approaches:

- *Social sustainability analysis*: This involves effectiveness and safety integrity analyses, which are critical as the primary function of barriers is to protect humans from injuries and fatalities in accidents. These methods have been discussed in previous chapters as the foundational to barrier analysis.
- *Economic sustainability analysis*: Techno-economic analysis aims to identify the key factors that affect the economic performance of different barriers and operational strategies. It is important to consider the lifecycle costs of barriers, which include initial capital expenditures (CAPEX), operational expenditures (OPEX), expenses in reliability-centered maintenance (RAMEX), and potential losses or expenses in case of accidents (RISKEX). Factors such as tax incentives and the long-term impact of inflation also play significant roles. In most cases, the economic impacts of activating barriers on the normal operation of the asset being protected are also needed to be taken into consideration.
- *Environmental sustainability analysis*: There are many methods for environmental sustainability analysis, and one of most used is the life cycle assessment (LCA). This method analyzes the environmental impact at each stage of the technical system lifecycle (Finnveden et al., 2009; Guinée et al., 2011). An LCA for a barrier system will include several key tasks:
  - Assessing the extraction of raw materials and resources needed in the installation, production, transportation, and decommissioning phases of entitative barriers;
  - Investigating alternative materials and technologies used for the same barrier function that can replace or reduce the reliance on scarce resources with the purpose of minimizing the environmental footprint;
  - Estimating emissions associated with production and operation, and identifying opportunities for optimizing energy efficiency; and
  - Evaluating waste management strategies, focusing on minimizing waste generation, promoting recycling.

The results of the above analysis can raise requirements for barrier specification, design, and operation. The requirements are diverse and sometimes conflicting. However, many decisions in barrier engineering have to be made considering these requirements. For instance, it is challenging to directly compare the following two barrier systems and decide which should be adopted: One that is low cost with a higher environmental impact, and another that is more expensive but environmentally friendly. Addressing these challenges needs multidisciplinary knowledge and the multi-criteria decision-making techniques.

In this chapter, we do not present the multidisciplinary knowledge required for assessing barrier sustainability, such as the detailed LCA process, due to the page limitation. It is also unreasonable to expect any one individual to possess all necessary competencies in practices. However, we will present some classical methods for guiding the comprehensively evaluation of multiple criteria.

### 13.3.1 Weighting and Scoring

The weighting and scoring method provides a straightforward but systematic approach for evaluating and prioritizing various options or alternatives based on multiple criteria. This method is particularly effective when assessing the sustainability and technical properties (TPs) of barrier systems, as detailed in a study by Zhao et al. (2024).

In this decision-making process, sustainability requirements (SRs) are linked with TPs of a barrier system. Each SR, such as $CO_2$ emissions, energy consumption, and social acceptance, is connected to specific TPs, which pertain to certain design or operational properties of the barrier system under review. The weight $w_{ij}$ represents the significance of a technical property $TP_i$ in meeting a A requirement for sustainability $SR_j$. Assuming there are $m$ requirements, for each requirement $SR_j$, we have

$$\sum_{i=1}^{n} w_{ij} = 1, \text{ for } j = 1,2,\ldots m \tag{13.1}$$

For a simplified model incorporating only the environmental-, social-, and economic pillars of sustainability, $m = 3$. A weight $w_{ij} = 0$ indicates that there is no relationship between $TP_i$ and $SR_j$, whereas $w_{ij} = 1$ implies that the $SR_j$ is entirely determined by $TP_i$. The performance of each $TP_i$ regarding $SR_j$ is denoted as $g_{ij}$, ranging from −1 to 1, where a negative value indicates a harmful effect on sustainability and a positive value indicates a beneficial impact.

The values of $w_{ij}$ and $g_{ij}$ are typically derived from technical and operational data, expert assessments, and stakeholder interviews and are updated as new information becomes available.

Additionally, the importance of each SR relative to the overall sustainability objectives is evaluated. This importance is quantified using $ISR_j$, as

$$\sum_{i=1}^{m} ISR_j = 1 \tag{13.2}$$

The overall sustainability score (OSS) of a barrier system or component is then calculated as follows:

$$OSS = \sum_{i=1}^{m} ISR_j \cdot \sum_{i=1}^{n} g_{ij} w_{ij} \tag{13.3}$$

Based on this formula, the OSS ranges from –1 to +1, where a score closer to –1 indicates a system that is less sustainable, and a score closer to +1 indicates a system that is more sustainable. The parameters in this formula can also be expressed as functions of time, allowing for dynamic sustainability assessments over time OSS($t$).

While this method offers a clear framework for sustainability evaluation, it relies heavily on subjective judgments and may come into totally different conclusion when different persons are asked. Such evaluation requires verification through empirical data or further validation studies.

### 13.3.2 Analytic Hierarchy Process

Developed by Thomas L. Saaty, the Analytic hierarchy process (AHP) is a structured technique for organizing and analyzing complex decision problems. An AHP process can structure a decision problem with decomposing it into a hierarchy, compare criteria and alternatives pairwise, and prioritize and ranking the alternatives (Saaty, 1980).

Rather than making a "correct" decision, the aim of AHP is to identify the most suitable option relative to the ultimate goal of decision-maker and her/his understanding of the problem. This approach is more flexible and practical than other scoring and weighting methods by allowing for inconsistencies in judgments and then resolving them.

To illustrate, we can consider the application of AHP in evaluating the overall sustainability of a barrier system. The main steps include the following:

1. *State the decision problem*: The AHP process starts by clearly defining the decision-making problem. Here in this illustrative example, the problem is selecting a safety barrier system for a process plant. The barrier system is expected to have an optimal performance in terms of the overall sustainability.
2. *Define the criteria/objectives*: The next step is to identify and categorize the criteria that can influence the decision. These criteria should comprehensively cover all considerations of the decision-maker. As mentioned in the last subsection, for the overall sustainability of a barrier system, we can consider three criteria or SRs:
   - *Economic sustainability*: Costs of installing, operating, and maintaining a barrier system, as well as some items that might be difficult to be quantified as monetary values, such as maintenance difficulty, and scalability.
   - *Social sustainability*: Impact on reducing the risks related to operator safety and well-being, as well as enhancing the confidence of neighborhood communities to the facility.
   - *Environmental sustainability*: Effects of the installation and maintenance of the barrier system on the local and broader environment, in comparison to its barrier function for reducing the potential damages and pollutions of accidents on environment.

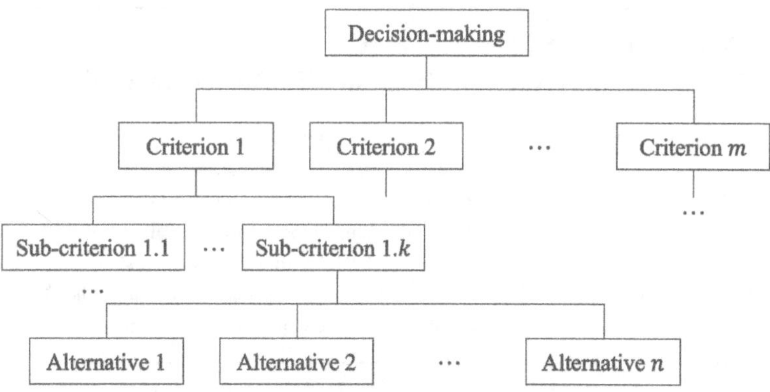


**FIGURE 13.1** An illustrative example of AHP process.

As shown in Figure 13.1, AHP can decompose such a decision problem into a hierarchy where the overall decision problem is at the top level, and criteria are at the second level.

If a criterion is regarded too broad, such as environmental sustainability in this example, it can be further divided into sub-criteria, like $CO_2$ emission, energy consumption, and pollution, as illustrated by the decomposition of Criterion 1 in Figure 13.1.

3. *Identify the alternatives*: Viable alternatives are listed at the bottom level. For example, we suppose to consider three types of barrier systems here:
    - *System A*: Low moderate maintenance cost, highly effective on safety (social impact), and low environmental friendliness.
    - *System B*: High cost, low maintenance, highly effective on safety, and high environmental friendliness.
    - *System C*: Moderate cost, high maintenance, moderate effective on safety, and moderate environmental friendliness.
4. *Conduct pairwise comparisons*: All the elements at each level of the hierarchy are compared in this step to assess their relative importance. This step includes the following tasks:
    - Compare criteria against each other to determine their relative importance to the decision.
    - Compare each alternative against others under each criterion.

    This is critical since it forms the basis for deriving priority scales. As illustrated in Table 13.1, numerical values can be used for describing the relative importance of one element to another.

    For example, we can establish a comparison matrix for the three aspects of overall sustainability, as shown in the following Table 13.2:

    It can be found that the judgments for relative importance of elements are not fully consistent. For example, the relative importance of social sustainability to economic one is 7, and relative importance of social sustainability to environmental one is 3, but the relative importance of economic

**TABLE 13.1**
**Description of Numerical Values in AHP Pair Comparison**

| Numerical Value | Description |
|---|---|
| 1 | The two elements are equally important. |
| 3 | One element is moderately more important than another. |
| 5 | One element is strongly more important than another. |
| 7 | One element is much more important than another. |
| 9 | One element is extremely more important than another. |
| 2, 4, 6, 8 | Intermediate values for compromising between judgments. |

one to environmental one is not equal to 3/7 in the table, but equal to 1/2. Such inconsistencies are acceptable in the AHP process.

Then, comparison matrices for sub-criteria and alternatives can be developed using the same principles.

5. *Rank alternatives*: The next step is to determine relative rankings of alternatives. These matrices developed in the last step are normalized and the priority vector is computed by averaging across the rows after normalizing each column. The alternative with the highest aggregating score across all criteria, determined by multiplying the priority vectors of alternatives by the priority vector of criteria, is selected as the most suitable option.

Saaty (1980) has demonstrated mathematically that the eigenvector approach is very effective in the AHP process. In linear algebra, an *eigenvector* of a square matrix is a vector that does not change its direction under the associated linear transformation. In other words, if $\mathbf{v}$ is a vector that is not zero, it is an eigenvector of a square matrix $A$ if $A\mathbf{v}$ is a scalar multiple of $\mathbf{v}$: $A\mathbf{v} = \lambda\mathbf{v}$, where $\lambda$ is a number known as the eigenvalue with the eigenvector $\mathbf{v}$.

Using the eigenvector approach, we can raise the pairwise matrix to powers that are successively squared each time, and the row sums are then calculated and normalized. The process stops when the difference between these sums in two consecutive calculations is smaller than a prescribed value.

Due to page limitation, we are unable to provide a numerical example of the AHP process in this section. However, interested readers can find a wealth of case studies and applications of the AHP method in various domains available in academic

**TABLE 13.2**
**Comparison Matrix of Three Criteria in the Example**

| Criterion | Economic | Social | Environmental |
|---|---|---|---|
| Economic | 1 | 1/7 | 2 |
| Social | 7 | 1 | 3 |
| Environmental | 1/2 | 1/3 | 1 |

and professional literature, such as those listed in the review paper by Russo and Camanho (2015).

### 13.3.3 Technique for Order Preference by Similarity to Ideal Solution

Technique for Order Preference by Similarity to Ideal Solution (TOPSIS) is another commonly used method in multi-criteria decision-making, which was first proposed by Hwang and Yoon (1981). TOPSIS identifies the ideal and anti-ideal solutions and ranks the alternatives based on their distances from these solutions, namely favoring alternatives closer to the positive ideal solution (PIS) and farther to the negative ideal solution (NIS). The steps involved in the TOPSIS process are designed to evaluate each alternative against a set of criteria, facilitating a straightforward comparative analysis.

Here are the main steps considering the overall sustainability of a barrier system:

1. *Construct a decision matrix*: The TOPSIS process starts by establishing a decision matrix comprising the alternatives and criteria. Each element in the matrix represents the performance of an alternative (with different barrier design and operation parameters) with respect to a specific criterion (SR).
2. *Construct the weighted normalized decision matrix*: The decision matrix is normalized to avoid the impact of differing scales among SRs. This can be achieved using the Euclidean norm method or other normalization techniques, ensuring that each criterion contributes equally to the decision-making process. The next task is to multiply each element of the normalized decision matrix by the weight of the corresponding criterion. Such a weight reflects the relative importance of each evaluation criterion.
3. *Determine the positive (ideal) and negative (anti-ideal) solutions*: This step is to identify the PIS and NIS. PIS is the solution that satisfies all SRs in the best way, namely maximizes the benefit criteria (such like safety integrity and public acceptance) and minimizes the cost criteria (such like operational cost and $CO_2$ emission), effectively representing the best possible outcome. Conversely, NIS maximizes the cost criteria and minimizes the benefit criteria, representing the worst possible outcome. These are not necessarily existing alternatives but theoretical extremes.
4. *Calculate the relative closeness to the ideal solution*: The distance of each alternative from the PIS and the NIS is calculated. This is typically done using the Euclidean distance formula. For each alternative, it is necessary to calculate the distance to the PIS (denoted as $d^+$) and the distance to the NIS (denoted as $d^-$). Then, the *relative closeness* (RC) of an alternative to the ideal solution is calculated by dividing the distance from the NIS by the sum of the distances from both the PIS and NIS:

$$RC = \frac{d^-}{d^+ + d^-} \tag{13.4}$$

RC ranges from 0 to 1, where a value closer to 1 indicates that the alternative is closer to the ideal solution and thus more desirable.

5. *Rank the alternatives*: Finally, all the alternatives are ranked based on their relative closeness scores. The alternative with the highest score is considered the best choice.

TOPSIS can provide a clear, logical analysis of complex decision problems involving multiple criteria and alternatives, offering a quantifiable basis for comparison and final selection. More details of TOPSIS can be found in the introductory article by Lai et al. (1994).

There are many more methods for multi-criteria decision-making that can be used in barrier specification and operation considering the overall sustainability goal. Readers who have interests can refer to the review papers by Triantaphyllou (2000), Wang et al. (2009), and Ceballos et al. (2016) for the applications, advantages, and limitations of different methods.

## 13.4 BARRIER ENGINEERING IN NEW APPLICATIONS FOR SUSTAINABILITY DEVELOPMENT

### 13.4.1 Some Challenges in New Applications

In terms of employing the principles of barrier engineering and deploying safety barriers in the new applications for sustainability development, such as in new energy sectors, pollution control, and $CO_2$ capturing and storage, a lot of barrier equipment can be re-used but the following challenges need more attention:

- *Novel requirements of new techniques, material properties, and processes*: New materials and energy formats have different requirements for the types, operational modes, and effectiveness of safety barriers. For instance, consider hydrogen as an emerging energy source and $CO_2$ that needs to be collected and stored, their physical properties differ significantly from those of traditional natural gas (primarily methane), as shown in Table 13.3. Many safety

**TABLE 13.3**
**Comparison of Physical Properties of Gaseous $H_2$, $CO_2$, and Natural Gas ($CH_4$)**

| Property | Hydrogen | $CO_2$ | Natural Gas ($CH_4$) |
|---|---|---|---|
| Molecular weight | 2.016 g/mol | 44.01 g/mol | 16.04 g/mol |
| Density (at STP) | 0.08988 kg/m³ | 1.98 kg/m³ | 0.717 kg/m³ |
| Flammability | 4%–75% in air | Non-flammable | 5%–15% in air |
| Ignition Temp. | 500°C | Non-flammable | 537°C |
| Odor | Odorless | Odorless | Odorless (odorant added for safety) |
| Solubility in Water | Low | High | Low |
| Toxicity | Non-toxic | Asphyxiant in high concentrations | Asphyxiant in high concentrations |

barriers have been deployed in the process industries to manage the potential events related to natural gas. However, the differences between various gases necessitate specific considerations in the specification and operation of the safety barriers that are used for similar equipment but handling different gases. For example, due to the broad flammability range of hydrogen, highly sensitive sensors or gas detectors are essential, and relying solely on gas dispersion as a safety barrier cannot reduce risks effectively. In contrast, because $CO_2$ is denser, leaks can create high concentrations at breathing levels, making venting channels as critical safety barriers for humans.

- *Lack of data and historical records*: The novelty of these sustainability technologies and processes also means that there is limited experience in handling, maintaining, and predicting their behavior under various conditions and thus limited data for barrier analysis. Without historical data in barrier design and performance estimation, engineers must rely on theoretical models, simulations, and extrapolations from related fields. The failure prediction methods based on reference data we have introduced in Chapter 7.3 can be used to provide useful information for barrier analysis. On the other hand, industries and relevant governmental agencies should be encouraged to develop common databases for the new applications, as the foundation of preventing accidents and ensuring safety in the emerging sectors.
- *High interdependencies and more complexities*: The integration of new technologies often involves complex systems where various components and processes are highly interdependent. This complexity can amplify the consequences of a single failure, as it may lead to cascading effects throughout the system. Understanding these interdependencies is important for designing effective barrier systems that can isolate problems and prevent them from escalating. We have discussed the interdependency issues in Chapter 10.
- *High uncertainty*: This uncertainty increases the risk of unexpected issues and intensifies the importance of tests and monitoring techniques during implementation. When coming to new applications, the uncertainty in predicting outcomes and assessing risks increases substantially. This uncertainty may come from incomplete knowledge about the interactions between different materials, the long-term stability of the technologies involved, or unforeseen environmental impacts. Robust analytical methods and models need to be developed to better predict these uncertainties and integrate them into the safety design and decision-making processes. We will have more discussions on dealing with this challenge in the next subsection.

### 13.4.2 Barrier Engineering under an Uncertain Environment

Although there are many arguments on the definition of uncertainty, as well as the relationship between risk and uncertainty, in this book, we understand risk as the result of risk assessment (Rausand & Haugen, 2020) and define uncertainty as

- *Uncertainty*: The measure of confidence in the risk assessment result considering the lack of own knowledge and the incompletely available information.

In literature, according to the causes, uncertainty can be categorized into two types:

- *Aleatory uncertainty*: This type of uncertainty arises from randomness or natural variability. An example is the direction of the wind at a specific future time.
- *Epistemic uncertainty*: This type of uncertainty is due to a lack of knowledge or information. For instance, the results of an examination will be released tomorrow.

Based on the definition, aleatory uncertainty cannot be reduced by more observations or additional information, but a more accurate estimation about the probability distribution of a certain result can be expected. On the other hand, epistemic uncertainty can diminish as more knowledge is available. The distinction between these two types is actually blurred. Increasing knowledge can reveal underlying patterns or rules in what was previously considered aleatory uncertainty, which is thus not further regarded as unpredictable. For example, events such as comets and solar eclipses were once thought to be random, but now their timings can be predicted with precision.

According to the types of missing information or knowledge, uncertainty in risk assessment can be further classified as

- *Model uncertainty*: This type of uncertainty arises from discrepancies between the model used in risk assessment and the reality it aims to represent. A model may diverge from reality or only partially reflect it (also known as *completeness uncertainty*). For instance, an integrity assessment model based on the exponential distribution for a technical barrier might not account for valve degradation.
- *Parameter uncertainty*: This type of uncertainty arises from differences between the estimated parameter values used in risk assessments and the actual values of those parameters. For example, the failure rate estimation by an exponential distribution model with the reference data from the industrial database may be underestimated if the valve is operating in a harsh environment.

There are existing methods for dealing with uncertainties in barrier engineering. For example, *fuzzy logic* can be a useful method for dealing with the uncertainty issue of lacking data. Fuzzy logic is a mathematical framework that allows for the representation and manipulation of imprecise or vague information. In the situations where traditional binary logic may not adequately capture the uncertainty of real-world phenomena, fuzzy logic offers a more flexible approach. For example, fuzzy logic allows us to define linguistic variables and terms to describe qualitative aspects of barrier performance, such as "very effective", "moderately effective", or "low effective". By quantifying these linguistic variables using fuzzy sets and membership functions, engineers can model and reason about subjective concepts more effectively. Readers who have interest can refer to the books by Hájek (2013) and Trillas and Eciolaza (2015).

Another approach for managing uncertainty is to conduct *sensitivity analysis*. This method is used to examine how the changes in an input parameter, an individual assumption, or the structure of the model influence the results of risk assessment. For instance, the reliability importance calculations discussed in Section 5.3.3 are a kind of sensitivity analysis. As an example, for a novel application of barrier, if the failure rate of a barrier component cannot be precisely estimated due to uncertainty, a range of values may be set instead of a single value. Then, the performance of the barrier and the risk to assets can be evaluated using both the maximum and minimum values of this range. This is a straightforward application of changing the failure rate value for sensitivity analysis. For a more comprehensive understanding of sensitivity analysis, interested readers can refer to the review article by Frey and Patil (2002) for an overview of sensitivity analysis methods and the review article by Borgonovo and Plischke (2016) for the mathematical principles of sensitivity analysis.

## REFERENCES

Borgonovo, E., & Plischke, E. (2016). Sensitivity analysis: A review of recent advances. *European Journal of Operational Research*, *248*(3), 869–887.

Ceballos, B., Lamata, M. T., & Pelta, D. A. (2016). A comparative analysis of multi-criteria decision-making methods. *Progress in Artificial Intelligence*, *5*, 315–322.

Finnveden, G., Hauschild, M. Z., Ekvall, T., Guinée, J., Heijungs, R., Hellweg, S., Koehler, A., Pennington, D., & Suh, S. (2009). Recent developments in life cycle assessment. *Journal of Environmental Management*, *91*(1), 1–21.

Frey, H. C., & Patil, S. R. (2002). Identification and review of sensitivity analysis methods. *Risk Analysis*, *22*(3), 553–578.

Guinée, J. B., Heijungs, R., Huppes, G., Zamagni, A., Masoni, P., Buonamici, R., Ekvall, T., & Rydberg, T. (2011). Life cycle assessment: Past, present, and future. *Environmental Science & Technology*, *45*(1), 90–96.

Hájek, P. (2013). *Metamathematics of Fuzzy Logic* (Vol. 4). Springer Science & Business Media.

Hwang, C. L., & Yoon, K. (1981). *Multiple Attribute Decision Making: Methods and Applications*. Springer-Verlag.

Lai, Y.-J., Liu, T.-Y., & Hwang, C.-L. (1994). TOPSIS for MODM. *European Journal of Operational Research*, *76*, 486–500.

Purvis, B., Mao, Y., & Robinson, D. (2019). Three pillars of sustainability: In search of conceptual origins. *Sustainability Science*, *14*, 681–695.

Rausand, M., & Haugen, S. (2020). *Risk Assessment: Theory, Methods and Applications*. Wiley.

Russo, R. d. F. S. M., & Camanho, R. (2015). Criteria in AHP: A systematic review of literature. *Procedia Computer Science*, *55*, 1123–1132.

Saaty, T. L. (1980). *The Analytic Hierarchy Process*. McGraw-Hill.

Triantaphyllou, E. (2000). Multi-criteria decision making methods. In *Multi-Criteria Decision Making Methods: A Comparative Study. Applied Optimization* (Vol. 44). Springer.

Trillas, E., & Eciolaza, L. (2015). *Fuzzy Logic: An Introductory Course for Engineering Students*. Springer Cham.

Wang, J.-J., Jing, Y.-Y., Zhang, C.-F., & Zhao, J.-H. (2009). Review on multi-criteria decision analysis aid in sustainable energy decision-making. *Renewable and Sustainable Energy Reviews*, *13*(9), 2263–2278.

Zhao, Y., Cai, B., Zeng, T., He, Z., & Liu, Y. (2024). Sustainability evaluation of multi-component subsea transmission system considering failure dependence and maintenance activities. *Ocean Engineering*, *296*, 116945.

# 14 Resilience Enhancer
## *Extension of Safety Barrier*

### 14.1 FROM RISK TO RESILIENCE

All the discussions about safety barriers presented in this book are within the framework of risk management, which considers potential negative events that might affect the asset of interest and employs diverse accidental models to analyze the likelihood and impact of these occurrences. If the asset is not harmed, the efforts of risk management can be regarded successful. However, these models are not used in the operation of the asset when no unexpected events occur, and effect of risk management is not accounted in normal operation performance of the asset. It can be interesting to integrate the performance evaluation of an asset under the normal conditions and the consideration on mitigating negative events into a single framework.

Another concern is that the risk-based approach often relies on a binomial assumption that an item can only exist in one of two states: functioning (normal) or failed. This simplification does not adequately capture the behavior of sociotechnical systems. Given the complex architecture and multiple components within a system, the occurrence of a single-point failure or an external shock typically does not lead to immediate bankruptcy. Instead, there is usually a time-lasting performance degradation. In certain scenarios, the functionality of the system, though impacted by a shock, does not completely vanish but decreases to a specific level. Furthermore, restoration is not only a swift transition from a completely failed state to full functionality, and it actually involves a gradual process of recovery.

Based on these considerations, the concept of *resilience* is employed to describe the performance of an asset. Figure 14.1 illustrates the distinctions between a risk

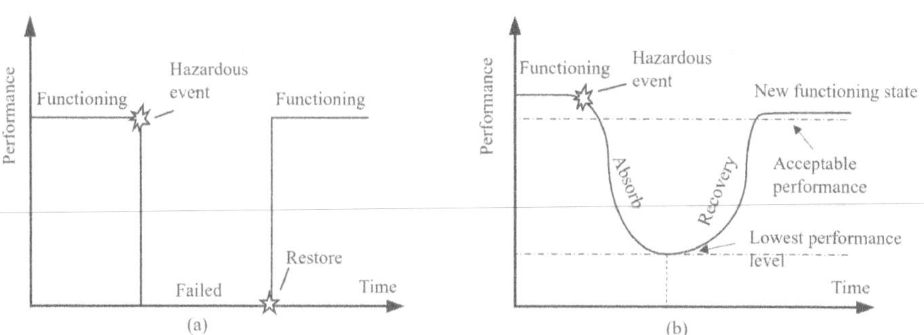


**FIGURE 14.1** System performance with time, (a) from a risk perspective, and (b) from a resilience perspective.


DOI: 10.1201/9781003245636-14

model (Figure 14.1(a)) and a resilience model (Figure 14.1(b)) in describing the performance of an item following a hazardous event. It can be found that in the resilience model, the item is in a stable, functioning state initially, or a state *resisting* a shock without performance loss. After a hazardous event or shock, the item undergoes an *absorbing* period with gradual decline of performance from the original level to a low level (not necessary zero). After reaching the lowest level (may sojourn at this level for a certain time), repair and other response actions can be initiated. It also takes some time in *recovery*, where the performance of the item is increased gradually. Eventually, the item may enter a new stable state with a performance level higher than the acceptable threshold, but the new stable performance level may differ from the original one before the hazardous event. This final state can be regarded as the *adapted* state.

The term "resilience" originally means "bouncing back", and it can be used to describe the capability of the item reflecting by whole process shown in Figure 14.1(b) from being shocked to returning to a new stability. The part above the resilience curve in Figure 14.1(b) is the performance loss of the item due to the hazardous event or shock. In overall, a more resilient item can have lower performance loss following the shock.

Resilience has been defined variously across different disciplines. For a comprehensive overview of these definitions, readers are encouraged to consult the article by Roostaie et al. (2019). In this chapter, however, we will not engage in further discussions due to the page limitation but will define resilience as follows:

- *Resilience*: A metric that evaluates how effectively a sociotechnical system can withstand and absorb shocks, restore itself, and adapt following a negative event.

Considering the four stages described in Figure 14.1(b), resilience can also be seen as a composite of four key capabilities within a sociotechnical system: *resistibility*, *absorbability*, *recoverability*, and *adaptability*. These capabilities influence the duration the system sojourns in each state and the corresponding performance levels:

- *Resistibility*: The ability of an item to survive without performance loss. Higher resistibility means that the system keeps its original functioning state for a longer period, even in the case of occurring hazardous and unexpected events.
- *Absorbability*: The ability of an item to keep performance or slow down the performance decline given that a hazardous event or shock has occurred on the item. The period from the hazardous event to the moment that the performance reaches the lowest level is the absorbing stage. Higher absorbability allows the item to endure the absorbing stage longer, achieving more graceful degradation following the hazardous event and maintaining a higher performance level at its lowest point.
- *Recoverability*: The ability of an item to bounce back to the functioning state after a hazardous event and performance decline. The recovery

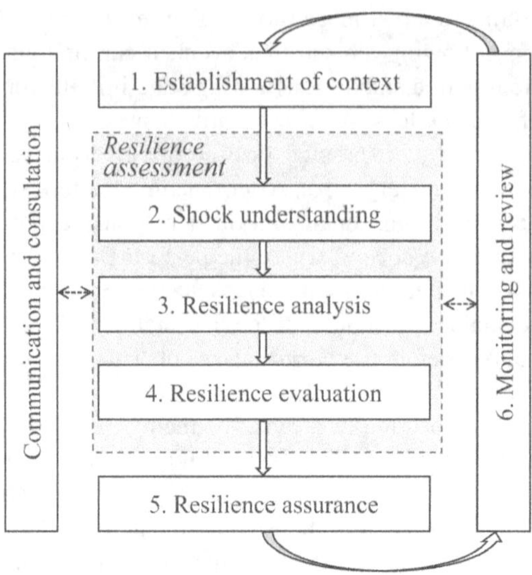


**FIGURE 14.2** Main steps of resilience management.

stage starts from the moment of entering the lowest performance level to the time that performance of the item becomes acceptable again. Higher recoverability implies a quicker restoration of the item and a shorter recovery stage.

- *Adaptability*: The ability of an item to perform well in the new stable state and prepare for future hazardous events. Higher adaptability results in a greater performance level in the new stable state and have less performance loss in case of the next hazardous event.

Therefore, *resilience management* of a sociotechnical system goes beyond risk management to address the complexities of such a system and the uncertainty of responses after a hazardous event/shock occurs. As the counterpart of risk management flowchart shown in Figure 1.4, Figure 14.2 illustrates the main steps of resilience management.

## 14.2 CONCEPTUALIZATION OF RESILIENCE ENHANCER

### 14.2.1 Definition of Resilience Enhancer

A sociotechnical system can be designed to be safe and also to be resilient. Some examples of effective design approaches for resilience include:

- *Redundancy*: This can involve duplicate critical components, backup resources, or alternative processes that can take over in case one of the components exposes to shocks.

- *Diversity*: This helps avoid system disruptions due to common-cause failures and makes the system less susceptible to specific threats or at least degrading slower.
- *Modularity*: This allows the failed parts to be isolated and repaired without disrupting the entire system and so that facilitate restoration.

It can be found that the above are also used in system reliability and safety design. In this chapter, we explore another strategy for bolstering resilience by integrating additional measures into the asset of interest. This strategy serves as a counterpart to safety barriers in resilience engineering. We refer to it as a *resilience enhancer*, defined as:

- *Resilience enhancer*: An entity and/or non-entitative means that is introduced in a sociotechnical system to reduce potential performance loss in case of occurrences of hazardous events, by improve the resistibility, absorbability, recoverability, or adaptability of this system.

Figure 14.3 compares the performance losses without and with a resilience enhancer to illustrate the expected functionality of the resilience enhancer. The solid curve is used to describe the performance changing with time following a shock given the inherent resilience of a system, like that in Figure 14.1(b), while the dotted curve is used to illustrate the improved performance given that a resilience enhancer is installed. The shadow area between the two curves is the avoided performance loss, or the expected functionality of the resilience enhancer.

It is also possible to some alternative terms, such like *resilience enabler*, *resilience driver*, or *resilience facilitator*, to express the almost same meanings. We choose resilience enhancer because while these added measures contribute to resilience, they are not the sole factors. In fact, resilience-oriented system design is also important and even more critical. The added measure enhances or strengthens resilience, by reducing performance loss in one or more stages of the resilience curve, but it does not act as the only enabler or driver.

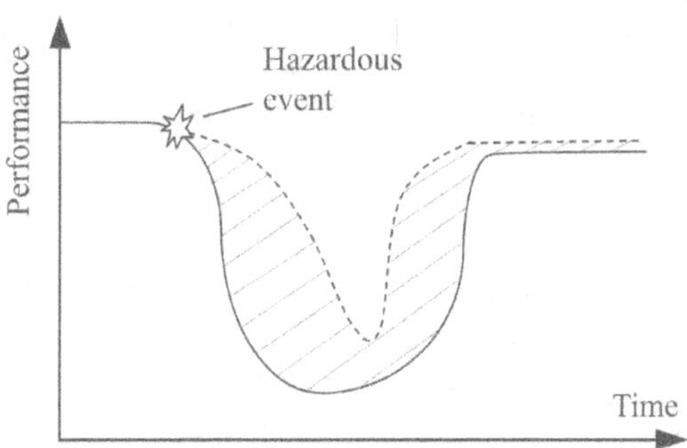


**FIGURE 14.3** The functionality of a resilience enhancer.

Rieger, Gertman et al. (2009) proposed a resilient control system with considerations similar to those of a resilience enhancer. A resilient control system is used to maintain state awareness and upholds an acceptable level of operational normality in response to disturbances. Another concept developed by Rieger and Villez (2012) is resilient agent, as a mechanism designed to detect anomalous events and initiate corrective actions.

Similar to the distinction between inherent safety and safety barriers, it is reasonable to differentiate a resilience enhancer from the integrated system elements that make the system resilient: A resilience enhancer does not directly affect system functionality under normal conditions, coming into play only during unexpected events.

In practices, examples of resilience enhancers can be commonly observed, such as:

- *Buffers in a production line*: These are used to manage fluctuations and interruptions in manufacturing processes (enhancing resistibility). By holding extra materials or products, buffers ensure that downstream operations continue smoothly even if upstream activities face delays, thus maintaining production flow and reducing system downtime (enhancing absorbability and recoverability).
- *Emergency lane in a traffic system*: An emergency lane acts as a dedicated pathway for emergency vehicles during crises, ensuring quick response times and maintaining a level of operational normalcy despite traffic disruptions. This lane also provides an escape route for all vehicles if there are incidents blocking the regular traffic lanes (enhancing recoverability).
- *Firewalls, fire detectors, and firefighters*: These form a comprehensive defense against fire-related incidents. Firewalls prevent the spread of flames, thereby extending the time before a fire-damaged plant collapses (enhancing absorbability). Fire detectors provide early warnings to facilitate a timely response (enhancing recoverability), and firefighters are trained to manage and mitigate fire outbreaks (enhancing recoverability).
- *Emergency training and first aid boxes*: Regular training on emergency procedures and accessible first aid resources prepare individuals to respond effectively to health emergencies. This proactive approach helps in minimizing the impact of accidents or sudden health issues, maintaining safety and resilience in workplaces and public spaces (enhancing resistibility and recoverability).
- *Aftercare and psychological counseling*: These services are vital for recovery after traumatic events, enhancing the resilience of individuals by providing emotional support and coping strategies. Such interventions help communities and organizations bounce back more effectively, fostering mental resilience and well-being (enhancing recoverability and adaptability).
- *Artificial inference on the reproduction of rare animals*: This involves using advanced technologies like artificial intelligence to predict and enhance the breeding patterns of endangered species (enhancing absorbability). By

ensuring the survival and growth of rare animal populations, such efforts contribute to the resilience of ecosystems, preserving biodiversity and ecological balance (enhancing recoverability).

### 14.2.2 Comparison of Safety Barrier and Resilience Enhancer

The concept of a resilience enhancer is derived from the notion of a safety barrier. Mathematically, referring back to Equation (1.3), the role of a barrier in risk management can be explained by the inequality $R[S|E(p, c), B] < R[S|E(p, c)]$, indicating that the installation of a barrier is expected to reduce the risk to the asset.

Similarly, let $U$ represent the utility or performance of a sociotechnical system, $E$ an unexpected event impacting the system, and $H$ signify a resilience enhancer, leading to the formulation,

$$U\left[E(p, c)\right] < U\left[E(p, c), H\right] \tag{14.1}$$

We could alternatively use Resilience ($RS$) instead of $U$ as the objective function, without altering the relationship in the inequality, as the overall utility is monotonically and positively correlated with resilience.

Safety barriers can be the proactive measures for preventing hazardous events and can thus be considered types of resilience enhancers that enhance system resistibility. On the other hand, reactive barriers are designed to mitigate the consequences of hazardous events once they occur, positioning them as resilience enhancers that improve absorbability. From this perspective, safety barriers constitute a subset of resilience enhancers.

However, it is important to note that the term resilience enhancer primarily focuses on whether a sociotechnical system can perform its required functions or whether its utility can be satisfied. Typically, a resilience enhancer is closely linked to a specific function within the system. In contrast, a safety barrier is associated with protecting the system against specific hazards or threats, which may not directly relate to the main functions of the system.

For instance, consider gas detectors in a chemical plant. If we assess these detectors as resilience enhancers, we can evaluate their effectiveness based on how much production time they save by detecting gas leaks early enough to prevent a fire. On the other hand, if we view them as safety barriers, their effectiveness would be measured by their ability to protect lives and property, such as the number of people in the surrounding area could be saved from a potential explosion, which is not directly related to the production function of the plant. In the latter case, it is difficult to regard the gas detectors as resilience enhancers when the productivity is the performance measure of the plant.

We have presented the role of safety barrier in risk management in Figure 1.6, and similarly, the role of resilience enhancer can be noted in resilience management, as illustrated in Figure 14.4. In this flowchart, we analyze resilience from the four perspectives of resistibility, absorbability, recoverability and adaptability, as discussed in the last section. Any improvement in these aspects can be regarded as resilience enhancement.

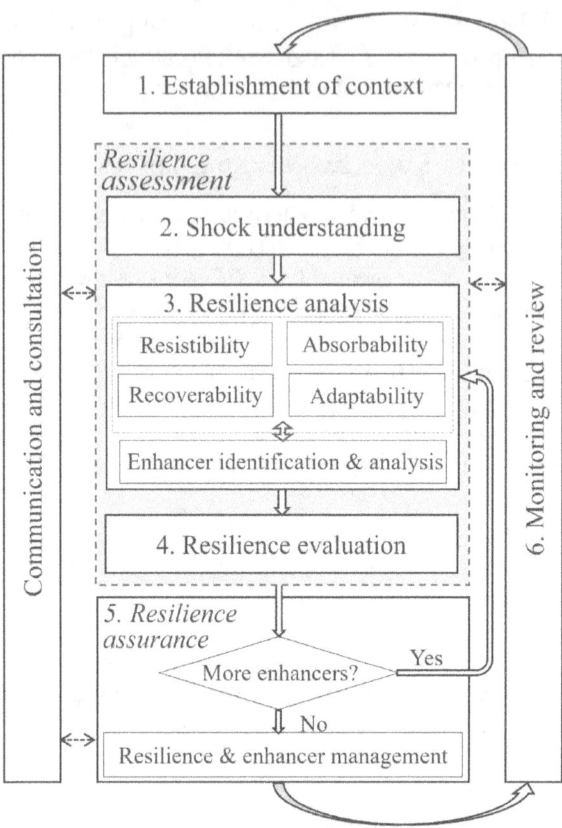


**FIGURE 14.4** Role of resilience enhancer in resilience management.

## 14.3 CLASSIFICATION OF RESILIENCE ENHANCERS

As what we have discussed on safety barriers in Chapter 1, we can similarly classify resilience enhancers into groups.

### 14.3.1 Classification According to the Phases

Based on our discussions in this chapter, this classification method considers resilience enhancers deploying at the various stages of system response to a hazardous event. Here, resilience enhancers are divided into four groups:

- *Resistibility enhancer*: An enhancer that is employed proactively to prevent exposure or increases resistance of a system to a shock, prior to the occurrence of an unexpected event. Examples include the barriers in the middle of road, and sensors and transmitters for monitoring system status.
- *Absorbability enhancer*: An enhancer operates after the hazardous event but before the situation worsens to its most severe state. It is used to moderate the

performance degradation gracefully by slowing down the rate of degradation or preventing a complete collapse. Examples include buffers in production line, the damping system equipped in high buildings for reduce vibrations caused by wind forces, and the surge protectors in electrical systems. The last example actually helps to manage and mitigate the impact of power surges on sensitive electronic equipment, allowing the system to absorb the shock.
- *Recoverability enhancer*: An enhancer that is designed to expedite the recovery process and enhance the performance of restoration activities following the worst phase but before new stable state is reached. Examples include the automated backup solutions for IT systems. These tools can quickly restore data, facilitating a swift return to operational status after an unexpected event, such as a cyberattack or software failure.
- *Adaptability enhancer*: An enhancer employed in the new stable state that helps the system adapt and be better prepared for future shocks by learning from past events. Examples can be adaptive traffic management systems that learn and evolve based on historical traffic jams. These systems adjust signals and manage flow based on real-time data and historical trends, making the traffic system more resilient to unexpected changes like accidents or sudden increases in traffic volume.

### 14.3.2 Classification According to the Approaches

Based on the classification of safety barriers in Section 1.4.2, resilience enhancers also can be classified as *entitative resilience enhancer* and *non-entitative resilience enhancer*.

Moreover, entitative resilience enhancer can be classified as:

- *Physical resilience enhancer*: Physical incorporation to improve resilience. Examples include firewalls and shear walls for earthquake-resistance.
- *Technical resilience enhancer*: An enhancer that involves technological elements or tools that improve the resilience of a system. Examples include the sensors and detectors for identifying the abnormalities.
- *Informative resilience enhancer*: This enhancer uses information to improve decision-making and response strategies. Examples include traffic lights and some data analytics systems that provide real-time insights to predict potential failures.

On the other hand, non-entitative resilience enhancer includes:

- *Procedural resilience enhancer*: Additional step(s) to existing procedures or protocols to improve resilience. For example, incorporating a time buffer in project planning, such as adding a two-week cushion in a construction project timeline to accommodate unforeseen delays from extreme weather or supply chain disruptions.
- *Organizational resilience enhancer*: The effort that adds or strengthens organizational practices to better manage and respond to disruptions.

Examples include cross-training employees to perform multiple roles, ensuring operational continuity if key staff are unavailable.
- *Perceptional resilience enhancer*: The additional efforts for changing the perception or awareness among individuals to improve the resilience of a sociotechnical system. Examples include emergency evacuation drills that improve the perception and skills of operators to response actual emergencies.

In addition, considering the complexities of large sociotechnical systems and diversity of performance influence factors, resilience enhancers are often the combinations of different types of elements. For example, to improve the air quality of a city (where poor air quality and resident dissatisfaction represent the performance losses), we can introduce the following resilience enhancers:

- *Natural-physical enhancer*: This enhancer uses natural solutions to mitigate pollution, such as increasing green spaces like parks and urban forests which absorb PM 2.5 and other pollutants.
- *Physical-technical enhancer*: This involves the implementation of advanced filtration and purification technologies in the industrial processes in this example.
- *Technical-informative enhancer*: This enhancer combines technological tools with information dissemination, such as deploying air quality monitoring apps that provide real-time air quality levels to residents. The integration of air quality forecasts into public event planning and school outdoor activity schedules.
- *Procedural-organizational enhancer*: This enhancer improves resilience by embedding procedures within organizational structures. For instance, the municipal government can introduce mandatory requirement for air quality assessment that can impact for all new urban development projects.
- *Organizational-perceptional enhancer*: This focuses on changing cultural norms and behaviors to support a cleaner environment. Campaigns that promote the use of public transport, cycling, and walking instead of driving personal vehicles can help reduce overall emissions.

### 14.3.3 Classification Based on Operational Features

Resilience enhancers also can be classified according to their operational features:

- *Passive resilience enhancer*: An enhancer that performs its function as an inherent property after it is in place, without requiring any external command, action, or energy. Examples of passive resilience enhancer include all the passive safety barriers.
- *Active resilience enhancer*: An enhancer that needs to be activated by a command, action, or some energy to perform their functions in response to certain events. Examples of active resilience enhancers include the active safety barriers and those elements in system operation, restoration, and repair, such as smart grid technology that can adjust flows based on

real-time data to optimize and maintain energy distribution and prevent overloads, requiring active management and monitoring.

There are other classification methods for resilience enhancers, in consideration of different factors, but we do not have more discussions here due to the page limitation.

## 14.4 ASSESSMENT OF RESILIENCE ENHANCERS

Based on the insights into safety barriers in Chapter 6 and resilience engineering literature, this section provides a brief overview on the assessment of resilience enhancers.

### 14.4.1 Resilience Metrics

To effectively evaluate resilience enhancers, it is essential first to comprehend how system resilience is measured. A significant amount of literature has proposed various metrics and methodologies for this purpose (for example, see the review by Hosseini et al. (2016)). Here, we discuss two fundamental metrics:

The first type is the deterministic-based metrics, where the performance loss following a shock event is linked to the resilience of the system in study: Greater losses indicate lower resilience, whereas smaller losses indicate higher resilience. In many models based on such metrics, the resilience curve is simplified into a trapezoidal shape as depicted in Figure 14.5.

In such a model, the transitions between critical phases are modeled linearly at a constant rate. In Figure 14.5, we use $\varphi(t)$ to represent the performance of an item (system) at time $t$, and in the originally normal operation state, $\varphi(t_0) = 100\%$. The other key time points are defined as follows:

- $t_e$: Occurrence time of the hazardous event,
- $t_w$: Time when the system reaches its lowest performance or degradation completes, and the performance level in the worst situation is $\varphi_w$,
- $t_s$: Time when repair or restoration starts,
- $t_r$: Time when the system stabilizes in a new equilibrium, and
- $t_f$: Time when observation is completed.

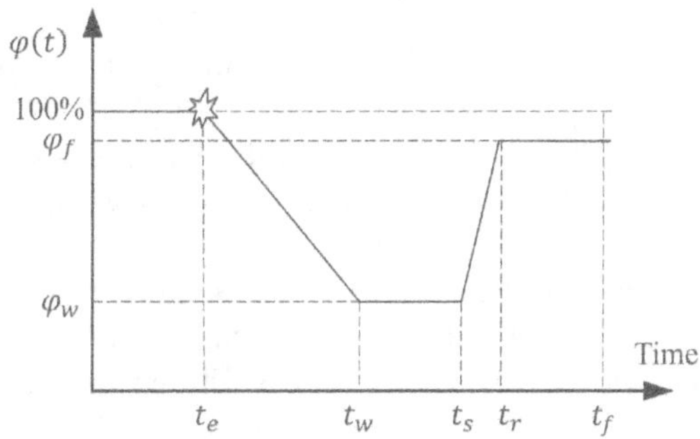


**FIGURE 14.5** A trapezoid model for quantitative resilience analysis.

Assuming that the performance of such an item in the new stable state is equivalent to its original state, we do not need to consider changes after $t_r$. However, if the new performance differs from the original, $\varphi_f$ can be used to represent the new performance level.

Referring to the model in Figure 14.5, the resilience ($RS$) of the item can be calculated as the ratio of the remaining performance after a shock to the expected performance in the absence of the hazardous event. This relationship is represented by the following equation:

$$RS \approx 1 - \frac{\frac{\left[(t_s - t_w) + (t_r - t_e)\right](1 - \varphi_w)}{2} + (t_f - t_e)(1 - \varphi_f)}{(t_f - t_e) \cdot 100\%} \tag{14.2}$$

The numerator of this equation reflects the performance loss, which is approximated by the sum of the areas of two geometric shapes in Figure 14.5: The trapezoid above the resilience curve from $t_e$ to $t_r$, and the rectangle approximated above the resilience curve in the new stable state from $t_r$ to $t_f$.

In some literature, the trapezoidal model in Figure 14.5 is simplified further into a triangular model, disregarding the duration spent at the worst performance level ($t_s - t_w$).

Resilience is also considered a time-dependent variable. The instantaneous resilience at time $t$ can be given by:

$$RS_{\mathrm{I}}(t) = \frac{\varphi(t) - \varphi_w}{\varphi(t_0) - \varphi_w} \tag{14.3}$$

Then, when $t > t_e$, the cumulative resilience performance $RS_{\mathrm{C}}(t)$ can be calculated as:

$$RS_{\mathrm{C}}(t) = \frac{\int_{t_e}^{t} \left[\varphi(u) - \varphi_w\right] du}{\int_{t_e}^{t} \left[\varphi(t_0) - \varphi_w\right] du} \tag{14.4}$$

These metrics primarily focus on the recoverability of the system. If the value of $RS_{\mathrm{C}}(t)$ ranges between [0, 1], when $\varphi(t) = \varphi(t_w)$, $RS(t) = 0$, indicating a disrupted state, when $\varphi(t) = \varphi(t_0)$, $RS(t) = 1$, indicating full recovery. It should be noted that this equation assumes that the performance level in the new stable state is same with the original.

Another type of resilience assessment metrics are probabilistic-based, where resilience is defined as the probability that performance after a shock is higher than a minimum acceptable level and recovery time is within an acceptable duration (e.g., see Chang & Shinozuka, 2004):

$$RS = \Pr\left(\varphi_w, T \mid e\right) = \Pr\left(\varphi_w > \varphi_w^* \text{ and } T < T^*\right) \tag{14.5}$$

where $\varphi_w$ and $T$ represent the lowest performance level and the time from the hazardous event to the new stability, respectively. $T$ can be explained as $(t_r - t_e)$ in Figure 14.5, while $\varphi_w^*$ is the minimum acceptable performance, and $T^*$ maximum acceptable recovery time considering the impact of event $e$.

### 14.4.2 Measures of Resilience Enhancers

To ensure the resilience of a system, it is natural to consider whether a resilience enhancer is effective, and thus *effectiveness* serves as a fundamental metric in the management of these enhancers within large systems. We still can quantify effectiveness using the *effectiveness coefficient* (EC) like in Chapter 6, defined as the ratio of the performance loss avoided by the enhancer to the expected total performance loss without the enhancer. From the resilience perspective, this is represented by the difference between the resilience with the enhancer ($RS^*$) and the resilience without the enhancer ($RS$), expressed as

$$EC = \frac{RS^* - RS}{RS} \tag{14.6}$$

For instance, using Equation (14.2), we can calculate the values of $RS^*$ and $RS$. Here, the equation not only quantifies the direct impact of an enhancer but also helps in strategic planning and optimization of resilience measures within the system. Resilience enhancers can be deployed to influence various metrics to improve system resilience:

- $t_w$: By employing methods to slow down the performance decline, the duration before a system reaches a critical degraded state ($t_w$) can be extended. Keeping other conditions constant, this approach directly boosts the resilience of the system. In a process plant, for example, implementing advanced corrosion-resistant painting materials in piping systems can be a resilience enhancer for slowing the rate of wear and tear.
- $t_s$: By planning and training for emergency scenarios in advance, the duration of critical disruptions ($t_s$), can be minimized. This preparation helps in swiftly managing and mitigating adverse events, thus enhancing overall resilience. The example in a process plant can be regular emergency drills and pre-defined response protocols. This preparation ensures rapid containment and cleanup, reducing the impact and duration of the incident.
- $\varphi_w$: By implementing robust system designs and preventive maintenance strategies to avoid complete disruption in an accident to leverage the value of $\varphi_w$, thus maintaining resilience. The example in a process plant can be backup power systems and generators, they can avoid shutdown of the total plant.
- $t_r$: By enhancing the capability and efficiency of recovery procedures, the time required to return to operational baseline $t_r$, can be significantly reduced, improving resilience. Upgrading control systems and training personnel for rapid troubleshooting can speed up the recovery process after a failure.
- $\varphi_f$: By fostering adaptability, a system can better perform under the new stable condition with the performance level of $\varphi_f$, adapting to changes and disturbances, thereby enhancing resilience. After a significant equipment

failure due to unexpected operational conditions, a process plant can invest in analyzing historical operation data and the specific failure modes, enabling the plant to identify patterns and predictive factors leading up to the failure.

On the other hand, effectiveness also can be regarded as a probabilistic measure as in Chapter 6. We can introduce the measure of *effectiveness in probability* (EP) for resilience enhancers as:

$$\begin{aligned}\text{EP} = \Pr\big(&\text{Resilience is sufficient with enhancer} \mid \text{Resilience is insufficient} \\ &\text{without enhancer}\big)\ \textit{or}\ \text{EP} = \Pr\big(\text{Resilience requirement is met}\big)\end{aligned} \tag{14.7}$$

If there is a minimum requirement for resilience enhancement ($\vartheta$), EP can be specified based on EC as:

$$\text{EP} = \Pr(\text{EC} \geq \vartheta) = \Pr\left[\frac{RS^* - RS}{RS} \geq \vartheta\right] \tag{14.8}$$

Then, EP by time $t$ can be expressed as:

$$\text{EP}(t) = \Pr\left[\frac{RS^*(t) - RS(t)}{RS(t)} \geq \vartheta\right] \tag{14.9}$$

Equations (14.8) and (14.9) can be integrated with Equation (14.2) or (14.5) to determine the values of critical parameters.

It is important to recognize that the concept of a resilience enhancer is very novel. Consequently, there is a limited literature available for reference, and the methods for assessing these enhancers are still under development. For instance, while the probability of failure on demand (PFD) is used to assess the integrity of safety barriers, a corresponding metric for resilience enhancers has not yet been fully established. A potential metric, based on Equation (14.5), can be the *probability of failure to the expected resilience*. However, this metric has not been clearly defined or empirically validated. Readers interested in exploring this topic further are encouraged to contact the author of this book for collaborative research opportunities.

## REFERENCES

Chang, S. E., & Shinozuka, M. (2004). Measuring improvements in the disaster resilience of communities. *Earthquake Spectra*, *20*(3), 739–755.

Hosseini, S., Barker, K., & Ramirez-Marquez, J. E. (2016). A review of definitions and measures of system resilience. *Reliability Engineering & System Safety*, *145*, 47–61.

Landucci, G., Argenti, F., Tugnoli, A., & Cozzani, V. (2015). Quantitative assessment of safety barrier performance in the prevention of domino scenarios triggered by fire. *Reliability Engineering & System Safety*, *143*, 30–43.

Rieger, C. G., Gertman, D. I., & McQueen, M. A. (2019). Resilient control systems: Next generation design research. 2nd Conference on Human System Interactions, Catania, Italy.

Rieger, C. G., & Villez, K. (2012, August). Resilient control system execution agent (ReCoSEA). *5th International Symposium on Resilient Control Systems*, Salt Lake City, UT, USA.

Roostaie, S., Nawari, N., & Kibert, C. J. (2019). Sustainability and resilience: A review of definitions, relationships, and their integration into a combined building assessment framework. *Building and Environment*, *154*, 132–144.

# Appendix A: Acronyms

| | |
|---|---|
| ABS | Anti-lock braking system |
| AHP | Analytic hierarchy process |
| AI | Artificial intelligence |
| ALARP | As low as reasonably possible |
| AOI | Asset of interest |
| ASIL | Automotive Safety Integrity Level |
| AUP | Asset under protection |
| BBN | Bayesian belief network |
| BCPS | Basic control process system |
| BFR | Binomial failure rate |
| BN | Bayesian network |
| BOP | Blowout preventer |
| BORA | Barrier and operational risk analysis |
| CAF | Cascading failure |
| CAPEX | Capital expenditure |
| CCF | Common-cause failure |
| CCPS | Center for Chemical Process Safety |
| CDF | Cumulative distribution function |
| CPT | Conditional probability table |
| DAG | Directed acyclic graph |
| DBN | Dynamic Bayesian network |
| DC | Diagnostic coverage |
| DD | Dangerous detected |
| DNV | Det Norske Veritas |
| DOE | U.S. Department of Energy |
| DSPN | Deterministic and stochastic Petri net |
| DT | Digital twin |
| DU | Dangerous undetected |
| EC | Effectiveness coefficient |
| E/E/PE | Electrical/electronic/programmable electronic |
| EFBA | Energy flow/barrier analysis |
| EP | Effectiveness in probability |
| ETBA | Energy trace and barrier analysis |
| ESP | Electronic stability program |
| ETA | Event tree analysis |
| EUC | Equipment under control |
| FMEA | Failure mode and effects analysis |
| FPSO | Floating production storage and offloading |
| FPT | Full proof test |
| FT | Fault tree |
| FTA | Fault tree analysis |

| | |
|---|---|
| GLM | Generalized linear model |
| GSPN | Generalized stochastic Petri net |
| HALT | Highly accelerated life test |
| HASA | highly accelerated stress audit |
| HASS | highly accelerated stress screen |
| HAZID | Hazard identification |
| HAZOP | Hazard and operability study |
| HIPPS | High integrity pressure protection system |
| HPP | Homogeneous Poisson process |
| IEC | International Electrotechnical Commission |
| IMV | Ice management vessel |
| IPL | Independent protection layer |
| ISO | International Organization for Standardization |
| LLM | Large-language model |
| LOPA | Layer of protection analysis |
| LOTO | Lockout/tagout |
| MBF | Multi-beta factor |
| MEF | Mitigated event frequency |
| MGL | Multiple Greek letter |
| ML | Machine learning |
| MORT | Management Oversight and Risk Tree |
| MRL | Mean residual lifetime |
| MTTF | Mean time-to-failure |
| NASA | National Aeronautics and Space Administration |
| NEA | Nuclear Energy Agency |
| NIS | Negative ideal solution |
| NLP | Natural language processing |
| OPEX | Operational expenditure |
| PCA | Principal component analysis |
| PCR | Principal component regression |
| PDF | Probability density function |
| PFD | Probability of failure on demand |
| PFH | Average frequency of dangerous failures per hour |
| PHA | Process hazard analysis |
| PIF | Performance influencing factor |
| PIS | Positive ideal solution |
| PLC | Programmable logic solver |
| PLSR | Partial least squares regression |
| PMF | Probability mass function |
| PN | Petri net |
| PPT | Partial proof test |
| PSA | Norwegian Petroleum Safety Authority |
| RBD | Reliability block diagram |
| RBDM | Risk-based decision-making |
| RBD-PN | Reliability block diagram driven Petri net |
| RIBM | Risk-informed barrier management |

| | |
|---|---|
| RIDM | Risk-informed decision-making |
| RIF | Risk influence factor |
| RPN | Risk priority number |
| RRF | Risk reduction factor |
| RUL | Remaining useful lifetime |
| SDRP | Similarity-difference-based reliability prediction |
| SEL/SL | Security level |
| SFF | Safety failure fraction |
| SIL | Safety integrity level |
| SIS | Safety instrumented system |
| SPC | Statistical process control |
| SPN | Stochastic Petri net |
| SR | Sustainability requirement |
| STAMP | Systems-theoretic accident model and process |
| STL | Spurious trip level |
| STPA | Systems-theoretic process analysis |
| TAAF | Test-analyze-and-fix |
| TEF | Tolerable event frequency |
| UEF | Unmitigated event frequency |

# Index

## C

## D

## E

## I

## P

## S

For Product Safety Concerns and Information please contact our EU representative GPSR@taylorandfrancis.com
Taylor & Francis Verlag GmbH, Kaufingerstraße 24, 80331 München, Germany

www.ingramcontent.com/pod-product-compliance
Lightning Source LLC
LaVergne TN
LVHW011259110826
845149LV00001B/184